ORGANOMETALLICS IN ORGANIC SYNTHESIS

Volume One

Organometallics in Organic Synthesis

EI-ICHI NEGISHI

Purdue University

VOLUME ONE

General Discussions

Organometallics of Main Group Metals in Organic Synthesis

A WILEY-INTERSCIENCE PUBLICATION

JOHN WILEY & SONS, New York • Chichester • Brisbane • Toronto

Library of Congress Cataloging in Publication Data:

Negishi, Ei-ichi, 1935–
  Organometallics in organic synthesis.

  "A Wiley-Interscience publication."
  Includes index.
  1. Organometallic compounds. 2. Chemistry,
Organic—Synthesis. I. Title.
QD411.N35   547′.05   79-16818
ISBN 0-471-03193-3

Printed in the United States of America

10 9 8 7 6 5 4 3 2 1

**To
Sumire**

PREFACE

Since the discovery of the Grignard reagents at the turn of the century, the organometallic compounds have played a major role in organic synthesis both in academic laboratories and in industry. Today it is not only unwise but rather difficult to accomplish an efficient and selective multistep synthesis without using organometallics.

Despite wide-spread use of organometallics, there still exists a strong tendency to view organic chemistry as a branch of science dealing primarily with C, H, N, O, S, and halogens. Except for a few classes of organometallics, such as organolithiums and the Grignard reagents, the organometallics are often viewed as exotic. Persuaded by the literature that promises an attractive route to the desired compound, chemists occasionally use some organometallics with much hesitation. The results are often disappointing and frustrating. Poor results appear to stem, at least in part, from lack of sufficient knowledge of the scope and limitations of the procedure and of the proper handling of organometallics in such a procedure.

Another frequent problem is one of selecting the most suitable procedure for the desired transformation. For example, most elementary organic chemistry textbooks describe in some detail the addition reaction of Grignard reagents to ketones. The corresponding reaction of organolithium compounds, which often is far superior to the Grignard reaction, however, is seldom adequately discussed. Thus chemists are asked to learn this simple fact the hard way. Since the number of synthetic procedures to be exploited is increasing at a rapid pace, this difficulty is becoming increasingly serious. Some remedial measures must be provided.

Over the past decade or so a number of excellent textbooks (*0.1* to *0.13*)* and monographs (*0.14* to *0.32*) covering the general area of organometallic chemistry

---

*Reference numbers beginning with 0 are located in Appendix I, "General References," at the end of this volume.

have been published. Unfortunately, however, these discussions focus mostly on the preparation and structure of organometallic compounds themselves and/or the inter-conversion of organometallic substances. While teaching a course on organic synthe-sis at the graduate and undergraduate senior level, I saw the obvious need for a textbook or a supplementary textbook that would fill the gap between the more usual organic textbooks and organometallic textbooks. This book is written in response to that need.

The main objective of this book is to provide, in a concise form, a current overall picture of the chemistry of organometallics in organic synthesis. One way to ac-complish such a task is to focus on a relatively small number of key features common to a large number of related reactions. While it may be ideal to classify various reactions according to mechanistic schemes, the present knowledge of organometal-lic reactions is not sufficiently developed to allow this approach. My discussion is therefore based primarily on the reactant-product relationship as well as on some tentative but reasonable interpretations provided both by the authors of individual papers and by myself. Such an approach is justified because (1) it is not inconsistent with experimental observations, (2) it helps to explain a large number of results (including even some that will be published in the future) in terms of a small number of common schemes, and, perhaps more importantly, (3) it provides some useful working hypotheses.

This is neither a reference book nor a comprehensive treatise. Thus the literature covered is highly selective. An effort has been made, however, to cite rather exten-sively reviews in which a number of pertinent original papers can be found. The following journals and review series provide the majority of reviews written in English in the area of organometallic chemistry.

Review Journals

1. *Acc. Chem. Res.*
2. *Angew. Chem. Int. Ed. Engl.*
3. *Chem. Rev.*
4. *Chem. Soc. Rev.*
5. *Intra-Sci. Chem. Rep.*
6. *J. Organometal. Chem.*
7. *Organometal. Chem. Rev. A*
8. *Pure Appl. Chem.*
9. *Quart. Rev.*
10. *Russ. Chem. Rev.*
11. *Synthesis*
12. *Tetrahedron Rep.*

Review Series

1. *Adv. Chem. Ser.*
2. *Adv. Inorg. Chem. Radiochem.*
3. *Adv. Org. Chem.*
4. *Adv. Organometal. Chem.*
5. *Org. React.*
6. *Organometal. React.*
7. *Progr. Inorg. Chem.*
8. *Top. Curr. Chem.*
   (Fortschr. Chem. Frosch.)

In discussing works published in or before 1975 extensive use was made of various textbooks, monographs, and collections of reviews (Appendix I).

The nomenclature and terminology in organometallic chemistry are still quite underdeveloped and often inconsistent. Mainly for the sake of simplicity and consistency, I have decided to adopt certain unconventional names and terms. For example, tricoordinate and tetracoordinate organoboron compounds are called organoboranes and organoborates, respectively. There is, however, no simple name that applies to both. I suggest "organo + element + s" to describe collectively any organometallic compounds containing the element directly bonded to carbon regardless of their structural details. Thus both organoboranes and organoborates as well as subvalent organoboron compounds may be called *organoborons*. Organomercury compounds have traditionally been called organomercurials. The *al* ending, however, does not appear to have been used in any other class of organometallic compounds. Here again, the term *organomercuries* may be used to refer collectively to all types of organomercury compounds.

I wish to express my sincere appreciation to my former mentor, Professor Herbert C. Brown of Purdue University. His guidance, assistance, and influence have been essential in the writing of this book. I am directly and indirectly indebted to my colleagues in the Chemistry Department of Syracuse University. Occasional friendly reminders and assistance provided by Drs. Gershon Vincow and Daniel J. Macero are appreciated. I am deeply indebted to members of my research group, particularly Drs. Shigeru Baba, Anthony O. King, Ronald E. Merrill, Nobushisa Okukado, David E. Van Horn, and Takao Yoshida, who made significant contributions to our research activities. Among my current graduate students, I would especially like to thank Michael J. Idacavage, Louis F. Valente, and Cinthia L. Rand, not only for their research efforts, but for their help in preparing this manuscript. I would also like to mention the enjoyable and fruitful collaborative research efforts with Dr. Augustine Silveira, Jr. of the State University of New York at Oswego. I am grateful to a number of my colleagues in the field of organometallic chemistry for their helpful assistance. My thanks to Dr. Alan B. Levy of SUNY at Stony Brook, who read the entire manuscript and assisted in proofreading. Dr. Richard C. Larock of Iowa State University and Dr. Phillip L. Fuchs of Purdue University provided useful comments. Finally I would like to thank my wife, Sumire, for her direct and indirect assistance, including the typing of the entire manuscript, and Virginia Ditz, David Ditz, Diane Piraino and last but not least Roberta "Tooti" Molander for preparing the camera ready copy.

Ei-ichi Negishi

Syracuse, New York

CONTENTS

PART I cf. Part III GENERAL DISCUSSIONS

1 Some Fundamental Properties of Metal Atoms, Bond to Metals, Carbon Groups, and Organometallic Compounds   **3**

    1.1   Metal Atoms, 4
    1.2   Bonds to Metals, 8
    1.3   Carbon Groups, 11
    1.4   Molecular Structure, 14
    1.5   Intermolecular Forces, 19
    1.6   Some Factors Controlling Chemical Reactions, 21

2 Methods of Preparation of Organometallic Compounds   **30**

    2.1   Oxidative Metallation of Organic Halides (Method I), 31
    2.2   Oxidative Displacement by Metallate Anions (Method II), 37
    2.3   Metal-Halogen Exchange (Method III), 38
    2.4   Oxidative Metallation of Active C-H Compounds (Method IV), 40
    2.5   Metal-Hydrogen Exchange (Method V), 41
    2.6   Hydrometallation (Method VI), 45
    2.7   Heterometallation (Method VII), 48
    2.8   Complexation (Method VIII), 50
    2.9   Oxidative Coupling (Method IX), 51
    2.10 Oxidative Metallation of Unsaturated Organic Compounds (Method X), 53
    2.11 Transmetallation (Method XI), 54
    2.12 Oxidative-Reductive Transmetallation (Method XII), 56

3. General Patterns of Organometallic Reactions   60

3.1 Reagent Versus Intermediate, 60
3.2 Organometallics as Intermediates, 61
3.3 Organometallics as Reagents, 83

PART II cf. Part III ORGANOMETALLICS OF MAIN GROUP METALS IN
ORGANIC SYNTHESIS

4 Organometallics Containing Group IA, Group IIA and Group IIB Elements
(Li, Na, K, Mg, Zn, Cd)   91

4.1 General Considerations, 91
4.2 Preparation of Organometallics Containing Group IA, Group IIA and
     Group IIB Metals, 95
4.3 Carbon-Carbon Bond Formative via Organoalkali and Organoalkaline
     Earth Metals, 104
4.4 Carbon-Hetero Atom Bond Formation via Organoalkali and
     Organoalkaline Earth Metals, 222
4.5 Other Synthetic Applications of Organometallics of Group IA and
     Group II Metals, 250

5 Organoborons and Organoaluminums (B, Al)   286

5.1 General Considerations, 286
5.2 Organoborons, 287
5.3 Organoaluminums, 350

6 Organosilicons and Organotins (Si, Sn)   394

6.1 Some Fundamental Properties of Silicon and Tin and the Bonds to
     These Elements, 394
6.2 Structure, 399
6.3 Preparation of Organosilicons and Organotins, 403
6.4 Reactions of Organosilicons and Organotins, 416

7 Organometallics of Heavy Main Group Metals (Hg, Tl, Pb)   455

7.1 General Considerations, 455
7.2 Preparation of Organomercuries, Organothalliums and Organoleads,
     460

7.3   Reaction of Organomercuries, 470
7.4   Reactions of Organothalliums and Organoleads, 479

Appendix I   General References   497

Appendix II   Periodic Table and Electronegativities of the Elements   500

Appendix III Electronic Configurations of the Elements   501

Appendix IV   Ionization Energies   504

Appendix V   Covalent and Ionic Radii   505

Appendix VI   Acidity of Brønsted Acids   506

Index   511

CONTENTS FOR VOLUME TWO

Part III   Organotransition Metal Compounds in Organic Synthesis

8   General Discussions of Organotransition Metal Compounds

9   Carbon-Carbon Bond Formation via Cross-Coupling, Conjugate Addition, and Related Reactions of Organotransition Metal Compounds

10   Insertion Reactions Involving Olefins and Acetylenes

11   Carbonylation and Related Reactions

12   Metal-Carbene Complexes

13   Miscellaneous Application of Organotransition Metal Compounds to Organic Synthesis

ORGANOMETALLICS IN ORGANIC SYNTHESIS

Volume One

PART I

GENERAL DISCUSSIONS

1
SOME FUNDAMENTAL PROPERTIES OF METAL ATOMS, BOND TO METALS, CARBON GROUPS, AND ORGANOMETALLIC COMPOUNDS

The physical and chemical properties of organometallic compounds, which are the subjects of this book, must be governed by the properties of the metal (M) as well as organic (R) and inorganic (L) ligands. They must also be influenced by the environmental conditions, such as solvents (S), concentration (C), temperature (T), and pressure (P), in addition to the reagents with which they react.

If we try to understand numerous organometallic reactions and attempt to predict some unknown but desirable organometallic reactions on at all a rational basis, it is essential to familiarize ourselves with some fundamental parameters affecting the properties of metals, organic and inorganic ligands, solvents, and so on. A detailed discussion of these parameters is beyond the scope of this book. The following brief discussion is intended merely to refresh our understanding of these parameters in order to establish a common platform for subsequent discussion. For more detailed treatment of the subject, the reader is referred to advanced textbooks (*1.1* to *1.7*).

1.1   METAL ATOMS

1.1.1   Metals

The definition of the metallic elements is very nebulous at best.
In our discussion, the metallic elements are those that can form
bonds to carbon in which the carbon atom is negatively polarized,
as in 1.1.

$$\overset{\delta+}{M} \underline{\quad\quad} \overset{\delta-}{C}$$

<u>1.1</u>

This requires that the electronegativity (EN) of the element in
question be smaller than that of carbon, which has been arbitrarily set at 2.5 by Pauling (*1.1*). In practice, it is more convenient to limit the electronegativity range to <2.0. Since there
is no such thing as a truly reliable electronegativity value, our
definition of the metallic elements is on precarious ground.
Nonetheless, we adopt it mainly for the sake of convenience. As
indicated in the Periodic Table (Appendix II), both boron (EN =
2.01) and silicon (EN = 1.74) are considered to be metallic elements here.

1.1.2   Electronic Configuration

Anyone who has taken an introductory course in chemistry knows
that the elements within the same group behave similarly in many
ways. This, of course, stems basically from the fact that they
possess the same number of valence electrons. After all, the
properties of elements, including their mutual similarities and
differences, can in principle be explained in terms of the numbers of subatomic entities, that is, protons, neutrons, and electrons. These numbers are, of course, indicated by the atomic
numbers and the atomic weights. Furthermore, these electrons are
distributed among orbitals surrounding the atomic nuclei according to the aufbau principle based on the Pauli exclusion principle and Hund's rule. The distribution of the electrons of an
atom is indicated by the electronic configuration (Appendix III).
It is therefore highly advisable to familiarize ourselves with
the electronic configurations of the elements of our concern.

1.1.3   Atomic Orbitals

Essential to our familiarization of the electronic configurations
is our understanding of atomic orbitals, especially their shapes.

The approximate shapes of the s, p, and d orbitals with nodal signs are shown in Fig. 1.1.

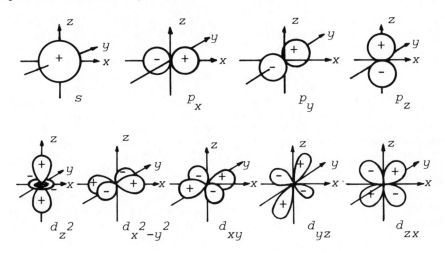

Fig. 1.1 Shapes of s, p, and d Atomic Orbitals

1.1.4 Electronegativity, Ionization Energy, Electron Affinity, and Atomic Size

Although the elements within the same group behave similarly in some ways, as stated earlier, they also behave very differently in many other ways. For example, whereas organoalanes (R_3Al) can react readily with aldehydes and ketones to form adducts, the corresponding organoboranes (R_3B) are generally inert under usual reaction conditions.

$$R_3^1Al \ + \ R_2^2C=O \ \longrightarrow \ R^1R_2^2COH \qquad\qquad (1.1)$$

$$R_3^1B \ + \ R_2^2C=O \ \longrightarrow \ \text{no reaction} \qquad\qquad (1.2)$$

As will become clear in subsequent discussions, the differences among the elements of the same group often overshadow their similarities. As already mentioned, we should, in principle, be able to explain these differences in terms of the numbers of the subatomic components. It is generally more convenient, however, to discuss them in terms of more composite parameters which can be determined for each element, that is, electronic parameters,

such as electronegativity, ionization energy, and electron affinity, as well as parameters of size, such as covalent and ionic
radii.

   Electronegativity was first defined by Pauling (*1.1*) as the
power of an atom in a molecule to attract electrons to itself.
It is a property of an atom in a molecule under the influence of
surrounding atoms. However, it has proven, in practice, to be
useful to assign an electronegativity value to each element.
Based on thermochemical data, Pauling first determined the electronegativities of various elements (*1.1*). Since then, a number
of attempts have been made to devise an improved set of electronegativity values. Interested readers are referred to an excellent discussion by Huheey (*1.3*). For our purposes, however, it
is sufficient to know only approximate electronegativity values.
We arbitrarily choose Allred-Rochow values (*1.8*) as well as the
original Pauling values (*1.1* Appendix II). Allred and Rochow
(*1.8*) defined electronegativity as the electrostatic force exerted by the nucleus on the valence electrons and obtained the for-
mula

$$EN = 0.359z^*/r^2 + 0.744 \tag{1.3}$$

where z^* and r are the effective nuclear charge obtained by Slater's rules and the covalent radius, respectively. The Allred-
Rochow and Pauling electronegativity values are very close to
each other for most elements. Their values for most of the d
block transition metals, however, are appreciably different from
each other. Our current knowledge of these elements suggests
that the Pauling values are much too high to be realistic. We
will therefore be using mainly the Allred-Rochow values in these
cases. In some other cases, EN of mercury, for example, neither
the Allred-Rochow values nor the Pauling values appear to be
realistic. Other values will therefore be considered in such
cases.

   It is useful to note that electronegativity increases as we
ascend the Periodic Table within an A-subgroup or go from left to
right among main group elements. The electronegativity values
range from 0.7 (Pauling scale) for francium to approximately 4
for fluorine.[†] These values can now be used to estimate the direction of bond polarization, as discussed earlier, and the ionicity of a polarized bond, as discussed later.

   Ionization energy (ionization potential) may be defined as
the energy needed to remove an electron from an atom, that is,

[†]The electronegativity of neon has been estimated at 5.1 by
Allred and Rochow (*1.8*).

the energy difference between the highest occupied energy level and that corresponding to the complete removal ($n = \infty$). The ionization process is always endothermic. Thus, the ionization energy (IE) is always indicated by a positive value. The first four ionization energies for various elements are summarized in Appendix IV.

For the main group elements, the ionization energy tends to decrease as we descend within an A-subgroup due to the combined effects of size and shielding. As we go from left to right among main group elements, the effective nuclear charge (z^*) increases, which tends to cause an increase in ionization energy.

As we go from Group IIA (s orbital) to Group IIIA (p orbital), however, we observe a decrease in ionization energy. This is due to the lower energy level of an s orbital relative to a p orbital of the same principal quantum number. The exchange energy between electrons of like spin stabilizes a system of parallel electron spins. This is responsible for the fact that the ionization energy of nitrogen is greater than that of oxygen.

The d-block transition atoms generally undergo the following ionization.

$$(n - 1)d^x ns^2 \xrightarrow{-e^-} (n - 1)d^x ns(I) \xrightarrow{-e^-} (n - 1)d^x(II) \quad (1.4)$$

This provides an explanation for the fact that the +2 oxidation state is very common among transition elements.

Electron affinity (EA) is defined as the energy released when an electron is added to the valence shell of an atom. Note that a positive electron affinity value corresponds to a negative enthalpy value, and vice versa. While the definition of electron affinity is clear, its experimental determination has been difficult. As a result, only a limited number of electron affinity values are available. As might be expected, electron affinity trends more or less parallel those of ionization energies. According to Mulliken (*1.9*), electronegativity, ionization energy, and electron affinity are correlated according to the following equation:

$$EN = 0.168(IE_{vs} + EA_{vs} - 1.23) \quad (1.5)$$

where IE_{vs} and EA_{vs} are valence state values in electronvolts.

Atomic size is another nebulous quantity associated with a given atom, and is very much dependent on the conditions under which the atom exists. It is, however, useful to discuss the following notable trends in atomic sizes.

1. Atomic sizes tend to increase, as we descend within the same group.
2. Atomic sizes decrease, as we go from left to right within a given series. It follows that a sudden increase in atomic size occurs, as we go from an inert gas to an alkali metal with the next higher atomic number.

In order to discuss atomic sizes in a quantitative manner, we must know the nature of the bonds formed by the atoms of our concern. Various types of atomic radii will be described in our discussion of bonds to metals in the following section.

1.2 BONDS TO METALS

1.2.1 Covalent Bond Versus Ionic Bond

Covalent bonds and ionic bonds are two representative chemical bonds. Although it is not our intention to discuss here any quantum mechanical theory of these chemical bonds, it is useful to review briefly the relation between the atomic properties discussed above and the nature of the chemical bond. A bond formed between two like atoms, for example, an H-H bond, may be viewed as being essentially covalent. On the other hand, there is no such thing as a 100% ionic bond. Thus all bonds between two or more unlike atoms are partially ionic and partially covalent.

Ionicity of a bond between two unlike atoms, A and B, is primarily governed by the difference between the electronegativity values of the two atoms. Pauling (1.1) proposed the following equation to estimate the percent ionicity of a bond between atoms A and B (Table 1.1).

$$\text{Ionicity (\%)} = 1 - e^{-\frac{1}{4}(EN_A - EN_B)} \qquad (1.6)$$

If we use this equation, we find that the bond between B (EN = 2.0) and C (EN = 2.5) is only approximately 6% ionic. Even the Cs-C bond, which is regarded as one of the most ionic metal-carbon bonds, is only approximately 50% ionic. We are not in a position to critically evaluate the validity of the above estimations. Furthermore, we do know that the ionicity of the metal-carbon bond is greatly affected by the nature of the carbon group, as discussed later. Nonetheless, it is important to realize that most of the metal-carbon bonds that concern us are likely to be relatively covalent.

Table 1.1 Relation Between Electronegativity Difference and
Ionicity of Single Bonds

$EN_A - EN_B$	Ionicity(%)	$EN_A - EN_B$	Ionicity(%)
0.2	1	1.8	55
0.4	4	2.0	63
0.6	9	2.2	70
0.8	15	2.4	76
1.0	22	2.6	82
1.2	30	2.8	86
1.4	39	3.0	89
1.6	47	3.2	92

Although it is generally difficult to correlate the ionicity
of a bond with some measurable quantities, the ionicity (q) of a
diatomic molecule can readily be estimated from its bond length
(r) and gas-phase dipole moment (μ) according to eq. 1.7.

$$q = \frac{\mu(D) \times 10^{-18} \, \text{Å-esu}}{r \, \text{Å} \times 4.8 \times 10^{-18} \, \text{esu}} \tag{1.7}$$

1.2.2 Polarizability

A common misconception is to think that a highly covalent bond is
very reluctant to participate in ionic reactions. We all know
that the essentially covalent iodine molecule readily participates in a variety of ionic reactions. This is qualitatively explained in terms of the high polarizability of the iodine-iodine
bond. Polarizability is yet another highly nebulous property
that is a function of a number of factors including the shape,
size, and other properties of the bond-forming orbitals. Whereas
the ionicity of a bond is related to its permanent bond dipole
moment, the polarizability of a bond is related to its induced
bond dipole moment. Despite the vagueness of our current understanding, however, the concept of polarizability is of essential
importance in our discussion of the chemistry of organometallic
compounds containing heavy atoms, such as Hg, Tl, and Pd.

1.2.3 Bond Strength and Bond Length

In addition to polarity and polarizability, bond strength and
bond length are important parameters of any bond affecting the
kinetics and thermodynamics of the reactions in which the bond in

question participates.

Covalent Bond Energies and Covalent Bond Lengths   The energy required to cleave a covalent bond to give the constituent radicals
is called the bond dissociation energy (DE). Although its definition is clear, its measurement is often difficult. Consequently, a more ambiguous but more readily obtainable quantity termed
"bond energy" has been introduced. This corresponds to the aver-
age of the bond dissociation energies for all bonds of one kind
present in a given molecule. In our discussions, we will be primarily dealing with bond energies (BE). The metal-carbon bond
energies for some relatively covalent organometallic compounds of
the Me_nM type are listed in Table 1.2.

Table 1.2   Metal-Carbon Bond Energy for Me_nM

Me_2M	BE	Me_3M	Be	Me_4M	BE
	(kcal/mole)		(kcal/mole)		(kcal/mole)
Me_2Be	–	Me_3B	87	Me_4C	83
Me_2Mg	–	Me_3Al	66	Me_4Si	70
Me_2Zn	42	Me_3Ga	59	Me_4Ge	59
Me_2Cd	33	Me_3In	–	Me_4Sn	52
Me_2Hg	29	Me_3Tl	–	Me_4Pb	37

A large number of bond energy and bond length values have been
compiled by Cottrell (1.10).
        The covalent bond length can be determined accurately by X-
ray analysis and microwave spectroscopy. It is, however, useful
to be able to estimate the bond length (r_{A-B}) between atoms A and
B from their atomic parameters, termed covalent radii (r_{cov}). If
we can determine the covalent bond lengths (r_{A-A}) of homonuclear
bonds, the covalent radii (r_A) can be obtained by dividing them
by two. If not, some indirect methods must be used. Schomaker
and Stevenson (1.11) proposed the following relation:

$$r_{A-B} = r_A + r_B - 0.09 \, (|EN_A - EN_B|) \tag{1.8}$$

Equation 1.8 can be used to estimate r_A from the heteronuclear
bond length (r_{A-B}), the covalent radius of atom B (r_B), and the
electronegativity difference ($|EN_A - EN_B|$). This equation sug-
gests that for highly covalent bonds the covalent atomic radii

are reasonably additive. The covalent radii for various elements
are listed in Appendix V.

Lattice Energy and Ionic Radii   The energy of the crystal lattice
of an ionic compound is represented by the lattice energy, which
can be determined experimentally. Although the distances between
two neighboring cations and anions in ionic crystals can be mea-
sured with a high degree of accuracy, it is difficult to correctly apportion them between the cations and the anions. Many
devices have been designed for this purpose. For our discussions, it will suffice to know general trends in ionic radii and
their approximate values (Appendix V).

1.3   CARBON GROUPS

1.3.1   Electronegativity and Basicity of Carbon Groups and Acidity of Carbon Acids

One of the most important properties associated with the organic
moiety (R) of an organometallic compound is its basicity. Since
the basicity of R is significantly modified by the metal-containing moiety, it is more convenient to discuss the acidity (pK_a) of
the conjugate carbon acid (RH), which is a parameter representing
the electronegativity of the R group. Detailed discussions of
this subject have been presented in many excellent monographs
(*1.12* to *1.16*). We shall briefly touch on some factors affecting
the acidity of carbon acids. The pK_a values of some representative carbon acids are summarized in Appendix VI.

1.3.1.1   Inductive Effect Due to Hetero Atoms

The pK_a values for methane and chloroform are 48 and 24, respectively. This difference is readily explained in terms of the
electron-withdrawing inductive effect of the highly electronegative Cl atom.

1.3.1.2   Resonance Effect and Resonance Inhibition

The resonance effect on the acidity of carbon acids is clearly
seen in the pK_a values of a series of phenylated methans:
$PhCH_3(pK_a = 41)$, $Ph_2CH_2(pK_a = 34)$, and $Ph_3CH(pK_a = 31-32)$. The
markedly higher acidity of toluene relative to that of methane is
largely due to the extensively delocalized nature of the corresponding benzyl carbanion. The acidity of diphenylmethane can be
interpreted in an analogous manner. On the other hand, the intro-

duction of the third phenyl group has a much smaller effect than
that of the first or second. This has been interpreted in terms
of steric inhibition of resonance due to the propeller-shaped
structure of the trityl anion. This explanation is plausible, as
triptycene (1.2), whose benzylic carbanion cannot be stabilized
through delocalization, is indeed a considerably weaker acid
(pK_a 42) than triphenylmethane.

1.2

1.3.1.3 Aromaticity

The relative pK_a values of cyclopentadiene and cycloheptatriene
cannot be explained in terms of the extent of delocalization a-
lone. The aromaticity ($4n + 2$ electrons) and anti-aromaticity
($4n$ electrons) of these carbanions must also be taken into consideration.

(pK_a = 15)   (pK_a = 36)

1.3.1.4 Hybridization Effect

The pK_a values of ethane (pK_a = 49), ethylene (pK_a = 44), and a-
cetylene (pK_a = 25) can be interpreted in terms of the differences in hybridization. The $2s$ atomic orbital energy level is lower
than that of the $2p$ orbital. Thus, electrons are closer and more
tightly bound to the nucleus in the $2s$ orbital than in the $2p$ orbital. In other words, the $2s$ orbital is more electronegative
than the $2p$ orbital. The electronegativity order for sp hybrid
orbitals as well as s and p orbitals is $s > sp > sp^2 > sp^3 > p$.

1.3.2 Bond Energies and Bond Lengths of Some Carbon-Carbon and Carbon Hetero Atom Bonds

In the preceding section, we considered only one bond energy value and one bond length value for a given bond (A-B). This, of course, is a gross oversimplification. We do know that the bond between two given atoms A and B can be significantly affected by neighboring and surrounding atoms, groups, molecules, and so on. While this must be true with all types of bonds, we shall consider only the hybridization effect on the carbon-carbon and some carbon-hetero atom (nonmetal) bond lengths as one of the most noticeable and significant examples.

Table 1.3 lists some carbon-carbon and carbon-hetero atom bond lengths. The pronounced effect of bond multiplicity is readily understood.

Table 1.3   Bond Lengths of Some Carbon-Carbon and Carbon-Hetero Atom Bonds

Bond Type	Length (Å)	Bond Type	Length (Å)
C-C		C-H	
sp^3-sp^3	1.54	sp^3, sp^2, or sp-H	1.10 ± 0.02
sp^3-sp^2	1.50	C-N	
sp^3-sp	1.46	sp^3-N	1.47
sp^2-sp^2	1.48	sp^2-N	1.36
sp^2-sp	1.43	C=N	
sp-sp	1.38	sp^2-N	1.28
C=C		C≡N	
sp^2-sp^2	1.34	sp-N	1.16
sp^2-sp	1.31	C-O	
sp-sp	1.28	sp^3-O	1.41
		sp^2-O	1.34
C≡C		C=O	
sp-sp	1.21	sp^2-O	1.20
		sp-O	1.16

The effect of hybridization has been interpreted in terms of partially increased bond multiplicity due to resonance and hyperconjugation, as indicated below.

These effects are also seen in the bond energies of these bonds. However, as we have already discussed earlier, the bond energy is a less well-defined parameter than the bond length. To indicate the effect of bond multiplicity, some bond energy data are shown in Table 1.4.

Table 1.4   Bond Energies for Some Carbon-Carbon and Carbon-Hetero Atom Bonds

Bond	Bond Energy (kcal/mole)	Bond	Bond Energy (kcal/mole)
C-C	83-85	C-O	85-91
C=C	146-151	C=O	173-181
C≡C	199-200	C-S	66
C-H	96-99	C-Cl	79
C-N	69-75	C-Br	66
C=N	143	C-I	52
C≡N	204		

1.4   MOLECULAR STRUCTURE

We have so far discussed some of the important atomic and bond parameters without dealing with the overall molecular shapes or structures. In this section, let us briefly review some factors governing (1) the elemental composition and the topology, that is, the atom-linking sequence and then (2) the three-dimensional molecular shape, without resorting to sophisticated modern quantum mechanical theories. For more rigorous and sophisticated treatments, the reader is referred to some of the textbooks and monographs cited earlier (*1.1* to *1.7*).

1.4.1   The Lewis Octet Rule

Despite the development of many sophisticated theories, the simple empirical rule of bonding developed by Lewis, called the Lewis octet rule, still provides a convenient basis for understanding and predicting molecular structures containing many im-

portant main group elements. It simply states that the first-row
atoms have a strong tendency to acquire eight valence electrons,
which the inert gas atom of the same row, namely neon, has. We
all know that we must apply the modified "two-electron rule" to
hydrogen and helium and that hypervalency or valence-shell expan-
sion can be observed with heavier atoms. In dealing with d-block
transition metals, the 16 or 18 electron rules discussed in Sect.
8.1 have proved useful.

The application of the Lewis octet rule to the organometallic compounds of the main group metals requires special comments.
If we examine the monomeric representations of LiMe, BeMe$_2$, and
BMe$_3$, we find that none of them is in accord with the octet rule.
They are a duet, a quartet, and a sextet, respectively. In light
of the low ionicity of the metal-carbon bonds of these compounds,
the ionic representations, such as $Be^{+2}Me_2^{-1}$, which would be in
agreement with the octet and two-electron rules, cannot be significant. Consequently, if the Lewis octet rule were at all mean-
ingful in such cases, it must imply either that these molecules
do not exist as monomers or that they have a strong tendency to
accept electrons so as to acquire a total of eight valence electrons. As discussed in Sect. 4.1, LiMe indeed exists as its
tetramer 1.3, and BeMe$_2$ as its polymer 1.4.

1.3   1.4

As discussed later in more detail the metal atoms in these
compounds are bonded to the methyl groups via multicenter bonding. Each lithium atom is simultaneously bonded to three methyl
groups, as shown in 1.5, and each methyl group is simultaneously
bonded to three lithium atoms, as shown in 1.6. Only BMe$_3$ appears to exist as a monomeric species under normal conditions.
In a formal sense, the Li atom in (LiMe)$_4$ and the B atom in BMe$_3$
are sextets, and the Be atom in (BeMe$_2$)$_n$ an octet. Moreover,
these electron-deficient species, BMe$_3$, for example, have a
strong tendency to acquire electrons, that is, Lewis acidic, so
as to form Lewis octets through either self-association or com-

plexation with electron donors (Lewis bases). In conclusion,
then, many commonly employed monomeric representations of organometallic compounds of the main group metals do not obey the Lewis
octet rule. However, they either exist as octet species through
association or display properties, which can be readily explained
within the context of the Lewis octet rule.

1.5   1.6   1.7

1.4.2 Coordinative Saturation and Unsaturation and Multicenter Bonds

Electron-deficient species, such as BMe_3, are said to be coordinatively unsaturated, since there is at least one empty valence
orbital available for extra coordination or bond formation. Con-
versely, chemical species, in which all valence orbitals are
doubly occupied, are coordinatively saturated. As previously indicated, a large number of organometallic compounds are either
coordinatively unsaturated or capable of generating coordinatively unsaturated species. This is not the case with organic compounds, which do not contain metals. Organic compounds rarely
exist as coordinatively unsaturated stable species. Carbonium
ions and singlet carbenes are coordinatively unsaturated, but
they are usually very short-lived. On the other hand, most triorganoboranes exist as thermally stable sextets. Similarly, di-
organoberylliums and triorganoalanes contain thermally stable
monomeric species in the vapor phase.
    One of the characteristics of these coordinatively unsatu-
rated species is their tendency toward self-association, as mentioned earlier. This association phenomenon should clearly be
separated from other much weaker intermolecular association phe-
nomena, such as the association of coordinatively saturated covalent molecules through very weak London forces. The metal-carbon and metal-hetero atom bonds that are responsible for forming
associated organometallic species are conveniently described in
terms of multicenter two-electron bonds, originally proposed by

Longuet-Higgins (*1.17*) for explaining the structure of diborane.
Thus, the Be-C-Be bond in $(BeMe_2)_n$ is a three-center two-electron
bond represented by 1.8, and the Li atom of $(LiMe)_4$ participates
in a four-center two-electron bond represented by 1.9. Only the
bonding orbitals are shown.

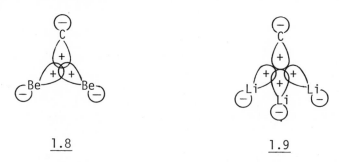

1.8   1.9

Note that the carbon atoms of $(BeMe_2)_n$ and $(LiMe)_4$ are pentava-
lent and hexavalent, respectively. The familiar but naive concept of tetravalency for carbon must be abandoned in describing
these multicenter bonds. What we do not have to discard is the
Lewis octet rule. It should be clearly recognized that, despite
their "abnormal" valencies, the carbon atoms of $(BeMe_2)_n$ and
$(LiMe)_4$ are all octets. The significance of the concept of multicenter bond in organometallic chemistry cannot be overemphasiz-
ed. Unfortunately, a detailed description of multicenter bonds
is beyond the scope of our discussion. Interested readers are
referred to advanced textbooks and reviews discussing this subject, such as those by Cotton (*1.7*) and Lipscomb (*1.18*).

1.4.3 Parameters Governing Molecular Shapes - Bond Angle, Hybridization, and Electron-Electron Repulsion

Although the Lewis octet rule is useful in understanding and predicting the atom-linking sequence in a molecule, it does not predict or help us understand the three-dimensional structure of the
molecule, except in some very simple cases, such as diatomic
molecules. For example, the Lewis octet rule itself does not
distinguish 1.10 from 1.11.
    Experimentally, what we need to know are some bond angles in
the molecule and information on some dynamic aspects of the molecule. Thus if we know the C-Si-C, H-C-H, and Si-C-H bond angles
and the freely rotating nature of the Si-C bond in addition to

1.10   1.11

the atom-linking sequence, we can unequivocally establish the three-dimensional structure of $Si(CH_3)_4$. X-ray and electron diffraction methods have been widely used for the determination of bond angles. The dynamic properties of molecules have been studied using various spectroscopic methods including NMR.

The concept of hybrid orbitals helps explain molecular shapes. Thus the tetrahedral arrangement about the Si atom of $SiMe_4$ is readily explained in terms of the sp^3 hybridization of the Si valence orbitals. The bond angles associated with some common hybridizations are summarized in Table 1.5.

Table 1.5   Relation Between Hybridization and Bond Angles

Hybridization	Bond Angle(s)	Hybridization	Bond Angle(s)
sp	180°	dsp^2	90°
sp^2	120°	dsp^3	90° and 120°
sp^3	109.5^o	d^2sp^3	90°

It should be emphasized, however, that it is not hybridization that dictates the three-dimensional structure. In fact, the bond angles listed in Table 1.5 are observed only with atoms that are bonded to one type of atoms or groups. In the other cases, the bond angles usually deviate from these listed values. It is the energetics of a molecule that determine its three-dimensional structure. In other words, with a given topology or atom-linking sequence, a molecule will shape itself up in three-dimensional space so as to minimize its potential energy.

What then controls the energy of a molecule with a given atom-linking sequence? While there can be various factors governing the bond angles or three-dimensional shape of a molecule, the

principal factor is presumably an electron-electron repulsion
force. Three types of electron-electron repulsions that must be
considered are: lone pair-lone pair > lone pair-bonding pair >
bonding pair-bonding pair. The order of the magnitude of repulsion energies is as indicated above (*1.19*). Although such a
crude and qualitative theory does not give precise bond angles,
it will permit us to predict the overall structure of a molecule
and the approximate bond angles with reasonable confidence. The
trigonal planar arrangement for the B atom of BMe_3, the pseudo
tetrahedral arrangement for the Be atoms of $(BeMe_2)_n$ and the tetrahedral arrangement for the Si atom of $SiMe_4$ can readily be predicted based on the electron-electron repulsion theory presented
above. In addition to the papers by Gillespie (*1.19*), Huheey
provides a detailed discussion of this subject in his textbook
(*1.3*).

1.5 INTERMOLECULAR FORCES

Highly covalent organic molecules are held together in liquid and
solid phases by weak van der Waals forces, such as London disper-
sion force, and the intermolecular distances are represented by
van der Waals radii. The relation between the covalent and van
der Waals radii in homonuclear diatomic molecules is shown in
Fig. 1.2.

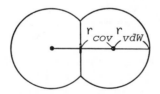

Fig. 1.2 The relation between the covalent and van der Waals
radii

As the ionicity of a bond increases, stronger electrostatic
interactions, such as hydrogen bonding, solvation, inclusion,
charge-transfer complex formation, and so on, become more and
more important. These interactions are often largely electro-

static in nature and can be classified into dipole-dipole, pole-(ion)-dipole, and pole(ion)-pole(ion) interactions. Even in these intermolecular interactions, however, orbital-orbital interaction, that is, covalent bonding, often plays a significant role. Moreover if we consider all types of intermolecular interactions, we soon recognize that it is impossible to classify all chemical bonds into two distinct classes, that is, intramolecular and intermolecular bonds. It is neither appropriate nor necessary to discuss all these different types of intermolecular bondings here.

It is very important, however, to fully realize that both the relatively high polarity of the metal-carbon bonds and the wide occurrence of coordinatively unsaturated organometallic species, as well as self-associated organometallic species, make intermolecular interactions that are stronger than van der Waals interactions widely observed and highly significant in organometallic chemistry. As a typical and significant example, consider solvation. The process of dissolving typical organic compounds, such as cyclohexanone, in a typical organic solvent, such as tetrahydrofuran (THF), is normally associated with weak electrostatic interactions. The IR carbonyl stretching frequency, for example, is only slightly affected by solvation. On the other hand, the process of dissolving organometallic compounds in organic solvents, such as THF, is often accompanied by much stronger interactions. For example, $(AlMe_3)_2$, a self-associated dimeric species, undergoes dissociation and forms a relatively stable 1:1 complex, that is, $AlMe_3 \cdot THF$, which gives totally different spectral data than $(AlMe_3)_2$. In other words, THF is acting as both a solvent and a reactant. In trying to understand and predict the course of an organometallic reaction, it is often of essential significance to consider the "solvent" as a reagent or reactant. For example, if we compare the rates of hydroboration of typical olefins, such as ethylene, with BH_3 and $ClBH_2$, we obtain the following results. In THF, the reaction is much faster with BH_3 than with $ClBH_2$, whereas the opposite is the case in ethyl ether. A priori, one might predict that the stronger Lewis acid of the two, that is, $ClBH_2$, would be the more reactive of the two. That is what one finds in ethyl ether, irrespective of their precise structures. In THF, however, their reactivity order is reversed, because the stronger Lewis acid, $ClBH_2$, forms a much tighter complex with THF than BH_3. In short, it is advisable to consider any solvents of electron-donating ability as reagents. For more systematic discussions of intermolecular interactions, the reader is referred to some of the textbooks cited earlier (*1.1* to *1.3*).

1.6 SOME FACTORS CONTROLLING CHEMICAL REACTIONS

Chemical reactions, including all organometallic reactions, may be arbitrarily classified into the following three general types: (1) generalized acid-base or electrophile-nucleophile reactions, that is, polar reactions, (2) free-radical reactions, and (3) concerted molecular reactions, which do not correspond to either (1) or (2), that is, pericyclic reactions. Within the past two decades, significant advances have been made in our understanding of the polar and pericyclic reactions. Especially noteworthy are the principle of conservation of orbital symmetry developed by Woodward and Hoffmann, called the Woodward-Hoffmann rule (1.6), and the principle of the hard and soft acids and bases (the HSAB principle) developed by Pearson (1.20), which has been significantly reinforced by Klopman's perturbation treatment (1.21). There are many excellent reviews on both the Woodward-Hoffmann rule and related theories (1.5, 1.6, 1.22 to 1.25) and the HSAB principle (1.20, 1.25). Consequently, it will suffice to present a very brief description of these principles and a few comments pertaining to their use in organometallic chemistry.

1.6.1 The Principle of Conservation of Orbital Symmetry

Based on the principle that orbital symmetry is conserved in concerted reactions, Woodward and Hoffmann (1.6) presented the following general rule for pericyclic reactions.

A ground-state or thermal reaction is symmetry allowed if the total number (Σ) of ($4q + 2$) suprafacial and $4r$ antarafacial components is odd (q and $r = 0$ or integers). The number of electrons involved in each component is indicated by either $4q + 2$ or $4r$. The reverse would be true for excited-state or photochemical reactions. The suprafacial and antarafacial modes of interactions are shown below for π, σ, and ω (nonbonding) orbitals.

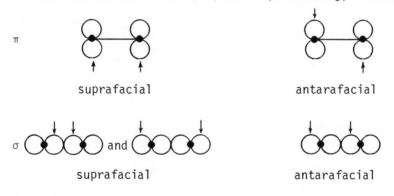

suprafacial   antarafacial

For example, the 1,2-migration reactions of organoboranes via organoborate intermediates discussed in Sect. 5.2.3 is a $[_\sigma 2_s + _\omega 0_s]$ process ($\Sigma = 1$) which is thermally allowed and should proceed with retention of configuration of the migrating group.

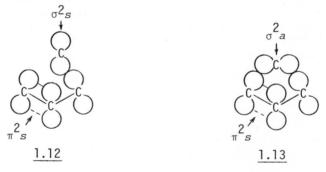 (1.9)

If one starts applying the Woodward-Hoffmann rule to organometallic reactions one soon recognizes that many organometallic reactions, which appear to be concerted, proceed readily under thermal conditions, and must therefore be thermally allowed, even when their organic counterparts do not occur readily. For example, the 1,3-shift or allylic rearrangement of an alkyl group (R) does not occur readily, presumably because the low strain process represented by 1.12 is a thermally disallowed $[_\pi 2_s + _\sigma 2_s]$ process ($\Sigma = 2$), whereas the thermally allowed $[_\pi 2_s + _\sigma 2_a]$ process represented by 1.13 is a high strain process.

$\sigma^2 s$

$\pi^2 s$

1.12

$\sigma^2 a$

$\pi^2 s$

1.13

On the other hand, a number of 1,3-metallotropic reactions are very facile, we further notice that the metal atoms of facile 1,3-metallotropic reactions either are coordinatively unsaturated or can readily be converted into coordinatively unsaturated atoms.

Indeed, many other seemingly concerted facile organometallic reactions, which might at first sight appear thermally disallowed by the principle of orbital symmetry conservation, are associated with the presence of an empty metal orbital. A wide variety of cis hydrometallation and carbometallation reactions fall into this category. The effect of an empty metal orbital can be explained most conveniently by Fukui's frontier orbital theory (*1.22*).

The highest occupied molecular orbital (HOMO) and the lowest unoccupied molecular orbital (LUMO) are called the frontier orbitals, which may be loosely equated to the filled and empty valence orbitals of atoms. If the symmetry of the HOMO of one reactant or component is such that it can overlap with the LUMO of the other reactant or component, the reaction between the two reactants or components is favored as a concerted process. The HOMO-LUMO interactions in the 1,3-metallotropy are shown in 1.14 and 1.15.

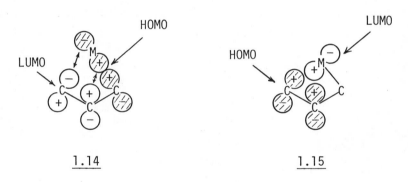

1.14   1.15

The interaction in 1.14 is equivalent to the $[_{\pi}2_s + _{\sigma}2_a]$ interaction shown in 1.13. Therefore, it appears the HOMO-LUMO interaction shown in 1.15 is mainly responsible for the ease with which the coordinatively unsaturated allylmetals undergo the 1,3-shift.

1.6.2 The Principle of Hard and Soft Acids and Bases

The Woodward-Hoffmann rule, which deals only with the symmetry aspects of orbitals, is a very rigid theory that has been successfully applied to most, if not all, of the pericyclic reactions. Its main limitation is that it cannot be applied to other types of reactions. Although no comparable theory that is applicable to other types of reactions has been developed, a more widely applicable but much more "soft" principle of the hard and

soft acids and bases (*1.20*) has proved to be highly useful.
In Klopman's quantum mechanical treatment (*1.21*), energy change (ΔE) in a chemical reaction is evaluated in terms of a charge-charge interaction representing electrostatic interactions, an orbital-orbital interaction, and a solvation shown in eq. 1.10

$$\Delta E = \frac{-q_r q_s e^2}{d\varepsilon} + \underset{\substack{m \quad n \\ \text{occ. unocc.}}}{\Sigma \quad \Sigma} \left[\frac{2(c_r^{\,m})^2 (c_s^{\,n})^2 \beta^2}{E_m^* - E_n^*} \right] + \Delta_{solv} \qquad (1.10)$$

where q_r and q_s are charges of the reactants R and S, respectively, d is the distance between these sites, ε is the dielectric constant, $c_r^{\,m}$, $c_s^{\,n}$ are the coefficients of the atomic orbital r in a frontier orbital m of energy E_m^* and of the atomic orbital s in a frontier orbital n of energy E_n^*, respectively, β is the resonance integral of the developing bond, $E_m^* - E_n^*$ is the energy difference of the frontier orbitals m and n, and Δ_{solv} is solvation energy. When the difference in energy between the frontier orbitals is large, the second term in eq. 1.10 is small, and the reaction is dominated by electrostatic interactions and is said to be charge controlled. When $|E_m^* - E_n^*|$ is small, the interactions are primarily covalent, and the reaction is said to be frontier controlled. A charge-controlled reaction is favored by species of small size and of low polarizability with high effective nuclear charge. Acids (electrophiles) and bases (nucleophiles) that have these properties are called "hard" acids and "hard" bases. On the other hand, the frontier-controlled reaction is favored by large, highly polarizable species of low or zero nuclear charge. These species are called "soft" acids and "soft" bases. The HSAB principle clearly indicates that it is intrinsically incorrect to relate a reaction rate to a particular reactivity index of a reactant. Thus there can be no such thing as the universal nucleophilicity or electrophilicity order. The relation between the ease of reaction and the hardness or softness of the acids and bases is shown in Table 1.6.
More than 20 years ago, Chatt (*1.26*) classified various metal ions and ligand bases into two classes, which correspond to Pearson's hard and soft species. The abilities of various ligand bases to complex with hard metal ions, such as Al^{+3}, Ti^{+4}, and Co^{+3}, and with soft metal ions, such as Pt^{+2}, Ag^+, and Hg^{+2}, are shown in Table 1.7. Only the atoms acting as a basic center are indicated.

Table 1.6 Rates of Reactions between Hard and Soft Reagents

Acid	Base	$|E_m^* - E_n^*|$	e^2/d	β	Rate
Hard	Hard	Large	Large	Small	High
Soft	Soft	Small	Very small	Large	High
Hard	Soft	Medium	Small	Very small	Low
Soft	Hard	Medium	Small	Very small	Low

Table 1.7 The Ability of Bases to Complex with Metal Ions

Ability to Complex with Hard Metal Ions	Ability to Complex with Soft Metal Ions
N >> P > As > Sb	N << P > As > Sb
O >> S > Se > Te	O << S < Se ~ Te
F > Cl > Br > I	F < Cl < Br < I

Table 1.8 Classification of Bases[a]

Hard Bases	NH_3, RNH_2, NH_2NH_2
	$H_2O(-10.73)$, ^-OH (-10.45), HOR, ^-OR, OR_2
	^-OAc, CO_3^{-2}, NO_3^-, PO_4^{-3}, SO_4^{-2}, ClO_4^-
	$F^-(-12.18)$, $Cl^-(-9.94)$
Borderline Bases	$ArNH_2$, C_5H_5N, N_2, N_3^-
	NO_2^-, SO_3^-
	$Br^-(-9.22)$
Soft Bases	$H^-(-7.37)$
	R^-, olefin, arene, $C{\equiv}O$, $CN^-(-8.78)$, RNC
	PR_3, $P(OR)_3$, AsR_3
	HSR, ^-SR, $^-SH(-8.59)$, SR_2, $S_2O_3^{-2}$
	$I^-(-8.31)$

[a] The numbers in parentheses are Klopman's softness parameters ($E^{\ddagger}$) in electronvolts.

Table 1.9 Classification of Metal Ions (Softness Parameter $E^{\ddagger}$, eV)

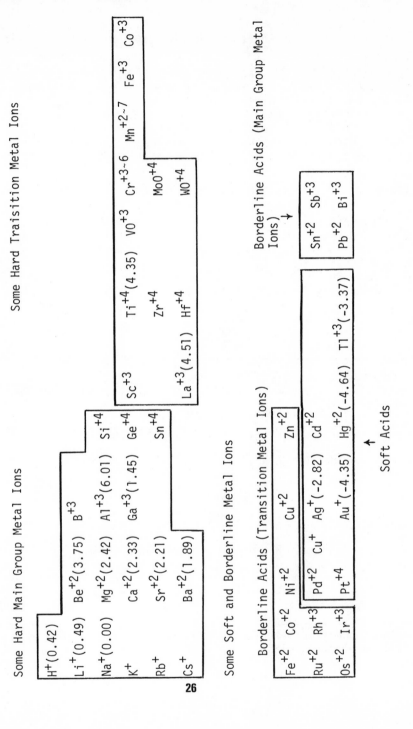

Some Hard Main Group Metal Ions

Some Hard Traisition Metal Ions

$H^+(0.42)$

$Li^+(0.49)$ $Be^{+2}(3.75)$ B^{+3}

$Na^+(0.00)$ $Mg^{+2}(2.42)$ $Al^{+3}(6.01)$ Si^{+4}

K^+ $Ca^{+2}(2.33)$ $Ga^{+3}(1.45)$ Ge^{+4}

Rb^+ $Sr^{+2}(2.21)$ Sn^{+4}

Cs^+ $Ba^{+2}(1.89)$

Sc^{+3} $Ti^{+4}(4.35)$ VO^{+3} $Cr^{+3\sim6}$ $Mn^{+2\sim7}$ Fe^{+3} Co^{+3}

Zr^{+4} MoO^{+4}

$La^{+3}(4.51)$ Hf^{+4} WO^{+4}

Some Soft and Borderline Metal Ions

Borderline Acids (Transition Metal Ions)

Fe^{+2} Co^{+2} Ni^{+2} Cu^{+2} Zn^{+2}

Ru^{+2} Rh^{+3} Pd^{+2} Cu^+ $Ag^+(-2.82)$ Cd^{+2}

Os^{+2} Ir^{+3} Pt^{+4} $Au^+(-4.35)$ $Hg^{+2}(-4.64)$ $Tl^{+3}(-3.37)$

Soft Acids

Borderline Acids (Main Group Metal Ions)

Sn^{+2} Sb^{+3}

Pb^{+2} Bi^{+3}

Table 1.10   Classification of Acids Other than Metal Ions

Hard Acids BF_3, BCl_3, $B(OR)_3$, AlH_3, AlR_3, $AlCl_3$

                       RCO^+, CO_2

                       N^{+3}, As^{+3}, SO_3, RSO_2^+, Cl^{+3}, Cl^{+7}, I^{+5}, I^{+7}

Borderline BR_3, GaH_3
Acids
                       R_3C^+, Ph^+

                       NO^+, SO_2

Soft Acids BH_3, GaR_3, TlR_3

                       CH_2

                       Quinones, $(NC)_2C=C(CN)_2$

                       HO^+, RO^+, RS^+, Br_2, Br^+, I_2, I^+, ICN

                       $Cl\cdot$, $Br\cdot$, $I\cdot$, $RO\cdot$, $RO_2\cdot$

                       Metal atoms and bulk metals

Pearson (*1.20*) has expanded this concept and classified qualitatively a variety of acids and bases according to their hardness or softness (Tables 1.8 to 1.10). Klopman introduced a parameter $E^\ddagger$, which is a measure of softness, based on his quantum mechanical perturbation treatment. The softness parameter ($E^\ddagger$) for some acids and bases is indicated in parentheses in Table 1.8 and 1.9. The high positive and negative $E^\ddagger$ values indicate hard acids and bases, respectively.

It is evident from these tables that the hardness or softness of a given element varies greatly depending on the charge and the nature of surrounding groups. For example, BF_3, BMe_3, and BH_3 have been classified as hard, borderline, and soft acids, respectively. This situation is quite analogous to that which we encountered earlier in our discussion of electronegativity. Indeed, species having high electronegativities are generally hard and those having low electronegativities are generally soft. In this connection, it should be recognized that, although Li metal is electropositive, the Li^+ ion is highly electronegative.

Unlike the Woodward-Hoffmann rule, the HSAB principle does not always predict in an unequivocal manner if a given reaction will be facile or not. It does usually permit us, however, to predict which of the two closely related reactions will be more facile. For example, it has been successfully employed in explaining the regiochemistry of a number of reactions which ambi-

dent species undergo. Thus the reaction of the enolate anion derived from acetophenone with EtI and $Et_3O^+BF_4^-$ gives the O-alkyl-
ation/C-alkylation ratio of 0.1 and 4.9, respectively (*1.27*)(eq.
1.11). This trend is readily explained by the HSAB principle.

$$PhC{=}CH_2 \xrightarrow{\;EtX\;} \underset{\text{O-alkylation}}{PhC{=}CH_2} + \underset{\text{C-alkylation}}{PhCCH_2Et} \qquad (1.11)$$

X = I (soft)   O-/C- = 0.1

X = $Et_2O^+BF_4^-$ O-/C- = 4.9
   (hard)

REFERENCES

1.1   Pauling, L., *The Nature of the Chemical Bond*, 3rd ed., Cor-
        nell University Press, Ithaca, N. Y., 1960, 644 pp.

1.2   Cotton, F. A., and Wilkinson, G., *Advanced Inorganic Chem-
        istry*, 3rd ed., Wiley-Interscience, New York, 1972, 1136 pp.

1.3   Huheey, J. E., *Inorganic Chemistry*, Harper & Row, New York,
        1972, 737 pp.

1.4   Ballhausen, C. J., and Gray, H. B., *Molecular Orbital Theo-
        ry*, Benjamin, New York, 1964, 273 pp.

1.5   Dewar, M. J. S., *The Molecular Orbital Theory of Organic
        Chemistry*, McGraw-Hill, New York, 1969, 484 pp.

1.6   Woodward, R. B., and Hoffmann, R., *The Conservation of Or-
        bital Symmetry*, Academic Press, New York, 1970, 178 pp.

1.7   Cotton, F. A., *Chemical Applications of Group Theory*, 2nd
        ed., Wiley-Interscience, New York, 1971, 271 pp.

1.8   Allred, A. L., and Rochow, E. G., *J. Inorg. Nucl. Chem.*, **5**,
        264 (1958).

1.9   Mulliken, R. S., *J. Chem. Phys.*, **2**, 782 (1934); **3**, 573
        (1935).

1.10 Cottrell, T. L., *The Strength of Chemical Bonds*, Butter-
        worths, London, 1954, 310 pp.

1.11 Schomaker, V., and Stevenston, D. P., *J. Am. Chem. Soc.*, 63, 37 (1941).

1.12 Cram, D. J., *Fundamentals of Carbanion Chemistry*, Academic Press, New York, 1965, 289 pp.

1.13 Ebel, H. F., *The Acidity of CH Acids*, Georg Thieme Verlag, Stuttgart, 1969.

1.14 Schlosser, M., *Struktur und Reaktivität polarer Organometalle*, Springer-Verlag, New York, 1973, 187 pp.

1.15 Jones, J. R., *The Ionization of Carbon Acids*, Academic Press, N. Y., 1973, 236 pp.

1.16 Buncel, E., *Carbanions: Mechanistic and Isotopic Aspects*, Elsevier, Amsterdam, 1975, 270 pp.

1.17 Longuet-Higgins, H. C., *J. Chem. Phys.*, 46, 268 (1949).

1.18 Lipscomb, W. N., *Acc. Chem. Res.*, 6, 257 (1973).

1.19 (a) Gillespie, R. J., *J. Am. Chem. Soc.*, 82, 5987 (1970); (b) Gillespie, R. J., *J. Chem. Educ.*, 40, 295 (1963); 47, 18 (1970).

1.20 (a) Pearson, R. G., *J. Am. Chem. Soc.*, 85, 3533 (1963); (b) Pearson, R. G., *Science*, 151, 172 (1966); (c) Pearson, R. G., *J. Chem. Educ.*, 45, 581, 643 (1968).

1.21 Klopman, G., *J. Am. Chem. Soc.*, 90, 223 (1968).

1.22 Fukui, K., *Acc. Chem. Res.*, 4, 57 (1971).

1.23 Lehr, R. E., and Marchand, A. P., *Orbital Symmetry*, Academic Press, New York, 1972, 190 pp.

1.24 Gilchrist, T. L., and Storr, R. C., *Organic Reactions and Orbital Symmetry*, Cambridge University Press, 1972, 271 pp.

1.25 Pearson, R. G., *Symmetry Rules for Chemical Reactions*, Wiley, New York, 1976, 548 pp.

1.26 Ahrland, S., Chatt, J., and Davies, N. R., *Quart. Rev.*, 12, 265 (1958).

1.27 Heiszwolf, G. J., and Kloosterziel, H., *Rec. Trav. Chim. Pays-Bas*, 89, 1153 (1970).

2
METHODS OF PREPARATION
OF ORGANOMETALLIC COMPOUNDS

Although numerous methods for the preparation of organometallic
compounds are known, the great majority of them can be classified
into a dozen or so categorically different methods (Table 2.1).
Organic halides, organic compounds containing active C-H bonds,
and unsaturated organic molecules, such as olefins, acetylenes,
and arenes, are the three most common groups of organic starting
materials. In addition to these compounds, certain organometallic compounds themselves often serve as precursors to organometallic compounds containing different metals. Transformations between two organometallics containing the same metal, however, are
not included in the following discussion, as these are better
viewed as reactions rather than preparations of organometallics.

Table 2.1   Methods of Preparation of Organometallic Compounds

Method	Description	General equation
From halides		
Method I	Oxidative metallation of organic halides	$RX + 2M \longrightarrow RM + MX$ or $RX + M \longrightarrow RMX$
Method II	Oxidative displacement of halogen by metallate anions	$RX + {}^{-}ML_n \longrightarrow RML_n + X^{-}$

Method III	Metal-halogen exchange	$RX + MR' \longrightarrow RM + R'X$

From active C-H Compounds

Method IV	Oxidative metallation of active C-H compounds	$RH + M \longrightarrow RM + \tfrac{1}{2}H_2$
Method V	Metal-hydrogen exchange	$RH + MR' \longrightarrow RM + HR'$

From unsaturated compounds

Method VI	Hydrometallation	$\text{C=C} + HM \longrightarrow H\text{-C-C-M}$
Method VII	Heterometallation	$\text{C=C} + XM \longrightarrow X\text{-C-C-M}$
Method VIII	Complexation	$\text{C=C} + M \longrightarrow \overset{}{\underset{M}{C\text{≡}C}}$
Method IX	Oxidative coupling	$\text{C=C} + M \longrightarrow \overset{}{\underset{M}{C\text{-}C}}$
Method X	Oxidative metallation of unsaturated compounds	$\text{C=C} + M \longrightarrow \cdot\text{C-C-M}$ and so on

From organometallics

Method XI	Transmetallation	$RM + M'X \longrightarrow RM' + MX$
Method XII	Oxidative-reductive transmetallation	$RM + M' \longrightarrow RM' + M$

2.1 OXIDATIVE METALLATION OF ORGANIC HALIDES (METHOD I)

$$RX + 2M(0) \longrightarrow RM(I) + M(I)X \qquad (2.1)$$

or

$$RX + M(0) \longrightarrow RM(II)X \qquad (2.2)$$

The reaction between a metal and an organic halide provides one of the most important and general methods of preparation of main group and transition metal derivatives. The method is especially suited for the preparation of organometallics of readily oxidizable or ionizable metals, that is, those of low ionization energies (IE), (Appendix IV). Among the main group metals lithium and magnesium participate in this reaction quite readily, thereby making organolithiums and Grignard reagents readily accessible. Other metals that can react readily with organic halides include Ca, Sr, and Ba. Although their reactivity appears roughly comparable to that of organolithiums, information about these organometallics is scanty. One of the often troublesome side reactions is the Wurtz coupling which involves the reaction of the product (RM) and the starting material (RX).

$$RM + RX \longrightarrow R-R + MX \qquad (2.3)$$

Thus the method tends to fail in cases where the product and/or the starting material are highly reactive. For this reason, organometallics of higher alkali metals, such as sodium and potassium, as well as allylic, propargylic, and benzylic organolithiums are not readily obtainable by this method. By slow addition of the halide to an excess of finely divided metal at low temperatures, however, this difficulty can be minimized. For example, the relatively stable arylsodiums are readily obtained by using fine dispersions of sodium or the liquid sodium-potassium alloy and stirring them efficiently.

$$C_6H_5Cl + 2Na \longrightarrow C_6H_5Na + NaCl \qquad (2.4)$$

In general, the reactivity of organic halides increases in the order Cl < Br < I. When the metal is of relatively low reactivity, only highly reactive organic halides may be used. Thus zinc reacts satisfactorily only with (1) alkyl iodides, (2) allylic and propargylic halides, (3) α-halocarbonyl derivatives (Reformatsky reagent), and (4) other reactive halides, such as α-halo ethers. As discussed in this section, however, zinc powder obtained via reduction reacts with alkyl and aryl bromides. Aluminum and beryllium are only of limited utility in this reaction. The reaction of allyl halides with aluminum produces the corresponding sesquihalides (sesqui = 3/2 or 1.5).

$$3\ CH_2=CHCH_2X + 2Al \longrightarrow (CH_2=CHCH_2)_2AlX + CH_2=CHCH_2AlX_2 \qquad (2.5)$$

Beryllium reacts only when heated with alkyl halides.

$$CH_3I + Be \xrightarrow{90°} CH_3BeI \qquad (2.6)$$

Cadmium and boron do not appear to participate readily in this reaction. At high temperatures Si, Sn, and Sb (also phosphorous and arsenic) react with alkyl halides to form the corresponding organometallic derivatives. For example, when volatile organic halides are passed over silicon, mixed with copper and heated at 250 to 400°C, organosilicons are formed. This reaction is of commercial importance and is often called "direct synthesis."

$$Si + 2RCl \longrightarrow R_2SiCl_2 \qquad (2.7)$$

Organometallic derivatives of Hg, Tl, Sn, Pb, Bi, and so on, can be prepared by the reaction of the alloys of these metals with alkali metals, such as sodium. These reactions must initially form the organoalkali metal compounds, which then undergo transmetallation reactions. This so-called "alloy method" is used in the manufacture of tetraethyllead, an antiknock additive in gasoline.

$$4EtCl + 4NaPb \longrightarrow Et_4Pb + 3Pb + 4NaCl \qquad (2.8)$$

$$2MeBr + Na_2Hg \longrightarrow Me_2Hg + 2NaBr \qquad (2.9)$$

$$3PhX + Na_3Bi \longrightarrow Ph_3Bi + 3NaX \qquad (2.10)$$

Certain low-valent transition metal complexes, such as those containing Ni(0) and Pd(0), undergo facile oxidative metallation (or oxidative addition) reactions with various organic halides. These reactions are discussed later in detail (Sect. 8.3).

2.1.1   Metal Powders

A potentially useful method for generating highly reactive metal powders through reduction of the metal salt with an alkali metal in an etherial or hydrocarbon solvent has been developed by Rieke (2.1).

$$MX_n + n M' \longrightarrow M + n M'X \qquad (2.11)$$

Potassium is an alkali metal often used because of its high reactivity and low melting point. Highly reactive metal powders including those of Mg, Zn, Cd, Al, In, Tl, Sn, Cr, Ni, Pd, Pt, U, Th, and Pu have been prepared by this method.

The preparation of magnesium and zinc powders using this method is of special synthetic significance in that the highly reactive magnesium and zinc powders can now be utilized in preparing those Grignard and organozinc reagents that are otherwise difficult to prepare, as exemplified by the following equations.

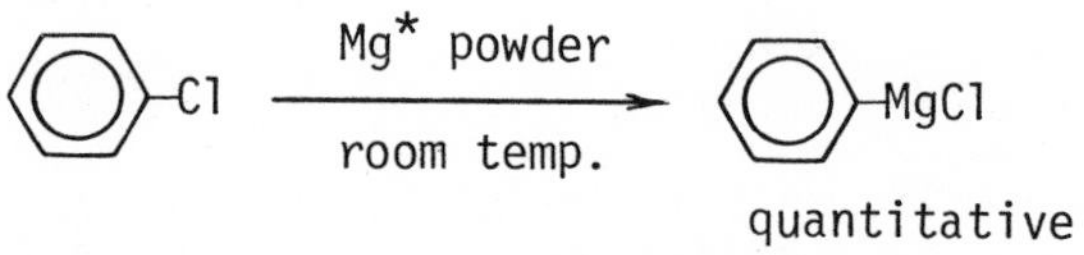

$$\text{Ph–Cl} \xrightarrow[\text{room temp.}]{\text{Mg}^* \text{ powder}} \text{Ph–MgCl} \qquad (2.12)$$

quantitative

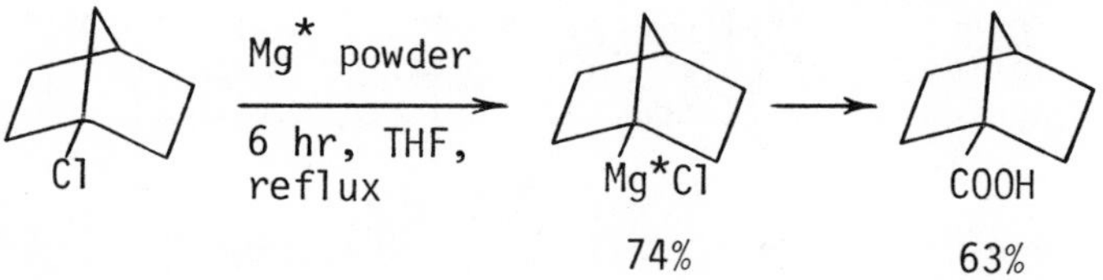

$$(2.13)$$

74%   63%

All previous efforts to prepare the Grignard reagent from fluorobenzene and magnesium had failed. On the other hand, magnesium powder generated by the reaction of $MgCl_2$ with potassium in the presence of KI reacted with fluorobenzene to form phenylmagnesium fluoride (2.2).

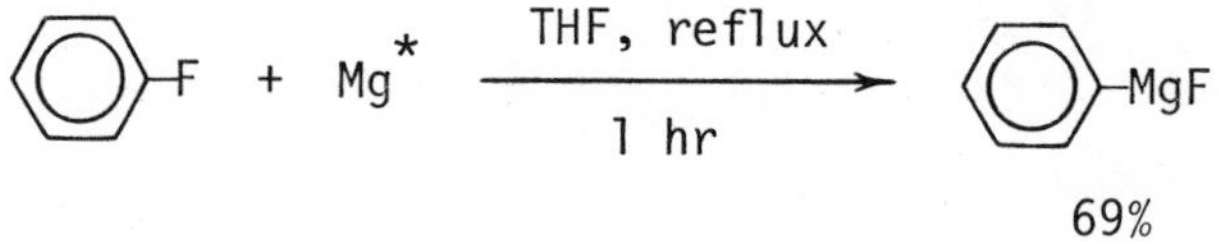

$$\text{Ph–F} + \text{Mg}^* \xrightarrow[\text{1 hr}]{\text{THF, reflux}} \text{Ph–MgF} \qquad (2.14)$$

69%

Whereas commercially available zinc does not readily react with organic bromides, Rieke's zinc powder generated from anhydrous $ZnCl_2$ and potassium in THF readily reacts even with alkyl and aryl bromides as shown in the following equations (2.3).

$$\text{RBr} \xrightarrow[\text{THF, reflux}]{\text{Zn}^*} R_2Zn \quad (R = \text{alkyl}) \qquad (2.15)$$

quantitative

$$ArX \xrightarrow{\quad Zn^* \quad} ArZnX \qquad (2.16)$$

X = I or Br

$$BrCH_2COOEt \xrightarrow[\text{room temp., 0.5 hr}]{\quad Zn^* \quad} BrZnCH_2COOEt \qquad (2.17)$$

The reaction of Ni, Pd, and Pt powders is discussed later (Sect. 8.3).

2.1.2   Metal Vapors as Reagents

So far we have dealt with either polymeric metals, as in the preparation of Grignard reagents, or metal-ligand complexes, such as $Ni(PPh_3)_4$ and $Pd(PPh_3)_4$. There is, however, at least one more way to utilize metals as reagents. Metals can be vaporized as mainly monoatomic species, and such species have been widely used to plate plastics in industry. Although these metal vapors had received little attention from synthetic chemists until recently, intensive studies by Skell (*2.4*), Timms (*2.5*), Klabunde (*2.6*), and others have clearly established that metal vapors are highly reactive and useful reagents in the preparation of organometallics, providing routes to even those organometallics that are otherwise inaccessible.

The technology of metal vaporization is well established (*2.7*), and a simple metal atom reactor is now commercially available from Kontes, Vineland, N.J. In short, metals are vaporized at high temperatures using resistive or inductive heating, electron guns, lasers, or arcs in a vacuum chamber (10^{-3} to 10^{-6} torr). Metal vapors deposit on the inside walls of the reactor kept at liquid nitrogen temperature, while vapors of organic substrates are directed through a shower head, codeposited in the same area, and allowed to react with the metal vapors.

Metals that have been utilized include both main group metals, such as Li, Mg, Ca, Zn, Cd, B, Al, In, Si, Ge, and Sn, and transition metals, such as Ti, Zr, V, Nb, Ta, Cr, Mo, W, Mn, Fe, Ru, Co, Ir, Ni, Pd, Pt, Cu, Ag, Au, U, and various lanthanide metals.

Although the method is quite general, it is not without some serious difficulties and limitations. Firstly, repolymerization of the metal vapors, which usually takes place readily even at 77° K, is always a competitive low activation energy process. Therefore, the use of a large excess (>100x) of organic substrates is desirable. Secondly, since organic substrates must enter the reactor as vapors, only those substrates that boil at about -80 to 200°C at atmospheric pressure can be utilized conveniently.

The reactivity of each metal-substrate pair depends on a number of factors, such as (1) acid-base properties, (2) availability of orbitals for complexation, and (3) electronic spin state of the metal atom. It can be said that nearly all substances with π or nonbonding electrons tend to react with atomic metals to form at least weak complexes. Among organic compounds, organic halides and unsaturated hydrocarbons have been most extensively utilized. Some noteworthy reactions observed with organic halides are discussed here, and those with unsaturated hydrocarbons are discussed later.

Nonsolvated Grignard reagents can now be prepared from magnesium and organic halides (*2.8*). These nonsolvated Grignard reagents exhibit some unusual properties. For example, they do not undergo the usual addition reaction with acetone, the only reaction observed being proton abstraction (eq. 2.18).

$$n\text{-PrMgX} + \text{CH}_3\text{COCH}_3 \begin{array}{c} \nearrow\!\!\!\!/\!\!\!\!\!\! \quad n\text{-Pr}\underset{\underset{\text{CH}_3}{|}}{\overset{\overset{\text{CH}_3}{|}}{\text{C}}}\text{-OH} \\ \\ \searrow \quad n\text{-PrH} + \text{CH}_3\underset{\underset{\text{OMgX}}{|}}{\text{C}}\text{=CH}_2 \end{array}$$

(2.18)

Calcium atoms react very efficiently with perfluoro olefins and hexafluorobenzene to yield C-F insertion products, see eq. 2.19 (*2.9*)

$$\underset{F}{\overset{F}{>}}\text{C=C}\underset{F}{\overset{F}{<}} + \text{Ca} \longrightarrow \underset{F}{\overset{F}{>}}\text{C=C}\underset{\text{CaF}}{\overset{F}{<}}$$

(2.19)

These appear to be the only examples of the metal insertion reaction involving the C-F bond.

The reaction of nickel and palladium vapors with organic

halides produces oxidative addition products of the RMX type
(M = nickel or palladium which are not stabilized with any ligands (*2.10*). These species may prove useful in examining various
catalytic reactions involving these metals. As might be expected, these products (RMX), which are still coordinatively unsaturated, are generally unstable even at room temperature except for
perfluorinated derivatives. The reaction of palladium atoms with
benzyl chloride, however, directly produces a dimeric η^3-benzyl-
palladium derivative (*2.1*) which is stable up to 100 to 110°C in
the solid state (*2.10b*).

$$\text{(structure with } CH_2Cl) + Pd \longrightarrow \text{(dimeric Pd structure)} \tag{2.20}$$

<u>2.1</u>

Although the scope of studies on the metal vapor-organic halide reaction is still limited, the following generalization can
tentatively be made. Firstly, the general order of reactivity of
organic halides is: RI > RBr > RCl. Secondly, aryl and benzyl
halides react much more efficiently than alkyl halides. These
results are in general agreement with those observed with either
polymeric metals or metal complexes. As a preparative method to
be used in practical organic synthesis, this new technique in
most cases does not yet provide advantages over conventional methods. It shows high promise, however, (1) as a tool for investi-
gating mechanistic details and (2) as a method for preparing otherwise inaccessible or less accessible species.

2.2 OXIDATIVE DISPLACEMENT BY METALLATE ANIONS (METHOD II)

$$RX + {}^-ML_n \longrightarrow RML_n + X^- \tag{2.21}$$

In the oxidative displacement reaction shown in eq. 2.21, the organic halide acts as an oxidizing agent, and the formal oxidation
number of the metal increases by 2. The metallate anions of the
IA to IIIA main group metals are essentially unknown. On the other hand, the IVA to VIA main group metals, such as Si, Sn, and
Sb, as well as a number of transition metals, such as Fe, Co, Ni,
and Cu, form various types of metallate anions that act as nucleophiles. In fact, some of the metallate anions containing transition metals are among the strongest nucleophiles known to chem-

ists (*2.11*). The metallate anions can react with a variety of electrophiles, such as organic halides, α,β-unsaturated carbonyl derivatives, and epoxides, thereby providing a convenient route to organometallics.

$$L_nM^- + \overset{|}{C}=C-C=O \longrightarrow L_nM-\overset{|}{C}-C^-_=C=O \qquad (2.22)$$

$$L_nM^- + \underset{O}{>C-C<} \longrightarrow L_nM-\overset{|}{C}-\overset{|}{C}-O^- \qquad (2.23)$$

The reaction of organic halides with metallate anions containing certain transition metals, copper and iron in particular has proven to be highly significant from the viewpoint of organic synthesis. The synthetic utility of the reaction of the Group IVA to VIA metallate anions, such as $KSiMe_3$ and $LiSnBu_3$, with organic halides is not clear, because the same products can be obtained generally more conveniently from the reaction of the halogen derivatives of these metals with a variety of organometallic reagents.

$$\begin{array}{c} RX + KSiMe_3 \\ \\ RLi + ClSiMe_3 \end{array} \searrow \begin{array}{c} \\ RSiMe_3 \\ \\ \end{array} \qquad (2.24)$$

Moreover, the metallate anion reaction occurs readily and selectively only with alkyl halides and sulfonates containing a C_{sp^3}-X bond. In such cases, the reaction presumably proceeds by the S_N2 mechanism. Although silanions and other metallate anions react with aryl halides to produce the arylated derivatives, such reactions evidently proceed via benzynes and are therefore associated with the well-known regiochemical and stoichiometric (R_3M^-/ ArX = 2/1) problems (*2.12*).

2.3   METAL-HALOGEN EXCHANGE (METHOD III)

$$RX + MR' \rightleftharpoons RM + R'X \qquad (2.25)$$

X = halogen

The reaction can proceed forward when group R is more electronegative or acidic than group R' (pK_a of RH < pK_a of R'H). It

provides one of the most widely used methods for preparing organolithiums; see Sec. 4.2.1.1 (*2.13*). The order of reactivity of
halogens is: I > Br > Cl >> F. The scope of the metal-halogen
exchange reaction, in practice, is limited to the preparation of
organolithiums. Other electropositive metals, such as Na, K, and
Mg, can also participate in the metal-halogen exchange reaction.
The Wurtz-type coupling reaction (eq. 2.26) and other side reactions, however, tend to compete with the desired metal-halogen
exchange reaction.

$$RX + MR' \\ \updownarrow \qquad\qquad\qquad \longrightarrow RR + RR' + R'R' + MX \qquad\qquad (2.26) \\ RM + XR'$$

Even in the preparation of organolithiums, the Wurtz-type
coupling reaction can be a troublesome side reaction. Benzylic,
allylic, and propargylic halides tend to undergo a competitive
Wurtz-type coupling reaction and are, therefore, of little use in
the metal-halogen exchange reaction. Whereas a variety of aryllithiums can be prepared according to eq. 2.27 using ethyl ether
as a solvent, the same reactants in THF or other polar solvents,
such as HMPA, tend to give mostly the cross-coupled produces, presumably via metal-halogen exchange.

$$n\text{-BuLi} + \text{ArX} \quad \xrightarrow{\text{Et}_2\text{O}} \quad n\text{-BuX} + \text{ArLi}$$
$$\xrightarrow{\text{THF}} \quad n\text{-BuAr} + \text{LiX} \qquad\qquad (2.27)$$

In fact, the latter reaction provides an excellent method for preparing alkylarenes (*2.14*), which appears superior to the classic
Wurtz reaction, as discussed later (Sect. 4.3.2.1).
    The lithium halogen exchange reaction can take place rapidly
even at very low temperatures, for example,-78°C. Surprisingly,
very little is known about the mechanism of this low activation
energy process. The reaction of alkenyl iodides with *n*-, *sec*-,
or *tert*-BuLi proceeds with retention, thereby providing a useful
route to stereodefined alkenyllithiums (*2.15*).

$$R^1\!\!\diagdown_{R^2}\!\!C=C\!\diagup^{R^3}_{\diagdown I} \quad \xrightarrow{\text{RLi}} \quad R^1\!\!\diagdown_{R^2}\!\!C=C\!\diagup^{R^3}_{\diagdown Li} \qquad\qquad (2.28)$$

Although not well established, when stereochemically rigid organolithiums are involved, retention of configuration seems to be
the rule. These stereochemical data do not tend to favor any
mechanism involving alkenyl free-radicals but do not rule out one-
electron processes. The following mechanisms have been suggested, but at the present time they appear to be no more than mere
speculations (1.14).

$$R\underset{X}{\overset{M}{\lessgtr}}R$$

$$RX + MR' \; \rightleftharpoons \; M^+[R\text{-}X\text{-}R'] \; \rightleftharpoons \; RM + XR' \qquad (2.29)$$

$$[R\text{-}X^+\text{-}M^-\text{-}R'] \; \rightleftharpoons \; \overset{M}{[R\text{-}X\bullet, \; \bullet R']}$$

The following radical chain mechanism has been proposed for the
reactions of alkyllithiums with alkyl halides (2.16).

$$RX + MR' \quad \dashrightarrow \quad R\bullet + R'\bullet + MX$$

$$R\bullet + MR' \quad \longrightarrow \quad RM + R'\bullet \qquad\qquad (2.29a)$$

$$R'\bullet + RX \quad \longrightarrow \quad R'X + R\bullet$$

2.4 OXIDATIVE METALLATION OF ACTIVE C-H COMPOUNDS (METHOD IV)

$$RH + M \quad \longrightarrow \quad RM + \tfrac{1}{2}H_2 \qquad\qquad (2.30)$$

The method is far more limited than the corresponding reaction of
organic halides (Method I). The reaction generally requires relatively acidic hydrocarbons ($pK_a \leq 25$ to 30), such as acetylenes
and cyclopentadiene, and highly electropositive metals, such as
alkali metals, as shown in the following example (eq. 2.31).

$$\text{(cyclopentadiene)} + Na \longrightarrow \text{(cyclopentadienide)} Na^+ + \tfrac{1}{2}H_2 \qquad (2.31)$$

In cases where favorable results are obtained, however, the reaction provides a convenient and direct route, the only byproduct
being hydrogen. Although not clear, the following dual-path
mechanism may tentatively be suggested as a plausible scheme.

$$RH + \bullet M \longrightarrow R\bullet + HM$$

$$R\bullet + \bullet M \longrightarrow RM \qquad\qquad (2.32)$$

$$RH + HM \longrightarrow RM + H_2$$

2.5 METAL-HYDROGEN EXCHANGE (METHOD V)

$$RH + MX \longrightarrow RM + HX \qquad\qquad (2.33)$$

Although represented by the same general equation (eq. 2.33), there are at least two distinct processes that should be discussed here. One involves a simple acid-base reaction in which a metal-containing reagent acts as a base, whereas in the other reaction the metal containing reagent acts as a Lewis acid.

2.5.1 Metal-Hydrogen Exchange with Basic Reagents

The metal-hydrogen exchange reaction with basic metal-containing reagents provides a very general route to organometallics. The reagent itself is frequently an organometallic compound. For the reaction shown in eq. 2.33 to proceed forward, the starting compounds (RH) must be more acidic than the conjugate acid (HX) of the basic reagent. A simple thermodynamic calculation tells us that a difference of only 2 to 3 pK_a units should be sufficient to drive the reaction to completion (>98%). In practice, however, a greater difference in pK_a is required for satisfactory results, because the basicity of MX seldom reaches the maximum value corresponding to the pK_a value of its conjugate acid (HX).
    In addition to the pK_a value of the conjugate acid (HX), the basicity of MX is also dependent on various other factors including the electronegativity value of the gegenion (M) and the solvent used. These factors exert marked influence on the metal-hydrogen exchange reaction.

2.5.1.1 Gegeion

With a given anionic species (X), the basicity of MX decreases as the electronegativity of M increases, the approximate order being: Cs > Rb > K > Ba > Sr > Na > Ca > Li > Mg. Other less electropositive metals are of little use in this reaction. Let us consider the reaction of the n-butylmetals of K, Na, Li, and Mg(X) with n-butane (pK_a = 50), benzene (pK_a = 43), triphenylmethane (pK_a = 30), and acetylene (pK_a = 25). n-Butylmagnesium bromide will metallate only acetylene, whereas n-butyllithium readily metallates triphenylmethane as well. Even benzene, but not n-butane, can be metallated using n-butylsodium. Finally, n-butylpotassium reacts with all these hydrocarbons.

2.5.1.2   Solvent

Strongly coordinating solvents, such as tetramethylethylenediamine (TMEDA), enhance the basicity of MX. For example, whereas n-butyllithium in nonpolar solvents does not metallate toluene, its TMEDA complex does, thereby providing one of the simplest routes to benzyllithium, even though a large excess (10x) of toluene is needed for satisfactory results (*2.17*). Alternatively, a stronger base, such as *sec*-BuLi can also be used for this transformation (eq. 2.34).

$$n\text{-BuLi} \cdot \text{TMEDA}$$

$$\text{(2.34)}$$

$$sec\text{-BuLi}$$

2.5.1.3   Scope

The scope of the reaction with respect to the metal was already discussed briefly. Typical saturated hydrocarbons are in most cases not acidic enough to form the corresponding organometallics by this method. As discussed later (Sect. 4.2), the metal-hydrogen exchange reaction has long been the method chosen for converting organic compounds containing relatively acidic C-H bonds ($pK_a \leq 30$) including terminal acetylenes, various carbonyl and cyano derivatives containing the α C-H bond, and cyclopentadiene derivatives. Arenes and alkenes can also be metallated readily. In cases where these compounds contain the benzyllic and allylic C-H bonds, metallation can take place at both sp^2 and sp^3 carbon atoms, although the benzylic and allylic C-H bonds tend to be somewhat more reactive than the $C_{sp}2$-H bonds.

2.5.1.4   Chemoselectivity, Regioselectivity, and Stereoselectivity

Although highly versatile and useful, the metal-hydrogen exchange reaction suffers from a few serious difficulties. Since strong bases readily react with a variety of electrophilic functional groups, the chemoselectivity of the method is inherently low. It represents a difficult problem to overcome. Controlling the regio- and/or stereochemistry of the reaction is also problematical in many cases. Recent developments, however, permit us to exert considerable regio- and/or stereochemical control. For example, limonene can be selectively metallated at only one of the five allylic positions (2.18).

$$(2.35)$$

The regiochemistry of the metallation reaction is also very much dependent on the nature of the bases used. For example, whereas KCH_2SiMe_3 abstracts the benzylic hydrogen of cumene, a combination of n-BuLi and t-BuOK attacks its meta and para hydrogens (2.19).

Although it still is difficult to predict accurately the point of attack in a given case, strongly basic reagents, such as KCH_2SiMe_3, tend to attack the most acidic hydrogen of a given molecule, whereas the reaction of weaker bases, such as n-BuLi and t-BuOK, can be strongly influenced by other factors, such as steric hindrance. In favorable cases, the metal-hydrogen exchange reaction proceeds in a remarkably stereospecific manner, as exemplified by the following conversions; see eqs. 2.36, 2.37 (2.20).

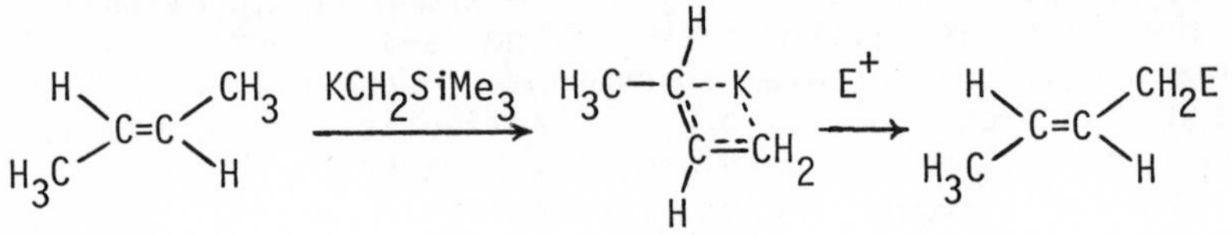

$$(2.36)$$

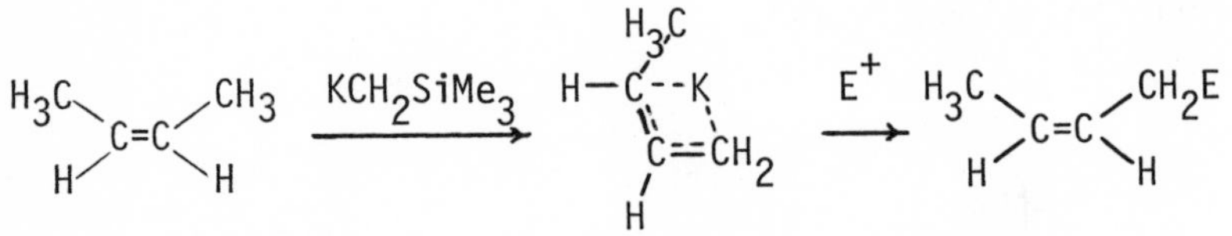

$$(2.37)$$

2.5.2 Metal-Hydrogen Exchange with Acidic Reagents

Certain metal-containing reagents that are Lewis acidic, such as BCl_3, $Hg(OAc)_2$, and $Tl(OOCCF_3)_3$, undergo the metal-hydrogen exchange reaction as shown in eq. 2.38. The method is most useful when RH is an arene. Many of these reactions can be best viewed as electrophilic aromatic substitution reactions involving metal-containing electrophiles, as exemplified by the following reaction (2.21, 2.22).

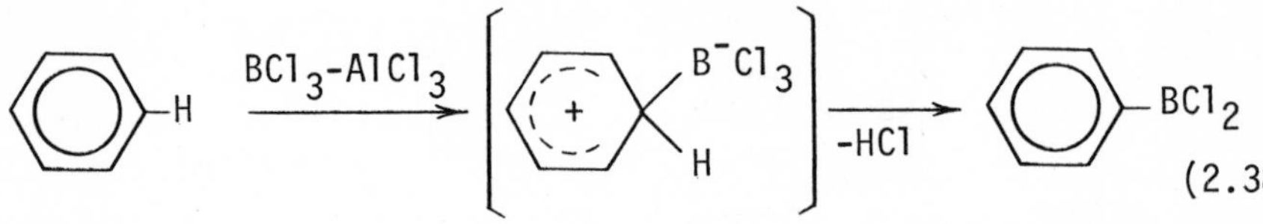

$$(2.38)$$

The reaction suffers from various difficulties and limitations associated with other electrophilic aromatic substitution reactions. Thus it often results in mixtures of regioisomers and arenes with electron-withdrawing, that is, deactivating, substituents often failing to undergo the metal-hydrogen exchange reaction. Despite these limitations, it represents one of the most direct and convenient routes to arylmercury and arylthallium derivatives (Sect. 7.2.3).

The corresponding reactions of alkenes and alkynes tend to undergo addition rather than substitution. Thus although mechanistically closely related to the aryl cases discussed above, these addition reactions are classified as the heterometallation reaction (Method VII).

2.6 HYDROMETALLATION (METHOD VI)

$$\text{>C=C<} \quad + \text{ HM} \quad \longrightarrow \quad \text{H-C-C-M} \qquad (2.39)$$

$$\text{-C≡C-} \quad + \text{ HM} \quad \longrightarrow \quad \text{H-C=C-M} \qquad (2.40)$$

Hydrometallation is a term devised to define a reactant-product relationship. Based on the stereochemistry of the products, the known hydrometallation reactions can be classified according to the following types: (1) stereoselective cis hydrometallation, (2) stereoselective trans hydrometallation, (3) nonstereoselective hydrometallation.

2.6.1 Stereoselective Cis Hydrometallation

In many cases, the addition of the M-H bond to the C=C or C≡C bond is essentially 100% cis. Those metals that participate in the cis hydrometallation include a few Group IIIA and Group IVA metals, such as B, Al, and Si, and a number of transition metals. Hydroboration (Sect. 5.2.2.2), hydroalumination (Sect. 5.3.2.3), hydrosilation catalyzed by a transition metal complex, such as H_2PtCl_6 (Sect. 6.3.3.2), and hydrozirconation (Sect. 10.1), however, are the only well-developed cis hydrometallation reactions that are currently useful as methods for the preparation of hydrometallated species as discrete reagents and intermediates. Each of these reactions is discussed in detail in the sections indicated.

One of the key requirements for the cis hydrometallation reaction is that the actual reacting M-H species be coordinatively unsaturated, that is, either a 6-electron or 16-electron (possibly 14-electron) species. Although the precise mechanism of the cis hydrometallation reaction has not been fully clarified, the presently available data are consistent with the following concerted, but probably non-synchronous, mechanism which may be classified as a thermally allowed $[_\pi 2_s + _\sigma 2_a + _\omega 0_s]$ reaction according to the Woodward-Hoffmann convension (1.6)

Hydrometallation

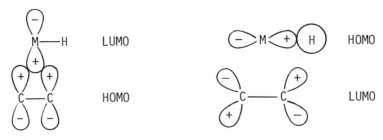

Carbene Addition (singlet)

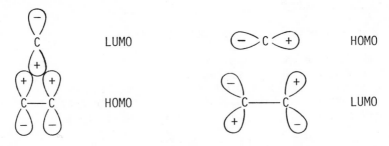

It might be instructive to compare this mechanism with that proposed for the singlet carbene reaction with olefins (*2.23*).

As discussed later in detail (Sect. 10.1), many other transition metal hydrides do undergo the hydrometallation reaction. In
most cases, however, the hydrometallated species cannot be isolated due to various competitive side reactions, of which dehydrometallation and reduction of the hydrometallated product with
the metal hydride are the most common.

2.6.2 Stereoselective Trans Hydrometallation

Although much more limited in scope, certain hydrometallation reactions involve a predominant, if not exclusive, trans addition
of the M-H bond, as shown in the following example; see eq. 2.41
(*2.24*).

$$R^1C\equiv CR^2 + LiAlR_3H \xrightarrow{\text{100 to 130°C}} \left[{}^{H}_{R^1}{>}C{=}C{<}^{R^2}_{Al^-R_3} \right] {}^+Li \qquad (2.41)$$

R^1, R^2 = alkyl (not H);   R = H or alkyl

The trans hydrometallation reaction with simple olefins and acetylenes is usually a slow reaction requiring high temperatures. In some favorable cases, however, it takes place even below 0°C. Thus the reaction of hexafluorobutyne with Cp_2MH_2 (M = molybdenum and tungsten) and Cp_2ReH gives the corresponding trans addition products at or below -30°C (2.25). The trans hydrometallation reaction of hexafluorobutyne with $HMn(CO)_5$ has also been reported (2.26). All of these metal hydrides are coordinatively saturated 18-electron species (Sect. 8.2). A concerted $[{}_\sigma 2_s + {}_\pi 2_a]$ mechanism has been proposed for the reaction of Cp_2MoH_2 and Cp_2WH_2 with $CF_3C\equiv CCF_3$ (2.25).

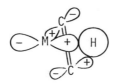

2.6.3   Nonselective Hydrometallation

Various metal hydrides including those of silicon and tin can undergo hydrometallation reactions that are not stereoselective. Some of these reactions appear to proceed at least partially by free-radical mechanisms. Hydrosilation and hydrostannation are discussed in detail in Sect. 6.3.3.

In some cases the stereochemistry of such reactions can be significantly improved by a suitable catalyst, such as H_2PtCl_6, which can participate in (1) a stereoselective hydrometallation and (2) a transmetallation with the metal hydride used, as shown in eq. 2.42 (Sect. 6.3.3.2).

$$RC\equiv CH \xrightarrow[\text{cat. } H_2PtCl_6]{HSiR_3'} \underset{\text{major}}{\overset{R}{\underset{H}{>}}C=C\overset{H}{\underset{SiR_3'}{<}}} + \underset{\text{minor}}{\overset{R}{\underset{R_3'Si}{>}}C=C\overset{H}{\underset{H}{<}}} \qquad (2.42)$$

100% cis

It should be clear that the applicability of such a principle is not restricted to the improvement of free-radical hydrometalla-tion reactions. Rather, it should be applicable to the modifica-tion of any type of hydrometalation reaction.

2.7   HETEROMETALLATION (METHOD VII)

Hydrometalation, which is discussed in the preceding section, is but one of many types of addition reactions which metal-contain-ing compounds undergo. Other bonds to metals that undergo addi-tion reactions to the C=C and C≡C bonds include M-C, M-O, M-X, (X = halogen), and M-M bonds. Addition of the M-C bond may con-veniently be termed carbometalation. Similarly oxymetalation (M-O), halometalation (M-X), and other related terms may be used to describe the corresponding addition reactions. In this book, all addition reactions of the M-X bonds other than hydrometalla-tion and carbometalation are collectively termed heterometalla-tion. Since carbometalation involves the transformation of one organometallic compound into another containing the same metal, it is not treated as a method for preparing organometallics.

2.7.1   Oxymetalation

Highly electropositive metals, such as Group IA and Group IIA me-tals, do not generally participate in heterometalation involving the addition of the M-O (oxymetalation) and M-halogen (halome-tallation) bonds. This is due mainly to the fact that these me-tals form thermodynamically strong bonds with electronegative elements, such as oxygen and chlorine. Thus even if an addition product containing such a metal were formed according to eq. 2.43, it would readily undergo the reverse reaction, that is, β-elimi-nation.

$$>C=C< \quad + M-X \quad \rightleftharpoons \quad M-\overset{|}{\underset{|}{C}}-\overset{|}{\underset{|}{C}}-X \qquad (2.43)$$

In fact, it is generally difficult by any method to prepare organometallics containing both electropositive metals and electronegative atoms which are β to the metal as species that can be
isolated.

On the other hand, certain, heavy metal compounds, such as
those containing Hg, Tl, Pb, Pd, and Pt, have exhibited a strong
tendency to undergo oxymetallation. One of the key requirements
for oxymetallation seems to be that the metal reagent should contain a highly polarizable and hence readily ionizable M-O bond
(2.27).

Both trans and cis oxymetallation reactions have been observed, although the former appears to be the more common of the
two. In cases where a highly stereoselective trans oxymetallation reaction is observed, the two step mechanism involving the
intermediacy of a π-complex has been proposed (2.27), as exemplified by the following.

$$\text{(2.44)}$$

Other nucleophilic reagents, such as water and alcohols used as
solvents, have often participated in the trans oxymetallation reaction. In some cases, however, the same metal-containing reagents exhibit the opposite stereoselectivity, producing cis addition products. This cis oxymetallation has been observed with
strained olefins, such as norbornene; see eq. 2.45 (2.28).

$$\text{(2.45)}$$

In such cases, one of the oxy ligands, rather than a solvent or
an added nucleophile, is competitively incorporated into the organometallic product, somewhat irrespective of the relative nucleophilicity and concentration. It appears highly probable that
the cis oxymetallation proceeds by an entirely different mechanism than that which operates in the trans oxymetallation. The

mechanistic details of oxymercuration will be discussed in Sect. 7.2.2.1. Oxymetallation involving mercury (oxymercuration) and thallium (oxythallation) represents one of the most significant routes to organometallics containing these metals and is further discussed in Sects. 7.2.2.1 and 7.4.1.3.

2.7.2 Halometallation and Other Heterometallation Reactions

Whereas oxymetallation appears unique to heavy metals, halometallation has been observed with both light metals, such as boron, and heavy metals, such as mercury and palladium. Heteropalladation is an important route to organopalladiums and is discussed in detail in Sect. 10.3.

As in the case of oxymetallation, both trans and cis halometallation reactions have been observed with heavy metals, and the trans addition appears to be the more common of the two. A few potentially useful examples of chloromercuration are shown below (*2.29*).

$$R_2\underset{\underset{OH}{|}}{C}-C\equiv CH \quad \xrightarrow{\ HgCl_2\ } \quad \underset{R_2C}{\overset{Cl}{\diagdown}}C=C\underset{HgCl}{\overset{H}{\diagup}} \qquad (2.46)$$
$$\hphantom{R_2C} {\scriptstyle OH}$$

$$RC\equiv CCOX \quad \xrightarrow{\ HgCl_2\ } \quad \underset{R}{\overset{Cl}{\diagdown}}C=C\underset{HgCl}{\overset{COX}{\diagup}} \qquad (2.47)$$

X = Me, OH, OMe, OEt, and so on

On the other hand, halometallation with haloboranes (Sect. 5.2.2.1) involves cis addition and appears to be mechanistically analogous to hydroboration. The coordinatively unsaturated nature of haloboranes, as well as the relatively high electronegativity of boron, which renders the haloborated products relatively stable to β-elimination, must be responsible for this unique cis halometallation of haloboranes.

Although less common, various other heterometallation reactions, including those of the M-N (aminometallation) and M-M bonds, have also been observed. These will be discussed as we encounter them in later chapters.

2.8 COMPLEXATION (METHOD VIII)

As we have already seen, coordinatively unsaturated metal-con-

taining species have a tendency to interact with electron donors
to form coordinatively saturated compounds. The reaction may be
viewed as a Lewis acid-Lewis base interaction, the equilibrium
position of which is a function of various electronic and steric
factors.

Olefins, acetylenes, and arenes are bases of low Brønsted
basicity. Thus these compounds do not form stable complexes with
the main group metal derivatives, although a number of π-complexes have been suggested as transient species in various reactions,
such as hydroboration and oxymercuration. A number of main group
metal derivatives, such as cyclopentadienyl sodium (2.2) and allyllithium (2.3), might be viewed as π-complexes of main group metals. The M-C bonds in these compounds, however, are essentially
ionic rather than covalent.

2.2   2.3

On the other hand, transition metal-containing compounds with
partially filled d orbitals can act simultaneously as both a donor and an acceptor. Thus they can form highly stable π-complexes with unsaturated organic compounds. Since this π-complexation
pertains only to the organotransition metal compounds, its detailed discussion is presented in Sect. 8.3.

2.9 OXIDATIVE COUPLING (METHOD IX)

Certain subvalent compounds of the main group metals (2.30), such
as B-X and SiX_2, and various coordinatively unsaturated organotransition metal compounds can react with unsaturated organic compounds to form addition products, which may be viewed as σ-complexes rather than π-complexes, as exemplified by 2.4 (2.31).
As might be expected, the difference between complexation and oxidative coupling can be vague. The true structure of organotransition metal compounds formed by the interaction of transition metal
complexes with olefins or acetylenes may, in some cases, be best
represented by a hybrid of two extreme structures representing
complexation and oxidative coupling. A more detailed discussion
of this subject is presented in Sect. 8.3.

2.4

This reaction is closely related to the complexation reaction discussed above, and distinction between the two is often not clear-cut. As in the case of complexation, the key requirement for the oxidative coupling reaction represented by eq. 2.48 appears to be that the metal compounds should possess both empty and filled nonbonding orbitals.

$$\text{>C=C<} \quad + \quad ML_n \quad \longrightarrow \quad \text{>C}\overset{ML_n}{-}\text{C<} \qquad (2.48)$$

Whereas a number of transition metal compounds satisfy this requirement, relatively few main group metal compounds of this type have been known. As a result, the scope of oxidative coupling as a method for preparing organometallics of the main group metals is still quite limited. The three-membered metallocyclic products containing main group metals tend to be highly unstable, and only a small number of compounds, such as 2.5, have been prepared and characterized as such (2.32).

2.5

It is also instructive to point out that the oxidative coupling reaction shown in eq. 2.48 is closely related to the well-known carbene addition reaction. Oxidative coupling involving transition metals is discussed in detail in Sect. 8.3.

2.10 OXIDATIVE METALLATION OF UNSATURATED ORGANIC COMPOUNDS (METHOD X)

Metals of low ionization potentials can undergo one-electron transfer reactions with unsaturated organic compounds with a low-lying unoccupied orbital (LUMO) to form ion pairs consisting of metal cations and anion radicals derived from the unsaturated organic compounds. In practice, only alkali metals and conjugated olefins including arenes and acetylenes participate readily in this reaction (2.33). A few representative examples follow.

$$(2.49)$$

green

$$Ph_2C=CH_2 \xrightarrow{Na} Na^+[Ph_2C-CH_2]^{\cdot-} \rightarrow 2Na^+[Ph_2C^-CH_2CH_2C^-Ph_2] \qquad (2.50)$$

$$PhC{\equiv}CPh \xrightarrow{Li} Li^+[PhC=CPh]^{\cdot-} \longrightarrow \qquad (2.51)$$

The radical anions formed by a single one-electron transfer can undergo various transformations. In the absence of any other reactive species, the radical anions tend to dimerize as shown in eqs. 2.50 and 2.51. In some cases, however, the radical anions participate in the second one-electron transfer reaction to form the corresponding dianions (eq. 2.52).

$$(2.52)$$

With a given substrate, formation of dianions is favored by using (1) lithium, which forms more covalent bonds to carbon than sodium and potassium, and (2) solvents of low dielectric constant

and poor solvating ability. In other words, the dianion formation is favored if the ionicity of the metal-carbon bond is low.
A typical example is the formation of the dilithio derivative of
naphthalene as shown in eq. 2.53 (*2.34*).

$$\text{green} \qquad \text{purple} \qquad (2.53)$$

Dianion formation is also favored if the dianion is highly stabilized as in the formation of the dianion of cyclooctatetraene;
see eq. 2.54 (*2.35*).

2.11   TRANSMETALLATION (METHOD XI)

$$RM + M'X \quad \rightleftharpoons \quad RM' + MX \qquad (2.55)$$

The transmetallation reaction shown in eq. 2.55, which does not
involve any oxidation-reduction, is probably the most general method for preparing those organometallics that cannot be obtained
directly from organic compounds. For this reaction to proceed in
a forward direction, M has to be more electropositive than M'.
It should be pointed out, however, that, even in cases where M is
less electropositive than M', it is still feasible to generate an
equilibrium quantity of RM' and utilize it in the subsequent step.
Thus certain organometallics containing mercury (EN = 1.9), for
example, have been successfully utilized in the vinylic hydrogen-
substitution reaction shown in eq. 2.56, in which organometallics
containing palladium (EN = 1.35) are products in the first step
(*2.36*).

$$RHgX + PdX_2 \longrightarrow RPdX + HgX_2$$

$$RPdX + CH_2=CHY \longrightarrow \underset{\underset{PdX}{|}}{RCH_2CHY} \tag{2.56}$$

$$\underset{\underset{H \ \ PdX}{| \ \ |}}{RCHCHY} \longrightarrow RCH=CHY + HPdX$$

Because of their ready availability and highly electropositive nature, organolithiums and Grignard reagents have been used most extensively as parent organometallics (RM). Organometallic compounds containing essentially all metals, whose electronegativities are greater than those of lithium and magnesium, have been prepared by this method.

As might be expected from the electronegativity values, organolithiums are generally more reactive than Grignard reagents. For example, EtLi reacts with $TlCl_3$ to form Et_3Tl, whereas EtMgCl replaces only two of the three chlorine atoms to form Et_2TlCl.

$$TlCl_3 \quad \begin{array}{c} \xrightarrow[\text{ether}]{\text{EtLi}} Et_3Tl \\ \\ \xrightarrow[\text{ether}]{\text{EtMgCl}} Et_2TlCl \end{array} \tag{2.57}$$

Despite their versatility, there are a few difficulties associated with organolithiums and Grignard reagents. Firstly, as these organometallic reagents react with practically all electrophilic functional groups, it is either impossible or difficult to prepare organometallics containing some of such functional groups by this method. In such cases, the use of less electropositive metals, such as zinc and aluminum, should be considered. Secondly, it is generally cumbersome to prepare organolithiums and Grignard reagents which are stereochemically defined at the metal-bound carbon atom. On the other hand, some of such species, *trans*-alkenylmetals, for example, can be readily prepared by hydrometallation of the corresponding alkynes (Sect. 2.6). It is therefore advantageous to use these readily available substances as parent organometallics, as shown in the following synthesis of alkenylmercuries (Sect. 7.2.1).

$$RC \equiv CH \xrightarrow{HB} \begin{array}{c} R \\ \diagdown \\ H \end{array} C=C \begin{array}{c} H \\ \diagup \\ B \diagdown \end{array} \xrightarrow{HgX_2} \begin{array}{c} R \\ \diagdown \\ H \end{array} C=C \begin{array}{c} H \\ \diagup \\ HgX \end{array} \qquad (2.58)$$

Group X in eq. 2.55 can be an organic group. In such a case, the MX is also an organometallic product. For example, a widely used method for preparing allyllithium involves the reaction shown in eq. 2.59 (Sect. 6.4.1).

$$\diagup\diagdown\diagup\diagdown SnPh_3 + LiPh \longrightarrow \diagup\diagdown\diagup\diagdown Li + SnPh_4 \qquad (2.59)$$

The more electronegative or less basic organic group, allyl, is nearly exclusively bonded to the more electropositive metal, lithium, in the product mixture. Thus as far as the synthesis of allyllithium is concerned, the starting allyl compound contains the less electropositive metal of the two, tin. It should be clear, however, that this fact does not contradict the general statement made above concerning eq. 2.55.

The scope of the transmetallation reaction shown in eq. 2.55 is too broad to discuss in detail, but is discussed in various sections throughout this book.

2.12   OXIDATIVE-REDUCTIVE TRANSMETALLATION (METHOD XII)

$$RM + M' \rightleftharpoons RM' + M \qquad (2.60)$$

This redox reaction can proceed forward when the free energy of formation of RM is more positive (or less negative) than that of RM'.

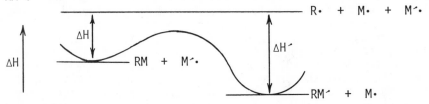

Since few entropy data for organometallic compounds are available, the heat of formation (ΔH) rather than free energy is used here. The ΔH values for the methyl derivatives of several main group metals are summarized in Table 2.2.

Table 2.2   Heat of Formation of Me_nM[a]

Me_nM	ΔH (kcal/mole)	Me_nM	ΔH (kcal/mole)
$Me_2Zn(g)$	13	$Me_3Al(g)$	-21
$Me_2Cd(g)$	26	$Me_4Si(g)$	-57
$Me_2Mg(g)$	22	$Me_4Sn(g)$	- 5
$Me_3B(g)$	-29	Me_3Bi	46

[a]These values are taken from Ref. 0.4, p. 8.

These data indicate that Me_3B and Me_4Si are stable with respect to their decomposition into the methyl and metal-free radicals (exothermic), whereas the methyl derivatives of Cd, Hg, and Bi are thermodynamically unstable (endothermic).

Since organomercuries can be directly prepared by either metal-hydrogen exchange and heterometallation (as thermodynamically unstable but kinetically stable, though toxic, compounds that can be easily handled in air at room temperature) they have been most widely used as the starting materials for this transmetallation reaction. Using organomercuries, a variety of organometallics including those that contain alkali and alkali earth metals, such as aluminum and tin, have been prepared (Sect. 7.3.1).

REFERENCES

2.1   Rieke, R. D., *Top. Curr. Chem.*, 59, 1 (1975), and references therein.

2.2   (a) Rieke, R. D., and Hudnall, P. M., *J. Am. Chem. Soc.*, 94, 7178 (1972); (b) Rieke, R. D., and Bales, S. F., *J. C. S. Chem. Comm.*, 789 (1973).

2.3   Rieke, R. D., Hudnall, P. M., and Uhm, S., *J. C. S. Chem. Comm.*, 269 (1973).

2.4   Skell, P. S., Havel, J. J., and McGlinchey, M. J., *Acc. Chem. Res.*, 6, 97 (1973), and references therein.

2.5   Timms, P. L., *Adv. Inorg. Radiochem.*, 14, 121 (1972).

2.6   Klabunde, K. J., *Acc. Chem. Res.*, 8, 393 (1975).

2.7   Reichelt, W., *Angew. Chem. Int. Ed. Engl.*, <u>14</u>, 218 (1975).

2.8   Skell, P. S., and Girard, J. E., *J. Am. Chem. Soc.*, <u>94</u>, 5518 (1972).

2.9   Klabunde, K. J., Low, J. Y. F., and Key, M. S., *J. Fluorine Chem.*, <u>2</u>, 207 (1972).

2.10 (a) Klabunde, K. J., and Low, J. Y. F., *J. Am. Chem. Soc.*, <u>96</u>, 7674 (1974); (b) Roberts, J. S., and Klabunde, K. J., *J. Organometal. Chem.*, <u>85</u>, C13 (1975).

2.11 King, R. B., *Acc. Chem. Res.*, <u>3</u>, 417 (1970).

2.12 Shippey, M. A., and Dervan, P. B., *J. Org. Chem.*, <u>42</u>, 2655 (1977).

2.13 (a) Jones, R. G., and Gilman, H., *Org. React.*, <u>6</u>, 339 (1951); (b) Applequist, D. E., and O'Brien, D. F., *J. Am. Chem. Soc.*, <u>85</u>, 743 (1963).

2.14 Merrill, R. E., and Negishi, E., *J. Org. Chem.*, <u>39</u>, 3452 (1974).

2.15 (a) Corey, E. J., and Beames, D. J., *J. Am. Chem. Soc.*, <u>94</u>, 7210 (1972); (b) Kluge, H. F., Untch, K. G., and Fried, J. H., *J. Am. Chem. Soc.*, <u>94</u>, 7827 (1972); (c) Cahiez, G., Bernard, D., and Normant, J. F., *Synthesis*, 245 (1976).

2.16 Russell, G. A., and Lampson, D. W., *J. Am. Chem. Soc.*, <u>91</u>, 3967 (1969).

2.17 Screttas, C. G., Estham, J. F., and Kamienski, C. W., *Chimia*, <u>24</u>, 109 (1970).

2.18 Crawford, R. J., Erman, W. F., and Broaddus, C. D., *J. Am. Chem. Soc.*, <u>94</u>, 4298 (1972).

2.19 Hartmann, J., and Schlosser, M., *Helv. Chim. Acta*, <u>59</u>, 453 (1976).

2.20 Schlosser, M., and Hartmann, J., *J. Am. Chem. Soc.*, <u>98</u>, 4674 (1976).

2.21 Bujwid, Z. J., Gerrard, W., and Lappert, M. F., *Chem. Ind.*, 1091 (1959).

2.22 Muetterties, E. L., *J. Am. Chem. Soc.*, <u>81</u>, 2597 (1959); <u>82</u>, 4163 (1960).

2.23 Anastassiou, A. G., *Chem. Commun.*, 991 (1968).

2.24 Zweifel, G., and Steele, R. B., *J. Am. Chem. Soc.*, <u>89</u>, 5185 (1967).

2.25 Nakamura, A., and Otsuka, S., in *Organotransition-Metal Chemistry*, Ishii, Y., and Tsutsui, M., Eds., Plenum, New York, 1975, p. 181.

2.26 Treichel, P. M., Pitcher, E., and Stone, F. G. A., *Inorg. Chem.*, 1, 511 (1962).

2.27 (a) Kitching, W., *Organometal. Chem. Rev. A*, 3, 61 (1968); (b) Kitching, W., *Organometal. React.*, 3, 319 (1972).

2.28 Traylor, T. G., *Acc. Chem. Res.*, 2, 152 (1969).

2.29 (a) Nesmeyanov, A. N., and Kochetkov, N. K., *Izv. Akad. Nauk SSSR, Otd. Khim. Nauk*, 74, 305 (1949); 77 (1950).

2.30 Timms, P. L., *Acc. Chem. Res.*, 6, 118 (1973).

2.31 Glanville, J. O., Stewart, J. M., and Grim, S. O., *J. Organometal. Chem.*, 7, 9 (1967).

2.32 Seyferth, D., *J. Organometal. Chem.*, 100, 237 (1975).

2.33 For reviews, see (a) Kalyanaraman, V., and George, M. V., *J. Organometal. Chem.*, 47, 225 (1973); (b) de Boer, E., *Adv. Organometal. Chem.*, 2, 115 (1964).

2.34 Smid, J., *J. Am. Chem. Soc.*, 87, 655 (1965).

2.35 Katz, T. J., *J. Am. Chem. Soc.*, 82, 3784 (1960).

2.36 Heck, R. F., *J. Am. Chem. Soc.*, 90, 5518 (1968).

3
GENERAL PATTERNS OF ORGANOMETALLIC REACTIONS

Most of the final products in organic synthesis do not contain the metal-carbon bond, although there are some biologically important organic compounds, such as vitamin B_{12}, which do contain the metal-carbon bond, and various synthetic organometallic compounds which are becoming increasingly important as final products. Thus the formation of organometallic compounds themselves seldom represents the completion of a synthesis. In the great majority of cases, they must be further transformed to produce organic compounds that do not contain the metal-carbon bond. In other words, the organometallic compounds usually act as either reagents or intermediates in organic synthesis.

3.1   REAGENT VERSUS INTERMEDIATE

The distinction between reagents and intermediates is rather vague. One unambiguous way of defining these terms is to view any organometallic compounds a part or the whole of the metal-bound carbon group of which is to be incorporated in the organic product as intermediates and the others as reagents. It is clear that a given organometallic compound may act as an intermediate and/or reagent depending on its specific role in a given reaction. A case in point is the reaction of Grignard reagents with ketones (Sect. 4.3.2.2), which can bring about addition, enolization, and reduction as well as other miscellaneous transformations. In the addition reaction, the Grignard reagent acts as an intermediate, whereas in the reduction reaction it acts as a reagent. In a more ambiguous but perhaps more widely used definition, the reagents include, in addition to the above-defined compounds, some carbon-group-incorporating compounds that are readily accessible and relatively inexpensive so that the excessive use of such

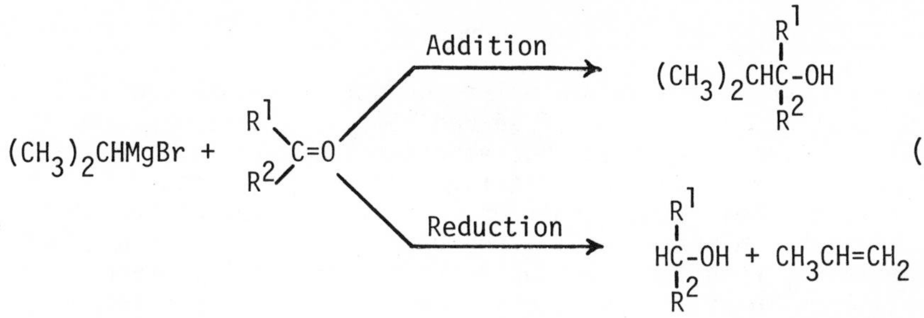

$$(CH_3)_2CHMgBr + \quad {}^{R^1}_{R^2}\!\!\diagdown C=O \quad \xrightarrow{\text{Addition}} \quad (CH_3)_2CH\overset{\displaystyle R^1}{\underset{\displaystyle R^2}{C}}\!-OH \qquad (3.1)$$

$$\xrightarrow{\text{Reduction}} \quad H\overset{\displaystyle R^1}{\underset{\displaystyle R^2}{C}}\!-OH + CH_3CH=CH_2$$

compounds is readily tolerated. Although the distinction between reagents and intermediates has some practical significance, it is of no theoretical or mechanistic significance. In this book we arbitrarily adopt the former definition in most cases simply for the sake of clarity.

3.2   ORGANOMETALLICS AS INTERMEDIATES

In cases where organometallic compounds are used as intermediates, the metal-carbon bond must be cleaved. In such cases, organic synthesis involving the use of organometallics follows either one of the two general paths shown below.

Scheme 3.1

Type A   M-R Formation + M-R Scission   M-R'

Type B   M-R Formation + M-R Interconversion + M'-R Scission

                                                M'-R'

          M or M' = metal or metal-containing group.
          R or R' = organic group.

As indicated above, the organometallic interconversion may in-volve changes in the organic moiety (R) and/or the metal-contain-ing moiety (M). Since we discuss some of the general patterns of the formation of organometallics in Chap. 2, we shall now be con-cerned with the scission of the metal-carbon bond and the organo-metallic interconversion.

3.2.1 Scission of the Metal-Carbon Bond

Inasmuch as carbon groups can exist and act as (1) carbanions, (2) carbocations, (3) carbon free-radicals, (4) carbenes, and (5) neutral and electron-paired species other than singlet carbenes, the organometallic compounds might, a priori, be expected to act as sources of these carbon species. It is, however, often difficult to establish unequivocally the precise mechanism of a given organometallic reaction. Fortunately, the successful use of organometallics in organic synthesis does not always require detailed mechanistic information. Even so, it may still be advantageous to attempt to interpret the observed results from the mechanistic viewpoint and to make use of the interpretations to formulate predictions.

In the following discussion, let us consider only a generalized situation, represented by eq. 3.2.

$$M-R \ + \ A-B \longrightarrow \ ? \tag{3.2}$$

The reactant A-B can represent a variety of compounds. It can be either an organic or an inorganic compound. Although not shown explicitly, groups A and B may be doubly or triply bonded. Interaction of M-R with A-B may even be preceded by decomposition of M-R and/or A-B.

3.2.1.1 Organometallics as Carbanion Sources

Probably the single most important application of organometallics to organic synthesis lies in their use as carbanion sources. By definition the metal-carbon bond is polarized in the sense indicated by 3.1. In cases where the contribution of the ionic resonance structure 3.2 is significant, the organometallic compound in question might be expected to act as a carbanion source or a nucleophile.

$$\overset{\delta+ \quad \delta-}{M-R}$$

$$M^+ \quad R^-$$

3.1 $\qquad\qquad\qquad\qquad\qquad$ 3.2

Both organic and inorganic electrophiles can react with such a nucleophilic organometallic compound. A representative scheme for the reactions of nucleophilic organometallic compounds with organic electrophiles is shown in Scheme 3.2.

Proton acids (H-X) and halogens (X_2), along with various metal-containing compounds (M-X), represent the most commonly encountered inorganic electrophiles. Other inorganic electrophiles include those that contain the S-halogen, S-O, O-halogen, O-O,

Scheme 3.2

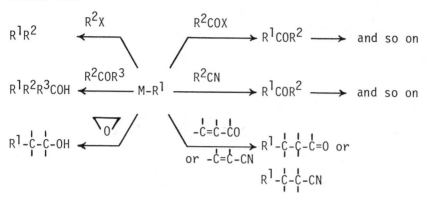

N-halogen, and N-O bonds. Since the reaction of a nucleophilic organometallic compound (M^1R) with a metal-containing compound (M^2X) would result in the formation of another organometallic compound either via complexation or transmetallation, it should be viewed as an organometallic interconversion reaction and is therefore discussed later.

The reactant-product relationships of these reactions are straightforward and can be represented by the following general equations (eqs. 3.3 to 3.5).

$$\overset{\delta+}{M}\text{---}\overset{\delta-}{R} + \overset{\delta+}{A}\text{---}\overset{\delta-}{B} \longrightarrow R\text{---}A + M\text{---}B \qquad (3.3)$$

$$\overset{\delta+}{M}\text{---}\overset{\delta-}{R} + \overset{\delta+}{A}\text{==}\overset{\delta-}{B} \longrightarrow R\text{---}A\text{---}B\text{---}M \qquad (3.4)$$

$$\overset{\delta+}{M}\text{---}\overset{\delta-}{R} + \overset{\delta+}{A}\text{≡}\overset{\delta-}{B} \longrightarrow R\text{---}A\text{==}B\text{---}M \qquad (3.5)$$

On the other hand, the mechanistic details of these reactions seem to vary considerably and can be far more complex than might be indicated by these equations. A detailed discussion of the mechanisms of all of these reactions is beyond the scope of this book. Let us discuss only the case of the cross-coupling reaction between organometallics (M-R^1) and organic halides or sulfonates (R^2X). The following mechanistic schemes appear to be worth considering.

Bimolecular Nucleophilic Substitution Mechanism (S_N2). Until recently, there had been a tendency among chemists to associate various cross-coupling reactions of organometallic compounds with organic halides and sulfonates with the S_N2 mechanism. Among other things, the S_N2 mechanism implies a colinear arrangement of the nonbonding orbital of the carbanion and the R^2-X bond during much of the bond-forming and bond-breaking process, including the transition state (eq. 3.6).

$$\tag{3.6}$$

It is clear that such a process would be favored by (1) highly electropositive metals, such as potassium and sodium, (2) highly electronegative or acidic R^1 groups, such as alkynyl groups, and (3) solvents of high coordination power, such as dimethoxyethane, which assist ionization of MR^1 through solvation of the M^+ cation.

More recent studies have indicated, however, that the s_N2 process probably represents a relatively small number of organometallic cross-coupling reactions and that it should be replaced, at least in part, with various alternate mechanistic schemes in a number of cases.

Substitution via One-Electron Transfer Process. In the s_N2 process, electron transfer and nuclear transfer take place at the same time. Alternatively, one of the nonbonding electrons of the carbanion R^1 or of the bonding electrons of M-R^1 may be transferred to R^2-X without being accompanied by any nuclear transfer. Such an electron transfer reaction will weaken not only the M-R^1 bond but also the R^2-X bond which now has one unshared electron in the LUMO. In either case, even the formation of the neutral free radicals $\cdot R^1$ and $\cdot R^2$ may occur under the reaction conditions. These radical ions and/or neutral radicals can combine to form R^1R^2 and other products (eq. 3.7).

$$M\text{-}R^1 + R^2\text{-}X \longrightarrow \left[\begin{array}{l} (M\text{-}R^1)^{\ddot{+}} + (R^2\text{-}X)^{\ddot{-}} \\ \text{or} \\ M^+ + \cdot R^1 + (R^2\text{-}X)^{\ddot{-}} \\ \text{or} \\ (M\text{-}R^1)^{\ddot{+}} + \cdot R^2 + X^- \\ \text{or} \\ M^+ + \cdot R^1 + \cdot R^2 + X^- \end{array} \right] \longrightarrow R^1\text{-}R^2 + M\text{-}X \tag{3.7}$$

The formation of free-radical species as transient intermediates can be detected either by ESR or by an NMR technique called "chemically induced dynamic nuclear polarization" (CIDNP), see Sect. 4.3.2.1 (3.1).

If an NMR spectrum is taken during the course of a reaction, certain signals may be enhanced, while others may be suppressed, either in a positive or negative direction. When this phenomenon, that is, CIDNP, is observed with the product of a reaction, it indicates that the product was formed either partially or entirely via a free-radical intermediate.

Many reactions of alkyllithiums with alkyl bromides, iodides, and geminal dichlorides in hydrocarbon solvents, such as the one in eq. 3.8, give products of coupling and disproportionation that show strong signal enhancements.

$$C_2H_5Li + C_2H_5I \longrightarrow n\text{-}C_4H_{10} + C_2H_6 + H_2C=CH_2 + LiI \quad (3.8)$$

It should be emphasized here that the occurrence of the one-electron process is, of course, not restricted to the cross-coupling reaction, and that any two-electron process can, in principle, have its one-electron counterpart which can produce the same products. In fact, recent studies indicate that a growing number of reactions, such as addition of Grignard reagents to ketones, see Sect. 4.3.2.2 (*3.2*), and conjugate addition of organocuprates, see Sect. 9.4 (*3.3*), which were thought to proceed by two-electron processes, in many cases evidently proceed by one-electron processes (*3.4*).

<u>Unimolecular Nucleophilic Substitution Process (s_N1) in Which the Organometallic Compound Acts as an *Amphophile*</u>. Both the s_N2 mechanism and the one-electron transfer process discussed above involve formation of the M^+ cation prior to the crucial carbon-carbon bond formation. Consequently, such processes become less and less operative, as the metal (M) becomes increasingly electronegative. Thus, for example, typical organic derivatives of zinc (EN = 1.66) and aluminum (EN = 1.47) do not react readily with typical primary alkyl halides, such as ethyl iodide. On the other hand, these organometallics can react to form cross-coupled products with hindered alkyl halides, such as t-butyl chloride and isopropyl mesylate (Chaps. 4 and 5), which would merely undergo β-elimination with more ionic and basic organometallic compounds. The order of reactivity of alkyl halides, that is, tertiary > secondary > primary, clearly indicates that the reaction involves ionization or extensive charge separation of the alkyl halides and sulfonates prior to the carbon-carbon bond formation (s_N1-like). Such a process may be induced by the interaction of the leaving group (X) and the empty orbital of the metal atom (M), where the organometallic compound acts as an electrophile rather than a nucleophile (eq. 3.9).

$$R^1\text{-}M + X\text{-}R^2 \longrightarrow R^1\text{-}M^-\text{-}X + {}^+R^2 \longrightarrow R^1\text{-}R^2 + MX \quad (3.9)$$

$$\underline{3.3}$$

The effect of the ionization or charge separation process is synergistic in that it clearly increases not only the electrophilicity of the R^2 group but also the nucleophilicity of the $M\text{-}R^1$ bond. A subsequent nucleophile-electrophile interaction, in which the organometallic compound now acts as a nucleophile, would then produce $R^1\text{-}R^2$.

Although $M\text{-}R^1$ acts as a source of a carbanionic species $(^-R^1)$, it is not entirely appropriate to say that this organometallic reagent is acting as a nucleophile, inasmuch as its role as an electrophile is of comparable significance. We therefore propose the term *amphophile* to denote a species that can act as both an electrophile and a nucleophile in a given reaction. Of course, any dipolar molecule or group is intrinsically *amphophilic* having both electrophilic and nucleophilic ends. Therefore the use of the term *amphophile* should be restricted to those cases in which both electrophilic and nucleophilic functions are essential and of comparable significance.

Substitution via Three-Center Interaction Involving the Metal-Carbon Bond (S_E2). In the discussion of the reaction shown in eq. 3.9, nothing is said about how the R^1 and R^2 groups might interact with each other. It is conceivable that the organometallic complex 3.3 undergoes either dissociation to produce the $^-R^1$ carbanion which can now couple with the $^+R^2$ cation or the one-electron transfer. In cases where the cross-coupled products $(R^1\text{-}R^2)$ are obtained in high yields, however, it is rather unlikely that the full-fledged carbanion is formed as an active species, since such a carbanion is expected by preference to undergo β-elimination with the $^+R^1$ cation.

Presently available information suggests that highly electrophilic species including carbon electrophiles may interact directly with the metal-carbon bond. Both the front-side attack (3.4) and the back-side attack (3.5) are conceivable.

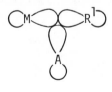

3.4   3.5

Front-side attack   Back-side attack

These processes are represented by the s_E2 mechanism (3.5, 3.6).
The back-side attack mechanism should not be confused with the
s_N2 mechanism. Although both are three-centered, the central
carbon atom in the s_E2 process is nucleophilic, whereas that in
the s_N2 process is electrophilic.

An important difference between the front-side attack and
the back-side attack is that the former proceeds with retention
of configuration of the R^1 group, while the latter proceeds with
its inversion. A recent study of iodination of organoborates,
for example, shows that the reaction proceeds with inversion. A
plausible mechanism, shown in eq. 3.10, involves the back-side
attack (3.7).

$$(3.10)$$

Although the number of well-established examples of the aliphatic carbon-carbon formation via organometallics that proceed
by the s_E2 mechanism is still limited, it appears likely that a
variety of reactions of organometallics having relatively covalent but reactive metal-carbon bonds, such as those containing
Hg, Al, Sn, and certain transition metals, actually involve the
s_E2 mechanisms. Although organoboranes themselves are quite reluctant to participate in these reactions, available information
suggests that organoborates are sufficiently nucleophilic to react with various electrophiles. Some of these reactions appear
to proceed by these mechanisms (Sect. 5.2.3).

Substitution via Addition-Elimination. Those metal-carbon bonds
that are highly covalent and short, hence not readily polarizable,
are generally quite resistant to the attack by electrophiles.
Such metal-carbon bonds are represented by the B-C (EN of boron =
2.0) and Si-C (EN of silicon = 1.8) bonds. Thus, for example, it
is exceedingly difficult to cleave the B-C bond of trimethylborane (but not of tetramethylborate) and the Si-C bond of tetramethylsilane with any electrophile including strong mineral acids
and halogens, although trimethylborane is reasonably reactive toward carboxylic acids (Sect. 5.2.3.2). It may safely be concluded that these metal-carbon bonds are intrinsically nonnucleophilic.

While the generalization made above is probably true, it

does not imply that all types of B-C and Si-C bonds are reluctant to participate in reactions with electrophiles. In fact,
various α,β- and β,γ-unsaturated derivatives of boron and silicon are generally far more reactive toward electrophiles than
their alkyl counterparts. For example, $PhSiMe_3$ reacts readily
with benzyl bromide to form diphenylmethane in the presence of
$AlCl_3$ (Sect. 6.4.2.1), whereas Me_4Si is completely unreactive.
What factors are responsible for such a marked difference?
Firstly, the unsaturated carbon-carbon bonds, such as alkenyl,
aryl, and alkynyl bonds, are reasonably good nucleophiles even in
the absence of any metal. Secondly, it has now been well established that the reaction of the unsaturated carbon-carbon bonds
with electrophiles is strongly facilitated by various metals,
such as Hg, Pb, Tl, Sn, and Si, which are either directly bonded
to or separated by one carbon atom from the unsaturated carbon-
carbon bonds (eqs. 3.11 and 3.12). Such an effect has been termed $\sigma-\pi$ conjugation effect, as discussed in Sect. 6.4.1.

$$-\overset{|}{\underset{|}{C}}=\overset{|}{\underset{|}{C}}-MX_n \xrightarrow{A^+} -\overset{|}{\underset{|}{C}}-\overset{\overset{+}{MX_n}}{\underset{|}{C}}-A \longleftrightarrow -\overset{|}{\underset{A}{C}}=\overset{\overset{+}{MX_n}}{\underset{|}{C}}-A \longrightarrow -\overset{|}{\underset{A}{C}}=\overset{|}{\underset{|}{C}}-A \quad (3.11)$$

$$-\overset{|}{\underset{|}{C}}=\overset{|}{\underset{|}{C}}-\overset{|}{\underset{|}{C}}-MX_n \xrightarrow{A^+} -\overset{|}{\underset{|}{C}}-\overset{\overset{A}{|}}{\underset{|}{C}}-\overset{\overset{+}{MX_n}}{\underset{|}{C}}- \longleftrightarrow -\overset{A}{\underset{|}{C}}-\overset{|}{\underset{|}{C}}=\overset{\overset{+}{M X_n}}{\underset{|}{C}}- \longrightarrow -\overset{A}{\underset{|}{C}}-\overset{|}{\underset{|}{C}}=\overset{|}{\underset{|}{C}}- \quad (3.12)$$

It should be noted that the $\sigma-\pi$ conjugation effect facilitates
not only the interaction between the π-system and an electrophile
(A^+) but also cleavage of the metal-carbon bond. This addition-
elimination sequence results in the substitution of a metal atom
or a metal-containing group with an electrophile.

The available data indicate that similar addition-elimination mechanisms probably operate in certain reactions of α,β-un-
saturated organoaluminums with electrophiles (Sect. 5.3.3.2). On
the other hand, the reactions of α,β-unsaturated organoborons
with electrophiles discussed in detail in Sect. 5.2.3 proceed
mostly by yet another mechanism which is unique to organoborons
(eq. 3.13).

$$-\overset{|}{\underset{|}{C}}=\overset{\overset{R}{|}}{\underset{|}{C}}-\overset{R}{\underset{}{B}}^-- \xrightarrow{A^+} A-\overset{|}{\underset{|}{C}}-\overset{\overset{R}{|}}{\underset{}{\overset{+}{C}}}-\overset{}{B}^-- \longrightarrow A-\overset{|}{\underset{|}{C}}-\overset{\overset{R}{|}}{\underset{|}{C}}-B- \quad (3.13)$$

This reaction, however, does not result in the cleavage of the
B-C bond, and should therefore be viewed as an organometallic interconversion reaction.

Substitution via Oxidative Addition-Reductive Elimination.   Certain organotransition metal compounds can bring about cross coupling through sequences that evidently are totally different from any of the mechanisms discussed above.   There appear to be at least two different paths.

  The reaction of organocuprates and related compounds with organic halides has been interpreted by some chemists in terms of the following oxidative addition-reductive elimination sequence, see eq. 3.14 (Sect. 9.1).

$$R_2^1Cu^- + R^2-X \xrightarrow[\text{addition}]{\text{oxidative}} R^1-\underset{\underset{R^1}{|}}{\overset{\overset{R^2}{|}}{Cu}}-R^1 \xrightarrow{\text{reductive elimination}} R^1-R^2 + R^1-Cu \quad (3.14)$$

$$\underline{3.6}$$

  On the other hand, the reaction of certain main group organometallics ($M-R^1$) with organic halides catalyzed by nickel and palladium complexes (eq. 3.15) has been interpreted in terms of the mechanism shown in Scheme 3.3, which involves an oxidative addition-transmetallation-reductive elimination sequence (Sect. 9.2).

$$M-R^1 + R^2-X \xrightarrow{\text{cat. } M'L_n} R^1-R^2 + M-X \qquad (3.15)$$

$M = Li, MgX, ZnX, Al$, and so on; $M'L_n = Ni(PPh_3)_4, Pd(PPh_3)_4$, and so on

Scheme 3.3

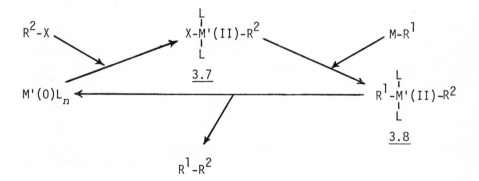

Although the overall equation shown in eqs. 3.14 and 3.15 clearly indicates that the R' group acts as a carbanionic moiety, it is not possible to classify one or the other of the two organic groups of 3.6 or 3.8 as either a nucleophile or an electrophile in the reductive elimination step which involves the crucial carbon-carbon bond formation at the expense of the cleavage of the metal-carbon bonds. These reactions are discussed in detail in Sect. 9.2.

In the above discussion, we arbitrarily chose cross coupling as a representative case of carbon-carbon bond formation via nucleophile-electrophile interaction. Similar considerations can also be made, however, with respect to various other nucleophile-electrophile reactions in which organometallics act as nucleophiles, as will become clear from various specific discussions presented in the following chapters.

3.2.1.2 Organometallics as Carbocation Sources

Since the metal-carbon bond is normally polarized as shown in 3.1, it might seem highly unlikely that any organometallic compound can act at all as a carbocation source, unless the organic group is oxidized either by added reagents or by some other means, such as electrolysis. While this generalization is valid in most cases, there are certain classes of organometallics that can readily act as carbocation sources.

One trick which organometallics can resort to in generating carbocationic species is shown in eq. 3.16.

$$R-M^{\delta+}-X^{\delta-} \longrightarrow R-M^{+} + X^{-} \longrightarrow R^{+} + M + X^{-} \qquad (3.16)$$

Such a process is facilitated if both the M-X and M-R bonds are labile. It is important to recognize that the second step of the reaction involves reduction of the metal by two electrons. Consequently, cleavage of the M-R bond will occur readily if it can be easily reduced by two electrons. Indeed, various organometallic compounds containing heavy, highly polarizable metals, such as Au, Tl, Pb, Pd, and Pt, can readily participate in the reaction shown in eq. 3.16.

The carbocationic species thus generated can then undergo various reactions represented by eqs. 3.17 to 3.20, which are characteristic of any carbocations irrespective of their origins. Whereas the first two reactions produce neutral products which can exist as stable compounds, the latter two reactions yield new carbocations which must eventually undergo either hetero coupling or elimination. These reactions of the carbocations derived from organometallics of heavy metals will be discussed in Chap. 7 and Sect. 10.3.

Hetero coupling: $R^+ + Y^- \longrightarrow R-Y$ (3.17)

Elimination: $-\overset{+}{\underset{|}{C}}-Y-Z \longrightarrow C=Y + Z^+$ (3.18)

Rearrangement: $-\overset{+}{\underset{|}{C}}-\overset{\overset{Y}{|}}{\underset{|}{C}}- \longrightarrow -\overset{\overset{Y}{|}}{\underset{|}{C}}-\overset{+}{\underset{|}{C}}$ (3.19)

Addition: $R^+ + Y=Z \longrightarrow R-Y-Z^+$ (3.20)

3.2.1.3 Organometallics as Carbon Free-Radical Sources

The presence or ready availability of empty orbitals that can act as low-lying LUMOs or acceptor sites characterizes a large number of organometallic compounds. These empty orbitals can promote the formation of carbon free-radicals in a few different ways, which is briefly discussed below.

Bimolecular Homolytic Substitution (S_H2). Whereas the S_H2 reaction has almost never been observed at the carbon center, various organometallics, such as those containing boron and tin, can readily participate in the S_H2 reaction (3.8). A number of synthetically useful organoboron reactions, in particular, have been shown to proceed by the S_H2 mechanism (3.9), as discussed in detail in Sect. 5.2.3.

$$R-M + \cdot Y \longrightarrow [R\text{---}M\text{---}Y]^{\cdot} \longrightarrow R\cdot + M-Y \quad (3.21)$$

α-Hydrogen Abstraction. The availability of an empty orbital can facilitate the α-hydrogen abstraction reaction shown in eq. 3.22. This reaction, however, does not involve cleavage of the metal-carbon bond.

$$\quad (3.22)$$

A typical example is the free-radical bromination of organoboranes discussed in Sect. 5.2.3.1 (*3.10*). At present, however,
the scope of the α-hydrogen abstraction reactions of organometallics does not appear to have been well delineated. There are
other ways of generating carbon free-radicals from organometallics which do not appear to require the availability of a low-lying empty orbital, as discussed below.

One-Electron Transfer. This process is discussed briefly in
Sect. 3.2.1.1.

$$R-M \xrightarrow{-e} R\cdot + M^+ \tag{3.23}$$

As mentioned earlier, this process is favored by highly electropositive and readily ionizable metals. It is conceivable, however, that organometallate anions, such as those containing boron
and aluminum, might also readily participate in this process,
which would produce stable neutral species along with carbon
free-radicals (eq. 3.24).

$$R-M^-X_3 \xrightarrow{-e} R\cdot + MX_3 \tag{3.24}$$

$$M = B \text{ or } Al$$

Thermal or Photochemical Excitation. Thermodynamically weak metal-carbon bonds can be cleaved thermally or photochemically (eq.
3.25).

$$R-M \xrightarrow{h\nu \text{ or } \Delta} R\cdot + \cdot M \tag{3.25}$$

This is a reaction characteristic of any thermodynamically weak
bond.

Ligand Abstraction-Reductive Cleavage. If an organometallic species contains a ligand that can be readily abstracted by a free
radical, the corresponding metal free-radical may decompose to
give a carbon free-radical and a reduced metal species, provided
that such a reduction is facile (eq. 3.27). This is the free
radical counterpart of the carbocation-forming reaction shown in
eq. 3.16.

$$R-M-X + \cdot Y \longrightarrow R-M\cdot + X-Y \longrightarrow R\cdot + M \tag{3.26}$$

Those metals that can be readily reduced by two successive one-
electron reductions, such as mercury, may participate in this re-

action, as shown in eq. 3.27 (Sect. 7.3.1).

$$RHgX \xrightarrow{NaBH_4} RHgH \xrightarrow{\cdot Y} R\cdot + Hg + HY \qquad (3.27)$$

There is no essential difference between the carbon free-radicals derived from organometallics and those obtained by other methods. Thus they can now participate in various reactions characteristic of the carbon free-radicals, which are represented by the following.

Radical coupling: $R\cdot + \cdot R' \longrightarrow RR + RR' + R'R'$ (3.28)

Disproportionation: $2 \underset{\underset{H}{|}}{-\overset{|}{C}-\overset{|}{C}\cdot} \longrightarrow \overset{\diagdown}{\diagup}C=C\overset{\diagup}{\diagdown} + H-\overset{|}{\underset{|}{C}}-\overset{|}{\underset{|}{C}}-H$ (3.29)

These two processes convert carbon free-radicals into electron-paired species, and are therefore known as termination processes. There are other processes that convert one free-radical into another.

Abstraction: $R\cdot + Y-R' \longrightarrow R-Y + \cdot R'$ (3.30)

Addition: $R\cdot + Y=C\overset{\diagup}{\diagdown} \longrightarrow R-Y-\overset{|}{\underset{|}{C}}\cdot$ (3.31)

Rearrangement: $-\overset{|}{\underset{|}{C}}-Z-\overset{|}{\underset{|}{C}}\cdot \longrightarrow \cdot\overset{|}{\underset{|}{C}}-Z-\overset{|}{\underset{|}{C}}-$ (3.32)

The free-radical rearrangement reaction may be viewed as the intramolecular version of the abstraction reaction. Although the free-radical 1,2-rearrangement is rather uncommon, the free-radical 1,5-rearrangement is a very important process (eq. 3.33).

$$(3.33)$$

3.2.1.4   Organometallics as Carbene Sources

A variety of α-heterosubstituted organometallics (<u>3.9</u>) have acted as carbene precursors and carbenoid species (*3.11*) (eq. 3.34).

$$
\begin{array}{c}
-\overset{|}{\underset{|}{C}}{}^{-} \quad + \quad M^{+} \\
X
\end{array}
$$

$$
\underline{3.10}
$$

$$
-\overset{|}{\underset{|}{C}}-M \qquad\qquad\qquad\qquad \downarrow \tag{3.34}
$$

$$
\underline{3.9} \qquad\qquad -\overset{|}{C}: \quad + \quad MX
$$

$$
\underline{3.11}
$$

All three carbon species in eq. 3.34, that is, $\underline{3.9}$, $\underline{3.10}$, and $\underline{3.11}$, are capable of acting as actual reactive species.
     The α-haloorganometals containing highly electropositive metals, such as lithium, can be generated as unstable species which can either act as carbanions $\underline{3.10}$ or generate carbenes $\underline{3.11}$. On the other hand, α-haloorganozincs, that is, Simmons-Smith reagents, evidently react as molecular species $\underline{3.9}$ (Sect. 4.3.3.5). Finally, α-haloorganomercuries, such as $PhHgCCCl_3$, can be thermally decomposed to produce directly the corresponding carbenes $\underline{3.11}$ (Sect. 7.3.2.2).
     The carbanion $\underline{3.10}$ normally acts as a typical nucleophile, unless it is converted to the corresponding carbene $\underline{3.11}$. The carbene $\underline{3.11}$ can participate in various reactions represented by eqs. 3.35 and 3.36, which are characteristic of the singlet carbenes.

Addition:
$$
ZYC: + \quad \overset{}{\underset{}{>}}C=C\overset{}{\underset{}{<}} \quad\longrightarrow\quad \overset{}{\underset{}{>}}C\!\!-\!\!C\overset{}{\underset{}{<}} \tag{3.35}
$$

Rearrangement:
$$
-\overset{..}{\underset{\underset{Y}{|}}{C}}-C- \quad\longrightarrow\quad \overset{}{\underset{Y}{>}}C=C\overset{}{\underset{}{<}} \tag{3.36}
$$

Insertion into the C-H bond does not appear to be a significant reaction of the carbenes generated by decomposition of α-haloorganometallics.
     Various transition metal complexes having both empty and filled valence-shell orbitals can stabilize carbenes by forming metal-carbene complexes (eq. 3.37). As discussed in detail in Chap. 12, the metal-carbene complexes can also be formed by various other means.

$$X_2C: + ML_n \longrightarrow X_2C=ML_n \longleftrightarrow X_2C^+-M^-L_n \longleftrightarrow X_2C^--M^+L_n \quad (3.37)$$

One mode of metal-carbene reaction which appears to be highly general is shown in eq. 3.38.

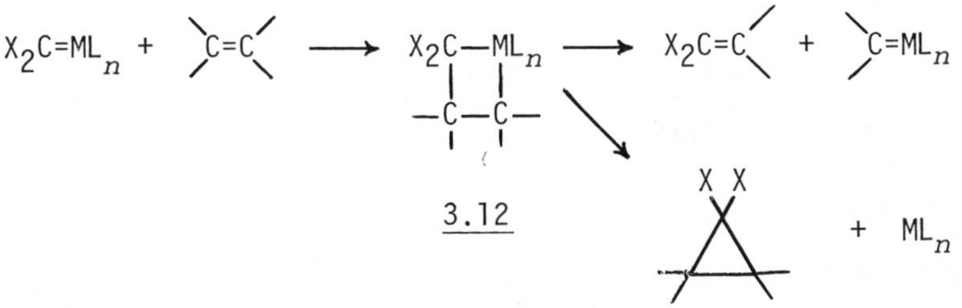

$$X_2C=ML_n + \;\; \overset{}{\underset{}{>}}C=C\overset{}{\underset{}{<}} \longrightarrow X_2C-ML_n \longrightarrow X_2C=C\overset{}{\underset{}{<}} + \;\; \overset{}{\underset{}{>}}C=ML_n$$

$$\underset{}{-\overset{|}{\underset{|}{C}}-\overset{|}{\underset{|}{C}}-}$$

$$\underline{3.12}$$

$$+ \; ML_n$$

$$(3.38)$$

The metallocyclic species $\underline{3.12}$ can undergo cycloreversion, reductive elimination, and other reactions. The formation and decomposition via cycloreversion of $\underline{3.12}$ appears to be one of the most plausible paths for olefin metathesis with tungsten and molybdenum catalysts (Sect. 12.3), and the formation and decomposition via reductive elimination of $\underline{3.12}$ can be invoked to explain the results obtained in the transition metal-catalyzed cyclopropanation (Sect. 12.1), as exemplified by the following reaction (*3.12*).

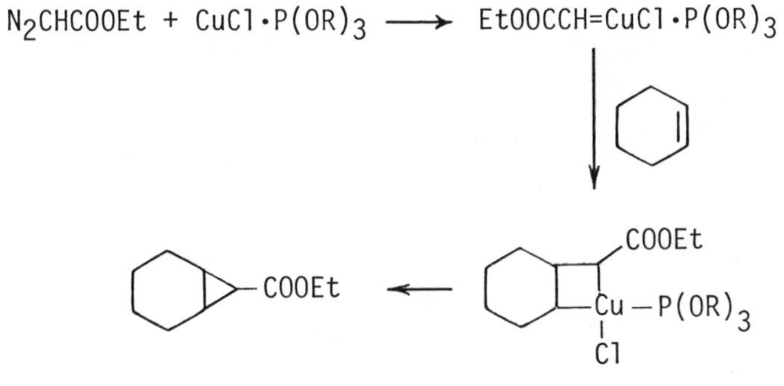

$$N_2CHCOOEt + CuCl \cdot P(OR)_3 \longrightarrow EtOOCCH=CuCl \cdot P(OR)_3$$

$$(3.39)$$

3.2.1.5 Scission of the Metal-Carbon Bond via Concerted Pericyclic Processes

In a wide variety of reactions involving scission of the metal-carbon bond, it is difficult or almost certainly inappropriate to classify them according to one of the four general types discussed in the preceding four sections. Many of these reactions appear best interpreted in terms of pericyclic processes (*1.6*). It should, of course, be clear that some of the metal-carbon bond-cleaving reactions, which can be represented by the four types of processes discussed above, can also be classified as pericyclic processes. For example, the formation of :CCl$_2$ by thermolysis of PhHgCCl$_3$ is best viewed as a carbene-forming pericyclic reaction (Sect. 7.3.2.2). Because it is impossible to discuss here all types of metal-carbon bond-cleaving pericyclic reactions, only a few representative examples are presented.

Two-Electron Pericyclic Processes. The formation of various olefin π-complexes of transition metals is best represented by the two-electron pericyclic process. Their thermal decomposition by the reversal of the above process must also be two-electron pericyclic reactions.

$$\overset{\overset{\textstyle ML_n}{\uparrow}}{\underset{\diagup}{\searrow}C \dotdiv C\underset{\diagdown}{\diagup}} \quad \rightleftharpoons \quad \underset{\diagup}{\searrow}C = C\underset{\diagdown}{\diagup} \; + \; ML_n \tag{3.40}$$

Four-Electron Pericyclic Processes. Perhaps far more intriguing are the ubiquitous thermal four-electron pericyclic processes, such as hydrometallation-dehydrometallation (eq. 3.41) and carbometallation (eq. 3.42), which at first sight might seem thermally forbidden.

$$\underset{\diagup}{\searrow}C = C\underset{\diagdown}{\diagup} \; + \; HML_n \quad \rightleftharpoons \quad \underset{\underset{\textstyle H \; ML_n}{|\;\;\;|}}{-\overset{|}{C}-\overset{|}{C}-} \tag{3.41}$$

$$\underset{\diagup}{\searrow}C = C\underset{\diagdown}{\diagup} \; + \; RML_n \quad \rightleftharpoons \quad \underset{\underset{\textstyle R \; ML_n}{|\;\;\;|}}{-\overset{|}{C}-\overset{|}{C}-} \tag{3.42}$$

In cases where the stereoselective cis addition is observed, these reactions most probably involve concerted, but not synchronous, pericyclic processes. As discussed in Sect. 2.6.1, the presence or ready availability of an empty orbital evidently makes otherwise thermally forbidden [2 + 2] reactions thermally allowed. These reactions appear to be best classified as $[_\pi 2_s + _\sigma 2_a + _\omega 0_s]$ reactions, although the mechanisms of these reactions have seldom been fully established. It also appears that various other organometallic reactions can be interpreted in terms of the same $[_\pi2_s + _\sigma2_a + _\omega0_s]$ processes. It is likely that certain oxidative addition and reductive elimination reactions (eq. 3.43) can also be represented by this process, as discussed in Sect. 8.4.

$$X-Y + ML_n \; \underset{\substack{\text{reductive}\\\text{elimination}}}{\overset{\substack{\text{oxidative}\\\text{addition}}}{\rightleftarrows}} \; \begin{array}{c} X \\ \diagdown \\ \diagup \\ Y \end{array} ML_n \qquad (3.43)$$

<u>Six-Electron Pericyclic Processes</u>. Despite the highly covalent nature of the B-C bond, allylic organoboranes can react readily with various polar and nonpolar compounds, such as water, aldehydes, ketones, and acetylenes, as discussed in Sect. 5.2.3. Moreover all of these reactions are accompanied by allylic rearrangement. These results are best interpreted in terms of the following six-electron pericyclic mechanism.

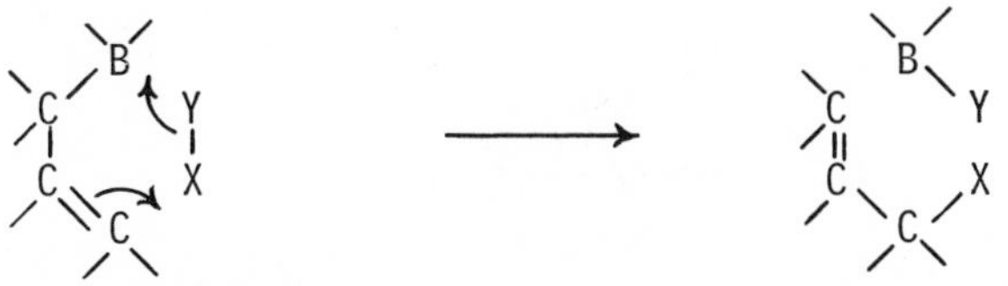

$$X-Y = H-OH, \quad RCH=O, \quad R_2C=O, \quad RC\equiv CR', \quad \text{and so on}$$

3.2.2 Organometallic Interconversion Reactions

3.2.2.1 Classification of Organometallic Interconversion Reactions

Organometallic compounds can undergo reactions which convert them into new organometallic compounds. These reactions are represented by the following general scheme.

Scheme 3.4

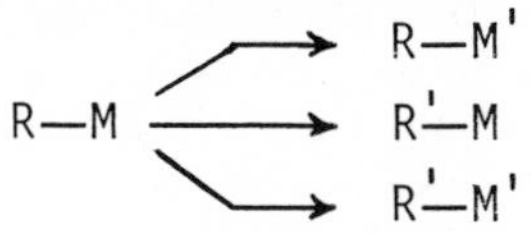

As indicated in Scheme 3.4, such reactions may involve changes in the metal-containing moieties and/or the organic moieties. Those reactions that involve changes in the metal-containing moieties are called transmetallation reactions, and are discussed in Sects. 2.11 and 2.12.

The organometallic interconversion reactions involving modification of the organic moieties may conveniently be classified as follows:

Type I   Organic ligand transformation not involving formation or cleavage of the metal-carbon bond.

Type II   Organic ligand transformation involving only formation or cleavage of the metal-carbon bond.

Type III Organic ligand transformation involving both formation and cleavage of the metal-carbon bonds.

The Type I reactions, exemplified by the reaction in eq. 3.45, can be better viewed as organic rather than organometallic reactions. No discussion of such reactions will therefore be attempted here.

$$Me\diagdown C=C\diagup H \qquad \xrightarrow{\quad\quad} \qquad Me\diagdown C\diagdown O\diagup C\diagup H \qquad (3.45)$$

Two examples of the Type II reactions are shown below.

$$i\text{-Bu}_2\text{AlH} \quad \xrightarrow{\text{HC}\equiv\text{CR}} \quad \begin{array}{c} H\diagdown C=C\diagup R \\ i\text{-Bu}_2\text{Al}\diagup \quad \diagdown H \end{array} \qquad (3.46)$$

$$\xrightarrow{\text{HOAc}} \qquad (3.47)$$

The first reaction involves formation but not cleavage of a metal-carbon bond, whereas the opposite is the case with the second reaction. These reactions should be viewed as metal-carbon bond-forming and metal-carbon bond-cleaving reactions, respectively.

We are now left with only the Type III reactions, which can be further divided into the following two types:

Type A   Organic ligand transformation unaccompa ied by skeletal changes in the organic ligands.

1. Metal migration (metallotropy).
2. Ligand exchange (disproportionation).
3. Ligand displacement.

Type B   Organic ligand transformation accompanied by skeletal changes in the organic ligands.

1. Carbometallation.
2. Ligand migration.

We briefly touch on these reactions in the following discussion.

3.2.2.2 Organic Ligand Transformation Unaccompanied by Skeletal Changes in the Organic Ligands

Metal Migration (Metallotropy). The $1,n$-shift reaction of an organometallic compound produces changes in the organic ligands, although such changes in some cases are not readily discernible. The 1,2-, 1,3-, and 1,5-rearrangements appear to be the most common modes of the $1,n$-rearrangements.

(a) Allylic rearrangement (1,3-shift). The 1,3-shift is commonly called the allylic rearrangement (3.13), and is ubiquitous in organometallic chemistry.

$$C=C-C^*-ML_n \rightleftharpoons L_nM-C-C=C^* \qquad (3.48)$$

A possible explanation for the ease with which a variety of coordinatively unsaturated allylmetals undergo the 1,3-metallotropy is presented in Sect. 1.6.

(b) 1,2- and 1,5-Rearrangements. The fluctional nature of cyclic allylic organometallics, such as 3.13 and 3.14 was discovered by Wilkinson (3.14) and extensively studied by Cotton and others (3.15).

3.13   3.14

Interestingly, detailed NMR studies of the metallotropic reactions involving η^1-cyclopentadienylmetals, such as 3.13, indicate that the 1,2- and/or 1,5-shift is favored over the 1,3-shift
(3.16).

In the case of the η^1-cyclopentadienylmetals the 1,2- and 1,
5-shifts are degenerate and therefore indistinguishable. Orbital
symmetry rules, however, predict that the observed processes
should be represented by the thermally allowed suprafacial 1,5-
shift rather than the 1,2-shift. At least in some cases, such as
those shown in eqs. 3.49 and 3.50, this prediction has been borne
out (3.17).

$$\text{and so on} \quad (3.49)$$

$$(3.50)$$

It appears that the cationotropic 1,2-metal shift is rare,
although certain metal-containing groups appear to participate
readily in the anionotropic 1,2-shift, as shown in eq. 3.51
(3.18).

$$(3.51)$$

(c) Others. There are various other types of metallotropic
rearrangements. For example, various ligand displacement processes discussed below, such as those involving dehydrometallation-hydrometallation and dissociation-complexation, can induce
metallotropic reactions. The metallotropic reactions of organotransition metal π-complexes are discussed in Sect. 8.5.

Ligand Exchange (Disproportionation). The ready availability of
an empty valence orbital makes various organometallics labile
with respect to ligand-ligand exchange (3.19). On the other hand,

those organometallics that satisfy the Lewis octet rule, such as organosilanes, and the 18-electron rule (Sect. 8.1), are quite stable with respect to ligand exchange.

The ligand exchange process can be most conveniently studied by NMR (*3.19b*). It appears that the mechanism of the reaction in most cases can be represented by eq. 3.52.

$$M-R^1 + R^2-M^* \rightleftharpoons \overset{R^1}{\underset{R^2}{M \cdots M^*}} \rightleftharpoons M-R^2 + R^1-M^* \qquad (3.52)$$

Although widely observed, this reaction does not usually seriously affect the course of organic synthesis via organometallics. It is, however, of critical importance in cases where the overall skeletal arrangement of an organometallic compound is important, as in some organoboron reactions. It is discussed later in appropriate sections.

Ligand Displacement. In the ligand-ligand exchange reaction, one ligand is displaced by another ligand. Alternatively, a ligand may be displaced by some group that initially is not bonded to a metal atom. Such reactions include (1) metal-halogen exchange (Sect. 2.3), (2) metal-hydrogen exchange (Sect. 2.5), and (3) displacement by unsaturated compounds via dehydrometallation-hydrometallation (eq. 3.53), dissociation-complexation (eq. 3.54), and other related processes.

$$R^1-\overset{H}{\underset{|}{\underset{|}{C}}}-\overset{|}{\underset{|}{C}}-M \xrightleftharpoons[]{R^1C=C-} MH \xrightleftharpoons[]{R^2C=C-} R^2-\overset{H}{\underset{|}{\underset{|}{C}}}-\overset{|}{\underset{|}{C}}-M \qquad (3.53)$$

$$R^1-\underset{\underset{M}{|}}{C}{=}\overset{|}{C}- \xrightleftharpoons[]{R^1C=C-} M \xrightleftharpoons[]{R^2C=C-} R^2-\underset{\underset{M}{|}}{C}{=}\overset{|}{C}- \qquad (3.54)$$

3.2.2.3   Organic Ligand Transformation Accompanied by Skeletal Changes in the Organic Ligands

Carbometallation.   This reaction may be represented by eqs. 3.55 and 3.56, and is closely related to hydrometallation (Sect. 2.6) and heterometallation (Sect. 2.7).

$$M-R \;+\; \text{\large$>$}C=C\text{\large$<$} \longrightarrow M-\underset{\displaystyle |}{\overset{\displaystyle |}{C}}-\underset{\displaystyle |}{\overset{\displaystyle |}{C}}-R \tag{3.55}$$

$$M-R \;+\; -C\equiv C- \longrightarrow M-\overset{\displaystyle |}{C}=\overset{\displaystyle |}{C}-R \tag{3.56}$$

As in the case of hydrometallation, those carbometallation processes that proceed readily appear to require coordinatively unsaturated organometallic species. It is probable that various hydrometallations and carbometallations are governed by some common mechanism, such as the one discussed in Sect. 2.6. The carbometallation process can repeat itself to produce oligomeric and polymeric products.

In general, the carbometallation reactions of organometallics containing main group metals are much more sluggish and tend to be less important than those of certain organotransition metals discussed in detail in Sect. 10.2. Carbolithiation (Sect. 4.3.2.5) and carboalumination (Sect. 5.3.2.1), however, have found significant industrial applications.

Ligand Migration.   While various types of ligand migration reactions are conceivable, by far the most common and significant is the anionotropic 1,2-shift. In this book, carbometallation is not viewed as a ligand migration reaction. The anionotropic 1,2-ligand shift of organometallics can be represented by the following general equations.

$$\overset{\displaystyle R}{\underset{\displaystyle }{M{-}Y{-}Z}} \longrightarrow \overset{\displaystyle R}{\underset{\displaystyle }{M{-}Y}} \;+\; Z \tag{3.57}$$

$$\overset{\displaystyle R}{\underset{\displaystyle }{M{-}Y{=}Z}} \longrightarrow \overset{\displaystyle R}{\underset{\displaystyle }{M{-}Y{-}Z}} \tag{3.58}$$

$$\overset{\displaystyle R}{\underset{\displaystyle }{M{-}Y{\equiv}Z}} \longrightarrow \overset{\displaystyle R}{\underset{\displaystyle }{M{-}Y{=}Z}} \tag{3.59}$$

For the sake of simplicity, no formal charge signs are shown. This convention is common in organotransition metal chemistry.

At present, organoborons are the only class of organometallics of main group metals that widely participate in this reaction. Their ligand migration is discussed in detail in Sect. 5.2.3. Organotransition metal compounds, on the other hand, readily participate in various ligand migration reactions. Particularly significant, from the viewpoint of organic synthesis, is the carbonylation reaction discussed in Chap. 11.

3.3 ORGANOMETALLICS AS REAGENTS

Transfer of the carbon group from the metal via cleavage of the metal-carbon bond is the key step in the use of organometallics as intermediates. On the other hand, in the use of organometallics as reagents, transfer of hydrogen and nonmetallic hetero atom groups, such as those containing N, O, P, S, and halogens, takes place via cleavage of the respective bond to metals (eq. 3.60).

$$R_nMY + Z \longrightarrow Y-Z + R_nM \qquad (3.60)$$

Y = H, halogen, N, O, P, and S groups, and so on

Z = organic and inorganic compounds

In some fortuitous circumstances, however, cleavage of the metal-carbon bond may also be involved.

3.3.1 Reduction via Hydride Transfer

Transfer of a hydride from a metal hydride to an organic compound brings about reduction, as represented by the following general equations.

$$-\overset{|}{\underset{|}{C}}-Y + HMR_n \longrightarrow H-\overset{|}{\underset{|}{C}}- + YMR_n \qquad (3.61)$$

$$\overset{\diagdown}{\underset{\diagup}{C}}=Y + HMR_n \longrightarrow H-\overset{|}{\underset{|}{C}}-Y-MR_n \qquad (3.62)$$

$$-C{\equiv}Y + HMR_n \longrightarrow H-C{=}Y-MR_n \qquad (3.63)$$

In cases where the Y group is a carbon group in eqs. 3.62 and 3.63, these equations, of course, represent hydrometallation. In other cases, the Y group is more electronegative than the carbon group, that is, $C^{\delta+}$-$Y^{\delta-}$. The regiochemical outcome shown in these equations stems primarily from two facts: (1) In most cases the metal atom is more electropositive than hydrogen, that is, $M^{\delta+}-H^{\delta-}$.   (2) A combination of the C-H and M-Y bonds general-ly is thermodynamically more favorable than a combination of the H-Y and M-C bonds, when Y = N, O, halogen, and so on.

Some of the most widely used metal hydrides bearing organic ligands are those that contain boron (Sect. 5.2.3.6), aluminum (Sect. 5.3.3.3), and tin (Sect. 6.4.3.3). Although various tran-sition metal hydrides are formed as reactive intermediates, it is often difficult to prepare them as discrete reagents that can be isolated. Despite their instability, however, many of them, such as those containing the Group VIII metals and copper, hold con-siderable promise as selective reducing agents (Chaps. 9 and 10).

In some cases, a hydrogen atom in an organic ligand can act as a hydride. It is usually one of the β-hydrogens, as shown in eq. 3.64. Certain α-hydrogens, however, can also act as hydrides, as shown in eq. 3.65 (Sect. 5.2.3.6).

$$(3.64)$$

$$(3.65)$$

3.3.2 Organometallics as Protecting Agents

Various active hydrogen-containing functional groups can be pro-tected by forming the corresponding organometallics.

$$HY \xrightarrow[\text{protection}]{R_nMX} R_nMY \xrightarrow[\text{deprotection}]{} HY + R_nMX' \qquad (3.66)$$

Y = NR_2, OR, SR, and so on

As might be expected, those organometallics that contain relatively covalent and stable metal-carbon bonds are suited for this
purpose. At present, triorganosilanes are by far the most useful
class of organometallic protecting agents (Sect. 6.4.2.4).
    Another significant mode of functional group protection in-
volves formation-decomposition of organotransition metal π-com-
plexes, as shown in the following examples; see eqs. 3.67 to 3.68
(3.20, 3.21).

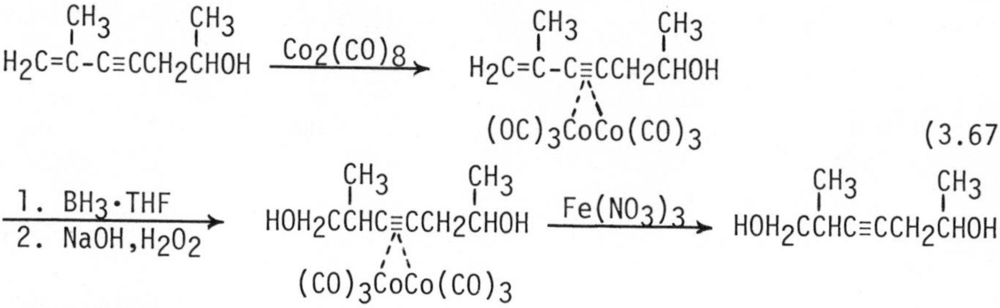

$$
\underset{\substack{|\\CH_3}}{H_2C=C}-\underset{\substack{|\\CH_3}}{C}\equiv CCH_2CHOH \xrightarrow{Co_2(CO)_8} \underset{\substack{|\\CH_3}}{H_2C=C}-\underset{\substack{|\\CH_3}}{C}\equiv CCH_2CHOH
$$

$$
(OC)_3CoCo(CO)_3 \qquad\qquad (3.67)
$$

$$
\xrightarrow[\text{2. NaOH,H}_2O_2]{\text{1. BH}_3\cdot\text{THF}} HOH_2C\underset{\substack{|\\CH_3}}{C}HC\equiv C\underset{\substack{|\\CH_3}}{C}CH_2CHOH \xrightarrow{Fe(NO_3)_3} HOH_2C\underset{\substack{|\\CH_3}}{C}HC\equiv C\underset{\substack{|\\CH_3}}{C}CH_2CHOH
$$

$$
(CO)_3CoCo(CO)_3
$$

$$
\xrightarrow{Fe_3(CO)_{12}} \qquad (3.68)
$$

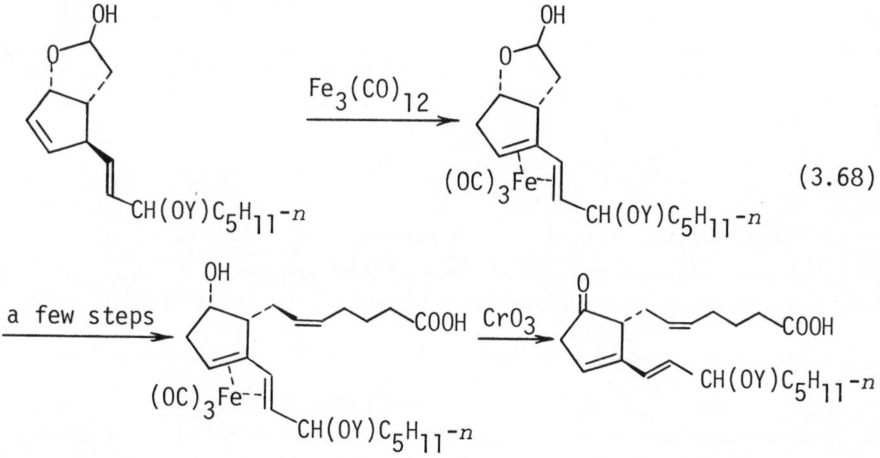

$(OC)_3Fe\!-\!|$ $CH(OY)C_5H_{11}\text{-}n$

a few steps $\longrightarrow$ ...COOH $\xrightarrow{CrO_3}$ ...COOH $CH(OY)C_5H_{11}\text{-}n$

$(OC)_3Fe\!-\!|$ $CH(OY)C_5H_{11}\text{-}n$

(Y = THP)

    There can be various other ways of protecting functional
groups by means of organometallics. The subject matter, however,
does not lend itself to systematic discussion. We therefore do
not attempt an extensive discussion of this subject.

REFERENCES

3.1   (a) Ward, H. R., *Acc. Chem. Res.*, 5, 18 (1972); (b) Lawler, R. G., *Acc. Chem. Res.*, 5, 25 (1972); (c) Lepley, A. R., and Closs, G. L., *Chemically Induced Magnetic Polarization*, John Wiley & Sons, Inc., New York, 1973.

3.2   Ashby, E. C., Laemmle, J., and Neumann, H. M., *Acc. Chem. Res.*, 7, 272 (1974).

3.3   House, H. O., *Acc. Chem. Res.*, 9, 59 (1976).

3.4   Kochi, J. K., *Acc. Chem. Res.*, 7, 351 (1974).

3.5   Hughes, E. D., and Ingold, C. K., *J. Chem. Soc.*, 244 (1935).

3.6   Jensen, F. R., and Rickborn, B., *Electrophilic Substitution or Organomercurials*, McGraw-Hill, New York, 1968, 203 pp.

3.7   Brown, H. C., De Lue, N. R., Kabalka, G. W., and Hedgecock, H. C., Jr., *J. Am. Chem. Soc.*, 98, 1290 (1976).

3.8   (a) Ingold, K. U., and Roberts, B. P., *Free Radical Substitution Reactions*, Wiley-Interscience, New York, 1971; (b) Davies, A. G., and Roberts, B. P., *Acc. Chem. Res.*, 5, 387 (1972).

3.9   Brown, H. C., and Midland, M. M., *Angew. Chem. Int. Ed. Engl.*, 11, 692 (1972).

3.10   Lane, C. F., *Intra-Sci. Chem. Rep.*, 7, 123 (1973).

3.11   For monographs, see (a) Jones, M., and Moss, R. A., *Carbenes*, 2 vols., John Wiley & Sons, Inc., New York, 1973 and 1975; (b) Kirmse, W., *Carbene Chemistry*, 2nd ed., Academic Press, New York, 1971, 615 pp.

3.12   Moser, W. R., *J. Am. Chem. Soc.*, 91, 1135, 1141 (1969).

3.13   For some earlier discussions of the allylic rearrangements of organometallics, see (a) De Wolf, R. H., and Young, W. G., *Chem. Rev.*, 56, 763 (1956); (b) Nordlander, J. E., Young, W. G., and Roberts, J. D., *J. Am. Chem. Soc.*, 83, 494 (1961); (c) Whitesides, G. M., Nordlander, J. E., and Roberts, J. D., *J. Am. Chem. Soc.*, 84, 2010 (1962).

3.14   (a) Wilkinson, G., and Piper, T. S., *J. Inorg. Nucl. Chem.*, 2, 32 (1956); (b) Piper, T. S., and Wilkinson, G., *J. Inorg. Nucl. Chem.*, 3, 104 (1956).

3.15   (a) Bernett, M. J., Cotton, F. A., Davison, A., Faller, J. W., Lippard, S. J., and Morehouse, S. M., *J. Am. Chem.*

Soc., <u>88</u>, 4371 (1966); (b) For a review, see Cotton, F. A., *Acc. Chem. Res.*, <u>1</u>, 257 (1968).

3.16 For a review, see Vrieze, K., and van Leeuwen, P. W. N. M., *Progr. Inorg. Chem.*, <u>14</u>, 1 (1971).

3.17 For a review, see Larrabee, R. B., *J. Organometal. Chem.*, <u>74</u>, 313 (1974).

3.18 Brook, A. G., *Acc. Chem. Res.*, <u>7</u>, 77 (1974).

3.19 For reviews, see (a) Lockhart, J. C., *Chem. Rev.*, <u>65</u>, 131 (1965); (b) Brown, T. L., *Acc. Chem. Res.*, <u>1</u>, 23 (<u>1</u>968); (c) Moedritzer, K., *Advan. Organometal. Chem.*, <u>6</u>, 1 (1968); (d) Oliver, J. P., *Adv. Organometal. Chem.*, <u>8</u>, 167 (1970); (e) Moedritzer, K., *Organometal. React.*, <u>2</u>, 1 (1971).

3.20 Nicholas, K. M., and Pettit, R., *Tetrahedron Lett.*, 3475 (1971).

3.21 Corey, E. J., and Moinet, G., *J. Am. Chem. Soc.*, <u>95</u>, 4449 (1973).

ORGANOMETALLICS OF MAIN GROUP METALS IN ORGANIC SYNTHESIS

PART II

ORGANOMETALLICS OF
MAIN GROUP METALS
IN ORGANIC SYNTHESIS

4

ORGANOMETALLICS CONTAINING GROUP IA, GROUP IIA, AND GROUP IIB ELEMENTS (Li, Na, K, Mg, Zn, Cd)

4.1 GENERAL CONSIDERATIONS

Organometallics containing Group IA and Group IIA metals are characterized by the presence of highly polar metal-carbon bonds. The Allred-Rochow electronegativity values (Appendix II) for these elements are as follows: Li(0.97), Na(1.01), K(1.04), Rb-(0.89), Cs(0.86), Be(1.47), Mg(1.23), Ca(1.04), Sr(0.99), Ba-(0.97), and Ra(0.97). The high polarity of the metal-carbon bonds of the organoalkali and organoalkaline earth metals is primarily responsible for both their high reactivity toward carbonyl and related polar carbon functional groups, that is, high nucleophilicity, as well as their high basicity. The terms *basicity* and *nucleophilicity* are very loosely defined in synthetic organic chemistry. The former generally refers to proton affinity, or Brønsted basicity, whereas the latter refers to the affinity toward other electrophilic centers, or Lewis basicity.

The application of francium and radium to organic synthesis is severely restricted due to their radioactivity and limited supply. Although it has become increasingly clear that organometallics containing rubidium and cesium can exhibit unique synthetic capabilities, their application to organic synthesis is still very limited. Nearly the same is true with the Group IIA metals other than magnesium. Whereas Cu, Ag, and Au can exist as d^9 and d^8 species and hence are regarded as transition metals, the Group IIB metals, that is, Zn, Cd, and Hg, do not exist in oxidation states higher than +2, and cannot therefore have incompletely filled d shells. Mercury is a heavy element of relatively high electronegativity, the chemistry of which has little resemblance to that of the Group IA and Group IIA metals. Its chemistry as well as that of thallium and lead is discussed in Chap. 7.

Because of the long history and wide-spread applications of the chemistry of organoalkali metals and Grignard reagents, it is impossible to present a comprehensive coverage of the subject. For more detailed discussions, the reader is referred to text-books and monographs on polar organometallics (*0.6*, *1.14*), or-ganolithiums (*4.1*), Grignard reagents (*0.8*, *4.2*), and organozincs and organocadmiums (*0.8*). Our discussion in this chapter is highly selective, emphasizing the chemistry of organolithiums and organomagnesiums.

4.1.1 Structure

Although organoalkali and organoalkaline earth metals represent the most ionic groups of organometallics, the metal-carbon bonds of these compounds generally have appreciable covalent character. The monomeric covalent structures of these organometallics, that is, M-R and R-M-X, are coordinatively unsaturated and do not gen-erally represent their actual structures. As discussed briefly in Sect. 1.4, organoalkali and organoalkaline earth metals usual-ly exist as oligomeric or polymeric species, in which the mono-meric units are held together via multicenter bonding.

4.1.1.1 Organolithiums

There are a few authoritative reviews on the structure of organo-lithiums by T. L. Brown (*4.3*). We have already learned that methyllithium is tetrameric in the crystalline state (Sect. 1.4). In nonpolar solvents, such as cyclohexane and benzene, sterically unhindered alkyllithiums, such as *n*-butyl- and *n*-octyllithiums, are usually hexameric, whereas sterically hindered members, such as isopropyl- and *t*-butyllithiums, are tetrameric. Benzyllithium in benzene is dimeric (*4.3*). In both tetrameric (1.3) and hexa-meric (4.1) structures, each lithium atom is bonded to three al-kyl groups, and each alkyl group to three lithium atoms, via four-center two-electron bonding. In a formal sense, the lithium atom is a sextet in these compounds. No [6]Li-^{7}Li coupling in ^{7}Li NMR has been observed, indicating that there is no Li-Li bond (*4.4*).

Organolithiums can form complexes with donor solvents, such as Et_2O, THF, TMEDA, and 1,4-diazabicyclo[2.2.2]octane (DABCO). These complexes are also often oligomeric. For example, the 1:1 complex between methyllithium and Et_2O is tetrameric.

Organolithiums derived from relatively acidic carbon acids, such as allyl-, benzyl-, and cyclopentadienyllithiums, have ap-preciable ionic character. For example, ^{1}H NMR spectrum of al-lyllithium in Et_2O or THF at low temperatures shows an AA'BB'C

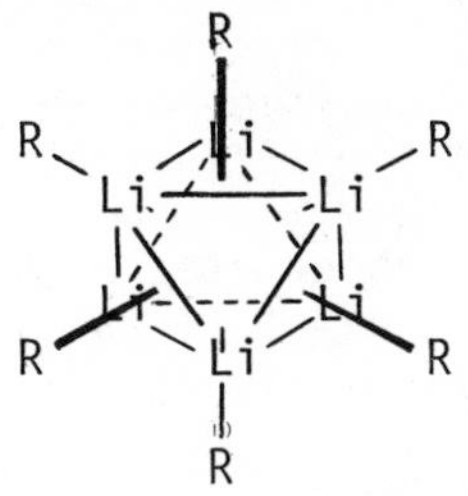

4.1

pattern in accordance with the equilibrium shown in eq. 4.1
(4.5).

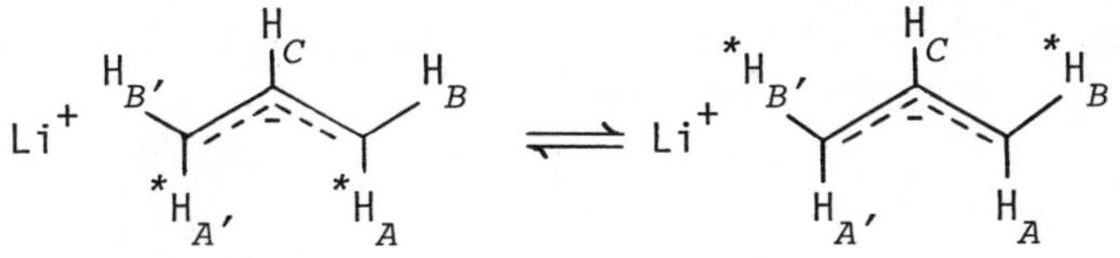

(4.1)

At higher temperatures, a rapidly interconverting AB_4 pattern is
observed. Covalent structures, such as 4.2 and 4.3, must be less
important.

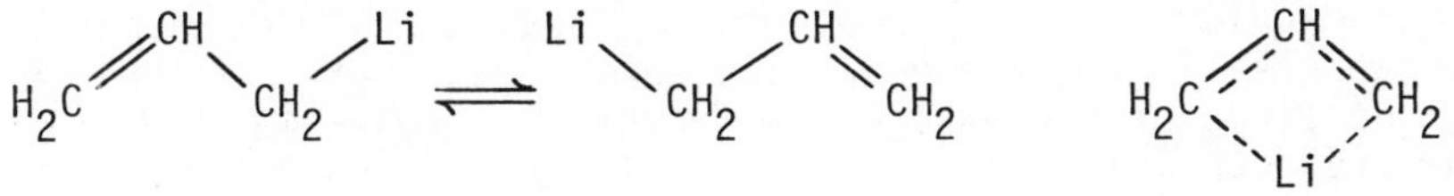

4.2   4.3

4.1.1.2 Organomagnesiums (4.6)

Typical Grignard reagents and diorganomagnesiums are monomeric in
THF, which exist as THF complexes. Similar complexes can form in

ether. The structures of $PhMgBr \cdot Et_2O$ (*4.7*) and $EtMgBr \cdot Et_2O$ (*4.8*) determined by X-ray diffraction indicate that the magnesium atom occupies the central position of an essentially tetrahedral configuration. Polymeric species having a $(-Mg-X-)_n$ polymer chain can also exist in ether.

Another complication associated with Grignard reagents is their disproportionation leading to the Schlenk equilibrium (eq. 4.2).

$$R_2Mg + MgX_2 \overset{K}{\rightleftharpoons} 2RMgX \qquad\qquad (4.2)$$

The Schlenk equilibrium constant (K) is much larger in ether (10^2 to 10^3) than in THF (*1.10*). It is reasonably clear that the solvation of MgX_2, the most Lewis acidic component of the three, governs the equilibrium and that it is more effective in THF than in ether.

^1H NMR examination of allylmagnesium bromide indicates that the α- and γ-protons are equivalent. On the other hand, iso-prenylmagnesium bromide does not rapidly interconvert with 3-methyl-1-buten-3-ylmagnesium bromide at $-40^{\circ}C$, supporting the co-valent σ-allylic structure **4.4** (*4.9*).

$$Me_2C=CHCH_2MgBr$$

$$\underline{4.4}$$

In summary, although a wide variety of structures occur in the chemistry of organoalkali and organoalkaline earth metals, the common driving force for the formation of associated and/or solvated species undoubtedly is stabilization of coordinatively highly unsaturated monomeric units, that is, RM and RMX, through multicenter bonding or solvation. From the viewpoint of organic synthesis, it is important to realize that the actual reactive species are not necessarily the same as those which are observed spectroscopically. While we may now say that the structures of many organoalkali and organoalkaline earth metals are reasonably clear, the structures of the actual reactive species in their re-actions are much less clear. Fortunately, this does not signifi-cantly hamper their applications to organic synthesis. For these reasons, in our subsequent discussions we adopt the conventional monomeric representations, RM and RMX, for these organometallics.

4.1.2 Configurational Stability

Since carbanions are isoelectronic with amines, free carbanions might be expected to undergo rapid inversion at room temperature.

On the other hand, covalent organometallic species are expected
to exist as configurationally stable species. The configurational stability of 3,3-dimethylbutyllithium was studied by ^{1}H NMR
(4.10). In ether, the α- and β-protons show an AA'BB' pattern at
-18°C, which collapse to an A_2B_2 pattern at 30°C. The activation
energy was estimated to be 62 ± 8 kj/mole. Secondary and tertiary alkyllithiums as well as alkenyllithiums are configurationally more stable than primary alkyllithiums. Many stereo-defined
organolithiums of these classes have been prepared and converted
into stereochemically pure organic products. These results indicate that free carbanions may not contribute significantly to the
constitution of simple organolithiums. Consideration of the relative electronegativities of various alkali and alkaline earth
metals suggests that organomagnesiums should be configurationally
more stable than organolithiums. By the same token, organosodiums and organopotassiums should be less stable than organolithiums. For further discussions of this subject, the reader is referred to Cram (1.12) and Buncel (1.16).

4.2 PREPARATION OF ORGANOMETALLICS CONTAINING GROUP IA, GROUP IIA, AND GROUP IIB METALS

4.2.1 Preparation of Organolithiums and Other Organoalkali Metals

Of the twelve general methods for preparing organometallics discussed in Chap. 2, oxidative metallation of organic halides
(Method I), metal-halogen exchange (Method III), metal-hydrogen
exchange (Method V), and transmetallation (Method XI) are of
practical significance as methods for the preparation of organoalkali metals. Oxidative metallation of active C-H compounds
(Method IV), oxidative metallation of unsaturated compounds
(Method X), and oxidative-reductive transmetallation (Method XII)
have also found some limited applications.

4.2.1.1 Preparation of Organolithiums

Oxidative Metallation of Organic Halides. The discovery that organolithiums can be prepared by the oxidative metallation of organic halides made in the early 1930s by Ziegler (4.11), Wittig
(4.12), and Gilman (4.13) has made organolithiums readily available.

$$RX + 2Li \longrightarrow RLi + LiX \qquad (4.3)$$

A large quantity of *n*-butyllithium, which is used primarily as an initiator in the polymerization of dienes, is produced by this method.

The scope of this method is broad, as indicated in Table 4.1. It is, however, less general than the corresponding method for the preparation of Grignard reagents in that those organic halides that readily undergo Wurtz coupling, such as allyl, propargyl, and benzyl halides, cannot successfully be converted into the corresponding organolithiums. Hydrocarbons, such as hexane and benzene, are preferred solvents. On the other hand, ether and THF may readily be cleaved by many organolithiums. Methyllithium and phenyllithium are insoluble in hydrocarbons, and their preparation usually requires the use of ethers. Fortunately, they are quite stable in ether. Alkenyllithiums are also relatively stable in ethers. Vinyllithium itself is commercially prepared in THF. Organic chlorides and bromides are preferred starting materials, whereas organic iodides tend to undergo Wurtz coupling extensively. Organic fluorides have almost never been used.

Various alkenyl halides have been converted to alkenyllithiums with retention of configuration. On the other hand, treatment of 4-*t*-butylcyclohexyl chloride with lithium produces a mixture of two epimers, which are themselves configurationally stable under the reaction conditions. Thus epimerization must occur during the formation of the organolithium compounds, suggesting that, at least in such a case, a one-electron transfer process appears to be operating.

Metal-Halogen Exchange. This method has been successfully used for the preparation of alkenyl- and aryllithiums (*2.13*). Alkyl halides are, in general, not reactive enough, while allyl, propargyl, and benzyl halides are prone to undergo Wurtz coupling. Its scope, stereochemistry, and possible mechanisms are discussed in Sect. 2.3. A few noteworthy recent developments should be briefly mentioned here, however.

Treatment of alkenyl iodides or bromides with either one equivalent of *n*-butyllithium or two equivalents of *t*-butyllithium (*2.15*) at low temperatures (<-50°C) produces the corresponding alkenyllithiums with retention of configuration. The second mole of *t*-butyllithium reacts with *t*-butyl halide to form isobutylene (eq. 4.4).

$$\text{R}\diagdown\text{C=C}\diagup\text{H} \xrightarrow[-120°]{t\text{-BuLi(2 equiv)}} \text{R}\diagdown\text{C=C}\diagup\text{H} + CH_2=C(CH_3)_2 + LiBr \qquad (4.4)$$

Table 4.1 Preparation of Organolithium Compounds

Method	⌒Li	≡⌒Li	PhCH₂Li	1°-RLi	2°-RLi	3°-RLi	⌒Li	ArLi	≡-Li
I. RX + M	−	−	−	+	+	+	+	+	
II. RX + MR'	−	−	−	−	−	−	+	+	
V. RH + MR'	+	+	+		−	−	−	−	+
XI. RM + M'X	+		+				+		

Table 4.2 Preparation of Grignard Reagents

Method	⌒MgX	≡⌒MgX	PhCH₂MgX	1°-RMgX	2°-RMgX	3°-RMgX	⌒MgX	ArMgX	≡-MgX
I. RX + M	+	+	+	+	+	+	+	+	
V. RH + MR'									+

Allenyl (4.4) and cyclopropyl halides (4.15) undergo similar transformations.

$$n\text{-}C_8H_{17}CH=C=CHX \xrightarrow[\text{2. MeI}]{\text{1. } n\text{-BuLi}} n\text{-}C_8H_{17}CH=C=CHMe \qquad (4.5)$$

$$90\% \ (X = Br); \quad 93\% \ (X = H)$$

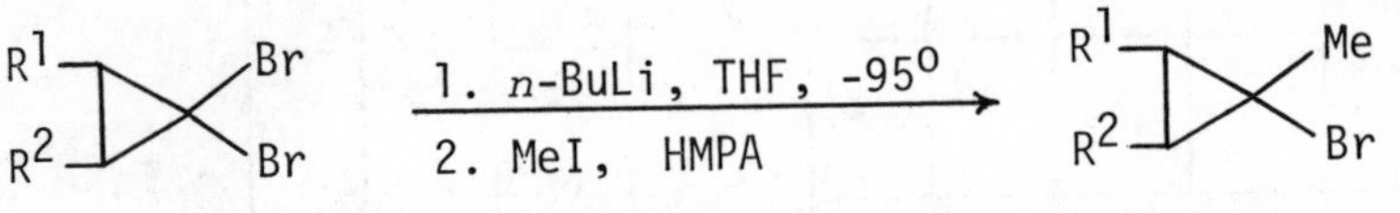

$\xrightarrow[\text{2. MeI, HMPA}]{\text{1. } n\text{-BuLi, THF, } -95^0}$ (4.6)

65 to 90%

It has recently been demonstrated that, at very low temperatures (<100°C), the metal-halogen exchange reaction can tolerate various electrophilic functional groups. Some aryl halides that have been successfully converted into aryllithiums are shown below.

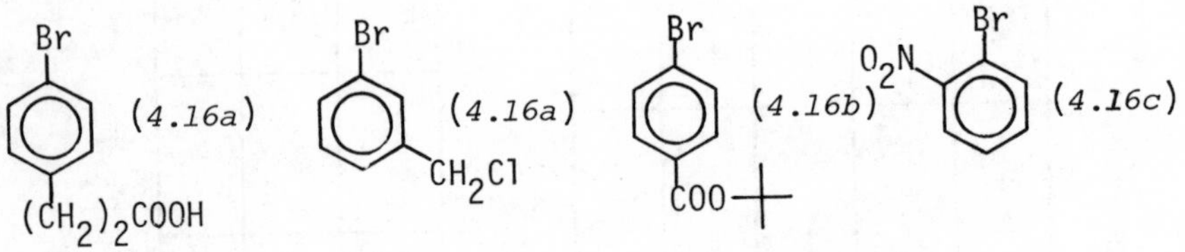

(4.16a)   (4.16a)   (4.16b)   (4.16c)

Metal-Hydrogen Exchange. The scope and limitations of this reaction as a method for preparing organolithiums are briefly discussed in Sect. 2.5.1 and summarized in Table 4.1. There are a number of extensive reviews on this subject, such as those by Wakefield (4.1) and Gilman (4.17). Consequently, a few recent noteworthy developments are briefly discussed here. At -78°C, acetylene can be converted to monolithioacetylene which can react with various electrophiles to form ethynylated products (4.18).

$$HC\equiv CH \xrightarrow[\text{THF, } -78^0C]{n\text{-BuLi}} HC\equiv CLi \xrightarrow[\text{2. } H_3O^+]{\text{1. } R^1COR^2} HC\equiv C-\overset{\overset{R^1}{|}}{\underset{\underset{R^2}{|}}{C}}-OH \qquad (4.7)$$

65 to 98%

Propyne can be lithiated in both the 1 and 3 positions. The dilithio derivative can act first as a propargyllithium derivative and then as an alkynyllithium, as shown in eq. 4.8 (*4.19*).

$$CH_3C{\equiv}CH \xrightarrow[\text{TMEDA}]{n\text{-BuLi}} LiCH_2C{\equiv}CLi \xrightarrow[\text{2. } CH_2O]{\text{1. } n\text{-BuBr}} n\text{-BuCH}_2C{\equiv}CCH_2OH \qquad (4.8)$$

<u>Transmetallation</u>. The transmetallation reaction between two organometallic species R^1M^1 and R^2M^2 would proceed forward, if their electronegativities (EN) are $EN_{R^1} > EN_{R^2}$ and $EN_{M^1} > EN_{M^2}$. The formation of R^1M^2 between the more electronegative R^1 group and the more electropositive M^2 must be the major driving force of the reaction (eq. 4.9).

$$R^1M^1 + R^2M^2 \longrightarrow R^1M^2 + R^2M^1 \qquad (4.9)$$

In practice, the choice of R^1 is very much limited by various factors, in particular "ate" complex formation in the case of the Group II and Group III metals. Only organotins have been shown, in practice, to be useful starting compounds for the preparation of organolithiums. Although alkyllithiums are difficult to prepare by this method, it provides a convenient route to organolithiums containing allyl, benzyl, alkenyl, aryl, and cyclopropyl groups, as briefly discussed in Sects. 2.11 and 6.4.3.1. Certain proximally hetero-substituted organolithiums have also been prepared by this method, as shown in eq. 4.10 (*4.20*).

$$(4.10)$$

Oxidative-reductive transmetallation can also be used to prepare organolithiums. Organomercuries (eq. 4.11) and organoleads (eq. 4.12), for example, (*4.21*) have been used as starting materials.

$$R_2Hg + 2Li \longrightarrow 2RLi + Hg \qquad (4.11)$$

$$(CH_2{=}CH)_4Pb + 4Li \longrightarrow 4CH_2{=}CHLi + Pb \qquad (4.12)$$

These methods are particularly suited for the preparation of halogen-free organolithiums. The scope of the preparation of organolithiums by exchange reactions has been expanded in recent years to include Li-B (Sect. 5.2.3.5), Li-S (*4.22* to *4.24*), Li-Se (*4.25*, *4.26*), and Li-Te (*4.27*) exchanges.

   Other methods of preparing organolithiums include ether cleavage by lithium metal (*4.28*) or organolithiums (*4.29*), see eq. 4.13, and reactions of olefins, arenes, and acetylenes with lithium metal or organolithiums. The latter reactions are discussed in Sects. 4.3.2.5 and 4.4.1.

$$\underset{\text{O}}{\bigcirc}\!\!-\text{H} \xrightarrow[25^\circ,16\ \text{hr}]{n\text{-BuLi}} \left[\underset{\text{O}}{\bigcirc}\!\!-\text{Li} \right] \longrightarrow CH_2{=}CHOLi + CH_2{=}CH_2 \quad (4.13)$$

$$\downarrow R_3SiCl$$

$$CH_2{=}CHOSiR_3$$

4.2.1.2 Preparation of Organosodiums and Organopotassiums

Two of the most widely used methods for preparing organolithiums, that is, oxidative metallation and metal-halogen exchange, are not well-suited for the preparation of other organoalkali metals due to highly competitive Wurtz coupling. Thus the applicability of oxidative metallation is largely limited to the preparation of aryl derivatives of sodium and potassium. The metal-halogen exchange reaction involving organosodiums and organopotassiums appears very facile. The Wurtz-coupling reaction, however, is also very fast. Consequently, the metal-halogen exchange does not provide, in practice, a useful method to prepare these organometallics.

   At present, metal-hydrogen exchange reactions and transmetallation reactions represent some of the most commonly employed routes to organosodiums and organopotassiums. The former reaction, which was discussed in some detail in Sect. 2.5.1, provides a convenient route to allyl, benzyl, aryl, and alkynyl derivatives of sodium and potassium.

   Perhaps the most common route to alkylsodiums and alkylpotassiums is the oxidative-reductive transmetallation reaction of alkylmercuries with sodium or potassium (eq. 4.14).

$$n\text{-Bu}_2\text{Hg} + \text{Na(excess)} \xrightarrow{\begin{array}{c}\text{petroleum}\\\text{ether}\end{array}} 2\ n\text{-BuNa} + \text{Na(Hg)} \quad (4.14)$$

Alkylpotassiums, such as n-BuK, react to a considerable extent, even with petroleum ether, within a few hours at room temperature, which seriously limits the synthetic usefulness of alkylpotassiums.

4.2.2 Preparation of Organometallics Containing Mg, Zn, and Cd

4.2.2.1 Preparation of Organomagnesiums

There are many extensive monographs and reviews on this subject, such as those by Kharasch (*4.2*), Coates (*0.6*), and Nesmeyanov (*0.8*). Furthermore, relatively little progress in this area has been made in recent years. Consequently, our discussion here is very brief. Of the 12 general methods of preparing organometallics, oxidative metallation (Method I) by far provides the most widely used route to organomagnesiums (eq. 4.15).

$$\text{RX} + \text{Mg} \xrightarrow{\text{ether}} \text{RMgX} \quad (4.15)$$

For the discovery and development of this reaction as well as the investigation of the chemistry of the organomagnesium products, that is, the Grignard reagents, V. Grignard was awarded the Nobel Prize for Chemistry in 1912.

As shown in Table 4.2, virtually all types of organomagnesium halides can be prepared by this method. Alkynylmagnesium halides are, however, usually prepared via metal-hydrogen exchange. Despite its wide applicability, many limitations also exist. Di-Grignard reagents, XMg-R-MgX, can present serious difficulties. 1,2-Dihaloethanes, such as $\text{BrCH}_2\text{CH}_2\text{Br}$, give ethylene and MgX_2. Similarly, 1,3-dibromopropane undergoes a complex reaction with magnesium which leads to cycopropane, propylene, and other products.

$$\text{BrCH}_2\text{CH}_2\text{Br} \xrightarrow{\text{Mg}} \text{BrCH}_2\text{CH}_2\text{MgBr} \longrightarrow \text{CH}_2\text{=CH}_2 + \text{MgBr}_2 \quad (4.16)$$

$$\text{Br(CH}_2)_3\text{Br} \xrightarrow{\text{Mg}} \text{Br(CH}_2)_3\text{MgBr} \longrightarrow \triangle + \text{MgBr}_2 \quad (4.17)$$

The latter problem has recently been solved by the development of the following indirect method, see eq. 4.18 (*4.30*).

$$H_2C=C=CH_2 \xrightarrow[\substack{2.\ Hg(OAc)_2 \\ 3.\ NaCl}]{1.\ BH_3} ClHg \diagup\!\!\diagup\!\!\diagdown HgCl \xrightarrow{MeLi} Li \diagup\!\!\diagup\!\!\diagdown Li$$

(4.18)

$$\xrightarrow{Mg,\ MgBr_2} BrMg \diagdown\!\!\diagup\!\!\diagdown MgBr$$

Alkenylmagnesium halides cannot be prepared readily in diethyl ether. This difficulty has been solved by using THF as a solvent (*4.31*). The preparation of allyl-, propargyl-, and benzylmagnesium halides is complicated by Wurtz coupling, which can be minimized, however, by very slowly adding the organic halides to magnesium suspended in a large quantity of ether. It is noteworthy that, in the conversion of propargyl bromide into propargylmagnesium bromide, abstraction of the acetylenic hydrogen atom does not occur to any serious extent (eq. 4.19).

$$HC\equiv CCH_2Br \xrightarrow[ether]{Mg} HC\equiv CCH_2MgBr \xrightarrow{slow} BrMgC\equiv CCH_2MgBr \quad (4.19)$$

high yield

The scope of the preparation of Grignard reagents via oxidative metallation has recently been expanded by the introduction of the active metal powder technique and the metal vapor technique (Sect. 2.1).

As the results shown in eq. 4.19 suggest, the metal-hydrogen exchange (Method V) is not a widely applicable method of preparing organomagnesiums. The active C-H compounds ($pK_a \leq 25$), however, can be converted to the corresponding Grignard reagents by treating them with highly basic Grignard reagents, such as *i*-PrMgCl, as shown in eq. 4.20.

$$RC\equiv CH \xrightarrow{i\text{-PrMgCl}} RC\equiv CMgCl \quad (4.20)$$

The metal-halogen exchange reaction (Method III) of Grignard reagents is generally a slow process which does not serve as a good synthetic method.

In addition to the oxidative metallation of organic halides, transmetallation reactions (Methods XI and XII) provide generally applicable routes to organomagnesiums. The transmetallation that does not involve any redox reaction generally requires organometallics that contain metals that are more electropositive than magnesium, such as alkali metals.

$$RLi \xrightarrow{MgX_2} RMgX \qquad (4.21)$$

Finally, the oxidative-reductive transmetallation (Method XII) provides a method for preparing halogen-free organomagnesiums.

$$R_2Hg + Mg \longrightarrow R_2Mg + Hg \qquad (4.22)$$

4.2.2.2 Preparation of Organozincs and Organocadmiums

Various methods for preparing organozincs and organocadmiums have been reviewed by Sheverdina and Kocheshkov (*4.32*).

Oxidative Metallation. In 1849, Frankland reported the synthesis of diethylzinc by the reaction of ethyl iodide with zinc metal (*4.33*). This appears to represent the first synthesis of an organometallic compound. Since then, the oxidative metallation reaction has been one of the main routes to organozincs along with transmetallation reactions.

In the conventional procedure, alkyl iodides are reacted with a Zn-Cu couple, which is prepared by heating a mixture of zinc metal (12 parts) and copper metal (ca. 1 part), CuO or a Cu salt, such as copper citrate, to high ($\geq 400^\circ$C) temperatures (*4.32*), either in the presence or in the absence of a solvent, such as ether.

$$2RI + 2Zn \longrightarrow 2RZnI \longrightarrow R_2Zn + ZnI_2 \qquad (4.23)$$

Allylic, propargylic, and benzylic derivatives of zinc are best prepared by reacting the corresponding bromides with zinc dust in THF (*4.34*).

$$CH_2=CHCH_2Br \xrightarrow{Zn, THF} CH_2=CHCH_2ZnBr \qquad (4.24)$$

$$HC\equiv CCH_2Br \xrightarrow[<20^\circ C]{Zn, THF} HC\equiv CCH_2ZnBr \qquad (4.25)$$

$$\downarrow \Delta$$

$$BrZnC\equiv CCH_3$$

$$PhCH_2Br \xrightarrow{Zn, THF} PhCH_2ZnBr \qquad (4.26)$$

It should be pointed out that the reactions shown in eqs. 4.24 to

4.26 are not complicated by Wurtz coupling.

The reaction of methylene iodide with the Zn-Cu couple gives a useful reagent for cyclopropanation, see eq. 4.27 (*4.35*).

$$CH_2I_2 + Zn(Cu) \longrightarrow ICH_2ZnI \qquad (4.27)$$
$$40\%$$

The reaction of α-bromo esters with zinc metal produces the Reformatsky reagents, see eq. 4.28 (*4.34, 4.36, 4.37*).

$$BrCH_2COOR + Zn \longrightarrow BrZnCH_2COOR \qquad (4.28)$$

As discussed in Sect. 2.1, the use of activated zinc powder prepared by treating $ZnCl_2$ with potassium metal in THF permits the conversion of alkyl and aryl bromides to organozinc products, thereby expanding the scope of the oxidative metallation.

The synthetic usefulness of the oxidative metallation as a route to organocadmiums appears to be rather limited. Alkyl iodides have been reacted with cadmium metal in HMPA to give good yields of dialkylcadmiums (*4.38*). A cadmium analogue of the Reformatsky reagent has been prepared in DMSO (*4.39*).

$$BrCH_2COOBu\text{-}t + Cd \longrightarrow BrCdCH_2COOBu\text{-}t \qquad (4.29)$$

Transmetallation. The reaction of organolithiums or Grignard reagents with zinc halides and cadmium halides, such as $ZnCl_2$ and $CdBr_2$, represents one of the most general routes to organozincs and organocadmiums, respectively. Some typical examples are shown below.

$$2PhLi \quad + ZnCl_2 \longrightarrow Ph_2Zn \quad (65\%) \qquad (4.30)$$

$$2i\text{-}BuMgBr + ZnCl_2 \longrightarrow i\text{-}Bu_2Zn \quad (62\%) \qquad (4.31)$$

$$2EtMgBr \quad + CdBr_2 \longrightarrow Et_2Cd \quad (90\%) \qquad (4.32)$$

On the other hand, the oxidative-reductive transmetallation reaction of organomercuries with zinc or cadmium is a generally sluggish equilibrium reaction, which does not appear to be of high synthetic significance.

4.3   CARBON-CARBON BOND FORMATION VIA ORGANOALKALI AND ORGANOALKALINE EARTH METALS

4.3.1 General Reaction Patterns

The great majority of selective carbon-carbon bond-forming reactions involving the use of organoalkali and organoalkaline earth metals proceed via nucleophile-electrophile interactions, which can conveniently be classified into several reaction types according to the structures of electrophiles, as summarized in Scheme 3.2 in Sect. 3.2.1.1. It is also convenient to classify organometallic nucleophiles according to the following three categories:

1. "Ordinary" or proximally nonfunctional organometallics, such as MeMgBr.

2. α-Hetero-substituted organometallics, such as $NaCCl_3$.

3. Metal enolates, such as $LiO(Ph)C=CH_2$.

Whether metal enolates should or should not be classified as organometallics is a matter of semantics. Since metal enolates can act as intermediates for the conversion of carbonyl compounds into α-alkylated derivatives, it is convenient to treat them as organometallic compounds regardless of their precise structures. In the following subsections, we discuss the reactions of each of the three classes of nucleophilic organometallics with several different types of electrophiles.

4.3.2 Carbon-Carbon Bond Formation via Proximally Nonfunctional Organometallics of Groups IA, IIA, and IIB Metals

4.3.2.1 Cross Coupling with Organoalkali Metals and Grignard Reagents

The cross-coupling reaction is arbitrarily defined as a process of a single carbon-carbon bond formation between two unlike carbon groups by the reaction of an organometallic species with an organic halide or a related electrophilic derivative (eq. 4.33).

$$R^1M + R^2X \longrightarrow R^1-R^2 + MX \qquad (4.33)$$

The reaction represents one of the most straightforward methods of carbon-carbon bond formation. It may even be said that, if one could achieve any type of cross coupling at will, most of the problems of organic skeletal construction would be solved. Despite its inherent simplicity, however, its synthetic utility had been quite limited until recently. Until the mid-1960s, alkali metals, lithium and magnesium in particular, had been used almost exclusively. Mainly within the past decade or so, the scope of cross coupling has been significantly broadened by the development of new cross-coupling procedures involving transition metals, particularly Cu, Ni, and Pd (Sects. 9.1 and 9.2). In this section, the scope and limitations of cross coupling involving organoalkali metals and Grignard reagents, as well as some mechanistic aspects of the reaction, are discussed.

A systematic and extensive survey of cross-coupling reactions has been presented by Mathieu and Weill-Raynal (*4.40*). The cross-coupling reactions of organolithiums and Grignard reagents have been reviewed by Wakefield (*4.1*) and Kharasch and Reinmuth (*4.2*), respectively.

Classification of Cross-Coupling Reactions. It is useful to classify the carbon groups to be considered in this section as follows:

1. sp^3 Hybridized carbon groups (Type I)
    primary, secondary, and tertiary alkyl groups.

2. sp^3 Hybridized carbon groups (Type II)
    allylic, propargylic, and benzylic groups.

3. sp^2 and sp Hybridized carbon groups (Type III)
    aryl, alkenyl, and alkynyl groups.

Let us now consider coupling any two of these carbon groups. The two unlike groups may be of the same type. The information available clearly indicates that each type of cross coupling presents unique characteristics and problems and hence deserves a separate discussion. For the sake of convenience, the cross-coupling reactions may be classified arbitrarily according to the following categories:

Category I $(C_{sp}3(I)-C_{sp}3(I)$ coupling).
Category II $(C_{sp}3(I)-C_{sp}3(II)$ or $C_{sp}3(II)-C_{sp}3(I)$ coupling).
Category III $(C_{sp}3(II)-C_{sp}3(II)$ coupling).
Category IV $(C_{sp}3(I)-C_{sp}2(sp)$ or $C_{sp}2(sp)-C_{sp}3(I)$ coupling).
Category V $(C_{sp}3(II)-C_{sp}2(sp)$ or $C_{sp}2(sp)-C_{sp}3(II)$ coupling).
Category VI $(C_{sp}2(sp)-C_{sp}2(sp)$ coupling).

In the following discussions, the first carbon group of each pair refers to that of an organometallic reagent (R^1M) and the second to the carbon group (R^2) of an organic halide or a related species (R^2X).

General Discussion of Cross Coupling with Organometallics Containing Group IA and Group II Elements. As pointed out earlier, the scope and synthetic utility of the cross coupling with organometallics containing Group IA and Group II elements are rather limited. This is mainly due to the following complications: (1) halogen-metal exchange (Sect. 4.2.1.1), (2) α- and β-elimination reactions (eq. 4.34), and (3) the general lack of chemoselectivity.

$$MR + \overset{\overset{H}{|}\ \overset{H}{|}}{-\underset{|}{C}-\underset{|}{C}-X} \quad \begin{cases} \xrightarrow{\alpha\text{-elimination}} \overset{\overset{H}{|}}{-\underset{|}{C}-\underset{|}{C}^--X} \longrightarrow -\underset{|}{C}-\overset{\cdot\cdot}{C}- \\[2em] \xrightarrow{\beta\text{-elimination}} -\underset{|}{C}^--\overset{\overset{H}{|}}{\underset{|}{C}}-X \longrightarrow {}^{\diagup}C=C^{\diagdown} \end{cases} \tag{4.34}$$

(a) Scope with respect to M. Of the Group IA and Group II metals, only lithium and magnesium have been used extensively in cross coupling. The higher alkali metals, such as sodium and potassium, are associated with several serious difficulties. Firstly, organometallics containing these alkali metals are generally much less readily available than those containing lithium and magnesium, since their preparation by either direct oxidative metallation (Sect. 2.1) or metal-halogen exchange (Sect. 2.3) is often accompanied by Wurtz homo coupling. Secondly, some of the undesirable side reactions, that is, halogen-metal exchange and α- and β-eliminations, take place more readily with the sodium and potassium derivatives than with the lithium and magnesium.

As a result, except in cases where organometallics containing sodium, potassium, and higher alkali metals are more readily available than those containing lithium and magnesium, these higher alkali metals do not generally offer any advantages over lithium and magnesium.

The nucleophilicity of organometallics containing Be, Zn, Cd, and Hg is generally too low to be of practical use. Likewise, the Group IIIA and Group IVA metals, such as B, Al, Tl, Si, Sn, and Pb, do not readily participate in the direct cross-coupling reaction. This may also be attributed to the relatively low ionicity of the M-C bond in these organometallic compounds. As discussed later, however, organoalanes and, to lesser extents, organozincs and related electrophilic organometallics have exhibited a unique capability to undergo clean substitution reactions with hindered alkyl halides. Perhaps more important is a

recent finding that organometallics containing moderately electronegative metals, such as Zn, Cd, Al, and Sn, can readily participate in nickel- or palladium-catalyzed cross coupling, as
discussed in detail in Sect. 9.2.

The nucleophilicity of organometallics containing calcium
and the higher Group IIA elements appears to be quite high. At
present, however, relatively little is known about their synthetic capability with respect to cross coupling.

(b) Scope with respect to the R^1 and R^2 groups. Before discussing specific cross-coupling reactions, it is worth discussing
the following general limitations:

1. In general, it is quite difficult to achieve successful
    cross coupling using highly hindered alkyl halides, such
    as tertiary and certain secondary alkyl halides, mainly
    due to competitive β-elimination. This seriously limits
    the scope of Category I, Category II, and Category IV
    cross-coupling reactions.

2. Unsaturated organic halides containing the $C_{sp}2$-X and
    C_{sp}-X bonds, that is, aryl, alkenyl, and alkynyl halides,
    are generally quite inert with respect to the direct
    carbon-carbon bond formation through their reaction with
    organometallics containing lithium, magnesium, and other
    main group metals. It is therefore either impossible or
    extremely difficult to achieve coupling of two unsatu-
    rated groups (Category VI) with these organometallics.
    The scope of Category IV and Category V cross-coupling
    reactions is also limited by the same fact. There are,
    however, a few solutions to this problem, as discussed
    later in this section.

3. As pointed out earlier, halogen-metal exchange presents
    a general difficulty. The difficulty is particularly
    serious when two groups to be coupled are of the same or
    of similar types, as in Category I and Category III cou-
    pling reactions.

4. Allylic and propargylic organolithiums and Grignard re-
    agents can undergo facile allylic rearrangements as dis-
    cussed in Sect. 4.1. This often presents difficulties
    regarding regio- and stereoselectivity in Category II,
    Category III, and Category V cross-coupling reactions.

Mechanistic Considerations. Despite the ubiquitous nature of the
reaction of organolithiums and Grignard reagents with organic

halides, relatively little is known about the mechanism of the
reaction in a clear-cut and definitive sense. As discussed
earlier, these organometallic compounds often exist as aggregat-
ed species. Thus the actual reactive species could be either ag-
gregated or monomeric. Ion pairing presents another major complication. Until recently, there was a widespread tendency to
view highly nucleophilic organometallics such as those containing
lithium, as carbanion sources and interpret the mechanism of
their reaction with alkyl halides within the framework of the S_N
mechanisms, involving the interaction of free carbanion species
with either alkyl halides themselves or carbocationic species
thereof in the rate-determining step. In principle, however,
these organometallics could react with alkyl halides either as
free carbanions or as ion pairs which may be either tight or solvent separated. Even in cases where the nature of the actual reactive species, with respect to aggregation and ion pairing phe-
nomena, is known, there still exists yet another major complication. Thus the reactive species could interact with organic halides via either two-electron or one-electron transfer (free-
radical intermediates). It is also possible that any given reaction may proceed by more than one mechanism operating concur-
rently. Because we are interested mainly in the synthetic aspects of these coupling reactions, only a few plausible mechanistic schemes proposed in recent years are briefly presented. Interested readers are referred to a recent review of this subject
(4.41).

(a) S_N2 and other related ionic mechanisms. The S_N2 mechanism shown in eq. 4.35 has been proposed primarily on the basis of
the stereochemistry of the reaction, that is, the inversion of
the C_B group.

$$(4.35)$$

Some representative stereochemical data are summarized in Table
4.3. These results indicate that some reactions of allyl- and
benzylmetals containing Li, Na, and Mg with certain secondary al-

kyl halides proceed with complete, or nearly complete, inversion of configuration of the R^2 group. The stereochemical results appear best accommodated by some mechanisms of the s_N2 type. It is important to note that these allyl- and benzylmetals are considerably more ionic than ordinary alkyl- and arylmetals containing the same metal atoms due to resonance stabilization. Consequently, they can better serve as sources of free carbanions required in the s_N2 mechanism depicted in eq. 4.35 than the alkyl and aryl derivatives.

Table 4.3 The Stereochemistry of the Reaction of Organometallics Containing Li, Na, and Mg with Organic Halides and Sulfonates

R^1M	R^2X	Yield of R^1R^2 (%)	Stereochemistry	Ref.
s-BuLi	2-OctI	–	80% racemization	4.42
n-BuLi	s-BuBr	37	98% racemization	4.43
n-BuNa	2-OctBr	35	racemization	4.44
PhLi	2-OctBr	30	racemization	4.45
AllylLi	2-OctX	59-95	$\geq$90% inversion	4.45
				4.46
	(X = Cl,Br,I, or OTs)			4.47
AllylNa	2-OctBr	83	80% inversion	4.48
AllylMgBr	2-OctBr	78	87% inversion	4.48
PhCH$_2$Li	s-BuBr	58	100% inversion	4.46
PhCH$_2$Li	2-OctX	–	93% inversion	4.46
	(X = Br,I, or OTs)			4.47

In marked contrast with the allyl and benzyl cases, the reactions of primary and secondary alkylmetals, as well as aryl-metals containing Li, Na, and Mg, with secondary alkyl halides and sulfonates tend to proceed with extensive racemization. It is clear that an s_N2-type mechanism cannot play a significant role in such cases.

In cases where the ionization of alkyl halides is facile for steric and electronic reasons, the s_N1 mechanism involving partial or complete racemization can compete with the s_N2 mechanism or even dominate the course of the reaction. In the reactions of allylic and propargylic halides, s_N2' and s_N1' mechanisms may also operate.

(b) One-electron transfer-radical recombination mechanism. Organoalkali metals and Grignard reagents can be readily oxidized via one-electron transfer producing radical cationic species. On the basis of product studies a number of workers have suggested the possible intermediacy of free-radical species in the reaction of organoalkali metals and alkyl halides. For example, the reaction of 6-bromo-1-phenyl-1-hexyne with n-BuLi proceeds as shown in eq. 4.36 (4.49).

$$PhC{\equiv}C(CH_2)_4Br \xrightarrow[\text{2. } H_2O]{\text{1. } n\text{-BuLi}}$$

(60%) + $CH_2{=}CHC_2H_5$

$$+ \; PhC{\equiv}CC_8H_{17}\text{-}n \; + \; PhC{\equiv}CC_4H_9\text{-}n$$

$$(20\%) \qquad\qquad (3\%) \qquad (4.36)$$

$$+ \; PhC{\equiv}C(CH_2)_2CH{=}CH_2 \; (1\%)$$

When D_2O was used in place of H_2O in the work-up, the cyclic product was only 25% deuterated. Most of the hydrogen must be incorporated into the product before the hydrolysis step. The results are in accord with a radical mechanism but not with a carbanionic mechanism, although some of the deuterated product may arise via the latter mechanism. The radical mechanism is further supported by the formation of 1-butene.

Clear-cut evidence for the existence of free-radical intermediates in reactions of this type was provided by Ward and Lawler (4.50). Thus the reaction of n-BuLi and n-BuBr produces 1-butene, which shows both emission and strongly enhanced absorption ^{1}H NMR signals, when the NMR measurement is made within several minutes of its formation. This abnormal NMR behavior has been called "chemically induced dynamic nuclear polarization" or CIDNP (3.1). The observed emission and enhanced absorption signals require population differences between nuclear spin energy levels that are more than 10 times greater than those of the Boltzmann population differences. The large population differences are attributable to the interaction of protons with the unpaired electrons of free-radical species. In the above example, the nuclear polarization presumably occurs in the butyl radical, which is then converted to 1-butene, which retains the imbalance of nuclear spin states decaying within several minutes via the normal spin relaxation process. The following free-radical mechanism has been proposed for the reaction (eq. 4.37).

$$R^1M + R^2X \longrightarrow [R^1{\cdot} + M^+ + X^- + R^2{\cdot}] \longrightarrow R^1R^2 + MX$$

$$(4.37)$$

Since then, many other examples of CIDNP observed with organolithiums and other organometallics have been reported. A few
representative reagent combinations are: n-BuLi + sec-BuI (4.51),
MeNa + MeI (4.53), n-BuMgCl + n-BuI (4.53), and t-BuMgCl + t-BuBr
(4.53). It should be noted that the proposed free-radical mechanism is consistent with the stereochemical results, that is,
extensive racemization (Table 4.3). A word of caution may be in
order. Observation of CIDNP signals does provide strong evidence
of the formation of radical intermediates. It does not readily
permit us, however, to establish the extent to which such free-
radical mechanisms operate.

(c) Radical chain mechanism. In a study of the reaction of
alkyllithiums with alkyl halides by ESR, Russell (4.54) obtained
direct evidence of the formation of radical intermediates. In
the reaction of n-BuLi with RI(R = Et, n-Oct, or t-Bu), only the
R• radicals arising from RI before metal-halogen exchange are observed. On the other hand, in the reaction of n-BuLi with certain alkyl bromides or benzyl chloride, only the signal of the n-
Bu• radical is observed, while both possible radicals were formed
in the reaction of sec-BuLi with EtBr and of n-BuLi with MeI. To
accommodate these results, Russell proposes the following radical
chain mechanism.

$$R^1M + R^2X \xrightarrow{k_1} R^1\cdot + R^2\cdot + MX$$
$$R^1\cdot + R^2X \xrightarrow{k_2} R^1X + R^2\cdot$$
$$R^2\cdot + R^1M \xrightarrow{k_3} R^2M + R^1\cdot \qquad (4.38)$$
$$R^1\cdot + R^2\cdot \xrightarrow{k_4} R^1R^2 + R^1H + R^2H + R^1(-H) + R^2(-H)$$

Depending on the relative magnitudes of k_2 and k_3, either the $R^1\cdot$
or the $R^2\cdot$ radical or both radicals may be observed. It has already been mentioned that the same scheme provides a plausible
mechanism for the metal-halogen exchange reaction (Sect. 2.3).

The Synthetic Utility of Cross Coupling with Organoalkali Metals
and Grignard Reagents. Since each of the ten or so different
classes of carbon groups tends to behave differently and present
unique synthetic problems, there are a hundred or so different
cases of cross coupling which should be considered separately. A
detailed discussion of all of these reactions is beyond the scope
of this book. Only a brief summary of the scope and a discussion
of certain cases of special interest are presented.

(a) Category I (alkyl-alkyl) coupling. The generally poor
alkyl-alkyl coupling reaction proceeds satisfactorily in a limit-

ed number of favorable cases, even when tertiary alkyl halides
are used; see eq. 4.39 (4.55).

$$MeMgBr + \underset{Br}{\overset{Me}{\bigtriangleup}}Me \xrightarrow{Et_2O} Me \underset{Me}{\overset{Me}{\bigtriangleup}}Me \quad (92\%) \quad (4.39)$$

In the great majority of cases, however, the use of the organocopper procedures (Sect. 9.1) should be considered.

   (b) Category II cross coupling. The reaction of allylmetal
derivatives with alkyl halides tends to give mixtures of regio-
and stereoisomers, see eq. 4.40 (4.56).

$$\diagdown\diagup\diagdown Li + \langle\rangle-Br \longrightarrow \quad (4.40)$$

Similar problems exist in the propargyl-alkyl coupling reaction;
see eq. 4.41 (4.57, 4.58).

$$HC\equiv CCH_2Li + RX \longrightarrow HC\equiv CCH_2R + H_2C=C=CHR \quad (4.41)$$

The regioselectivity problem can be solved by the use of silyl-
protected propargyllithium derivatives, see eq. 4.42 (4.57, 4.59).

$$\quad + LiCH_2C\equiv CSiMe_3 \longrightarrow \quad (4.42)$$

77%

   The cross-coupling reactions of substituted allylic and propargylic halides may also be complicated by the allylic rearrangement. Another side reaction associated with these and benzylic

halides is α-elimination. Thus, for example, the reaction of *n*-BuLi with benzyl chloride in THF gives *trans*-stilbene (*4.60*). At -100°C, 1-chloro-1,2-diphenylethane is obtained in 80% yield, indicating that *trans*-stilbene is formed via cross coupling-elimination rather than dimerization of phenylcarbene (eq. 4.43). The same reaction, however, produces *n*-pentylbenzene in high yield, when carried out in hexane.

$$
PhCH_2Cl \xrightarrow[THF]{n\text{-BuLi}} \left[PhCH \diagdown_{Cl}^{Li} \right] \xrightarrow{PhCH_2Cl} \underset{\underset{\displaystyle PhCH=CHPh}{\downarrow}}{\underset{\displaystyle Cl}{PhCHCH_2Ph}}
$$

$$
\left[PhCH: \right] \dashrightarrow PhCH=CHPh \tag{4.43}
$$

(c) Category III cross coupling. Allyl-allyl coupling is very important, since a number of terpenoids and other natural products contain the 1,5-diene unit that can be synthesized via allyl-allyl coupling. When ordinary allylmetal derivatives are used, regiochemical, stereochemical, and cross-homo scrambling reactions tend to accompany the desired reaction. These difficulties have been solved largely by the use of α-thioallyllithiums (Biellmann coupling); see eq. 4.44 (*4.61*).

$$\tag{4.44}$$

Allylic sulfones can also be used in place of allylic sulfides

(4.62). The regiospecificity of the reaction of allylic sulfones appears higher than the corresponding reaction of allylic sulfides. The reductive removal of the sulfone moiety, however, is generally considerably more difficult than that of the sulfide group. In some fortunate cases, the allyl-allyl coupling has been successfully achieved with ordinary allyllithiums (4.63) and allylmagnesium halides (4.64).

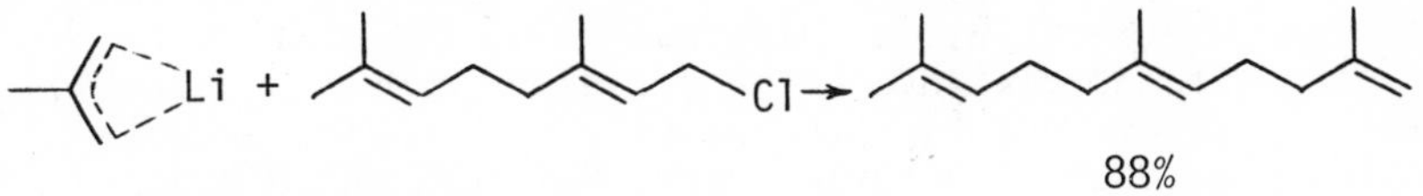

(4.45)

88%

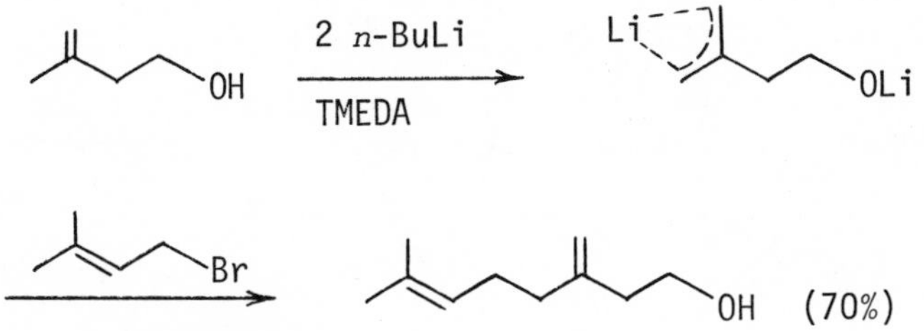

(4.46)

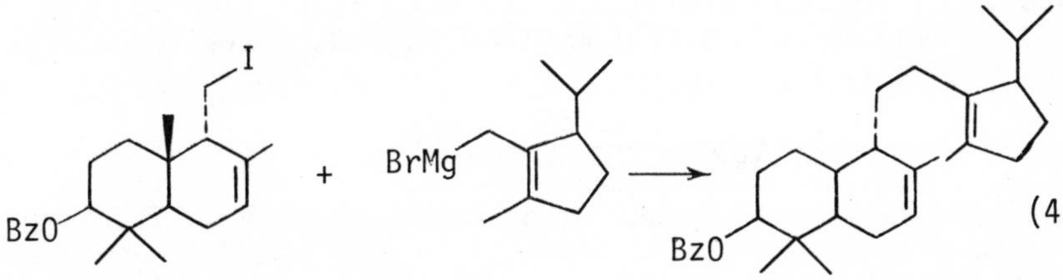

OH   (70%)

(4.47)

>50%

Propargylmagnesium bromide and 1-trimethylsilyl-3-lithiopropyne react with allylic and benzylic halides (4.57, 4.58) as well as with propargylic halides (4.65) to give the desired cross-coupled products along with minor amounts of allenic products. These products can be further converted into 1,5-dienes.

Generally speaking, Category III cross coupling tends to be

complicated by metal-halogen exchange and allylic rearrangement. These difficulties can, in principle, be overcome by the use of α-thioorganolithium reagents. This possibility, however, does not appear to have been tested in the benzyl-benzyl case.

(d) Category IV cross coupling. The direct carbon-carbon bond formation between alkyllithiums or alkylmagnesium halides and aryl, alkenyl, or alkynyl halides does not usually take place under ordinary reaction conditions. This difficulty has been largely overcome by the development of cross-coupling reactions of organocuprates (Sect. 9.1). Alternatively, the difficulty could be circumvented via the charge affinity inversion operation, which, in some cases, occurs spontaneously. Thus, for example, the reaction of n-BuLi with various aryl halides, which gives aryllithiums via metal-halogen exchange in ether, produces n-butylarenes in high yields in THF. The reaction presumably proceeds via metal-halogen exchange; see eq. 4.48 (*4.66*).

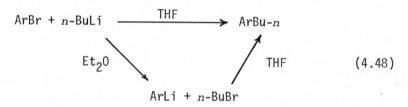

$$\text{ArBr} + n\text{-BuLi} \xrightarrow{\text{THF}} \text{ArBu-}n \tag{4.48}$$

with Et_2O and THF paths via $\text{ArLi} + n\text{-BuBr}$

As anticipated, *sec*-BuLi or *tert*-BuLi does not give cross-coupled products, but produces aryllithiums. These results can be exploited in developing the following primary alkyl-aryl coupling procedure (*4.66*).

$$\text{ArBr} + \text{RBr} \xrightarrow{\text{sec-BuLi}} \text{ArR} \tag{4.49}$$

Alkenyllithiums likewise react with primary alkyl bromides and iodides to give cross-coupled products in high yields. The use of polar solvents, such as THF and DMF, is essential. Primary alkyl tosylates and chlorides are poor substrates; see eq. 4.50 (*4.67*). The reaction appears highly stereospecific.

The reaction of alkynylmetals containing Li, Na, K, and Mg with primary alkyl halides generally proceeds satisfactorily (*4.40*). Treatment of terminal acetylenes with NaNH_2, n-BuLi, and i-PrMgCl represents a few of the most commonly used methods of generation of alkynylmetals. Here again, the use of polar solvents, such as THF, DMF, DMSO, and HMPA, is advantageous.

(e) Category V cross coupling. As in Category IV cross

$$\text{CH}_2\text{=CHCH}_2\text{Li} + n\text{-OctX} \longrightarrow \text{CH}_2\text{=CHCH}_2\text{Oct-}n \qquad (4.50)$$

X	Solvent	%
I	THF	100
Br	THF	100
OTs	THF	0
I	Et$_2$O	0

coupling, the direct carbon-carbon bond formation between allyl-, propargyl-, or benzylmetals containing alkali metals or Mg and aryl, alkenyl, or alkynyl halides does not occur readily. Here too, this problem can, in principle, be solved via charge affinity inversion. There is, however, one significant difference between Category IV and Category V coupling reactions. The products of Category V coupling contain, in the great majority of cases, highly acidic C-H bonds which could be competitively abstracted by organometallic species. Consequently, the use of highly basic organometallics, such as those containing Li, Na, and K, tends to produce less satisfactory results than those of Grignard reagents or other organometallics of lower basicity. In some favorable cases, even organoaluminums (Sect. 5.3.3.2) and organoborates (Sect. 5.2.3.3) can react with allyl, propargyl, or benzyl halides to give cross-coupled products. At present, however, organocoppers appear to be the most satisfactory class of compounds for achieving Category V cross coupling (Sect. 9.1). Another method that is highly suited for the preparation of allylated and benzylated arenes involves the reaction of aryllithiums with α,β-unsaturated and arylated carbonyl compounds, respectively, followed by reduction with lithium in NH$_3$ in the presence of NH$_4$Cl; see eq. 4.51 (4.68). When the reactants are compatible with the reduction procedure, the desired products are formed in high yields (>80%). Unfortunately, however, alkynyl and conjugated dienyl and polyenyl groups are incompatible with the reduction procedure.

$$(4.51)$$

(f) Category VI cross coupling. As discussed earlier, it is difficult to achieve the direct carbon-carbon bond formation between two sp^2 and/or sp hybridized carbon groups using organometallics containing alkali metals and magnesium. Fortunately, the problems associated with the $Csp^2(sp)$-$Csp^2(sp)$ coupling have been solved largely by the development of organocopper chemistry (Sect. 9.1) and nickel- or palladium-catalyzed cross coupling (Sect. 9.2). It is also worth mentioning that aryl-alkynyl, alkenyl-alkynyl, and alkynyl-alkynyl cross coupling can also be achieved via organoborates (Sect. 5.2.3.3).

4.3.2.2 Reactions of Organoalkali Metals and Grignard Reagents with Carbonyl and Related Compounds

The reaction of organometallics with carbonyl and related compounds may arbitrarily be divided into substitution and addition reactions, as shown in eq. 4.52.

$$R^1M + R^2\text{-}\overset{\overset{\displaystyle O}{\|}}{C}\text{-}Y \quad \begin{cases} \xrightarrow{\text{substitution}} R^1\text{-}\overset{\overset{\displaystyle O}{\|}}{C}\text{-}R^2 \\ \\ \xrightarrow{\text{addition}} R^1\text{-}\underset{\underset{\displaystyle R^2}{|}}{\overset{\overset{\displaystyle OH}{|}}{C}}\text{-}Y \longrightarrow R^1\text{-}\underset{\underset{\displaystyle R^2}{|}}{\overset{\overset{\displaystyle OH}{|}}{C}}\text{-}R^1 \end{cases} \tag{4.52}$$

The substitution reaction produces ketones, and it may be viewed as a special case of cross coupling. The addition reaction gives various types of alcohols.

<u>The Preparation of Ketones by the Reaction of Organometallics of Group IA and Group II Metals with Carboxylic Acids and Their Derivatives</u>. The reaction of organometallics of Group IA and Group II metals with carboxylic acids and their derivatives appears to involve the initial attack of the carbonyl group by organometallics. If the presumed intermediate <u>4.5</u> is sufficiently stable under the reaction conditions, it can subsequently be converted into the corresponding ketone via hydrolysis. Quite often, however, <u>4.5</u> presumably decomposes under the reaction conditions, producing the ketone in the presence of the organometallic reactant (R^1M). In such cases, the corresponding tertiary alcohol tends to be formed competitively (eq. 4.53). The synthesis of the tertiary alcohol <u>4.6</u> does not present any major difficulty. Even if it is not formed readily, it can be obtained from

$$R^1M + R^2COY \longrightarrow \underset{\underset{R^2 \ (\underline{4.5})}{|}}{\overset{O-M}{\underset{|}{R^1-\overset{|}{C}-Y}}} \xrightarrow{\ H_2O\ } \overset{O}{\overset{\|}{R^1-C-R^2}}$$

$$\Big\downarrow -MY \hspace{5cm} (4.53)$$

$$\overset{O}{\overset{\|}{R^1-C-R^2}} \xrightarrow{\ R^1M\ } \underset{\underset{R^1}{|}}{\overset{OH}{\underset{|}{R^1-\overset{|}{C}-R^2}}}$$

$$\underline{4.6}$$

$Y = OH, OR, SR, NR_2, CN,$ and so on

the ketone (R^1COR^2) in a separate step. We shall therefore focus our attention on the synthesis of ketones. The early results have been reviewed by various authors (*4.1, 4.2, 4.69, 4.70*). Numerous recent developments, however, make it desirable to update our knowledge of this subject in this section.

As is well known, the approximate order of reactivity of various carbonyl groups toward organometallic nucleophiles is: $COX(X = $ halogen$) > COOR > CONR_2 > CN$. The reaction of acyl halides and esters with organolithiums and Grignard reagents tends to produce tertiary alcohols as major products. To circumvent this difficulty, various alternate procedures have been developed. They tend to fall into either one of the following two categories:

1. Use of less reactive organometallics.
2. Use of less reactive carboxylic acid derivatives.

The use of organozincs and organocadmiums is well known and thoroughly reviewed; see eq. 4.54 (*4.32, 4.36*).

$$RCOCl \xrightarrow{\ R_2^1Zn \text{ or } R_2^1Cd\ } RCOR^1 \hspace{2cm} (4.54)$$

In addition to these organometallics, it has recently been shown that a variety of other organometallics react readily with acyl halides to produce ketones without the concomitant destruction of the ketones. They include organometallics containing B (Sect. 5.2.3.3), Al (Sect. 5.3.3.2), Si, and Sn (Sect. 6.4.2.1), Hg (Sect. 7.3.2.3), Cu (Sect. 9.5), Zr (Sect. 9.5), Fe (Sect. 11.3), Rh (Sect. 9.5), and Pd (Sect. 11.3). Although their relative advantages and disadvantages are not very clear, these new procedures seem to offer one distinct advantage over the use of organolithiums and Grignard reagents in that they can tolerate various electrophilic functional groups better than the lithium or magnesium procedure.

Some carboxylic acid derivatives that have been successfully employed in the ketone synthesis via organometallics containing lithium and magnesium are summarized in Table 4.4. The courses of the reaction of Grignard reagents with acyl derivatives can also be controlled by the use of highly polar aprotic solvents, such as HMPA, as shown in eq. 4.55 (4.74).

$$RCOY \xrightarrow[\text{ether or } C_6H_6]{R^1MgX-3HMPA} RCOR^1 \qquad (4.55)$$
$$63 \text{ to } 95\%$$

$$Y = Cl, OAc, OMe, NMe_2$$

The Carbonyl Addition Reactions of Organometallics of Group IA and Group II Metals. The synthetic usefulness of the addition reactions of organometallics with acyl derivatives which lead to alcohols is limited by the fact that they produce symmetrically substituted alcohols (eq. 4.52). We shall therefore focus our attention on the addition reactions of organometallics with aldehydes and ketones (eq. 4.56).

$$R^1COH(R^2) \xrightarrow[\text{2. } H_3O^+]{\text{1. } RM} R-\overset{R^1}{\underset{H(R^2)}{\overset{|}{\underset{|}{C}}}}-OH \qquad (4.56)$$

Although the addition reaction shown in eq. 4.56 provides a reasonably general and highly useful method of carbon-carbon bond formation, the successful use of this reaction requires clear knowledge of its scope, limitations, and anticipated difficulties. There are at least three major side reactions that can complicate the desired addition reaction, (1) enolization (eq. 4.57), (2) reduction (eq. 4.58), and (3) condensation (eq. 4.59).

$$R^1-\overset{H}{\underset{|}{\overset{|}{C}}}-\overset{O}{\overset{\|}{C}}-H(R^2) \xrightarrow{RM} R^1-\overset{OM}{\underset{|}{\overset{|}{C}}}=CH(R^2) + RH \qquad (4.57)$$

$$R^1\overset{|}{\underset{O}{\overset{|}{C}}}H(R^2) \xrightarrow{R-\overset{|}{\underset{|}{C}}-\overset{|}{\underset{|}{C}}-M} \qquad (4.58)$$

$$\xrightarrow{\hspace{3cm}} R^1-\overset{H}{\underset{|}{\overset{|}{C}}}-OM + R-C=C-\underset{H(R^2)}{}$$

Table 4.4   Preparation of Ketones by the Reaction of Carboxylic Acid Derivatives with Organometallics Containing Lithium and Magnesium

Carboxylic Acid Derivative	Organometallic Reactant	Possible Intermediate	Typical Yields (%)	Ref.
RCOOH (RCOOLi)	R^1Li	$\begin{array}{c}R\\\ \ \ C\\R^1\end{array}\!\!\!<\!\!\begin{array}{c}OLi\\OLi\end{array}$		4.70
RCOS-(2-pyridyl)	R^1MgBr	$\begin{array}{c}R\\R^1\end{array}\!\!>\!\!C\!\!<\!\!\begin{array}{c}S\\O\end{array}$ (cyclic with pyridine N–Mg–X)	80 to 100	4.71
RCOO-(2-pyrazinyl)	R^1MgX	$\begin{array}{c}R\\R^1\end{array}\!\!>\!\!C\!\!<\!\!\begin{array}{c}O\\O\end{array}$ (cyclic with pyrazine N–Mg–X)	44 to 97	4.72
RCOCN	R^1MgX	$\begin{array}{c}R\\\ \ \ C\\R^1\end{array}\!\!\!<\!\!\begin{array}{c}CN\\OMgX\end{array}$	70 to 84	4.73
$RCONR'_2$	R^1Li	$\begin{array}{c}R\\\ \ \ C\\R^1\end{array}\!\!\!<\!\!\begin{array}{c}NR'_2\\OLi\end{array}$	40 to 85	4.1
RCN	R^1Li	$\begin{array}{c}R\\R^1\end{array}\!\!>\!\!C=NLi$	25 to 90	4.1
RCN	R^1MgX	$\begin{array}{c}R\\R^1\end{array}\!\!>\!\!C=NMgX$		4.2

$$R^1-\overset{\overset{\displaystyle H}{|}}{\underset{|}{C}}-\overset{\overset{\displaystyle O}{\|}}{C}-H(R^2) \longrightarrow R^1-\overset{\overset{\displaystyle HO}{|}}{\underset{\underset{\displaystyle H(R^2)}{|}}{C}}-\overset{\overset{\displaystyle R^1}{|}}{\underset{|}{C}}-\overset{\overset{\displaystyle O}{\|}}{C}-H(R^2) \longrightarrow \text{ and so on }\quad (4.59)$$

It is important to note that, in general, the Grignard reagents are far more susceptible to these side reactions than the corresponding organolithiums. The scope and limitations of the Grignard addition reaction may be indicated by the results summarized in Table 4.5 (*4.75*).

Table 4.5   The Reaction of Grignard Reagents with Diisopropyl Ketone

Grignard Reagent	Addition	Reduction	Enolization	Total
MeMgBr	95	0	0	95
EtMgBr	77	21	2	100
n-PrMgBr	36	60	2	98
i-PrMgBr	0	65	29	94
i-BuMgBr	8	78	11	97
t-BuMgBr	0	65	0	65

In contrast to the Grignard addition reaction, the scope of which is rather limited, the corresponding reaction of organolithiums is quite general, as indicated by the following examples; see eqs. 4.60, 4.61 (*4.76, 4.77*).

(88%)

(4.60)

(81%)

(50%)

(4.61)

The stereochemistry of the organometallic addition to ke-

tones presents interesting synthetic and mechanistic problems.
The stereochemistry of the reactions of simple acyclic ketones
may be predicted on the basis of Cram's rule (4.78), which states
that the conformational requirements of the carbonyl group are
generally greater in the transition state than those of either of
the two carbon groups of the ketone and that attack by a nucleophile takes place predominantly from the less hindered side of
the carbonyl group, as shown in 4.7.

direction of
preferred attack

4.7

Certain proximal polar groups can seriously affect the stereochemistry of the reaction. The results observed with ketones
having an α-oxide anion and an α-halogen atom can be rationalized
in terms of electronic effects depicted in 4.8 and 4.9, respectively (4.79).

direction of
preferred attack

repulsion

4.8   4.9

    The stereochemistry observed with cyclic ketones is more
difficult to rationalize. Some representative results obtained
with 4-t-butylcyclohexanone are summarized in eq. 4.62 (4.80).
Although many different explanations have been proposed, this author has not yet been convinced by any presented in the past.
Interested readers are referred to pertinent references cited in
Ref. 4.80. It is, however, clear that the equatorial side of the
C=O group of 4-t-butylcyclohexanone is sterically less hindered
than the axial side. In general, very bulky reactants tend to
attack preferentially from the equatorial side.
    From the viewpoint of organic synthesis, it is probably more

$$(4.62)$$

RM	Solvent	Equational Attack (%)	Axial Attack (%)
MeLi	Et_2O	65	35
PhLi	Et_2O	58	42
$HC{\equiv}CNa$	$Et_2O\text{-}NH_3$	12	88
Me_2Mg	Et_2O	62	38
$t\text{-}BuMgBr$	Et_2O	100	0
Me_3Al (0.5 eq)	C_6H_6	80	20
Me_3Al (2 eq)	C_6H_6	17	83
$2MeMgI + ZnCl_2$	Et_2O	38	62

important to be able to improve the isomer ratio. In this connection, recent results obtained with a mixture of MeLi and
$LiCuMe_2$ are noteworthy; see eq. 4.63 (*4.81*).

$$(4.63)$$

The mechanism of the carbonyl addition reaction is not very
well established. Some reasonable speculations can be made, however, based on the available information. The reaction of MeLi
with 2,4-dimethyl-4'-mercaptomethylbenzophenone is first order in
ketone and 0.25 order in MeLi (*4.82*). These results are consistent with the following mechanism involving the interaction of
the MeLi monomer with a ketone in the 1:1 molar ratio (eq. 4.64).
The Grignard addition reaction appears to involve a similar
four-center mechanism. Although methylmagnesium halides generally add to ketones via two-electron transfer, highly hindered

$$(MeLi)_4 \rightleftharpoons 4\ MeLi$$

$$MeLi + \begin{array}{c} R^1 \\ \diagdown \\ R^2 \diagup \end{array} C{=}O \longrightarrow \begin{array}{c} R^1 \\ \diagdown \\ R^2 \diagup \end{array} \overset{\text{----}}{C}{=}O \longrightarrow Me\text{-}\underset{\underset{R^2}{|}}{\overset{\overset{R^1}{|}}{C}}\text{-}OLi \qquad (4.64)$$
$$\underset{Me\text{-}Li}{\big|\big|}$$

Grignard reagents, such as t-BuMgCl, seem to react, at least in some cases, via one-electron transfer (*4.83*). For further discussions of this subject, the reader is referred to reviews by Ashby (*4.80*, *4.84*).

4.3.2.3   Reactions of Organoalkali Metals and Grignard Reagents with Epoxides

In an epoxide molecule, the carbon and oxygen atoms of the oxirane ring provide electrophilic and nucleophilic centers, as indicated in 4.10. In addition to the carbon atoms of the oxirane ring, the β-hydrogen atoms can also act as electron acceptors.

4.10

Various organolithiums and Grignard reagents are capable of forming carbon-carbon bonds by attacking the ring carbon atoms. Grignard reagents, however, tend to act as electrophiles as well. The epoxide-electrophile interaction can induce (1) isomerization of epoxides into aldehydes and ketones, (2) formation of halohydrins, and (3) polymerization of epoxides (eq. 4.65).

$$(4.65)$$

In general, the reaction of epoxides with organometallics, which are more Lewis acidic than Grignard reagents, such as organozincs and organoalanes, tends to be complicated by one or more of these side reactions, although some synthetically useful reactions of organoalanes are known (Sect. 5.3.3.2).

The reaction of epoxides with organolithiums is much less prone to manifest the above-mentioned side reactions. Highly basic organolithiums, however, such as n-BuLi, tend to act as Brønsted bases and abstract a β-hydrogen atom, which leads to the formation of allylic alcohols (eq. 4.66).

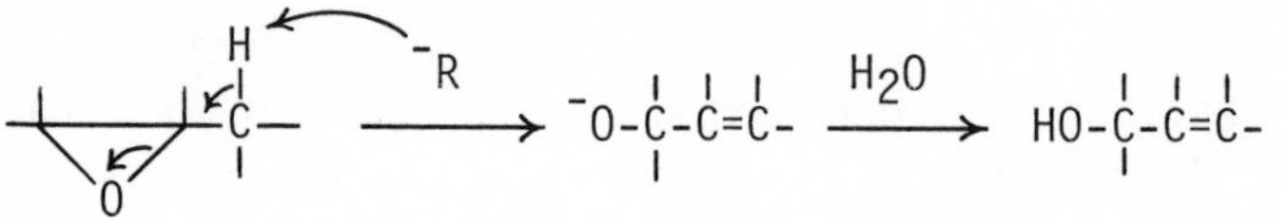

$$\text{(epoxide)} \xrightarrow{\text{-R}} \text{-O-C-C=C-} \xrightarrow{H_2O} \text{HO-C-C=C-} \quad (4.66)$$

If more basic organometallics, such as those containing sodium and potassium, are used, this side reaction might be more serious, but this is not clear. Although Grignard reagents have been used most widely in the past, it appears that organolithiums and organocuprates (Sect. 9.5.2) will prove at least as useful as and probably more useful than Grignard reagents. Some representative examples of epoxide opening reactions with organometallics containing lithium and magnesium are shown below. The early results on the organomagnesium-epoxide reaction have been summarized in an extensive review (4.85).

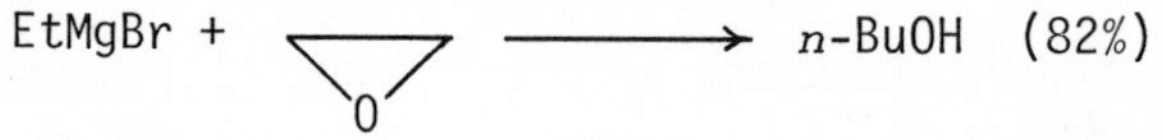

$$\text{EtMgBr} + \text{(epoxide)} \longrightarrow n\text{-BuOH} \quad (82\%) \quad (4.67)$$

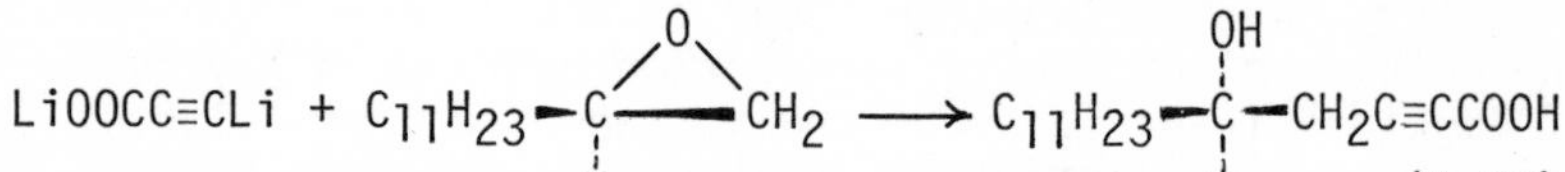

$$\text{LiOOCC}\equiv\text{CLi} + C_{11}H_{23}-\overset{H}{\underset{}{C}}-CH_2 \text{(epoxide)} \longrightarrow C_{11}H_{23}-\overset{OH}{\underset{H}{C}}-CH_2C\equiv CCOOH \quad (4.68)$$

$$\text{PhLi} + \text{(epoxide)}-\text{Ph} \longrightarrow \text{PhCh}_2\overset{}{\underset{OH}{C}}HPh \quad (70\%) \quad (4.69)$$

$$H_2C=\text{CHLi} + \text{(epoxide)} \longrightarrow \text{(cyclohexane)}\overset{CH=CH_2}{\underset{OH}{}} \quad (79\%) \quad (4.70)$$

Organolithiums tend to attack the less hindered carbon atom of the oxirane ring; see eqs. 4.68, 4.69 (4.86, 4.87). The regiochemistry of the epoxide opening reaction with Grignard reagents is much more complicated. With alkyl-substituted epoxides, the s_N2 opening appears to be favored. With aryl-substituted epoxides, however, the s_N1 opening has often been observed (eq. 4.71). The regiochemistry can be further complicated by the isomerization-addition path shown in eq. 4.72.

$$\text{PhCH}\underset{\displaystyle \diagdown_{\displaystyle O}\diagup}{\text{———}}\text{CH}_2 + \text{Me}_2\text{Mg} \longrightarrow \underset{\displaystyle |}{\text{PhCHCH}_2\text{OH}} \qquad (4.71)$$
$$\text{Me}$$

$$\text{PhCH}\underset{\displaystyle \diagdown_{\displaystyle O}\diagup}{\text{———}}\text{CH}_2 + \text{MeMgI} \longrightarrow \text{PhCH}_2\text{CHO}$$
$$\downarrow \text{MeMgI} \qquad\qquad (4.72)$$
$$\underset{\displaystyle \underset{\displaystyle \text{OH}}{|}}{\text{PhCH}_2\text{CHCH}_3}$$

The reaction usually involves a trans opening of the oxirane ring (eq. 4.70) (*4.88*).

4.3.2.4   Conjugate Addition Reactions of Organometallics of Group IA and Group II Metals

α,β-Unsaturated carbonyl derivatives can act as either carbonyl or alkenyl (alkynyl) compounds.

$$-\underset{|}{C}=\underset{|}{C}-\underset{|}{C}=O \longleftrightarrow -\overset{+}{\underset{|}{C}}-\underset{|}{C}=\underset{|}{C}-O^- \longleftrightarrow -\overset{+}{\underset{|}{C}}-\overset{-}{\underset{|}{C}}-\underset{|}{C}=O \longleftrightarrow -\underset{|}{C}=\underset{|}{C}-\overset{+}{\underset{|}{C}}-O^-$$

Conjugation with the carbonyl group makes the β-carbon atom a good electrophilic center. Attack at the carbonyl carbon atom by a nucleophile gives a 1,2-addition product, whereas attack at the β-carbon atom provides a 1,4-addition (or conjugate addition) product (eq. 4.73).

$$-\underset{|}{C}=\underset{|}{C}-\underset{|}{C}=O + M^{\delta+}-R^{\delta-} \quad\overset{\displaystyle 1,2-}{\nearrow}\quad \underset{\displaystyle \underset{\text{OH}}{|}}{-\underset{|}{C}=\underset{|}{C}-\overset{\overset{\text{R}}{|}}{C}}$$

$$\qquad\qquad\qquad\qquad\qquad\qquad\qquad (4.73)$$

$$\overset{\displaystyle 1,4-}{\searrow}\quad R-\underset{|}{C}-\overset{\overset{\text{H}}{|}}{\underset{|}{C}}-\underset{|}{C}=O$$

Despite the fact that numerous data on this reaction have been accumulated, no unifying interpretation of the available data appears to have been presented. Nonetheless, it is becoming increasingly clear that many different types of mechanisms operate in this reaction and that the 1,4/1,2 ratio is a function of a variety of factors. The following brief discussion is intended to facilitate our understanding of some factors, which presumably govern the 1,4/1,2 ratio, in a qualitative manner.

   (1) The formation of the 1,4-addition product should, in general, be thermodynamically more favorable than that of the 1,2-addition product, since the former product is stabilized through delocalization of either an unpaired electron or a negative charge, while the latter is not (eqs. 4.74, 4.75).

$$
\text{R}\cdot\ +\ -\underset{|}{\text{C}}=\underset{|}{\text{C}}-\underset{|}{\text{C}}=\text{O}
\quad
\begin{cases}
\xrightarrow{\ 1,2-\ } & -\underset{|}{\text{C}}=\underset{|}{\text{C}}-\overset{\overset{\text{R}}{|}}{\underset{|}{\text{C}}}-\text{O}\cdot \\[2ex]
\xrightarrow{\ 1,4-\ } & \text{R}-\underset{|}{\text{C}}-\overset{\cdot}{\underset{|}{\text{C}}}-\underset{|}{\text{C}}=\text{O} \longrightarrow \text{R}-\underset{|}{\text{C}}-\underset{|}{\text{C}}=\underset{|}{\text{C}}-\text{O}\cdot
\end{cases}
\tag{4.74}
$$

$$
\text{R}^-\ +\ -\underset{|}{\text{C}}=\underset{|}{\text{C}}-\underset{|}{\text{C}}=\text{O}
\quad
\begin{cases}
\xrightarrow{\ 1,2-\ } & -\underset{|}{\text{C}}=\underset{|}{\text{C}}-\overset{\overset{\text{R}}{|}}{\underset{|}{\text{C}}}-\text{O}^- \\[2ex]
\xrightarrow{\ 1,4-\ } & \text{R}-\underset{|}{\text{C}}-\overset{-}{\underset{|}{\text{C}}}-\underset{|}{\text{C}}=\text{O} \longrightarrow \text{R}-\underset{|}{\text{C}}-\underset{|}{\text{C}}=\underset{|}{\text{C}}-\text{O}^-
\end{cases}
\tag{4.75}
$$

It is not surprising therefore that free-radical addition reactions of α,β-unsaturated carbonyl compounds proceed by 1,4-addition. The great majority of the known conjugate addition reactions of organoboranes (Sect. 5.2.3.3) appear to fall into this category.

   (2) Based on simple considerations of electronegativity, it may be said that the carbonyl group is "harder" than the alkenyl group. As the ionicity or "hardness" of the organometallic reactants increases, the 1,2-addition, which involves a hard-hard interaction, can be kinetically more and more favored. The effect of the gegenion of the organometallic reactant on the 1,4/1,2 ratio is indicated by the data summarized in eq. 4.76 (4.89).

   (3) There is, however, a major pitfall in the above discussion. In the reaction of 2-cyclohexenone with Grignard reagents containing Me, Et, i-Pr, and t-Bu groups, the 1,4/1,2 ratio increases as the steric requirement of the Grignard reagent increases, as summarized in eq. 4.77 (4.90).

$$PhMX_n + PhCH=CHCOPh \longrightarrow PhCH=CHCPh_2 + Ph_2CHCH_2COPh$$

$$\underset{\displaystyle OH}{|}$$

$$(4.76)$$

MX_n	1,2(%)	1,4(%)
Na	39	4
Li	69	13
MgBr	0	high
ZnPh	0	91

$$(4.77)$$

R	1,2(%)	1,4(%)
Me	38	15
Et	52	24
i-Pr	10	44
t-Bu	0	70

Since the ionicity of the Grignard reagent should increase as the number of Me groups increases, one might expect t-BuMgX to exhibit the highest 1,2/1,4 ratio. In reality, however, t-BuMgX does not undergo the 1,2-addition at all. The results shown in eq. 4.77 may be rationalized as follows. The high carbanionic character of t-BuMgX renders it highly susceptible to one-electron transfer reactions involving the intermediacy of the t-Bu radical which should favor the 1,4-addition over the 1,2-addition (eq. 4.78).

$$t\text{-BuMgX} + \text{-C=C-C=O} \longrightarrow t\text{-Bu}\cdot + \text{-C=C-C=O}^{\overline{\cdot}} + MgX^+$$

$$\longrightarrow t\text{-Bu-}\overset{|}{C}\text{-C=C-OMgX} \xrightarrow{H_2O} t\text{-Bu-}\overset{|}{C}\text{-}\overset{H}{\underset{|}{C}}\text{-C=O}$$

$$(4.78)$$

Recently it has been proposed that the conjugate addition reaction of $LiCuR_2$ proceeds via one-electron transfer, as discussed in Sect. 9.4.1. The mechanism shown in eq. 4.78 suggests that the extent to which a given organometal-enone reaction undergoes the 1,4-addition via one-electron transfer should be a function of the reduction potential of the enone.

(4) In cases where the formation of kinetically favored 1,-2-addition products is reversible, the 1,2-addition products can be isomerized into the 1,4-addition products via equilibration. Weakly basic carbanions might participate in such a reaction. At least in some cases, the reaction of cyanoalanes with enones has been shown to proceed in this manner (Sect. 5.3.3.2).

(5) Certain highly covalent organometallics are capable of undergoing six-center pericyclic conjugate addition reactions with cisoid enones. The ready availability of an empty coordination site should facilitate such a reaction. Some conjugate addition reactions of organoboranes appear to proceed via a pericyclic mechanism (Sect. 5.2.3.3).

$$RM + \ -\underset{|}{C}=\underset{|}{C}-\underset{|}{C}=O \longrightarrow \ \ \longrightarrow \ R-\underset{|}{C}-\underset{|}{C}=\underset{|}{C}-OM \quad (4.79)$$

(6) The inductive, resonance, and steric effects can also affect the 1,4/1,2 ratio in a significant manner. As might be expected, steric hindrance around the carbonyl carbon and β-carbon atoms retards the 1,2- and 1,4-addition reactions, respectively. For example, whereas the *s*-butyl ester of crotonic acid undergoes predominantly a 1,4-addition reaction with *n*-BuMgBr, less hindered esters react at the carbonyl group; see eq. 4.80 (*4.91*).

$$
\begin{array}{c}
\xrightarrow[\text{2. } H_3O^+]{\text{1. } CH_3CH=CHCO_2Bu\text{-}s} \quad n\text{-}BuCHCH_2CO_2Bu\text{-}s \\[2pt]
n\text{-}BuMgBr \qquad\qquad\qquad\qquad\qquad\qquad\quad \underset{CH_3}{|} \qquad (4.80) \\[2pt]
\xrightarrow[\text{2. } H_2O]{\text{1. } CH_3CH=CHCO_2Me} \quad CH_3CH=CHC(Bu\text{-}n)_2 \\[2pt]
\qquad\qquad\qquad\qquad\qquad\qquad\qquad \underset{OH}{|}
\end{array}
$$

The fact that α,β-unsaturated aldehydes almost never undergo 1,4-addition reactions with organolithiums and Grignard reagents can, at least in part, be attributed to the low steric requirements of the formyl group.

An interesting and potentially useful example of electronic manipulation of the 1,4/1,2 ratio is shown in eq. 4.81 (4.92).

$$RLi + \underset{\underset{\overset{|}{PPh_3}}{+}}{-C=C-\overset{\overset{\displaystyle O}{\|}}{C}-C^- -COOEt} \longrightarrow R-\overset{|}{\underset{|}{C}}-\overset{|}{\underset{|}{C}}\!\!-\!\!\overset{\overset{\displaystyle Li\ O}{\ \ \|}}{\underset{\underset{\overset{|}{PPh_3}}{+}}{C}}-C^- -COOEt$$

with H$_2$O and R^{1}X branches   (4.81)

$$R-\overset{|}{\underset{|}{C}}-\overset{\overset{\displaystyle H\ O}{|\ \ \|}}{\underset{\underset{\overset{|}{PPh_3}}{+}}{C}-C}-C^- -COOEt \qquad R-\overset{|}{\underset{|}{C}}-\overset{\overset{\displaystyle R^1\ O}{|\ \ \|}}{\underset{\underset{\overset{|}{PPh_3}}{+}}{C}-C}-C^- -COOEt$$

The 1,2-addition evidently is not favored by the presence of the negatively charged ylide carbon atom. The requisite enones are readily prepared as stable species by treating α,β-unsaturated acyl chlorides with Ph$_3$P=CHCOOEt. The 1,4-addition products can be treated with alcohols and concentrated HCl to remove the ylide moiety (eq. 4.82).

$$\underset{\underset{\overset{|}{PPh_3}}{+}}{RCOC^-COOEt} \xrightarrow[\text{reflux}]{H^+, HOR'} RCOOR' \qquad (4.82)$$

The presence of the dipolar isocyanide group in the α-position of α,β-unsaturated carbonyl compounds electronically favors the 1,4-addition; see eq. 4.83 (4.93).

$$\underset{\overset{|}{Me_2C=C-COOEt}}{N^+\!\!\equiv C^-} \xrightarrow[\text{Et}_2\text{O}]{EtMgX} \underset{\overset{|}{Me_2EtC-CHCOOEt}}{N^+\!\!\equiv C^-} \quad (63\%)$$

$$(4.83)$$

Manipulation of detailed aspects of the conjugate addition reaction through modification of steric and electronic features of α,β-unsaturated carbonyl compounds, as exemplified by the two reactions shown above, seems to represent a promising area for future investigations. Especially noteworthy is a highly asymmetrical conjugate addition reaction of organolithiums with chiral vinyloxazolines which produces highly optically active products, though in modest yields; see eq. 4.84 (4.94).

$$(4.84)$$

31 to 76% (92 to 100% ee)

R^1 = Me, Et, c-Hex, or Ph; R^2 = Et, n-Bu, or Ph

(7) The 1,4/1,2 ratio can also be affected by other reaction conditions, such as the solvent used and the temperature applied. A recent report indicates that even readily polymerizable ethyl acrylate can successfully be reacted with ordinary Grignard reagents at low temperatures to produce 1,4-addition products in good yields; see eq. 4.85 (4.95).

$$CH_2=CHCOOEt \xrightarrow[\text{Et}_2\text{O, -50}^\circ\text{C}]{\text{RMgX}} RCH_2CH_2COOEt \qquad (4.85)$$

R = n-Pent (80%), —(68%), PhCH$_2$ (69%), Ph (54%)

Through manipulation of the various reaction parameters discussed above, we can now control the 1,4/1,2 ratio in many of the reactions of α,β-unsaturated carbonyl compounds with ordinary organolithiums and Grignard reagents. Even so, these reactions are not yet highly predictable and dependable. Fortunately, it has been fully established in recent years that the corresponding 1, 4-addition reactions of organocoppers, especially organocuprates, are not only more dependable but generally more chemoselective than the lithium or magnesium reactions, as discussed in detail in Sect. 9.4.1. Although more limited in scope, the 1,4-addition reactions of organoborons (Sect. 5.2.3.3) and organoaluminums (Sect. 5.3.3.2), as well as certain 1,4-addition reactions cata- lyzed by transition metal complexes, such as Ni(acac)$_2$ (Sect. 9.4.2), have exhibited a number of unique synthetic capabilities.

4.3.2.5 Reactions of Organometallics of Group IA and Group II Metals with Olefins and Acetylenes

The conjugate addition reaction of organometallics discussed in the preceding section formally involves addition of organometal- lics across carbon-carbon double or triple bonds. These carbon- carbon multiple bonds are electron deficient and, hence, highly activated toward nucleophiles. On the other hand, "ordinary" olefins and acetylenes are generally quite inert to organoalkali and organoalkaline earth metals under mild reaction conditions. In some special instances, however, facile carbometallation reac- tions of organometallics containing Group IA and Group II metals have been observed. The following reactions are worth mentioning here.

(1) Ethylene reacts with EtLi or n-BuLi under pressure to give simple addition products and oligomers (4.96).

(2) The corresponding reaction of secondary and tertiary al- kyllithiums turned out to be more facile than that of primary al- kyllithiums; see eq. 4.86 (4.97).

$$RLi + CH_2=CH_2 \longrightarrow RCH_2CH_2Li \qquad (4.86)$$
$$R = i\text{-Pr},\ c\text{-}C_6H_{11},\ t\text{-Bu, and so on}$$

Similar reactions of t-Bu$_2$Zn have also been reported (4.98).

(3) Although alkyl-substituted olefins are generally inert, highly strained olefins, such as cyclopropene (4.99) and norborn- ene (4.100), can undergo an addition reaction with alkyllithiums.

(4) The intramolecular version of the carbometallation of organoalkali and organoalkaline earth metals can be considerably more facile than the corresponding intermolecular reactions, as in the following examples.

$$(4.87)$$

$$(4.88)$$

$$(4.89)$$

$$Ph_2C=CHCH_2CH_2Li \underset{ether}{\overset{THF}{\rightleftharpoons}} Ph_2\underset{Li}{\overset{}{C}}\!\!-\!\!\triangleleft$$

$$(4.102)$$

(5) Certain hetero-substituted olefins and acetylenes undergo a facile carbometallation reaction, which presumably is anchimerically assisted by the hetero substituents (eqs. 4.90, 4.91).

$$CH_2=CHCH_2OH \xrightarrow[\,(4.103)\,]{2RLi,\ TMEDA} Li CH_2\underset{}{\overset{R}{C}}HCH_2OLi \qquad (4.90)$$

$$\swarrow H_2O \qquad \searrow CO_2$$

$$\underset{}{\overset{R}{}}CH_3\overset{|}{C}HCH_2OH \qquad\qquad R\!-\!\!\underset{O}{\diagdown}\!\!=\!O$$

$$n\text{-}BuC\equiv CCH_2X \xrightarrow[\,(4.104)\,]{\begin{array}{l}1.\ RMgX\\2.\ H_2O\end{array}} \underset{H}{\overset{n\text{-}Bu}{\diagup}}C=C\underset{CH_2X}{\overset{R}{\diagdown}} + \underset{R}{\overset{n\text{-}Bu}{\diagup}}C=C=CH_2$$

X = NR_2' or OH;   R = Me, Et, i-Pr, or Ph (4.91)

(6) Triorganosilyl and organoselenyl groups that are directly bonded to alkenyl groups activate the latter groups toward organometallics.

$$CH_2=CHSiMe_3 \xrightarrow[\,(4.105)\,]{RLi} RCH_2\underset{Li}{\overset{|}{C}}HSiMe_3 \qquad (4.92)$$

$$CH_2=CHSePh \xrightarrow[\,(4.106)\,]{\begin{array}{l}1.\ RLi\\2.\ H_2O\end{array}} RCH_2CH_2SePh \xrightarrow{Br_2} RCH_2CH_2Br \quad (4.93)$$

(7) Allylmetals containing Group IA and Group II metals can react readily with olefins and acetylenes. This reaction appears to proceed, at least partially, via six-centered transition states, such as 4.11, and appears to be closely related to the corresponding reaction of allylboranes (Sect. 5.2.3.5).

$$RC{\equiv}CH \xrightarrow[(4.107)]{\text{ZnBr}} \begin{array}{c} R\text{-}C{\equiv}CH \\ H_2C \diagdown \\ \diagdown CH{\vdots}CH_2 \end{array} ZnBr \longrightarrow \begin{array}{c} R \diagdown \\ C{=}CH \\ H_2C \diagdown \\ CH{=}CH_2 \end{array} ZnBr \quad (4.94)$$

4.11

$$R^1CH{=}CH_2 \xrightarrow[(4.108)]{\begin{array}{c}R^2\\ \diagup\!\diagdown\!MgX\end{array}} \quad CH_3\overset{R^1}{\underset{|}{C}}HCH_2\overset{R^2}{\underset{|}{C}}{=}CH_2 \quad (4.95)$$

(8) The currently most important carbometallation reaction of organometallics containing Group IA and Group II metals is probably the organolithium-initiated polymerization of conjugated dienes. The discovery (4.109) that isoprene can be polymerized by organolithiums to produce cis-poly(isoprene), whose properties closely resemble those of natural rubbers, led to the first large-scale industrial utilization of organolithiums. The course of the polymerization is, however, very much dependent on the metal atoms of organometallic initiators and solvents. The cis-1,4-addition occurs only when the reaction is carried out in a nonpolar solvent using either an organolithium or lithium metal as an initiator (4.110). The scope of this book does not permit a detailed discussion of this subject. Interested readers are referred to appropriate reviews on this subject (4.111).

Despite numerous papers and patents, relatively little is known definitively about the mechanism of the polymerization reaction. The cis stereochemistry may, however, be rationalized in terms of a six-centered transition state (4.12).

$$\begin{array}{c} H_3C \diagdown \\ C\text{-}CH \\ H_2C \diagup \diagdown CH_2 \end{array} \xrightarrow{\text{LiR}} \begin{array}{c} H_3C \diagdown \\ C{=\!=}CH \\ H_2C \diagdown \diagdown CH_2 \\ \diagdown Li\text{-}\text{-}\text{-}R \diagup \end{array} \longrightarrow \begin{array}{c} H_3C \diagdown \\ C{=}CH \\ LiCH_2 \diagdown CH_2R \end{array} \quad (4.96)$$

4.12

The following stereo- and regioselective dimerization of isoprene; see eq. 4.97 (4.112), represents a useful modification of the polymerization reaction discussed above.

(9) Finally, the nucleophilic substitution reaction of naphthalene with t-BuLi has been shown to proceed via an addition-elimination mechanism; see eq. 4.98 (4.113).

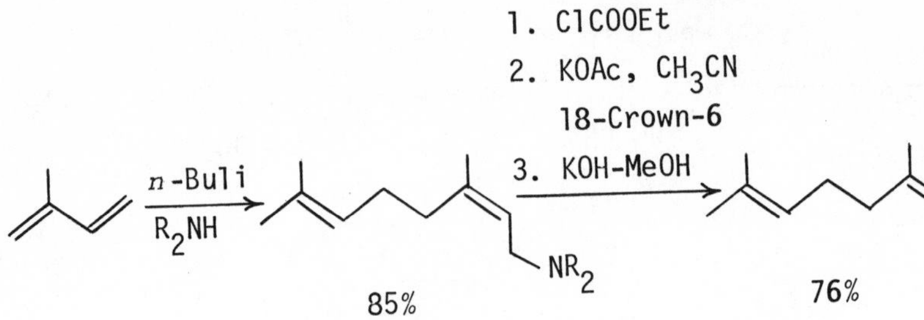

$$\text{(4.97)}$$

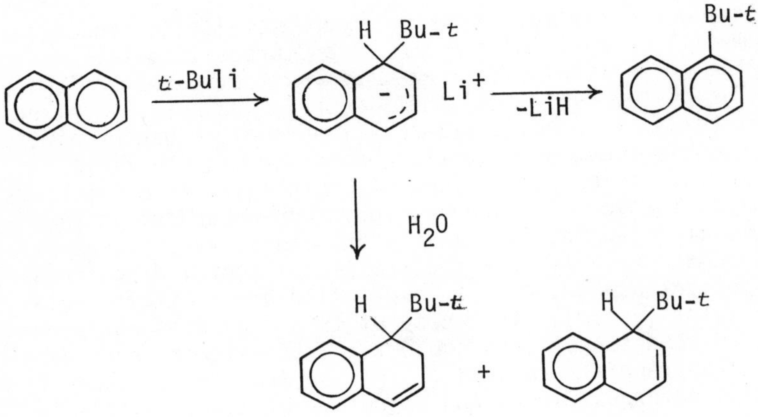

$$\text{(4.98)}$$

4.3.3 Carbon-Carbon Bond Formation via α-Hetero-Substituted Organometallics of Group IA and Group II Metals

The acidity of a C-H bond can be significantly enhanced by either electron-withdrawing hetero substituents or carbonyl and related unsaturated groups directly bonded to the carbon atom. This makes the corresponding organometallics containing alkali and alkaline earth metals readily accessible via metal-hydrogen exchange.

Although a limited number of α-hetero-substituted organometallics, such as KCN and NaCCl₃, have been known for many decades, the synthetic significance of α-hetero-substituted organometallics in general had not been fully recognized until recently.

With the advent of highly basic reagents including organolithiums, such as n-BuLi, a wide variety of α-hetero-substituted organometallics have become readily available. Investigations over the past few decades have clearly established that α-hetero-substituted organometallics are a class of uniquely useful species in organic synthesis.

Hetero atoms that have been incorporated in the α-positions of these organometallics include the following: halogens (F, Cl, Br, I), oxygen (OR), sulfur (SR, SOR, SO_2R, S^+R_2), selenium (SeR, SeOR, $^+SeR_2$), nitrogen (NO_2, $^+N{\equiv}C^-$, NR(NO), NR(CSR), NR(POX_2), NR_2, and the nitrogen atom of the $C{\equiv}N$ or $C{=}NR$ group), phosphorus (P^+R_3, PR_2, $PO(OR)_2$, POR_2), silicon (SiR_3), tin (SnR_3), and boron (BR_2, $B(OR)_2$). The metal-bound carbon atom can be sp^3 (alkyl, allyl, propargyl, benzyl), sp^2 (alkenyl, C=O, C=S, C=NR, heteroaromatic), or $sp(CN)$ hybridized, and may be bonded to one, two, or three hetero atoms, which may be the same or different. Some representative examples of α-hetero-substituted organometallics containing Group IA and Group II metals are listed in Tables 4.6 to 4.12.

A detailed discussion of all of these α-hetero-substituted organometallics is beyond the scope of this book. Their chemistry, therefore, is discussed in a highly condensed manner in the following several sections. For further details the reader is referred to pertinent reviews and original articles cited in the following discussions.

General Reaction Patterns. The α-hetero-substituted organometallics are often thermally unstable. One of the most commonly encountered and useful modes of their decomposition is α-elimination leading to the formation of carbenes (eq. 4.99).

$$M-\overset{\displaystyle |}{\underset{\displaystyle |}{C}}- \;\longrightarrow\; :\overset{\displaystyle |}{C}- + MX \qquad\qquad (4.99)$$
$$X$$

Organoalkali metals and organomercuries containing α-halogen atoms have proved to be especially useful as precursors to carbenes, as discussed in Sects. 4.3.3.5 and 7.3.2.2, respectively. Under appropriate conditions, α-hetero-substituted organometallics containing Group IA and Group II metals can react with various organic and inorganic electrophiles in the manner of their nonhetero-substituted analogues (Scheme 4.1). Many of the products shown in Scheme 4.1 are stable. In such cases, the presence of α-hetero substituents may not induce any transformation unique to the α-hetero substituents. In general, however, their presence permits us to achieve some synthetically useful manipulations, such as "umpolung" (Sect. 4.3.3.1) and regiochemical

Scheme 4.1

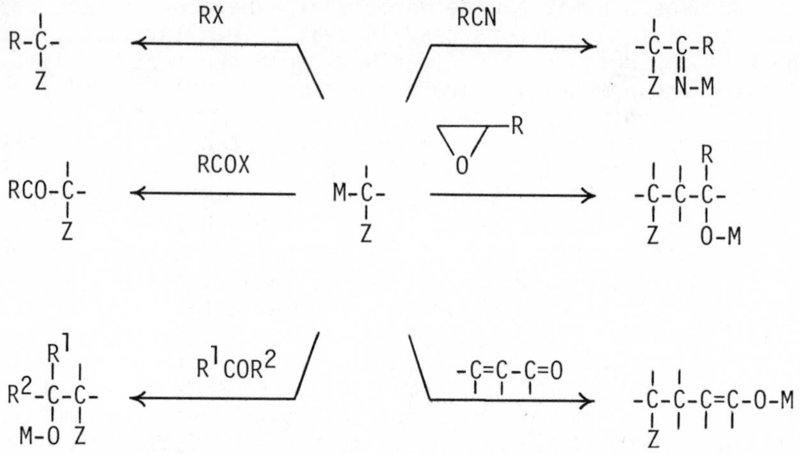

control (Sect. 4.3.3.2).

Quite frequently, the reaction of α-hetero-substituted organometallics with carbon electrophiles leads to the formation of unstable intermediates, which undergo subsequent spontaneous or assisted transformations under the reaction conditions. Some of the most useful reactions of this type are olefination (Sect. 4.3.3.3), epoxidation (Sect. 4.3.3.4), and cyclopropanation (Sect. 4.3.3.5). These reactions are shown in eqs. 4.100 to 4.102.

<u>Olefination</u>

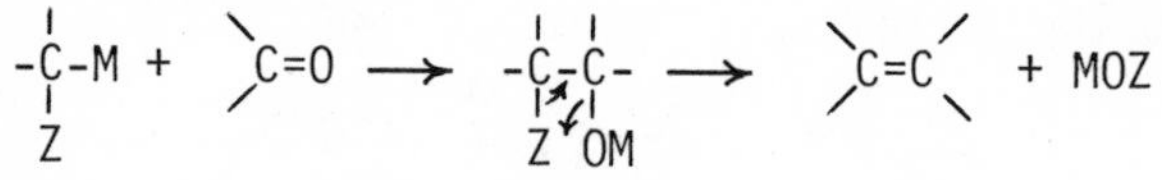

$$\text{(4.100)}$$

<u>Epoxidation</u>

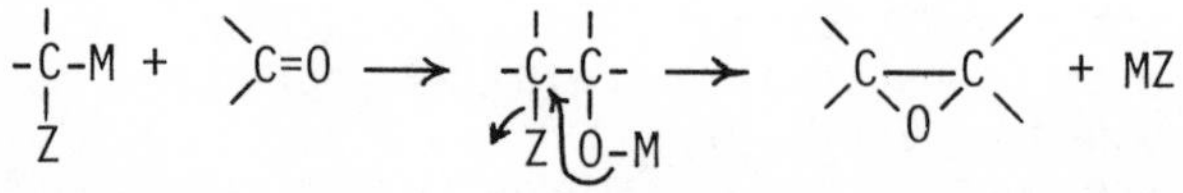

$$\text{(4.101)}$$

<u>Cyclopropanation</u>

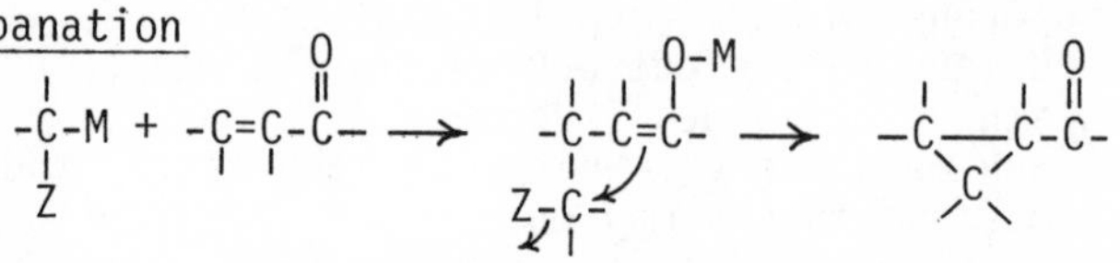

$$\text{(4.102)}$$

It should also be mentioned here that α-hetero-substituted organometallics can undergo various synthetically useful carbon-

carbon forming reactions with organoboranes (Sect. 5.2.3.4).

Table 4.6 α-Halo-Substituted Organometallics of Group IA and Group II Metals[a]

Compound	Method of Preparation	Use or Electrophile	Product	Ref.
$LiCH_2Cl$	n-BuLi (Li ⇌ Br)	Carbene precursor		4.114
$LiCHCl_2$	n-BuLi, LiN(Hex-c)$_2$	Carbene precursor		4.115
		R^1COR^2	$R^1R^2C(OH)CHCl_2$	4.116
		MeCOR	$RMeC(Cl)CHO$	4.117
		RCOOEt	$RCOCHCl_2$	4.118
$MCCl_3$ (M = Li, Na or K)	(M ⇌ H)	Carbene precursor		4.119
$LiC\!\!\underset{Cl}{=}\!\!C\!-$	n-BuLi			4.120
LiCBrHR	Li (metallation)	R^1COR^2	$R^1R^2\overset{\displaystyle\,}{C}\!\!\!\underset{O}{-}\!\!\!CHR$	4.121
$LiCHBr_2$	LiN(Hex-c)$_2$	R^1COR^2	$R^1R^2C(OH)CHBr_2$	4.116
$LiCBr_2R$	LDA	R^1X	RR^1CBr_2	4.122
$LiCBr_3$	n-BuLi	Carbene precursor		4.123
$LiC\!\!\underset{Br}{=}\!\!C\!-$	n-BuLi(Li ⇌ Br)			4.124
Li—(cyclopropyl)—Br	n-BuLi(Li ⇌ Br)	R^1COR^2	$R^2\!\!\underset{\displaystyle\;}{\overset{R^1}{\rceil}}\!\!=\!O$	4.125
$IMgCH_2I$	Mg (metallation)	R^1COR^2	$R^1R^2C=CH_2$	4.126
$IZnCH_2I$	Zn(Cu)	$\!\!\!>\!\!C=C\!\!<$	$>\!\!C\!\!\underset{CH_2}{-}\!\!C\!\!<$	4.127
$LiCHI_2$	n-BuLi(Li ⇌ I)	⟍⟍—Br	⟍⟍—CHI_2	4.128
	LiN(Hex-c)$_2$	R^1COR^2	$R^1R^2C(OH)CHI_2$	4.116
$Li\underset{Cl}{CHOR}$	LiTMP	$\!\!>\!\!C=C\!\!<$	$>\!\!C\!\!\underset{CHOR}{-}\!\!C\!\!<$	4.129
$K\underset{Cl}{CHSC_7H_7}$	KOBu-t	RCHO	$RHC\!\!\underset{O}{-}\!\!CHSC_7H_7$	4.130
$Li\underset{Cl}{CHSOPh}$	n-BuLi	R^1COR^2	$R^1R^2C\!\!\underset{O}{-}\!\!CHSOPh$	4.131

(continued...)

Table 4.6 continued

Compound	Method of Preparation	Use or Electrophile	Product	Ref.
		RCHO	RCOCH$_2$Cl	*4.132*
LiCHSO$_2$N⟨O⟩ | Cl	n-BuLi	R^{1}COR^2	R^{1}R^{2}C-CHSO$_2$N⟨O⟩ | | HO Cl	*4.133*
BrMgCHSO$_2$C$_7$H$_7$ | Cl	EtMgBr	CO$_2$	C$_7$H$_7$SO$_2$CHClCOOH	*4.134*
		R^{1}COR^2	R^{1}R^{2}C-CHSO$_2$C$_7$H$_7$ | | HO Cl	
LiCHPPh$_3$ | Cl	n-BuLi			*4.135*
LiCCl$_2$P(OEt)$_2$	n-BuLi	R^{1}COR^2	R^{1}R^{2}C=CCl$_2$	*4.136*
LiCBr$_2$P(OEt)$_2$		R^{1}COR^2	R^{1}R^{2}C=CBr$_2$	*4.137*
LiCMeSiMe$_3$ | Cl	s-BuLi	R^{1}COR^2	R^{1}R^{2}C——CMeSiMe$_3$ \\O/	*4.138*
LiC(SiMe$_3$)$_2$ | Br		RCHO	RCH=C(Br)SiMe$_3$	*4.139*
LiCRCOOEt | Br	LiN(SiMe$_3$)$_2$	R^{1}COR^2	R^{1}R^{2}C——CRCOOEt \\O/	*4.141*
KCRCOOEt | Cl (R = H, alkyl, or aryl)	KOBu-*t*	R^{1}COR^2	R^{1}R^{2}C——CRCOOEt \\O/	*4.140*
NaCHCOPh | Cl	NaOH	RCHO	RHC——CHCOPh \\O/	*4.142*
KCHCONEt$_2$ | Cl	KOBu-*t*	R^{1}HC=NR^2	R^{1}HC——CHCONEt$_2$ \\NR^2/	*4.143*
NaCHCN | Cl	NaOH	R^{1}COR^2	R^{1}R^{2}C——CHCN \\O/	*4.144*
LiCHCH=CHCOOMe | Cl	KOBu-*t*	ArCHO	ArHC——CHCH=CHCOOMe \\O/	*4.145*
NaCH-⟨C$_6$H$_4$⟩-NO$_2$ | Cl	NaOH	ArCHO	ArHC——CH-Ph-NO$_2$-*p* \\O/	*4.146*

[a] Unless otherwise specified, the listed compounds are prepared via metal-hydrogen exchange. Metal-halogen exchange is shown by (M ⇌ X).

Table 4.7 α-Oxy-Substituted Organometallics of Group IA and Group II Metals[a]

Compound	Method of Preparation	Electrophile	Product	Ref.
$\underset{\overset{\mid}{OEt}}{Li\overset{\mid}{C}=CH_2}$	t-BuLi	PhCHO	$\underset{HO\ \ OEt}{Ph\overset{\mid}{C}H\overset{\mid}{C}H=CH_2}$	4.147
		R^1COR^2 then H_3O	$\underset{HO}{R^1R^2\overset{\mid}{C}COCH_3}$	4.148
$\underset{\overset{\mid}{OSiMe_3}}{Li\overset{\overset{R^1}{\mid}}{C}CN}$	LDA	1. R^2X 2. HCl 3. NaOH	R^1COR^2	4.149
$\underset{\overset{\mid}{OTHP}}{Na\overset{\mid}{C}Ph(CN)}$	NaH	RX then H_3O^+	PhCOR	4.150
$LiCH_2O\overset{\overset{\displaystyle O}{\|}}{C}Ar$	s-BuLi TMEDA	RX then Hydrolysis	RCH_2OH	4.151
$\underset{\overset{\mid}{OR}}{Li\overset{\mid}{C}HR^1}$	BuLi (Li $\rightleftharpoons$ Sn)	R^2X	R^1R^2CHOR	4.152
$\underset{\overset{\mid}{OSiR_3}}{Li\overset{\mid}{C}HCH=CH_2}$	s-BuLi	R^1X	$R^1CH_2CH=CHOSiR_3$ (major)	4.153
		R^1COR^2	$\underset{HO\ \ OSiR_3}{R^1R^2\overset{\mid}{C}-\overset{\mid}{C}HCH=CH_2}$	4.154
$\underset{\overset{\mid}{OBu\text{-}t}}{Li\overset{\mid}{C}HCH=CH_2}$	s-BuLi	RX	$RCH_2CH=CHOBu\text{-}t$ (major)	4.155
$\underset{\overset{\mid}{OMe}}{Li\overset{\mid}{C}HC\equiv CSMe}$	LDA	RX	$\underset{\overset{\mid}{R}}{MeOCH=C=CSMe}$	4.156
$\underset{\overset{\mid}{OMe}}{K\overset{\mid}{C}HSO_2C_7H_7}$	KOBu-t	R^1COOR	$\underset{\overset{\mid}{OMe}}{R^1CO\overset{\mid}{C}HSO_2C_7H_7}$	4.157
$Li-\underset{\diagdown O}{\overset{S\diagup}{\langle}}$		RX	$R-\underset{\diagdown O}{\overset{S\diagup}{\langle}}$	4.158

[a] See footnote a of Table 4.6.

141

Table 4.8 Organometallics of Group IA and Group II Metals Containing α-Sulfur Substituents[a]

Compound	Method of Preparation	Electrophile	Product	Ref.
$LiCH_2SCH_3$	n-BuLi TMEDA	RCHO	$RCH(OH)CH_2SMe$	*4.159*
$LiCH_2SAr$	n-BuLi DABCO	1. RI 2. MeI, NaI, DMF	RCH_2I	*4.160*
		1. R^1COR^2 2. H_3O^+ 3. RX then base	$R^1R^2C\overset{\displaystyle CH_2}{\underset{O}{\diagdown}}$	*4.161*
$Li\underset{SPh}{CR^1R^2}$	n-BuLi (Li ⇌ Se)	R^3COR^4	$PhSCR^1R^2CR^3R^4\underset{OH}{\vert}$	*4.162*
$Li\underset{SPh}{CHCH{=}CR^1R^2}$	n-BuLi	1. $R^3R^4C{=}CHCH_2X$ 2. Li, $EtNH_2$	$R^1R^2C{=}CHCH_2$ $R^3R^4C{=}CHCH_2$	*4.61*
		R^3COR^4	see eq. 4.134	*4.163*
$Li\underset{SLi}{CHCH{=}CH_2}$	n-BuLi	1. $MgBr_2$ 2. R^1COR^2 3. MeI	$R^1R^2C\text{-}CHCH{=}CH_2$ $HO\ \ SMe$	*4.164*
$Li\underset{SLi}{CHPH}$	n-BuLi	PhCHO	$PhCHCHPh$ $HO\ SH$	*4.165*
$Li\text{—}\underset{SPh}{\triangleleft}$	n-BuLi	R^1COR^2	$R^1R^2C\text{—}\underset{HO\ \ \ SPh}{\triangleleft}$	*4.166*
$Li\underset{SEt}{C{=}CH_2}$	s-BuLi THF-HMPA	1. RBr 2. $HgCl_2$, H_2O MeCN	$RCOMe$	*4.167*
$Li\underset{SEt}{C{=}CH_2}$	s-BuLi THF-HMPA	1. $RCH\overset{\diagup CH_2}{\underset{O}{\diagdown}}$ 2. $HgCl_2$, H_2O MeCN	$RCH{=}CHCOMe$	*4.167*
$Li\underset{SPh}{C{=}CHOEt}$	t-BuLi	RI	$R\underset{SPh}{C{=}CHOEt}$	*4.168*
		R^1COR^2	$R^1R^2C\text{-}CHOEt$ $HO\ SPh$	*4.168*

142

(continued...)

Table 4.8 continued

Compound	Method of Preparation	Electrophile	Product	Ref.
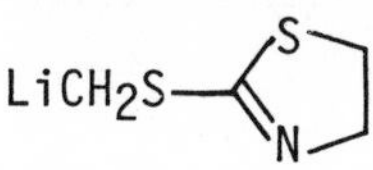 LiCH$_2$S—[thiazoline structure]	n-BuLi	1. RX 2. MeI, DMF	RCH$_2$I	4.169
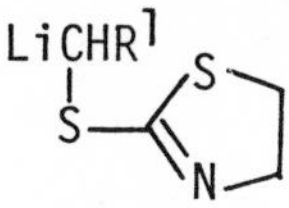 LiCHR^1 [S-C(=N)S thiazoline]	n-BuLi	1. R^{2}COR^3 2. H$_3$O$^+$ 3. Δ then base	R^{1}CH—CR^{2}R^3 (bridged by S)	4.170
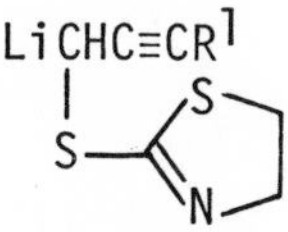 LiCHC≡CR^1 [thiazoline]	n-BuLi	R^{2}X	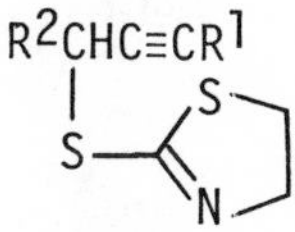 R^{2}CHC≡CR^1 [thiazoline]	4.171
LiCHCH=CH$_2$ SCH=CH$_2$	s-BuLi	1. RBr 2. DME, H$_2$O CaCO$_3$, Δ	RCH=CH(CH$_2$)$_2$CHO	4.172
LiCHC(OEt)=CH$_2$ SCH=CH$_2$	s-BuLi	1. RBr 2. DME, H$_2$O, Δ	RCH$_2$CO(CH$_2$)$_2$CHO	4.173
LiCHR^1 S-C(S)NMe$_2$	LDA	MeI	MeCHR^1 S-C(S)NMe$_2$	4.174
LiCHCR^1=CH$_2$ S-C(S)Me	s-BuLi	1. R^{2}Br 2. MeI	$\begin{array}{c}R^2R^1\\ \diagdown C=C \diagup\\ H(CH_2)_2\overset{S}{\overset{\|}{C}}SMe\end{array}$	4.175
LiCHCR^1=CH$_2$ SC(S)NMe$_2$	LDA	1. R^{2}X 2. Δ(PhH) 3. LDA 4. MeSSMe 5. HgX$_2$, MeCN	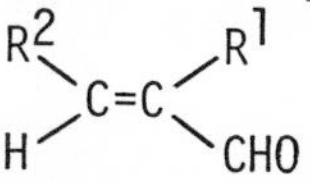 $\begin{array}{c}R^2R^1\\ \diagdown C=C \diagup\\ HCHO\end{array}$	4.176
LiCHC(OMe)=CH$_2$ SC(S)NMe$_2$	LDA	RX	RCHC(OMe)=CH$_2$ SC(S)NMe$_2$	4.176
LiCH$_2$SOMe NaCH$_2$SOMe KCH$_2$SOMe	n-BuLi NaH KOBu-t	E$^+$	ECH$_2$SOMe	4.177
LiCH$_2$SOC$_7$H$_7$	LiNEt$_2$	PhCH=NPh	PhCHNHPh CH$_2$SOC$_7$H$_7$	4.178

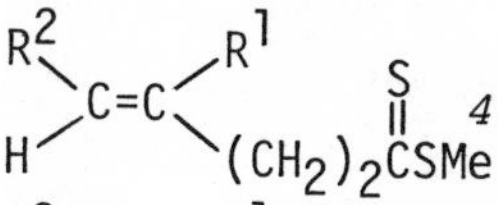

143

(continued...)

Table 4.8 continued

Compound	Method of Preparation	Electrophile	Product	Ref.
LiCHR[1] | SOPh	LiICA	1. R^2CH_2X 2. Δ	$R^1CH=CHR^2$ (R =Ar, SPh, SOPh, or CN)	4.179
LiCH$_2$SONAr | Li	n-BuLi	1. RCOOR' 2. H_2O	RCOMe	4.180
		RCHO then Δ	$RCH=CH_2$	4.180
LiCHR[1] | SOBu-t	MeLi	1. R^2COR^3 2. NCS or SO_2Cl_2	$R^1CH=CR^2R^3$	4.181
LiCHCH=CR^{1}R^2 | SOPh	LDA	1. R^3X 2. $P(OMe)_3$	$R^1R^2CCH=CHR^3$ | HO͡OH	4.182
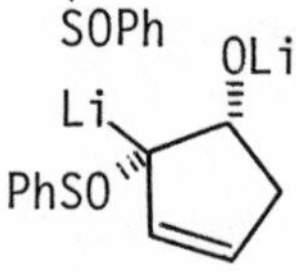	LDA	1. RX 2. $HNEt_2$	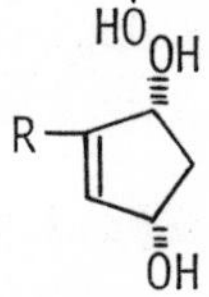	4.183
$NaCH_2SO_2CH_3$	$NaCH_2SOCH_3$	RCOOR'	$RCOCH_2SO_2CH_3$	4.177a
$KCH_2SO_2CH_3$	KOBu-t	RCOOR'	$RCOCH_2SO_2CH_3$	4.184
$BrMgCH_2SO_2Ar$	EtMgBr	RCOOR'	$RCOCH_2SO_2Ar$	4.185
$LiCH_2SO_2NMe_2$	n-BuLi	RCOOR'	$RCOCH_2SO_2NMe_2$	4.177a
LiCHCH$_2$—C—R | PhSO$_2$ O O	n-BuLi	1. R^1X 2. aq HOAc 3. base		4.186a
		1. R^1COOR^2 2. Al(Hg)	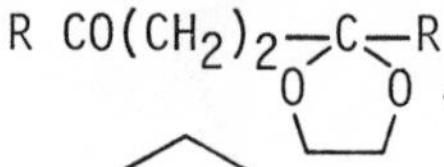	4.186b
		1. 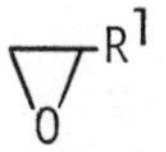 R^1 2. Na(Hg) 3. H_3O^+	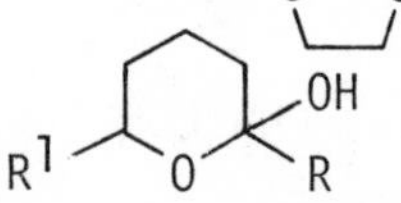	4.186c
LiCHCH=CR^{1}R^2 | SO$_2$C$_7$H$_7$	n-BuLi	$R^3R^4C=CHCH_2Br$	$R^1R^2C=CHCHSO_2C_7H_7$ $R^3R^4C=CHCH_2$	4.187
KCHC≡CR | SO$_2$Ph	KOBu-t	$CH_2=CHCOOMe$		4.188

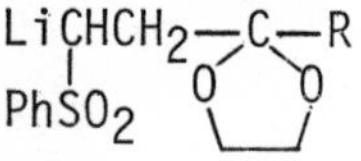

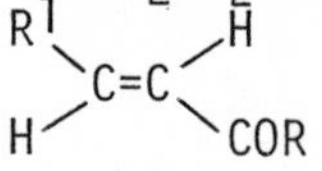

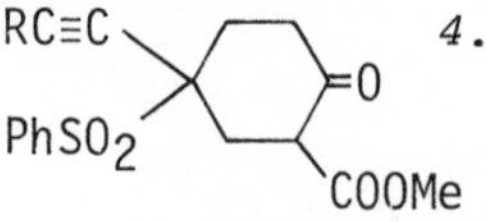

(continued...)

Table 4.8 continued

Compound	Method of Preparation	Electrophile	Product	Ref.
Li—⟨ring⟩ SO₂N‑R	n-BuLi	R^1X	R^1—⟨ring⟩ SO₂N‑R	4.189
		R^1COR^2	R^1R^2C—⟨ring⟩ SO₂N‑R, HO	4.189
LiCHR / Ph(O)S=NMe, O	n-BuLi	1. R^1COR^2 2. Al(Hg), H₂O, HOAc	$R^1R^2C=CHR$	4.190
LiR¹R²CSR / NTs	n-BuLi	R^3COR^4	R^1R^2C—CR^3R^4 (O)	4.191
LiCHCOOR / S⊥	LiNH₂	R^1X	R^1CHCOOR / S⊥	4.192
NaCHCOOEt / SPh	NaH	1. BrCH₂CH(OMe)₂ 2. NaH 3. RX 4. LAH 5. TsOH, Δ	⟨furan⟩—R	4.193
LiCHCOOEt / SLi	LDA TMEDA	1. R^1COR^2 2. ClCOOEt	$R^1R^2C=CHCOOEt$	4.194
Li—⟨lactone ring⟩ EtOCS(=S), O	LDA	R^1COR^2	$R^1R^2C=$⟨lactone ring⟩, O	4.195
LiCHCOOLi / SPh		R^1COR^2	R^1R^2C-CHCOOH / HO SPh	4.196
NaCHCOR / SOMe	NaH	1. R^1X 2. Al(Hg)	$RCOCH_2R^1$	4.197
		1. BrCH₂COOEt 2. Zn-HOAc	$RCO(CH_2)_2COOEt$	4.198
MCMeCOCH₂M / SOPh (M = Li and/or Na)	NaH then n-BuLi	1. RX 2. Δ	$CH_2=CHCOCH_2R$	4.199

(continued...)

Table 4.8   continued

Compound	Method of Preparation	Electrophile	Product	Ref.
NaCRCOR[1] $\\ $ SO$_2$Me	NaH	BrCH$_2$COOEt	R[1]COC=CHCOOEt (R on C)	4.200a
		1. BrCH$_2$COR[2] $\\$ 2. Zn, HOAc	R[1]COCHRCH$_2$COR[2]	4.200b
KCHN≡C $\\ $ SO$_2$Ar	KOBu-t	1. R[1]COR[2] $\\$ 2. H$_3$O$^+$	R[1]R[2]CHCOOH	4.201a
KCHN≡C $\\ $ SO$_2$Ar	KOBu-t	1. R[1]COR[2] $\\$ 2. NaOMe MeOH reflux	R[1]R[2]CHCN	4.201b
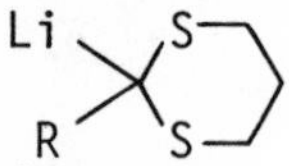 Li, R — (1,3-dithiane) (R = H or alkyl)	n-BuLi	See Scheme 4.2		4.202, 4.203
 Li, COOEt — (1,3-dithiolane)	LDA	RC=CHCOOR' $\\$ Cl	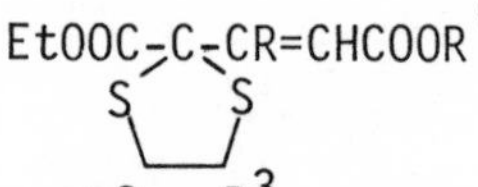 EtOOC-C-CR=CHCOOR' (dithiolane)	4.204
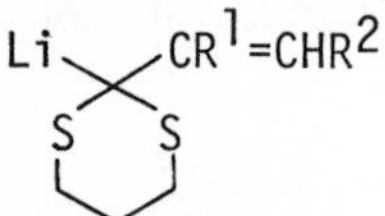 Li, CR[1]=CHR[2] — (1,3-dithiane)	LDA	R[3]X	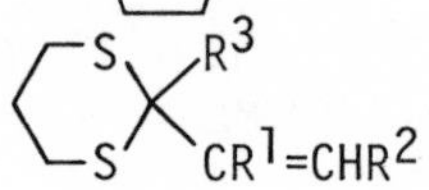 dithiane with R[3] and CR[1]=CHR[2]	4.205
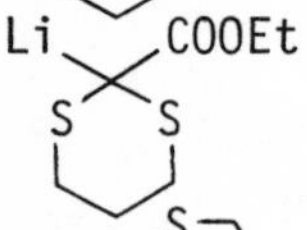 Li, COOEt — (1,3-dithiane)	NaH	1. RBr $\\$ 2. NBS	RCOCOOEt	4.206
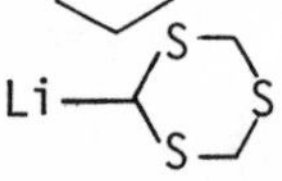 Li — (1,4-dithiane ring)	n-BuLi	RX	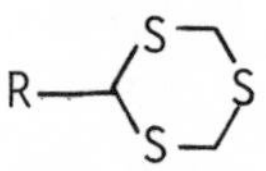 R — (1,4-dithiane ring)	4.207
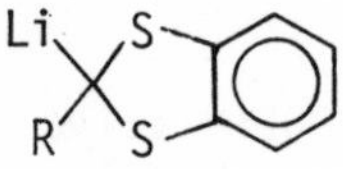 Li, R — (benzodithiole)	n-BuLi	1. E$^+$ $\\$ 2. HgO BF$_3\cdot$Et$_2$O	ECOR	4.208
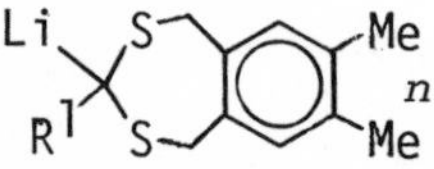 Li, R[1] — (benzodithiepine, Me,Me)	n-BuLi	R[2]X	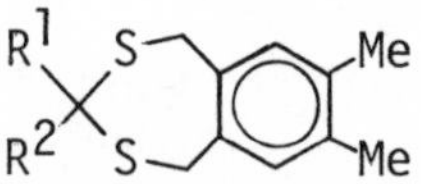 R[1], R[2] — (benzodithiepine, Me, Me)	4.209
NaCH(SPh)$_2$	NaNH$_2$	RX	RCH(SPh)$_2$	4.210

(continued...)

Table 4.8 continued

Compound	Method of Preparation	Electrophile	Product	Ref.
$NaC(SEt)_2COOMe$	NaH	$CH=CHCOY$ $(Y = OR \text{ or } Me)$	$MeOOCC(SEt)_2CH_2$ $\quad\quad\quad\quad YCOCH_2$	4.211
$Li\underset{\underset{SMe}{\vert}}{C}HSC(S)NMe_2$	n-BuLi	1. RI 2. HgX_2, MeOH	$RCH(OMe)_2$	4.212
$Li\underset{\underset{SMe}{\vert}}{C}R^1SOMe$ $(R^1 = H \text{ or alkyl})$	n-BuLi or LDA	1. E^+ 2. H_3O^+	$ECOR^1$	4.213
$Li\underset{\underset{SEt}{\vert}}{C}R^1SOEt$ $(R^1 = H \text{ or alkyl})$	n-BuLi or LDA	1. E^+ 2. H_3O^+	$ECOR^1$	4.214
$LiC(SPh)_3$	n-BuLi	RI R^1COR^2	$RC(SPh)_3$ $R^1R^2\underset{\underset{HO}{\vert}}{C}C(SPh)_3$	4.215
		$H_2C=C(SR)_2$	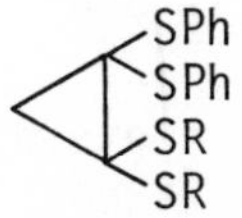 cyclopropane with SPh, SPh, SR, SR	4.215
		$R^1CH=CHCOR^2$	$(PhS)_3CR^1CHCH_2COR^2$	4.216
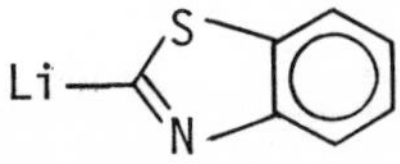 Li—benzothiazole	n-BuLi	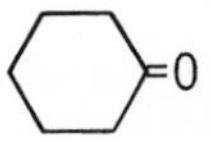 cyclohexanone (=O)	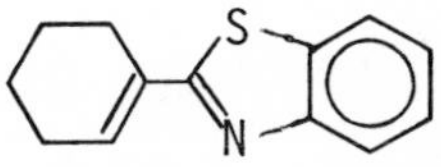 cyclohexenyl-benzothiazole	4.217

a See footnote a of Table 4.6.

Table 4.9 Organometallics of Group IA and Group II Metals Containing α-Selenium Substituents[a]

Compound	Method of Preparation	Electrophile	Product	Ref.
$Li\underset{\underset{SeR}{\vert}}{C}R^1R^2$	n-BuLi ($Li \rightleftharpoons Se$)	R^3COR^4	$R^1R^2\underset{\underset{RSe}{\vert}}{C}\!-\!\underset{\underset{OH}{\vert}}{C}R^3R^4$	4.218

147

(continued...)

Table 4.9 continued

Compound	Method of Preparation	Electrophile	Product	Ref.

$LiCR^1CH_2R^2$
  $\|$
 $Se(O)R$
	LDA	1. R^3COR^4
		2. HOAc
		3. heat
		  R^3X
			$R^2CH{=}CR^1CR^3R^4$
			           $\|$
			           OH
			$R^1R^3CCH_2R^2$
			    $\|$
			    $Se(O)R$
				4.219

$LiCR^1CH{=}CR^2R^2$
  $\|$
 $SePh$
		1. RX
		2. H_2O_2,Py
			$RR^1C{=}CHCR^2R^3$
			          $\|$
			         OH
				4.220

$LiCHC{\equiv}CLi$
  $\|$
 $SePh$
	LDA	1. RX
		2. E^+
			$RCHC{\equiv}CE$
			  $\|$
			 $SePh$
				4.221

$LiCHSiMe_3$ $\|$ $SePh$		1. RCH_2X 2. H_2O_2	RCH_2CHO	4.222

a See footnote a of Table 4.6.

Table 4.10 Organometallics of Group IA and Group II Metals Containing α-Nitrogen Substituents[a]

Compound	Method of Preparation	Electrophile	Product	Ref.

$LiCH_2N(NO)R$	LDA	R^1COR^2	$R^1R^2CCH_2N(NO)R$
			       $\|$
			      OH
				4.223

		R^1X	$R^1CH_2N(NO)R$	4.223

$LiCHPh$
  $\|$
$MeNPO(NMe_2)_2$
	n-BuLi	RX	$PhRHCNPO(NMe_2)_2$
			          $\|$
			         Me
				4.224

$LiCHCH{=}CH_2$ $\|$ $NMePh$	n-BuLi	RX	$CH{=}CHCH_2R$ $\|$ $NMePh$	4.225

$LiCHCH{=}CH_2$
  $\|$
 N (pyrrolidine)
	s-BuLi	RX	$RCHCH{=}CH_2$
			  $\|$
			 N (pyrrolidine)
				4.226

$LiCHR^1$
  $\|$
 NO_2
		1. $CH_2{=}CHC(O)R^2$	$R^1C(CH_2)_2CR^2$
			  $\|\|$ $\|\|$
			  O O
				4.227

(continued...)

Table 4.10   continued

Compound	Method of Preparation	Electrophile	Product	Ref.
		2. $TiCl_3$ 1. $CH_2{=}CH{-}C(O){-}R^2$	$R^1\underset{\parallel}{C}(CH_2)_2\underset{\parallel}{C}R^2$ ($C{=}O$, $C{=}O$)	4.228
$LiCHR^1$ $\mid$ $N{\equiv}C$		2. O_3,base $R^2R^3C{=}NR^4$	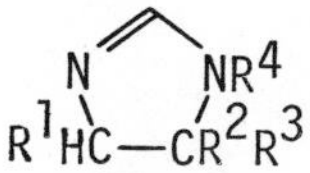 imidazoline ring: $N{=}\,\,NR^4$ / $R^1HC{-}CR^2R^3$	4.229
$LiC{=}CR^1R^2$ $\mid$ $N{\equiv}C$	n-BuLi	E^+	$E\underset{\mid}{C}{=}CR^1R^2$ $N{\equiv}C$	4.230
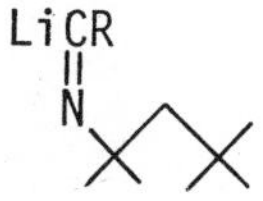 $Li\underset{\parallel}{C}R$, $\underset{N}{}$ (bicyclic)	$C_8H_{17}N{\equiv}C$ and RLi	$PhCHO$	$Ph\underset{\mid}{C}HCOR$ OH	4.231
		epoxide${-}R^1$	$R^1\underset{\mid}{C}HCH_2COR$ OH	4.231
$Li\underset{\mid}{C}{=}O$ $N(Pr\text{-}i)_2$	LDA	$Ph_2C{=}O$	$Ph_2\underset{\mid}{C}CON(Pr\text{-}i)_2$ HO	4.232
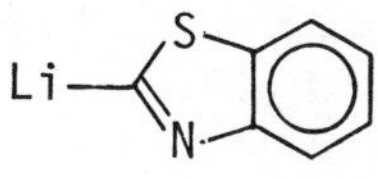 $Li{-}$benzothiazole (S, N ring)	n-BuLi	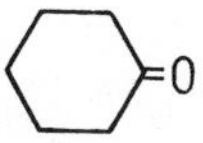 cyclohexanone	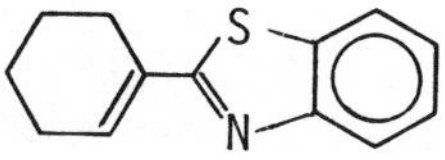 cyclohexenyl-benzothiazole	4.217

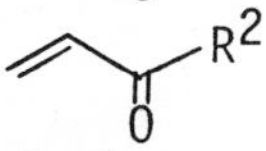

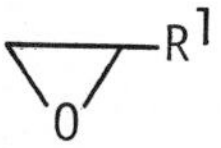

a   See footnote a of Table 4.6.

Table 4.11   Organometallics of Group IA and Group II Metals Containing α-Phosphorus Substituents[a]

Compound	Method of Preparation	Electrophile	Product	Ref.
$NaCRCOOEt$ $\mid$ $PO(OEt)_2$	NaH	R^1COR^2	$R^1R^2C{=}CRCOOEt$	4.233
$NaCHCN$ $\mid$ $PO(OEt)_2$	NaH	R^1COR^2	$R^1R^2C{=}CHCN$	4.234

(continued...)

Table 4.11 continued

Compound	Method of Preparation	Electrophile	Product	Ref.
LiCHCH=CR^{1}R^2 $\mid$ PO(OEt)$_2$	n-BuLi	1. R^{3}X 2. LiAlH$_4$	R^{1}R^{2}CHCH=CHR^3	4.235
LiCR(SMe) $\mid$ PO(OEt)$_2$	n-BuLi	R^{1}CHO	R^{1}CH=CR(SMe) $\downarrow$HgCl$_2$,H$_2$O,CH$_3$CN R^{1}CH$_2$COR	4.236
LiCHSOMe $\mid$ PO(OEt)$_2$	n-BuLi	R^{1}COR^2	R^{1}R^{2}C=CHSOMe	4.237
LiCHSO$_2$Ar $\mid$ P(OEt)$_2$	n-BuLi	R^{1}COR^2	R^{1}R^{2}C=CHSO$_2$Ar	4.238
LiCR(SePh) $\mid$ PO(OEt)$_2$	n-BuLi	R^{1}COR^2	R^{1}R^{2}C=CR(SePh)	4.239

[a] See footnote a of Table 4.6.

Table 4.12 Organometallics of Group IA and Group II Metals Containing α-Silicon Substituents[a]

Compound	Method of Preparation	Ref.	Compound	Method of Preparation	Ref.
LiCHPh $\mid$ SiMe$_3$	n-BuLi	4.240	LiCHP(S)Ph$_2$ $\mid$ SiMe$_3$	n-BuLi	4.240
LiCHR^1 $\mid$ SiR$_3$	n-BuLi (Li $\rightleftharpoons$ Se)	4.241	LiCHPO(OEt)$_2$ $\mid$ SiMe$_3$	n-BuLi	4.252
LiCHCH$_2$R$_1$ $\mid$ SiR$_3$	R$_3$SiCH=CH$_2$ +RLi	4.242	NaCH(SiMe$_3$)$_2$	NaOMe HMPA	4.250
LiCHCH=CH$_2$ $\mid$ SiMe$_3$	n-BuLi	4.243	LiC(SiMe$_3$)$_3$	LiOMe, MeLi	4.250 4.244
LiC=CH$_2$ $\mid$ SiMe$_3$	t-BuLi (Li $\rightleftharpoons$ Br)	4.244	LiCR(SiMe$_3$)$_2$		4.251

150

(continued...)

Table 4.12   continued

Compound	Method of Preparation	Ref.	Compound	Method of Preparation	Ref.
LiCHCOOR[1] $\|$ SiR$_3$	LDA	4.245	LiCHSMe $\|$ SiMe$_3$	n-BuLi	4.252
LiCHCOOLi $\|$ SiMe$_3$		4.246	LiCHSPh $\|$ SiMe$_3$	n-BuLi	4.253
LiCHCN $\|$ SiMe$_3$		4.247	LiCHSOPh $\|$ SiMe$_3$	n-BuLi	4.240
LiCHCONR$_2$ $\|$ SiMe$_3$	BuLi or LDA	4.248	LiC(SR)$_2$ $\|$ SiMe$_3$	n-BuLi	4.254
LiCMeCH=NR $\|$ SiMe$_3$	LDA	4.249	Li⟨S–S⟩ SiMe$_3$	n-BuLi	4.255
LiCHPPh$_2$ $\|$ SiMe$_3$	n-BuLi	4.240			

$\underline{a}$ See footnote $\underline{a}$ of Table 4.6.

4.3.3.1   "Umpolung" -- Charge Affinity Inversion

The carbonyl and neighboring carbon atoms of carbonyl compounds
normally act as either nucleophilic or electrophilic centers as
shown in 4.13. This charge affinity pattern can be reversed, as
shown in 4.14 through the use of various "masked" carbonyl compounds. The charge affinity inversion operation has been termed
"umpolung" by Seebach (4.256).

$$
\begin{array}{c}
\overset{\displaystyle O}{\overset{\|}{-C}}-C-C-C-C-C \text{-------} \\
+ \ - \ + \ - \ + \ - \\
\mathbf{4.13}
\end{array}
\qquad
\begin{array}{c}
\overset{\displaystyle O}{\overset{\|}{-C}}-C-C-C-C-C \text{-------} \\
- \ + \ - \ + \ - \ + \\
\mathbf{4.14}
\end{array}
$$

<u>Masked Carbonyl Compounds or Acyl Anion Equivalents</u>. Our brief
discussion of this important subject may be supplemented by re-
ferring to extensive reviews by Seebach (4.256, 4.257) and Lever
(4.258).

(a) Cyanides, nitronates, and other classical acyl anion equivalents. For many decades, metal cyanides and metal nitronates have been used as acyl anion equivalents. Note that these compounds are, in a formal sense, α-hetero-substituted organometallic species. A few representative carbon-carbon bond-forming reactions of metal cyanides are shown in eq. 4.103. The cyano group can be readily converted into various carbonyl groups either by reduction or by hydrolysis.

$$
MCN \quad
\begin{cases}
\xrightarrow{\quad RX \quad} & RCN \\[2em]
\xrightarrow{\quad R^1COR^2 \quad} & R^1\!\!\diagdown\!\!\overset{OM}{\underset{R^2\diagup}{C}}\!\!\diagup\!\!CN \\[2em]
\xrightarrow{\quad -\overset{|}{C}=\overset{|}{C}-\overset{|}{C}=O \quad} & NC-\overset{|}{\underset{|}{C}}-\overset{|}{\underset{|}{C}}-\overset{|}{C}=O
\end{cases}
\qquad (4.103)
$$

One practical difficulty in the use of alkali metal cyanides, such as NaCN and KCN, lies in their extreme insolubility in typical organic solvents. It has, however, been at least partially solved by the development of (1) polar aprotic solvents, such as DMF, DMSO, and HMPA, and (2) phase transfer catalysts (4.259). The formation of cyanohydrin can be facilitated by Me$_3$SiCN (4.260), and various difficulties encountered in the conjugate addition reaction of alkali metal cyanides can be overcome by Et$_2$AlCN, as discussed in Sect. 5.3.3.2.

As might be expected from the relatively high acidity of nitroalkanes (e.g., pK_a of CH$_3$NO$_2$ = 10), metal nitronates (4.15) readily undergo conjugate addition reactions with α,β-unsaturated carbonyl compounds (eq. 4.104).

$$
RCH_2NO_2 \xrightarrow{MX} RCH=N^{+}\!\!\underset{O^{-}}{\overset{O^{-}}{\diagup\!\!\diagdown}} M^{+} \xrightarrow{-\overset{|}{C}=\overset{|}{C}-\overset{|}{C}=O} O_2N-\overset{R}{\underset{H}{C}}-\overset{|}{\underset{|}{C}}-\overset{|}{\underset{|}{C}}-\overset{|}{C}=O
$$

$$
\underline{4.15} \qquad\qquad (4.104)
$$

The δ-carbonyl nitroalkane products can be converted into 1,4-dicarbonyl compounds by various methods. The classical Nef reaction (4.261), which involves treatment with NaOH followed by mineral acids, is often incompatible with the carbonyl and other functional groups present in the products. Consequently, a number of alternate procedures have been developed. Treatment of nitroalkanes with aqueous TiCl$_3$ (4.227), which presumably proceeds as shown in eq. 4.105, appears particularly promising.

$$R^1R^2CHNO_2 \xrightarrow[H_2O-THF]{TiCl_3} R^1R^2CHN^+{=}^{O}_{O-Ti^{2+}} \longrightarrow R^1R^2CHN{=}O \quad (4.105)$$

$$\longrightarrow R^1R^2C{=}NOH \xrightarrow{TiCl_3} R^1R^2C{=}NH \xrightarrow{H_2O} R^1R^2C{=}O$$

Although no detailed discussion is intended, it may be pointed out here that alkynylmetals and alkenylmetals can also act as acyl anion equivalents, as exemplified by eqs. 4.106 and 4.107.

$$RC{\equiv}CLi \xrightarrow{R^1COR^2} RC{\equiv}C{-}\underset{OH}{C}R^1R^2 \xrightarrow{[O]} RCH_2\underset{O\ OH}{C}{-}CR^1R^2 \quad (4.106)$$

$$R^1R^2C{=}CHLi \xrightarrow{R^3X} R^1R^2C{=}CHR^3 \xrightarrow[2.\ [O]]{1.\ HB} R^1R^2CHCOR^3 \quad (4.107)$$

(b) Other α-hetero-substituted organometallics used as acyl anion equivalents. Extensive investigations pioneered by Corey, Seebach, and others over the past decade or two have established that a wide variety of α-hetero-substituted organometallics can serve as acyl anion equivalents. Some representative examples are shown in Table 4.13.

Table 4.13   Some Representative Acyl Anion Equivalents

Compound	Ref.	Compound	Ref.
(Li,R / S,S ring)	4.202	$LiC(SPh)_3$	4.215
	4.203	$Li,R^1{-}C{-}SR,COOH$	4.196
(Li,R / S,O ring)	4.158		
(Li,R / S,S ring)	4.204	$Li\underset{SEt}{C}{=}CH_2$	4.167
		$Li\underset{OEt}{C}{=}CH_2$	4.148
(Li,R / S,S-S ring)	4.207	$M,R{-}C{-}OX,CN$	4.150

(continued...)

Table 4.13   continued

Compound	Ref.	Compound	Ref.
R, Li–C(S)$_2$–benzene (dithiolane)	4.208	$\underset{R}{\overset{Li}{>}}C\underset{SePh}{\overset{SiMe_3}{<}}$	4.222
R, Li–C(S-CH$_2$)$_2$–dimethylbenzene	4.209	$LiCH_2N(NO)R$	4.223
		$Li\overset{}{C}=NR$, R^1	4.231
Li–benzothiazole (N,S)	4.217	$Li\overset{}{C}=CH_2$, $SiMe_3$	4.244
$\underset{R^1}{\overset{Li}{>}}C(SR)_2$	4.210		
$\underset{R^1}{\overset{Li}{>}}C\underset{SR}{\overset{SOR}{<}}$	4.213 4.214		
$\underset{R}{\overset{Li}{>}}C\underset{SMe}{\overset{S-C(S)NMe_2}{<}}$	4.212		

Before we discuss the chemistry of masked carbonyl anions, a brief discussion of "unmasked" carbonyl anions is presented here. It has recently become feasible to deprotonate nonenolizable formyl derivatives to give metal carbonylates of alkali metals. Thus Schöllkopf (4.262) and Seebach (4.263) have treated DMF and its thio derivative, respectively, with LDA at low temperatures (<-78°C) and obtained unstable lithio derivatives which undergo carbonyl addition reactions (eq. 4.108).

$$\underset{X}{\overset{}{HCNMe_2}} \xrightarrow[\underline{<}-78°C]{LDA} \underset{X}{\overset{}{LiCNMe_2}} \xrightarrow{R^1COR^2} \underset{HO \;\; X}{\overset{}{R^1R^2C-CNMe_2}} \quad (4.108)$$

$$X = O \text{ and } S$$

Alternatively carbonylation of certain lithium amides has provided products which behave as lithioformamides, as shown in eq. 4.109 (4.264).

$$LiNHBu\text{-}t \xrightarrow{CO} LiCONHBu\text{-}t \xrightarrow{MeSiCl} Me_3SiCONHBu\text{-}t \quad (4.109)$$

While these metal carbonylates of main group metals are interesting, their synthetic values are severely limited due to their extreme instability. On the other hand, acyltransition metal compounds, such as $NaRCOFe(CO)_4$ and $LiRCONi(CO)_3$, have proved to be highly useful intermediates, as discussed in detail in Sect. 11.2.

Dithiane and related cyclic and acylic S,S-acetal derivatives. Dithiane (4.16) is probably the most widely used acyl anion equivalent. Since its use as an acyl anion equivalent was reported by Corey and Seebach (4.202a) in 1965, it has been applied to the synthesis of a wide variety of organic compounds including natural products. These results have been extensively reviewed by Seebach (4.203).

Conversion of dithiane (pK_a = 31) and its organo-substituted derivatives into the corresponding 2-lithiodithianes is usually achieved by treating them with n-BuLi in THF or other organic solvents at -30 to $0^{\circ}C$ (eq. 4.110).

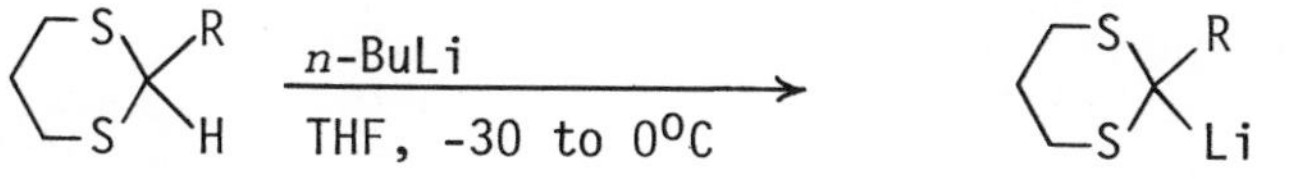

$$\text{(4.110)}$$

4.16

Some representative reactions of 2-lithiodithiane derivatives are summarized in Scheme 4.2.

Scheme 4.2

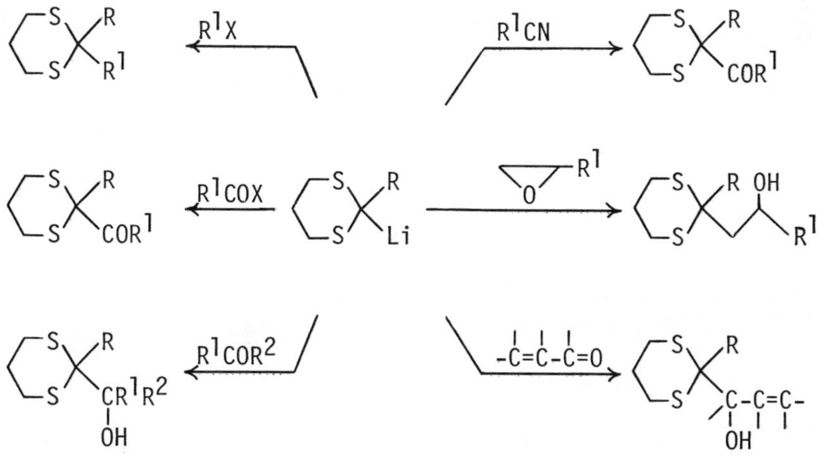

It should be noted that 2-lithiodithiane and its derivatives normally undergo only 1,2-addition reactions with α,β-unsaturated carbonyl compounds.

The dithiane ring is relatively stable and can withstand various conditions required for other reactions. This often represents an advantage as well as a disadvantage. Its conversion to the carbonyl group has presented a difficulty in some cases. Over the past several years, a number of methods for this transformation have been developed. The reagents used in these procedures include HgO-BF$_3$·OEt$_2$ (4.265), CuCl$_2$-CuO-aq acetone (4.266), TiCl$_4$-HOAc-H$_2$O-CHCl$_3$ (4.267), Tl(OOCCF$_3$)$_3$ (4.268), Ce(NH$_4$)$_2$(NO$_3$)$_6$ (4.269), chloramine T-H$_2$O-MeOH (4.270), mesityl-SO$_2$ONH$_2$ (4.271), NCS and NBS (4.272), and various alkylating agents, such as MeI, MeOSO$_2$F, and Meerwein salts (4.273).

As shown in Table 4.13, a number of other cyclic and acyclic S,S-acetals have been suggested as dithiane substitutes. Relative advantages and disadvantages of these acyl anion equivalents are not very clear, however, because experimental data for comparison are usually not available. Lithiodithiolane is too unstable to be of practical use. One noteworthy advantage of using acyclic S,S-acetals is being able to achieve the 1,4-addition with α,β-unsaturated carbonyl compounds which tend to undergo the 1,2-addition reaction with the corresponding dithiane derivatives, as shown in eq. 4.111 (4.274).

(4.111)

Both cyclic and acyclic S,S-acetals can be totally desulfurized by Raney nickel (4.275), LiAlH$_4$-CuCl$_2$-ZnCl$_2$ (4.276), LiAlH$_4$-TiCl$_4$ (4.277), and hydrazine (4.278). On the other hand, treatment with calcium in liquid NH$_3$ converts S,S-acetals into sulfides; see eq. 4.112 (4.279).

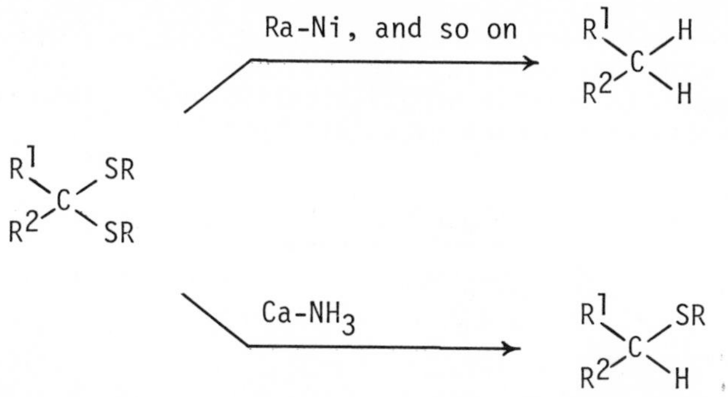

$$\xrightarrow{\text{Ra-Ni, and so on}} \quad \begin{array}{c} R^1 \\ \diagdown \\ R^2 \diagup \end{array} C \begin{array}{c} H \\ \diagup \\ \diagdown H \end{array}$$

$$\begin{array}{c} R^1 \\ \diagdown \\ R^2 \diagup \end{array} C \begin{array}{c} SR \\ \diagup \\ \diagdown SR \end{array}$$

(4.112)

$$\xrightarrow{\text{Ca-NH}_3} \quad \begin{array}{c} R^1 \\ \diagdown \\ R^2 \diagup \end{array} C \begin{array}{c} SR \\ \diagup \\ \diagdown H \end{array}$$

Methyl methylthiomethyl sulfoxide and related S,S-acetal S-oxides. Methyl methylthiomethyl sulfoxide (MMTS) (4.213) and its ethyl homologue (4.214) have been developed as useful alternatives to dithiane and related S,S-acetals (eq. 4.113). Although the ethyl derivative is less readily available of the two, it is reported to be superior to MMTS in some reactions presumably due to the absence of competitive methyl hydrogen abstraction.

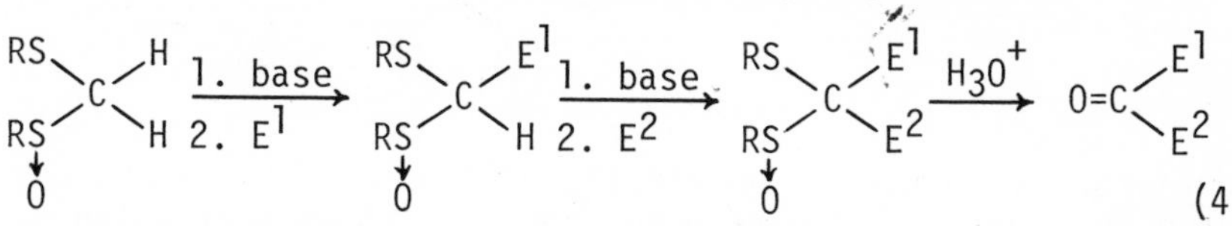

$$\begin{array}{c} RS \\ \diagdown \\ RS \diagup \\ \downarrow \\ O \end{array} C \begin{array}{c} H \\ \diagup \\ \diagdown H \end{array} \xrightarrow[\text{2. } E^1]{\text{1. base}} \begin{array}{c} RS \\ \diagdown \\ RS \diagup \\ \downarrow \\ O \end{array} C \begin{array}{c} E^1 \\ \diagup \\ \diagdown H \end{array} \xrightarrow[\text{2. } E^2]{\text{1. base}} \begin{array}{c} RS \\ \diagdown \\ RS \diagup \\ \downarrow \\ O \end{array} C \begin{array}{c} E^1 \\ \diagup \\ \diagdown E^2 \end{array} \xrightarrow{H_3O^+} \quad O=C \begin{array}{c} E^1 \\ \diagup \\ \diagdown E^2 \end{array}$$

(4.113)

R = Me or Et

These S,S-acetal S-oxides are considerably more acidic than S,S-acetals. Consequently, generation of their metallated derivatives can be achieved with a variety of bases, such as n-BuLi, LDA, NaH, and KH.

The reactions of these metallated species are generally quite analogous to those of lithiodithianes. The following differences should be clearly noted, however. Whereas metallated S, S-acetals in general are quite reluctant to undergo the 1,4-addition reaction with α,β-unsaturated carbonyl compounds, the corresponding S,S-acetal S-oxides readily undergo the 1,4-addition reaction (4.280). The 1,4/1,2 ratio of a given reaction, however, is affected by various factors. Thus it has been reported that the lithio derivative of MMTS undergoes predominantly the 1,2-addition reaction with cyclohexenone, whereas the 1,4-addition predominates in the case of cyclopentenone (4.280b). Another difference to be noted is that, whereas the reaction of lithiodithiane with cyanides is often complicated by α-hydrogen abstraction, the corresponding reaction of S,S-acetal S-oxides is straightforward.

Since sulfoxides are chiral, S,S-acetal S-oxides are often obtained as diastereomeric mixtures, which could represent an inconvenience. Conversion of S,S-acetal S-oxides into carbonyl compounds can be achieved by acidic hydrolysis.

$$(4.114)$$

Others. As shown in Table 4.13, various other α-hetero-substituted organometallics have been shown to act as acyl anion equivalents. Several notable examples are shown in eqs. 4.115 to 4.119.

$$(4.115)$$

$$(4.116)$$

$$\underset{\underset{PO(OEt)_2}{|}}{CHR(SMe)} \xrightarrow[\text{Ref. } 2.236]{n\text{-BuLi}} \underset{\underset{PO(OEt)_2}{|}}{LiCR(SMe)} \xrightarrow{R^1CHO} R^1CH=CR(SMe) \qquad (4.117)$$

$$R^1CH=CR(SMe) \xrightarrow[\text{MeCN}]{HgCl_2,\ H_2O} R^1CH_2COR$$

$$C\equiv N\!\!\diagdown\!\!\diagup\!\!\diagdown\!\!\diagup \xrightarrow[\text{Ref. } 4.231]{RLi} \underset{\underset{N}{\|}}{LiCR}\diagdown\!\!\diagup\!\!\diagdown\!\!\diagup \qquad (4.118)$$

$$\xrightarrow[\text{2. } H_2O]{\text{1. PhCHO}} \underset{\underset{OH}{|}}{PhCHCOR}$$

$$\xrightarrow[\text{2. } H_2O]{\text{1. } \triangle R^1} \underset{\underset{OH}{|}}{R^1CHCH_2COR}$$

$$\underset{\underset{SePh}{|}}{LiCHSiMe_3} \xrightarrow{RCH_2Br} \underset{\underset{SePh}{|}}{RCH_2CHSiMe_3} \xrightarrow{30\%\ H_2O_2} RCH_2CHO \qquad (4.119)$$

$$\underset{\underset{SePh}{|}}{CH_2SiMe_3} \xrightarrow[\text{Ref. } 4.222]{LDA} \ \ \uparrow$$

$$\Big\downarrow 30\%\ H_2O_2 \qquad\qquad \uparrow H_2O$$

$$[RCH=CHSiMe_3] \xrightarrow{30\%\ H_2O_2} \Big[RCH\overset{O}{\triangle}CHSiMe_3\Big]$$

It is instructive to note that, except in eq. 4.118, hetero-substituted alkenes act as key intermediates, which are converted into the corresponding carbonyl compounds via oxidation and/or hydrolysis (Scheme 4.3). Similarly, alkenylboranes can be oxidized to give aldehydes and ketones (Sect. 5.2.3.2).

As mentioned earlier, it is often difficult to identify the most suitable masked carbonyl reagent for a given synthetic task. Nevertheless, various reagents discussed above seem to add considerable flexibility to the carbon-carbon bond formation via acyl anion equivalents.

Scheme 4.3

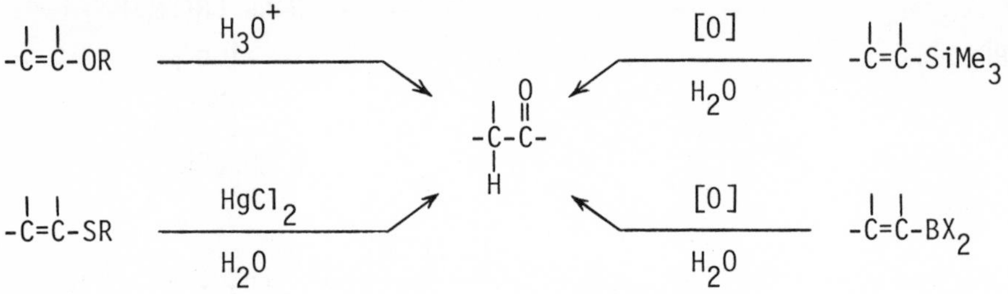

$$-\overset{|}{C}=\overset{|}{C}-OR \xrightarrow{\text{H}_3\text{O}^+}$$

$$-\overset{|}{C}=\overset{|}{C}-SR \xrightarrow[\text{H}_2\text{O}]{\text{HgCl}_2}$$

$$-\overset{|}{\underset{|}{C}}-\overset{O}{\overset{||}{C}}-$$
$$H$$

$$\xleftarrow[\text{H}_2\text{O}]{[O]} -\overset{|}{C}=\overset{|}{C}-SiMe_3$$

$$\xleftarrow[\text{H}_2\text{O}]{[O]} -\overset{|}{C}=\overset{|}{C}-BX_2$$

1,3 Charge Affinity Inversion Operations. Generation and use of acyl anion equivalents represent but one of many possible types of charge affinity inversion operations. To classify various types of normal and charge-affinity-inverted reagents, Seebach (4.257) has proposed the terms E^n and N^n to represent electron-donating and electron-accepting species, respectively, that have a reactive center on the nth carbon atom. Several examples of normal and charge-affinity-inverted reagents are shown below.

Normal reagent:

$$N^1 \qquad\qquad E^2 \qquad\qquad N^3$$

$$RC^{\delta+}-X^{\delta-} \qquad RC-\overset{|}{C}^{\delta-}-M^{\delta+} \qquad RC-C^{\delta-}=C^{\delta+}-$$
$$\overset{||}{O} \qquad\qquad \overset{||}{O}\,\overset{|}{} \qquad\qquad \overset{||}{O}\,\overset{|}{}\,\overset{|}{}$$

Charge-affinity-
inverted reagent:

$$E^1 \qquad\qquad N^2 \qquad\qquad E^3$$

$$RC^{\delta-}-M^{\delta+} \qquad RC-\overset{|}{C}^{\delta+}-X^{\delta-} \qquad RC-\overset{|}{C}-\overset{|}{C}^{\delta-}-M^{\delta+}$$
$$\overset{||}{O} \qquad\qquad \overset{||}{O}\,\overset{|}{} \qquad\qquad \overset{||}{O}\,\overset{|}{}\,\overset{|}{}$$

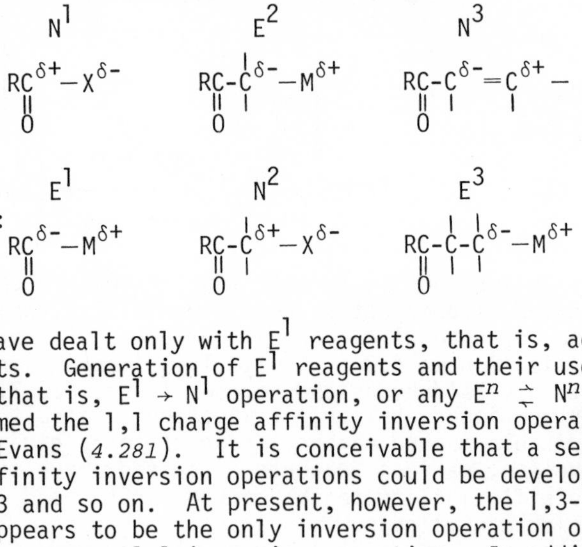

So far we have dealt only with E^1 reagents, that is, acyl antion equivalents. Generation of E^1 reagents and their use as N^1 substitutes, that is, $E^1 \to N^1$ operation, or any $E^n \rightleftharpoons N^n$ operation may be termed the 1,1 charge affinity inversion operation, as suggested by Evans (4.281). It is conceivable that a series of 1,n charge affinity inversion operations could be developed, where n = 1, 2, 3 and so on. At present, however, the 1,3-inversion operation appears to be the only inversion operation other than the much more common 1,1-inversion operation. In addition to the reviews by Seebach (4.256, 4.257) and Lever (4.258) cited above, a review by Evans (4.281) presents a lively discussion of the 1,3-inversion operation.

Alkylation on the alkenyl carbon atom of 2-cyclopenten-1,4-diol with an alkyl halide is not a normal synthetic operation. It can, however, be achieved via a 1,3-inversion operation, as shown in eq. 4.120 (4.183).

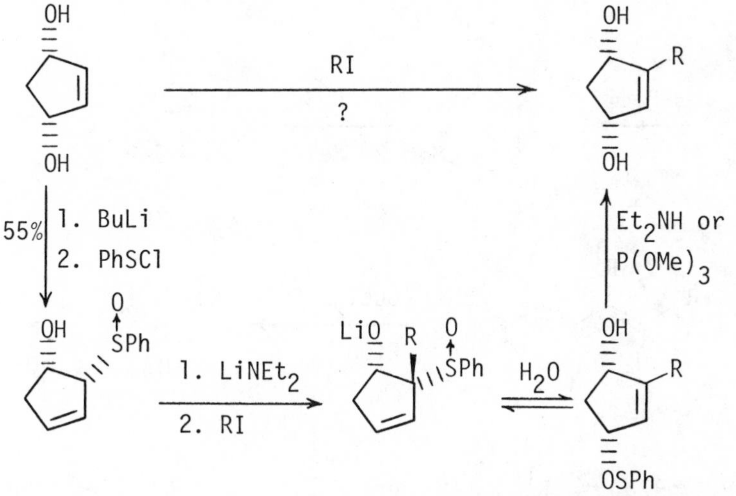

(4.120)

This principle has been applied to the synthesis of a number of γ-substituted allylic alcohols. It should be clearly noted that the 2,3-shift of allylic sulfoxides and sulfenates plays a crucial role in the sequence shown in eq. 4.120.

Other Carbon-Carbon Bond-Forming Reactions of α-Hetero-Substituted Organometallics Involving Sigmatropic Shifts. Other 2,3- and 3,3-shifts have been combined with the alkylation of α-oxy (4.282), α-thio (4.172, 4.175, 4.176), and α-seleno (4.220) organolithiums to develop stereoselective procedures for the synthesis of olefins.

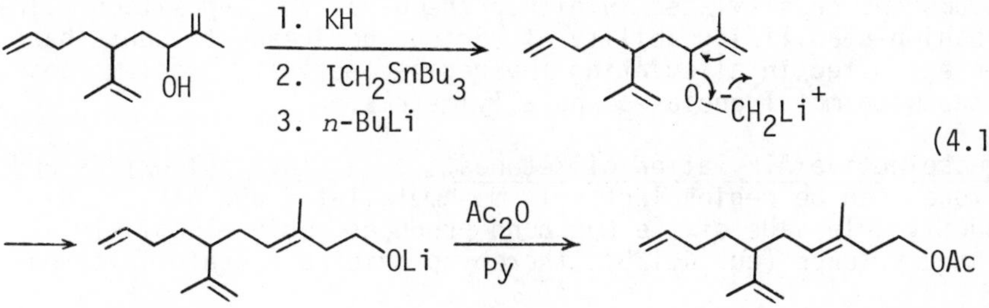

(4.121)

$$(4.122)$$

$$(4.123)$$

$$(4.124)$$

4.3.2.2 Regioselective Carbon-Carbon Bond Formation via α-Hetero-Substituted Organometallics of Group IA and Group II Metals

Alkylation of ketones can take place on either of the two α-carbon atoms. In fact, it is often difficult to achieve a highly regioselective alkylation. Likewise, allyllithiums and allylmagnesiums can be alkylated in either the α- or the γ-position. The carbanion-stabilizing ability of various heterosubstituents has been exploited in alleviating the regioselectivity problems associated with metal enolates and allylmetals.

Regioselective Alkylation of Ketones. β-Carbonyl sulfoxides and sulfones can be regioselectively monometallated and alkylated. Reductive cleavage of the C-S bond produces regioselectively alkylated ketones (eq. 4.125), thereby providing a useful alterna-

tive to the classic alkylation reaction of malonic and acetoacetic esters.

$$R^1CH_2COCH_2SO_nR \xrightarrow[\text{or LDA}]{n\text{-BuLi}} R^1CH_2COCHSO_nR \xrightarrow[\text{2. Al(Hg)}]{1.\ R^2X} R^1CH_2COCH_2R^2$$
$$\underset{Li}{\mid}$$

$(n = 1$ or $2)$ (4.125)

The corresponding dianions permit monoalkylation on the less
acidic α-carbon atom (eq. 4.126).

$$R^1CH_2COCH_2SO_2R \xrightarrow[\text{2. }n\text{-BuLi}]{1.\ \text{LDA}} R^1\underset{Li}{\underset{\mid}{C}}HCO\underset{Li}{\underset{\mid}{C}}HSO_2R \xrightarrow[\text{2. H}_2\text{O}]{1.\ R^2X} R^1R^2CHCOCH_2SO_2R$$

$$\xrightarrow{\text{Al(Hg)}} R^1R^2CHCOCH_3$$

                                                        (4.126)

It is also possible to dialkylate the dilithio derivatives, as
shown in eq. 4.127 (*4.283*).

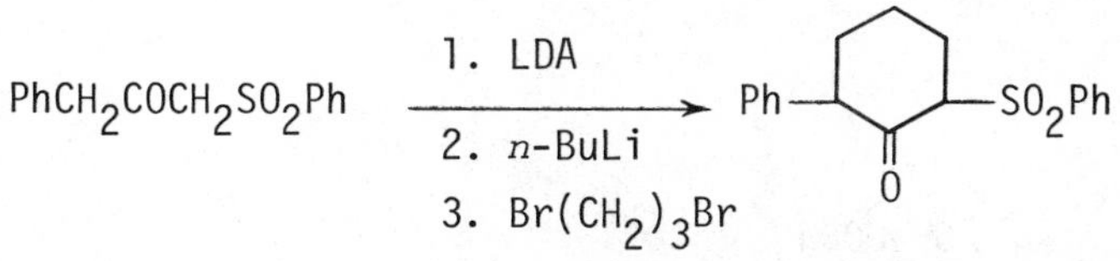

$$PhCH_2COCH_2SO_2Ph \xrightarrow[\substack{\text{2. }n\text{-BuLi} \\ \text{3. Br(CH}_2)_3\text{Br}}]{1.\ \text{LDA}}$$

                                    (4.127)

<u>Regioselective Carbon-Carbon Bond Formation Involving Allylic</u>
<u>Organoalkali Metals</u>. Several extensive reviews on this subject
are available (*4.20*, *4.257*, *4.258*, *4.281*). As already indicated
in the preceding section, α-hetero-substituted allyllithiums can
undergo, in favorable cases, highly regioselective carbon-carbon
bond formation (eq. 4.128).

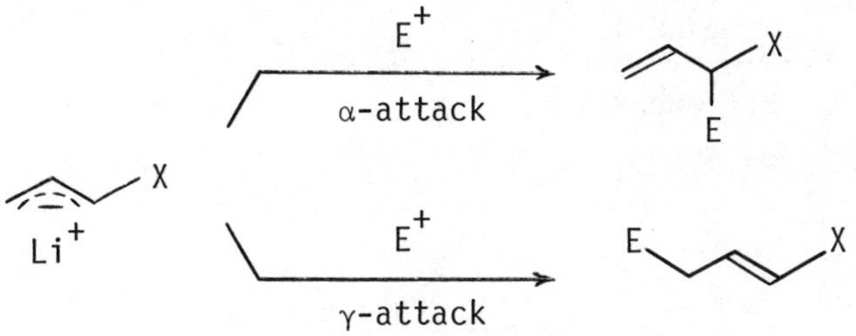

                                    (4.128)

The α/γ ratio, however, is a function of a number of factors including the nature of X, and the nature of other substituents,
gegenions, electrophiles, solvents, and additives. Unlike the
alkylation of metallated β-carbonyl sulfoxides and sulfones, it
is often difficult to attain a high regioselectivity in the reaction of α-hetero-substituted allylmetals. The following tentative generalizations concerning the α/γ ratio are presented here
as a crude empirical guide.

    (a) α-Hetero substituents. The heteroatom groups SR, SOR,
SO_2R, SeR, and POX_2 tend to direct electrophilic attack to the
α-position, whereas the OR and NR_2 groups favor attack at the γ-
position. The α/γ ratio tends to increase in the order: SR <
SOR < SO_2R. Thus whereas the lithium derivative of an allylic
sulfide reacts with aldehydes and ketones at both α- and γ-posi-
tions, the corresponding sulfone tends to react exclusively at
the α-position, for example, eqs. 4.129 and 4.130 (*4.284*).

$$32\% \qquad 49\% \quad (4.129)$$

$$(4.130)$$
$$71\%$$

Replacement of the phenyl group of phenyl sulfides with nitrogen-
containing heteroaromatic rings increases the α/γ ratio, presumably through chelation; see eq. 4.131 (*4.281*).

$$(4.131)$$

X	Yield (%)	α (%)	γ (%)
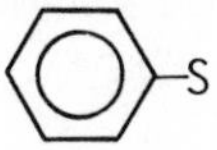 Ph–S	84	75	25
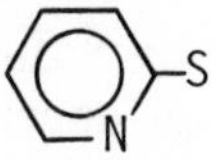 pyridyl–S	79	90	10
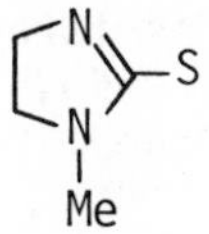 imidazolidinyl–S (N-Me)	86	92	8

(b) Substituents on allylic atoms.   The α/γ ratio depends on the substitution pattern in the allylic moiety, as summarized below for the ethylation reaction of allylic sulfoxides (*4.281*).

PhSO⌒⌒
α/γ = 2.9

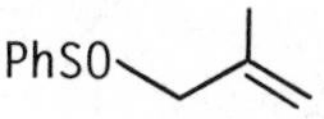

PhSO⌒⤙
α/γ = 2.4

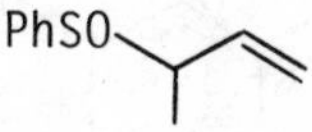

PhSO⌒⌒
α/γ = 2.5

PhSO⌒⌒⌒
α/γ = 6.7

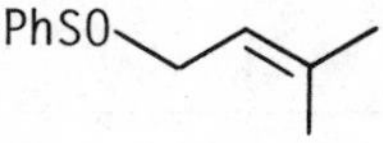

PhSO⌒⌒⤙
α/γ >> 10

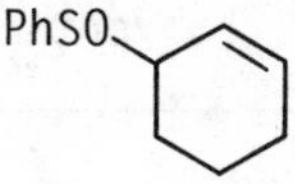

PhSO⬡
α/γ >> 10

(c) Gegenions.   The dilithio derivative of allylmercaptan reacts with electrophiles to give mixtures of α- and γ-products, with the latter predominating (eq. 4.132), whereas the magnesium derivative gives the α-products exclusively; see eq. 4.133 (*4.257*).

(4.132)

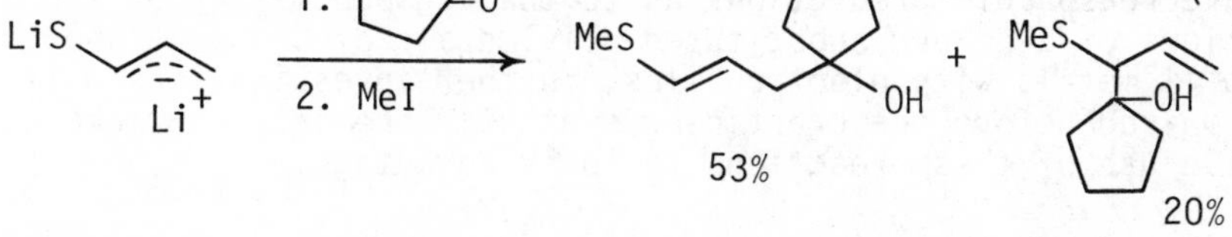

$$Mg^{++}S^- \diagup\!\!\!\diagdown\!\!\!\diagup \xrightarrow[\text{2. MeI}]{\text{1.} \quad \square\!\!=\!\!O} \quad MeS\diagup\!\!\!\diagup \text{ only (76\%)} \quad (4.133)$$

(d) Electrophiles. The α/γ ratio is often very much dependent on the nature of electrophiles. For example, whereas the
alkylation of the lithium derivatives of allylic sulfides takes
place predominantly at the α-position, for example, eq. 4.44,
their reaction with carbonyl compounds tends to occur predominantly in the γ-position, for example, eq. 4.129.

(e) Solvents and additives. The α/γ ratio in the reaction
of lithium derivatives of allylic sulfides with carbonyl compounds also depends on the solvent used as well as strongly coordinating additives, as exemplified by the results shown in eq.
4.134 (*4.285*).

$$\begin{array}{c} \diagup\!\!\!\diagdown\!\!\!\diagup^{SPh} \\ \underset{Li}{\diagdown} \end{array} \xrightarrow{CH_3COCH_3} \begin{array}{c} \diagup\!\!\!\diagdown\!\!\!\diagup^{SPh} \\ \diagdown_{OH} \end{array} + \begin{array}{c} \diagup\!\!\!\diagdown\!\!\!\diagup^{SPh} \\ \diagdown_{OH} \end{array} \quad (4.134)$$

Solvent	Additive	α (%)	γ (%)
THF	-	25	75
THF	DABCO	0	100
THF	HMPA	40	60
THF	[2.2.2] cryptate	100	0

In summary, while the currently available data may allow us
to make reasonable predictions as to the regiochemistry of the
reactions of α-hetero-substituted allylmetals of Group IA and
Group II metals with electrophiles, further investigations will
be required before the practicing synthetic chemists can make extensive use of these reactions to their advantage.

4.3.3.3   Olefination Reactions Involving α-Hetero-Substituted Organometallics of Group IA and Group II Metals

Some applications of α-hetero-substituted organolithiums to the synthesis of olefins via olefin interconversion are discussed in Sect. 4.3.3.1. In this section, applications of α-hetero-substituted organometallics to the synthesis of olefins via olefination are briefly discussed. A few synthetic strategies that have been exploited in the olefination of α-hetero-substituted organoalkali metals are shown below:

1. Carbonyl olefination reactions

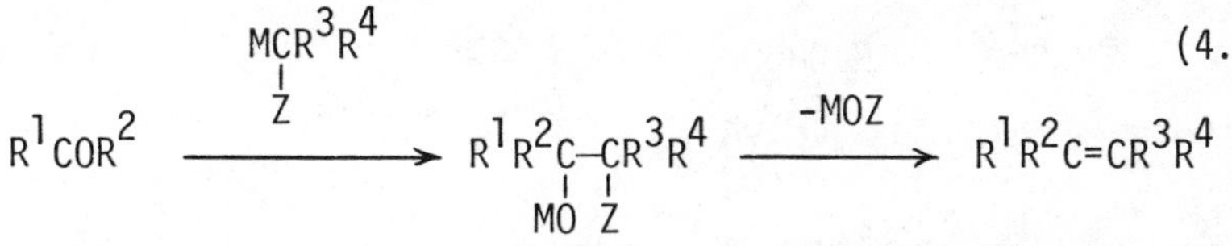

$$R^1COR^2 \xrightarrow[\substack{Z}]{\substack{MCR^3R^4}} R^1R^2C-CR^3R^4 \xrightarrow{-MOZ} R^1R^2C=CR^3R^4 \qquad (4.135)$$
$$\underset{MO \; Z}{}$$

Z = P, Si, and S groups,   M = alkali metals

2. Olefination via 2,3-shift reactions

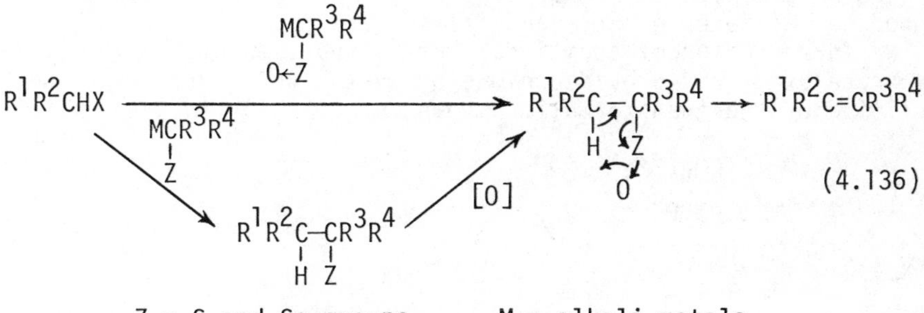

$$R^1R^2CHX \xrightarrow[\substack{MCR^3R^4 \\ | \\ Z}]{\substack{MCR^3R^4 \\ | \\ O \leftarrow Z}} R^1R^2C-CR^3R^4 \rightarrow R^1R^2C=CR^3R^4$$

$$\searrow R^1R^2C-CR^3R^4 \xrightarrow{[O]} \qquad (4.136)$$
$$\underset{H \; Z}{}$$

Z = S and Se groups,   M = alkali metals

3. Olefination via sulfur extrusion

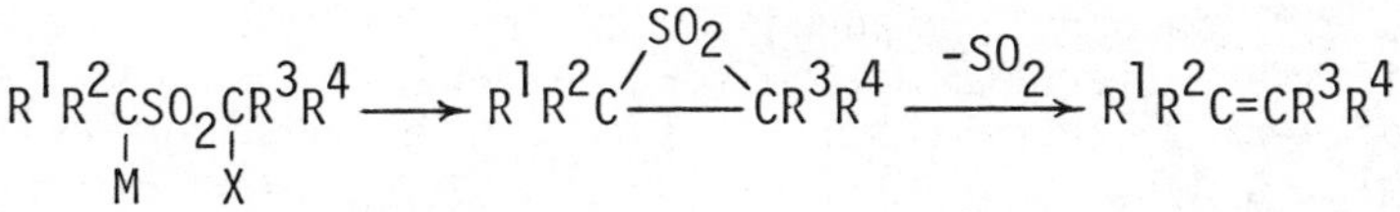

$$R^1R^2CSO_2CR^3R^4 \longrightarrow R^1R^2C\overset{SO_2}{-}CR^3R^4 \xrightarrow{-SO_2} R^1R^2C=CR^3R^4 \qquad (4.137)$$
$$\underset{M \quad X}{}$$

M = alkali metals

Carbonyl Olefination Reactions. Although the Wittig reaction is probably the most commonly used carbonyl olefination reaction, it is not an organometallic reaction, and is therefore outside the scope of this book. The reader is referred to reviews by Maercker (*4.286*), Schlosser (*4.287*), and Johnson (*4.288*).

(a) The Wadsworth-Emmons reaction. The carbonyl olefination reaction of α-metallated phosphonates, called the Wadsworth-Emmons reaction, is a useful modification of the Wittig reaction. Some representative examples are summarized in Table 4.11. In cases where the carbonyl compound is an aldehyde, the Wadsworth-Emmons reaction often, but not always, leads to the formation of the (E)-isomer in a highly stereoselective manner (~95% selectivity), as shown in eq. 4.138 (*4.233*).

$$CH_3CHO \;+\; \underset{\underset{PO(OEt)_2}{|}}{NaCHCOOMe} \;\xrightarrow{\;\;DME\;\;}\; \underset{96\% \text{ E}}{\overset{CH_3}{\underset{H}{>}}C{=}C\overset{H}{\underset{COOMe}{<}}} \qquad (4.138)$$

The stereoselectivity of the Wadsworth-Emmons reaction appears to be generally higher than that of the corresponding Wittig reaction. The reaction of α-metallated phosphonates with ketones, however, in general, is not stereoselective. Even in the olefination of aldehydes, the use of sterically hindered phosphonates produces mixtures of (E)- and (Z)-isomers (*4.233d*).

Introduction of additional hetero substituents in the phosphonate ester can provide unique synthetic procedures as exemplified by the transformation shown in eq. 4.139 (*4.289*).

$$\bigcirc{=}0 \;\xrightarrow[PO(OEt)_2]{LiCMe(SMe)}\; \bigcirc{-}C\overset{Me}{\underset{SMe}{<}} \;\xrightarrow[\underset{MeCN}{H_2O}]{HgCl_2}\; \bigcirc{-}COCH_3 \qquad (4.139)$$

(b) The Peterson reaction. The reaction of α-silylorganoalkali metals with carbonyl compounds, known as the Peterson reaction, provides an alternate method for the carbonyl olefination. A variety of α-silylorganoalkali metals, such as those summarized in Table 4.12, undergo this olefination reaction. Several representative examples are shown in Scheme 4.4.

Unlike the Wadsworth-Emmons reaction, the Peterson reaction is seldom highly stereoselective. It does, however, provide a convenient route to certain hetero-substituted olefins, as shown

Scheme 4.4

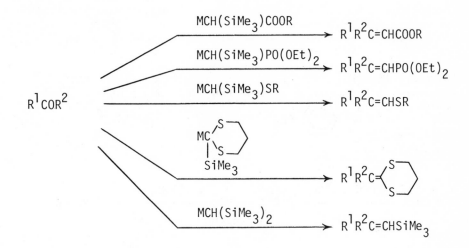

in Scheme 4.4. It is important to notice that the olefin-forming ability of a trialkylsilyl group is greater than that of a phosphonate group. It should also be noted that the reaction of α-halo-α-silylorganometals with carbonyl compounds can form either olefins or epoxides; see eqs. 4.140, 4.141 (*4.139*, *4.138*).

$$RCHO \xrightarrow[-115°C, \; THF]{LiCBr(SiMe_3)_2} \underset{\overset{|}{Br}}{RCH=CSiMe_3} \qquad (4.140)$$

$$R^1COR^2 \xrightarrow{\overset{Me}{\overset{|}{\underset{\underset{Cl}{|}}{LiCSiMe_3}}}} R^1R^2C\overset{Me}{\underset{O}{\diagdown\diagup}}CSiMe_3 \qquad (4.141)$$

(c) Miscellaneous carbonyl olefination reactions. Certain organoalkali metals containing α-sulfur substituents can undergo carbonyl olefination reactions similar to the Wadsworth-Emmons and Peterson reactions. A few representative examples follow.

$$\text{RCHO} \xrightarrow[\substack{(4.180)}]{\substack{1.\ \text{LiCH}_2\text{SON}\!\!-\!\!\bigcirc\!\!-\!\!\text{CH}_3 \\ \qquad\qquad\quad \text{Li} \\ 2.\ \text{H}_2\text{O}}} \text{RCHCH}_2\text{SONH}\!\!-\!\!\bigcirc\!\!-\!\!\text{CH}_3 \qquad (4.142)$$
$$\underset{\text{OH}}{|}$$

$$\xrightarrow[\text{reflux}]{\text{toluene}} \begin{array}{c} \text{RCH--CH} \\ |\qquad\ | \\ \text{O--S-OH} \\ \qquad | \\ \text{NH}\!-\!\bigcirc\!\!-\!\!\text{CH}_3 \end{array} \longrightarrow \text{RCH}=\text{CH}_2 + \text{SO}_2 + \text{H}_2\text{N}\!\!-\!\!\bigcirc\!\!-\!\!\text{CH}_3$$

$$R^1\text{COR}^2 \xrightarrow[\substack{(4.190)}]{\substack{\text{LiCHR} \\ | \\ \text{Ph(O)}\overset{\shortmid}{S}=\text{NMe}}} \begin{array}{c} R^1R^2\text{C}\!-\!\!-\!\text{CHR} \\ |\qquad\quad | \\ \text{LiO} \quad \text{O}=\overset{\shortmid}{S}=\text{NMe} \\ \qquad\qquad | \\ \qquad\qquad \text{Ph} \end{array} \xrightarrow[\text{H}_2\text{O, HOAc}]{\text{Al(Hg)}} R^1R^2\text{C}=\text{CHR}$$

$$(4.143)$$

$$R^1\text{COR}^2 + \begin{array}{c} \text{Li} \\ \diagdown \\ \text{EtOCS} \quad \text{O} \\ \ \ \| \\ \ \ \text{O} \end{array} \xrightarrow{\substack{(4.195)}} R^1R^2\text{C}=\!\!\diagup\!\!\searrow_{\text{O}} \qquad (4.144)$$

Olefination via 2,3-Shift Reactions. Alkyl sulfoxides and selen-oxides undergo a thermal 2,3-sigmatropic shift reaction to form olefins. Thus the combination of the alkylation of α-sulfur- or α-selenium-substituted organoalkali metals and the 2,3-shift pro-vides a method for the conversion of alkyl halides into olefins, as shown in eq. 4.136. The following examples are representa-tive.

$$\underset{\text{PhSe}\!\rightarrow\!\text{O}}{\text{LiCHR}^1} \xrightarrow[\substack{(4.219)}]{R^2\text{CH}_2\text{X}} \underset{\text{PhSe}\!\rightarrow\!\text{O}}{R^1\text{CH-CH}_2R^2} \xrightarrow{\Delta} R^1\text{CH}=\text{CHR}^2 \quad (4.145)$$

$$\underset{\text{PhS}\!\rightarrow\!\text{O}}{\text{LiCHR}^1} \xrightarrow[\substack{(4.179)}]{R^2\text{CH}_2\text{X}} \underset{\text{PhS}\!\rightarrow\!\text{O}}{R^1\text{CH-CH}_2R^2} \xrightarrow{\Delta} R^1\text{CH}=\text{CHR}^2 \quad (4.146)$$

R^1 = Ar, SPh, SOPh, COR, COOR, CN

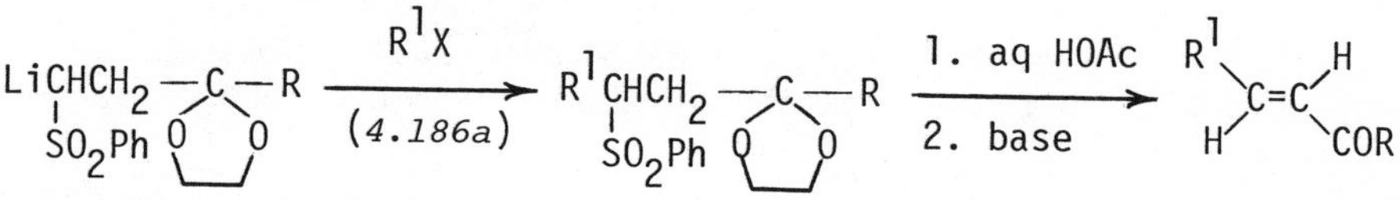

(4.147)

The formation of α,β-unsaturated carbonyl compounds and related conjugated olefins is stereoselective, producing the more stable (E)-isomers. For further details of olefination via organoseleniums and organosulfurs, the reader is referred to reviews by Clive (4.290) and Trost (4.291), respectively.

Olefination via Sulfur Extrusion. Although more limited than the other olefination reactions of α-hetero-substituted organometallics discussed above, the Ramberg-Bächlund reaction (4.292), the sulfur analogue of the Favorskii rearrangement, provides a unique route to olefins (eq. 4.137).

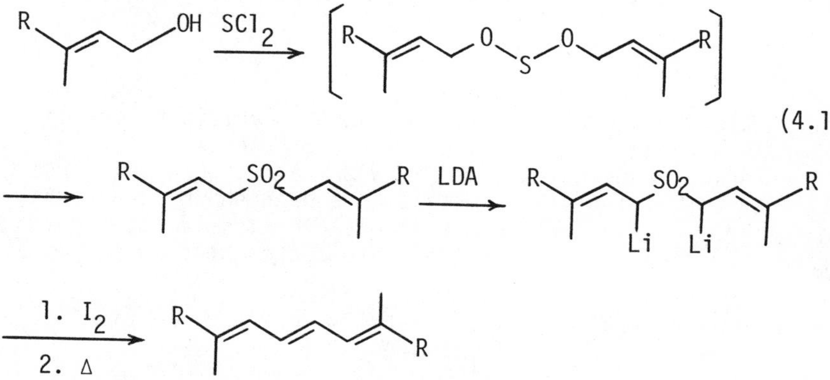

(4.148)

all-*trans*-β-carotene

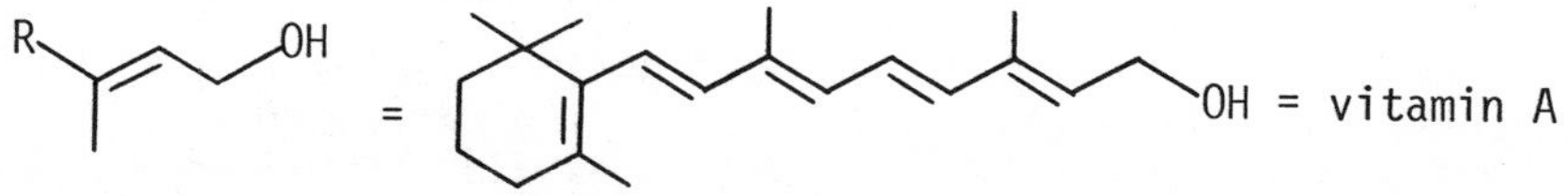

OH = vitamin A

4.3.3.4   Epoxidation Reactions Involving α-Hetero-Substituted Organometallics of Group IA and Group II Metals

The Darzens glycidic ester condensation (*4.294*) represents a classic example of epoxidation of aldehydes and ketones with α-haloorganometallics (eq. 4.149).

$$R^1COR^2 \xrightarrow{\begin{array}{c} KCHCOOEt \\ | \\ Cl \end{array}} R^1R^2C\!\!-\!\!CHCOOEt \longrightarrow R^1R^2C\!\!-\!\!\!-\!\!CHCOOEt$$

(4.149)

Although no detailed discussion of the Darzens reaction is intended here, it should be pointed out that, in many cases, the reaction produces (E)-isomers stereoselectively. The main synthetic use of the Darzens reaction lies in the carbonyl interconversion shown in eq. 4.150.

$$R^1COR^2 \xrightarrow{\text{Darzens}} R^1R^2C\!\!-\!\!\!-\!\!CR^3COOEt \xrightarrow[\substack{2.\ H_3O^+ \\ 3.\ \Delta}]{1.\ NaOH} R^1R^2CHCOR^3$$

(4.150)

Unfortunately, however, the conversion of glycidic esters into carbonyl compounds is usually a low-yield process.

As summarized in Table 4.6, a wide variety of α-haloorganometallics containing alkali metals have been shown to undergo epoxidation reactions which presumably are closely related to the Darzens reaction. These reactions are represented by eq. 4.151.

$$R^1COR^2 + \begin{array}{c} MCYZ \\ | \\ X \end{array} \longrightarrow R^1R^2C\!\!-\!\!CYZ \longrightarrow R^1R^2C\!\!-\!\!\!-\!\!CYZ$$

(4.151)

X = Cl, Br, or I;   Y, Z = H, alkyl, aryl;

CN, SR, SOR; SO_2R, SiR_3, and so on

Related to these epoxidation reactions is the reaction of carbonyl compounds with either sulfonium of sulfoxonium ylides (*4.295*). Some recent results (*4.161*, *4.191*) indicate that it is often more advantageous to use the corresponding sulfur-containing organometallics and generate the required sulfur betain intermediates via alkylation, as shown in eq. 4.152, especially in cases where the carbonyl compounds are sterically hindered.

$$(n = 0 \text{ or } 1) \qquad (4.152)$$

The presence of an extra alkyl substituent in sulfur ylides relative to the corresponding organometallics presumably exerts
steric hindrance to their addition to carbonyl compounds.

4.3.3.5 Cyclopropanation Reactions Involving α-Hetero-Substi-tuted Organometallics of Group IA and Group II Metals

As discussed briefly in Sect. 3.1.1.4, organometallics containing
good leaving α-substituents, such as halogens, can act as precursors of carbenes and related carbenoids. If the metal-carbon
bond of such organometallics is highly ionic, they may decompose
first to the corresponding carbanions and then to carbenes, as in
the well-established case of generation of dichlorocarbenes by
the reaction of chloroform with a base, such as NaOH (4.119).
Although methylene, that is, :CH$_2$ and its alkyl-substituted homo-
logues are more stable as triplets (4.17) than as singlets (4.18),
halocarbenes, such as :CCl$_2$, have singlet ground states, presumably due to the stabilization of the singlet states through the
p_π-p_π interaction between the empty p orbital and lone-pair electrons of halogens.

4.17   4.18

A detailed discussion of carbene chemistry is not intended
here. The reader is referred to monographs by Kirmse and by
Jones and Moss (3.11). It may be pointed out here, however, that
cyclopropanation of olefins and carbene rearrangement are their
two most important reactions. Insertion of carbenes into the C-H
bond does not appear to be an important process for those car-

benes generated from α-heteroorganometallics. As originally proposed by Skell (*4.296*), the cyclopropanation reaction of singlet
carbenes is believed to be a concerted process and should therefore be stereospecific, whereas the corresponding reaction of
triplet carbenes must be a stepwise, nonstereospecific process.
In this section, some carbene or carbenoid reactions of α-hetero-
substituted organometallics of Group IA and Group II metals is
discussed briefly. The reaction of α-haloorganomercuries is discussed in Sect. 7.3.2.2, and the chemistry of transition metal-
carbene complexes is discussed in Chap. 12.

Halocarbenes. Haloforms are sufficiently acidic so that their
treatment with a variety of bases can produce the corresponding
dihalocarbenes; see eq. 4.153 (*4.119*, *4.123*).

$$X_3CH + MY \xrightarrow[-HY]{} X_3CM \xrightarrow[-MX]{} X_2C: \qquad (4.153)$$

$$Y = alkyl,\ NR_2,\ OR,\ OH,\ and\ so\ on$$

In addition to simple haloforms in which all three halogen atoms
are the same, various "mixed" haloforms, such as $CHFI_2$(*4.297*)
and $CHClBr_2$(*4.298*), can also serve as carbene precursors. The
carbene products generated from mixed haloforms can be predicted
on the basis of the relative leaving ability of halogens, that
is, I > Br > Cl > F. Alternatively, trihaloacetic acids and
their esters can be treated with bases to produce carbenes, as
shown in eq. 4.154 (*4.299*).

$$\underset{\substack{\parallel \\ EtOCCl_3}}{O} \xrightarrow{NaOR} \underset{\substack{| \\ RO}}{EtO\overset{O^-}{\underset{|}{C}}-CCl_3} \longrightarrow {}^-CCl_3 \longrightarrow :CCl_2 \qquad (4.154)$$

Generation of monohalocarbenes from dihalomethanes requires highly basic reagents, such as organolithiums (*4.300*) and alkali
metal amides, such as $NaN(SiMe_3)_2$ (*4.301*).

Other Hetero-Substituted Carbenes. Various hetero-substituted
carbenes can be generated by treating appropriate α-hetero-substituted organic halides and related derivatives with strong
bases, as shown in eqs. 4.155 and 4.156.

$$CH_2ClOR \xrightarrow[(4.302)]{MeLi} \underset{\substack{| \\ Cl}}{LiCHOR} \longrightarrow :CHOR \qquad (4.155)$$

$$CH_2ClSR \xrightarrow[\text{(4.303)}]{\text{NaOH}} Na\underset{Cl}{C}HSR \longrightarrow :CHSR \qquad (4.156)$$

Methylene and Its Derivatives Substituted with Carbon Groups.
Although methyl halides have not successfully been converted into methylene by treatment with bases, certain dihalomethanes can
be treated with alkyllithiums at low temperatures to form methylene via metal-halogen exchange, as shown in eq. 4.157 (4.114).

$$CH_2BrCl \xrightarrow{\text{n-BuLi}} LiCH_2Cl + \text{n-BuBr} \longrightarrow :CH_2 \qquad (4.157)$$

In general, the metal-halogen exchange reaction of polyhalogen-
ated compounds, especially those containing iodine and bromine,
provides a useful alternative to the metal-hydrogen exchange
route (4.300).

Higher alkyl halides containing β-hydrogen atoms preferentially undergo β-elimination when treated with strong bases. On
the other hand, benzylic halides can be converted into benzylic
carbenes, which can undergo the expected cyclopropanation reaction (4.304). The corresponding reaction of allylic halides
leads to the formation of cyclopropenes via intramolecular cyclopropanation, as shown in eq. 4.158 (4.305).

$$CH_2\!=\!\underset{\overset{|}{Et}}{C}CH_2Cl \xrightarrow{\text{NaNH}_2} CH_2\!=\!\underset{\overset{|}{Et}}{C}\text{-}CH: \longrightarrow \underset{Et}{\triangle} \qquad (4.158)$$

The Simmons-Smith Reaction. The reaction of the organozinc reagent prepared from methylene iodide and Zn-Cu couple with olefins, known as the Simmons-Smith reaction, has proved to be a
convenient method for the synthesis of cyclopropanes (4.127,
4.306). The active species is believed to be iodomethylzinc iodide rather than free methylene. The following concerted mechanistic scheme has been proposed (4.306, 4.307), and is consistent
with the observed data.

$$CH_2I_2 \xrightarrow{\text{Zn-Cu}} ICH_2ZnI \xrightarrow{\substack{>C=C<}} \underset{>C=====C<}{\overset{I-----ZnI}{\underset{CH_2}{}}} \longrightarrow >C\!\!\underset{}{\overset{CH_2}{-\!\!-\!\!-}}\!\!C< \qquad (4.159)$$

Thus the reaction is first-order in both olefin and ICH_2ZnI. Although electron-donating substituents of olefins, such as alkyl groups, tend to facilitate this reaction, there is a delicate balance between electronic and steric effects, as reflected in the following reactivity order (4.306).

In general, attack by the Simmons-Smith reagents from the less hindered side of the double bond is preferred. Proximal oxy groups, such as OH and COOR, however, strongly favor attacking on the cis side of the oxy substituents, as shown in eq. 4.160 (4.308).

$$(4.160)$$

Another synthetically useful as well as mechanistically revealing feature of the reaction is that it proceeds with retention of the configuration of olefins, as shown in eq. 4.161 (4.309).

$$(4.161)$$

The reaction of the Simmons-Smith reagent with silyl enol ethers provides a convenient route to cyclopropanols; see eq. 4.162 and Sect. 6.4.2.2 (4.310).

$$(4.162)$$

The Furukawa Modification of the Simmons-Smith Reaction.   An important modification of the Simmons-Smith reaction has been reported by Furukawa (4.311).   In this modification, CH_2I_2 is treated with dialkylzinc, such as Et_2Zn, to generate the iodomethylzinc reagent via metal-halogen exchange.   Unlike the generation
of the Simmons-Smith reagent, this reaction is homogeneous and
proceeds readily under mild conditions, thereby preventing various difficulties encountered in the original Simmons-Smith reaction.

$$CH_2I_2 + Et_2Zn \longrightarrow ICH_2 \overset{I}{\underset{ZnEt}{\diamond}} Et \longrightarrow ICH_2ZnEt + EtI \quad (4.163)$$

$$\begin{array}{ccc}
 & CH_3CHI_2 \\
 & \xrightarrow{\hspace{1.5cm}} & \text{>C}\mathbin{\underset{CHCH_3}{\triangle}}\text{C<} \\
 & Et_2Zn \\
 & ArCHI_2 \\
\text{>C=C<} & \xrightarrow{\hspace{1.5cm}} & \text{>C}\mathbin{\underset{CHAr}{\triangle}}\text{C<} \qquad (4.164) \\
 & Et_2Zn \\
 & CHXYI \\
 & \xrightarrow{\hspace{1.5cm}} & \text{>C}\mathbin{\underset{CHX}{\triangle}}\text{C<} \\
 & Et_2Zn
\end{array}$$

X,Y = F, Cl, Br, or I

As indicated in eq. 4.164, the Furukawa procedure appears to
be much more general than the Simmons-Smith reaction.   For example, whereas the Simmons-Smith reaction is not readily adaptable
to alkylidene transfer, in many cases the modified procedure is
readily applicable to this type of transformation.   Cationically
polymerizable olefins, such as vinyl ethers, are generally not
very suitable for use in the Simmons-Smith reaction.   On the
other hand, the Furukawa reagents can react smoothly with such
olefins.   It is important to note that the modified procedure
seems to retain essentially all advantageous features, such as
stereospecificity, of the original procedure.

4.3.3.6   Miscellaneous Synthetic Applications of α-Hetero-Substi-
          tuted Organometallics of Group IA and Group II Metals

Although the great majority of synthetic applications of α-hetero-substituted organometallics of Group IA and Group II metals
are represented by the reactions discussed in the preceding sections, their applications to organic synthesis are by no means
limited to those discussed above.   The primary products in their
reactions with carbon electrophiles (Scheme 4.1) and nucleophiles, such as olefins, can be further transformed into a myriad
of organic compounds by taking advantage of the presence of α-
hetero substituents.   The synthetic scope of such transformations
is limited only by creativity and imagination.   A systematic discussion of this subject is not possible here.   Only a few specific examples are discussed merely to indicate their synthetic
potential yet to be fully explored.   The reaction of alkyl halides, such as RI, with α-thiomethyllithium derivatives followed
by treatment with an approproate alkylating agent, such as MeI,
permits a one-carbon homologation of alkyl halides; see eq. 4.165
(*4.160, 4.169*).

$$RI \xrightarrow{\text{LiCH}_2\text{SR}'} RCH_2SR' \xrightarrow{\text{MeI}} RCH_2 \overset{+}{S}R'Me \longrightarrow RCH_2I \quad (4.165)$$

$$I^-$$

$$R' = Me, Ph, \text{—}\overset{S}{\underset{N}{\langle}}, \text{ and so on}$$

One inspiring example of imaginative explorations in this
area is that of Trost (*4.312*).   Taking advantage of the presence
of an α-thio substituent as well as of the strained cyclopropyl
ring, the products of the reaction of α-thiophenoxycyclopropyl-
lithium (4.19) with carbonyl compounds have been converted into a
variety of organic compounds, as exemplified by Scheme 4.5.

4.3.4   Carbon-Carbon Bond Formation via Alkali and Alkaline Earth
        Metal Enolates

4.3.4.1   General Considerations

Until recently, the chemistry of carbonyl compounds had dominated
the carbon-carbon bond formation via polar reactions, and metal
enolates had been major sources of carbanionic species.   Although

Scheme 4.5

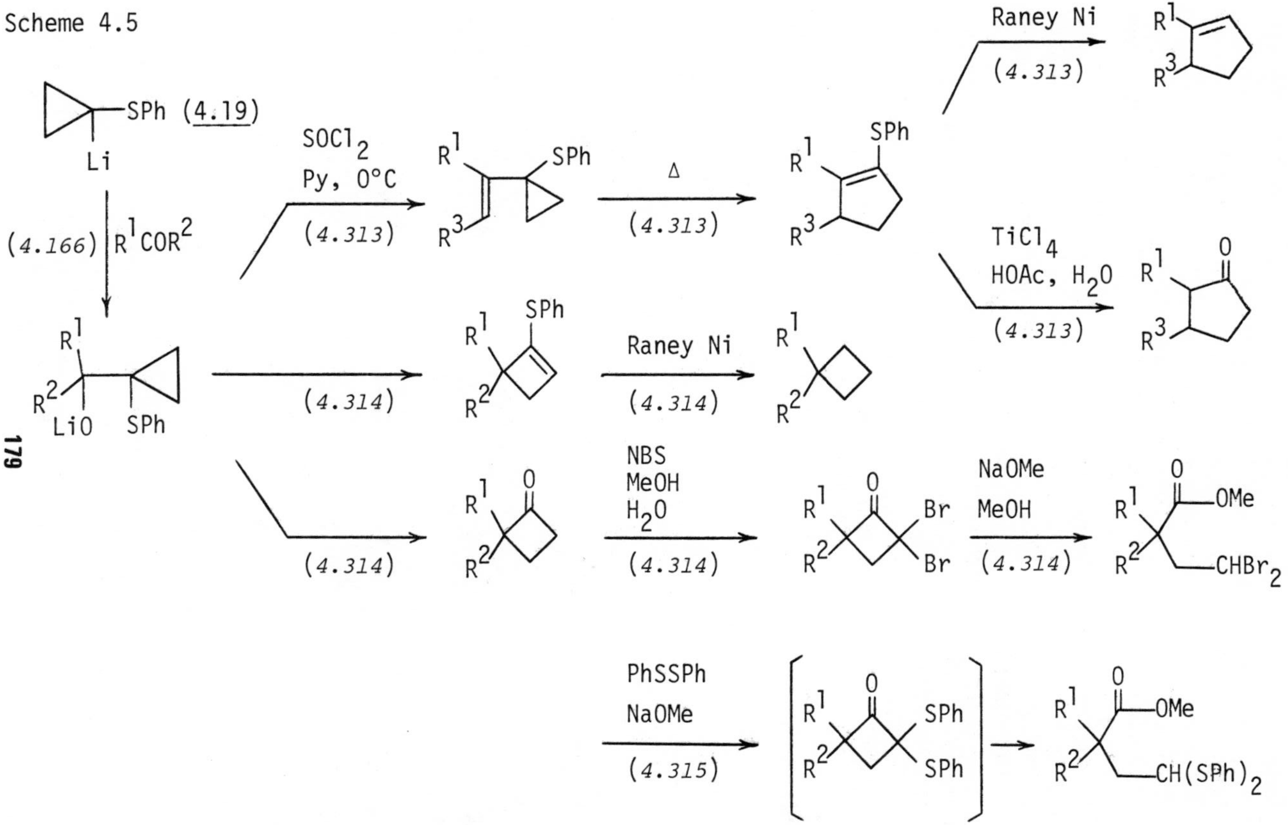

SPh (4.19)
Li
(4.166)
R^1COR^2
R^1
R^2
LiO
SPh
$SOCl_2$
Py, 0°C
(4.313)
(4.314)
(4.314)
R^1
R^3
SPh
Δ
(4.313)
SPh
R^1
R^2
Raney Ni
(4.314)
O
NBS
MeOH
H_2O
(4.314)
R^1
R^3
SPh
Raney Ni
(4.313)
$TiCl_4$
HOAc, H_2O
(4.313)
Br
Br
NaOMe
MeOH
(4.314)
OMe
$CHBr_2$
PhSSPh
NaOMe
(4.315)
SPh
SPh
OMe
$CH(SPh)_2$

recent developments in other areas, such as those discussed in preceding sections, have drastically changed this situation, the chemistry of metal enolates remains highly significant. In this and in several sections that follow, both new and old carbon-carbon bond-forming reactions involving enolates of Group IA and Group II metals are briefly discussed with emphasis on new developments. Our brief discussion may be supplemented with many excellent discussions presented previously. House (*4.316*) provides one of the most extensive discussions on this subject and covers the literature through 1970. For more recent results, the reader is referred to some recent reviews by Jackman (*4.317*), d'Angelo (*4.318*), and Seebach (*4.20*), in addition to various other reviews and original papers cited in this and in several sections that follow.

It is useful to note that the enolates of Group IA and Group II metals undergo the carbon-carbon bond-forming reactions summarized in Scheme 3.2, as do many other types of polar organometallics. The enolate versions of these reactions are summarized in Scheme 4.6.

Scheme 4.6

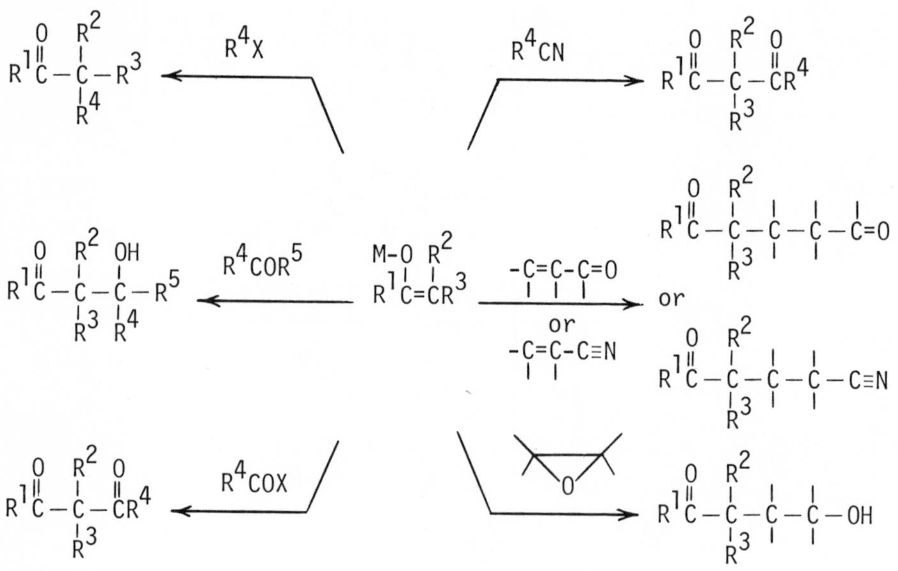

Before discussing these carbon-carbon bond-forming reactions, however, a very brief discussion of the structure and preparation of metal enolates is in order.

<u>Structure</u>. Although some β-carbonyl organometallics represented by <u>4.20</u> containing relatively electronegative metals, such as Hg, Si, Ge, and Sn (*4.319*), are known, structure <u>4.20</u> appears to be unimportant in metal enolates containing relatively electropositive metals, such as Li, Na, K, and Mg.

$$R^1-\overset{\overset{\displaystyle O}{\|}}{C}-\overset{\overset{\displaystyle M}{|}}{\underset{\underset{\displaystyle R^3}{|}}{C}}-R^2 \qquad\qquad R^1-\overset{\overset{\displaystyle O}{\|}}{C}\overset{M^+}{=}\overset{}{\underset{\underset{\displaystyle R^3}{|}}{C}}-R^2 \qquad\qquad R^1-\overset{\overset{\displaystyle O-M}{|}}{C}=\overset{}{\underset{\underset{\displaystyle R^3}{|}}{C}}-R^2$$

<u>4.20</u>   <u>4.21</u>   <u>4.22</u>

If the ionicity of the bonds to the metal atoms of metal enolates were very high, ionic structure <u>4.21</u> would be the most accurate representation of the monomer form of metal enolates. In such cases, the question of the M-O bonds versus the M-C bonds is rather meaningless. In reality, however, spectroscopic studies of certain lithium and sodium enolates (*4.320, 4.321*) suggest that, at least in these cases, structure <u>4.22</u> is most consistent with the available data. Thus, for example, the lithium enolate of phenylacetone exhibits a pair of strong IR absorptions in the 1550-1585 cm^{-1} region (*4.320*). Likewise, the 'H NMR spectrum of lithioisobutyrophenone indicates two distinct methyl signals (*4.321*).

Another question of fundamental significance is that which concerns the degree of aggregation, as discussed in Sect. 4.1. The degree of aggregation of lithioisobutyrophenone has been studied by measuring the ^{13}C NMR relaxation times (*4.322*). The results indicate that it is tetrameric in THF, dioxane, and tetrahydropyran. In our subsequent discussion of metal enolates, in most cases we simply adopt the monomeric representation <u>4.22</u> for metal enolates regardless of their exact structures.

<u>Preparation</u>. By far the most important method for preparing metal enolates is metal-hydrogen exchange (Method V), although other methods commonly used for the preparation of "simple" organolithiums and Grignard reagents, such as oxidative metallation (e.g., eq. 4.166), metal-halogen exchange (e.g., eq. 4.167), and transmetallation (e.g., eq. 4.168), have also been used on occation.

$$BrCH_2COOEt \xrightarrow{\text{Zn}} H_2C=\overset{\overset{\displaystyle }{}}{\underset{\underset{\displaystyle OZnBr}{|}}{C}}COOEt \qquad (4.166)$$

$$R^1COCHR^2_{}\overset{LiCuMe_2}{\underset{(4.323)}{\xrightarrow{\hspace{2cm}}}}\; R^1C=CHR^2 \qquad (4.167)$$

$$\underset{Br}{|}\qquad\qquad\qquad\qquad \underset{OM}{|}$$

$$M = Li \text{ or } CuL_n$$

$$R^1\underset{OSiMe_3}{\overset{R^2}{C=CR^3}} \overset{MeLi}{\underset{(4.324)}{\xrightarrow{\hspace{2cm}}}} R^1\underset{OLi}{\overset{R^2}{C=CR^3}} + SiMe_4 \quad (4.168)$$

In addition to these methods, the conjugate addition of either organometallics (e.g., eq. 4.169) or metal hydrides (e.g., eq. 4.170) and the dissolving metal reduction of α,β-unsaturated carbonyl compounds (e.g., eq. 4.171) have provided regioselective routes to enolates.

$$M = Li \text{ or } CuMe \; Li \qquad\qquad 63\% \quad (4.169)$$

$$(4.170)$$

$$(4.171)$$

Dissolving metal reduction is also applicable to the generation of enolates from cyclopropyl ketones; see eq. 4.172 (*4.328*).

$$(4.172)$$

In the preparation of metal enolates via metal-hydrogen exchange, unsymmetrical ketones having both α- and α'-hydrogen atoms can give two regioisomeric enolates (eq. 4.173).

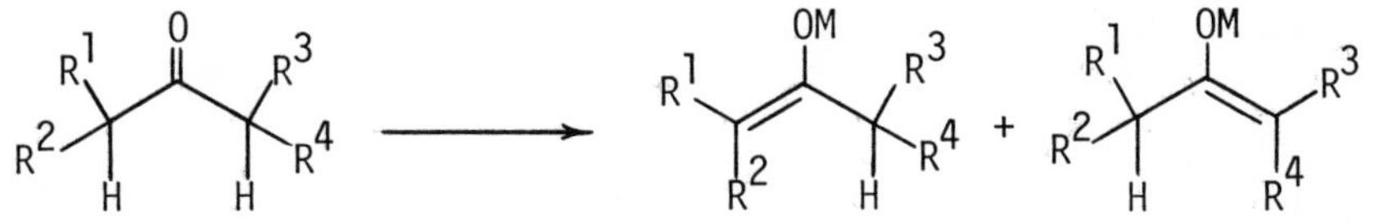

(4.173)

It is, of course, very desirable to be able to generate regioselectively either one of these two regioisomers.

Recent developments of nonnucleophilic but highly basic reagents, such as $LiN(Pr-i)_2$ (LDA), $LiNPr-i(Hex-c)$ (LICA), $LiN(SiMe_3)_2$, and $KN(SiMe_3)_2$, have made it possible to generate regioselectively "kinetic" enolates at low temperatures, $-78°C$, for example, in aprotic solvents, such as THF. Under these conditions, the least hindered hydrogen atom is abstracted, as shown in the following examples.

Scheme 4.7

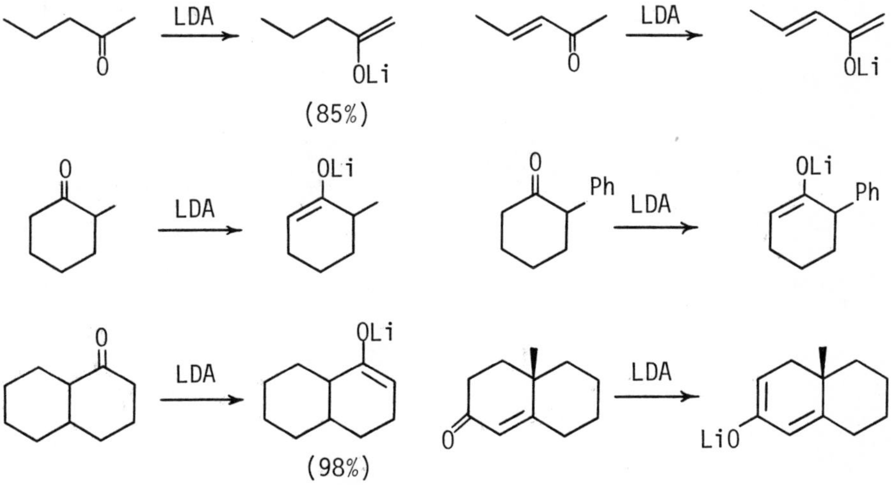

Ketones must be added slowly to a solution of a base to minimize the amount of free ketones that can act as proton donors capable of inducing undesirable scrambling.

In most cases, "kinetic" and "thermodynamic" enolates are different. Equilibration of enolates, which is strongly catalyzed by proton donors including the ketones used for the generation of enolates, usually leads to mixtures of two or more regioisomers in which "thermodynamic" enolates predominate. This is usually accomplished by generating enolates at or above room temperature in protic solvents. The equilibrium is dependent on both gegenions and temperature. Some typical results are summarized in Scheme 4.8.

Scheme 4.8

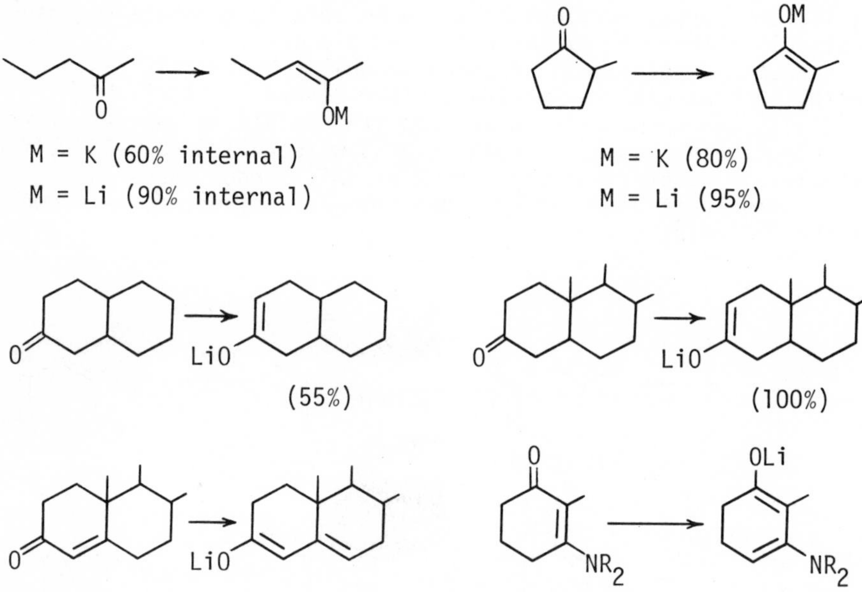
O
OM
M = K (60% internal)
M = Li (90% internal)
O
OM
M = K (80%)
M = Li (95%)
O
LiO
(55%)
O
LiO
(100%)
O
LiO
O
NR2
OLi
NR2

In some cases, "kinetic" and "thermodynamic" enolates are identical, as shown in eq. 4.174.

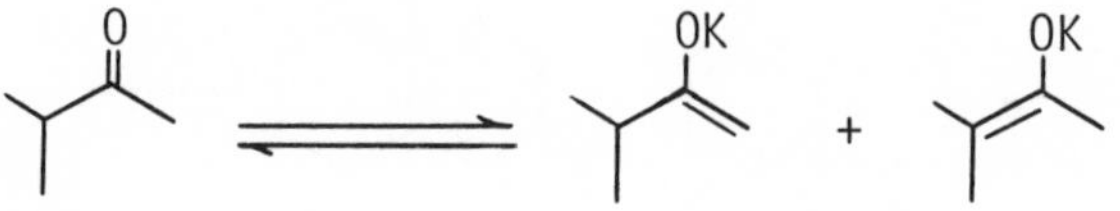
O
OK
+
OK

(4.174)

88 : 12

Although less general and less convenient, the other methods of generating enolates discussed above are regiospecific. Therefore, under nonequilibrating conditions, regiochemically pure enolates can be generated.

Even in cases where no regiochemical problem exists, the extent of enolate formation is a matter of serious concern. Before the advent of strong nonnucleophilic bases, the synthetic organic chemist often dealt with enolates formed as equilibrium mixtures containing both the starting carbonyl compounds and their enolates. While the use of such equilibrium mixtures may be advantageous in some cases, they present a serious problem of self-condensation. Here again, however, the use of strong nonnucleophilic bases, such as LDA and LICA, has solved many problems associated with the enolates of carboxylic acids and their esters by permitting complete metal enolate formation (*4.329*).

An important addition to the list of strongly basic but non-nucleophilic bases is KH (*4.330*). Using KH as a base, not only ketones (*4.330*) but aldehydes; see eq. 4.175 (*4.331*), can be converted into enolates without the complication due to self-aldolization. On the other hand, esters cannot be converted into enolates with KH, since the Claisen condensation is highly competitive (*4.332*).

$$\underset{\underset{R^2}{|}}{\overset{\overset{O\ \ H}{||\ \ |}}{HC-C-R^1}} \xrightarrow{\text{KH, THF}} \underset{\underset{R^2}{|}}{\overset{\overset{O-K}{|}}{HC=C-R^1}} \xrightarrow[\text{2. } H_3O^+]{\text{1. } R_3X} \underset{\underset{R^2}{|}}{\overset{\overset{O\ \ R^3}{||\ \ |}}{HC-C-R^1}} \quad (4.175)$$

4.3.4.2 Alkylation of Metal Enolates

The alkylation of metal enolates has been reviewed by many authors (*4.316* to *4.318*). Although the reaction has been widely used, it has also been plagued with several difficulties, such as those listed below:

1. Regiochemistry,

   1a. C-alkylation versus O-alkylation,
   1b. α-Alkylation versus α'-alkylation,

$$(4.176)$$

2. Mono- versus poly-alkylation,

$$\underset{\overset{\displaystyle O}{\overset{\parallel}{}}}{-C-CH_3} \quad \xrightarrow[\text{2. RX}]{\text{1. base}} \quad \underset{\overset{\displaystyle O}{\overset{\parallel}{}}}{-C-CH_2R} \quad \xrightarrow[\text{2. RX}]{\text{1. base}} \quad \underset{\overset{\displaystyle O}{\overset{\parallel}{}}}{-C-CHR_2}$$

(4.177)

$$\xrightarrow[\text{2. RX}]{\text{1. base}} \quad \underset{\overset{\displaystyle O}{\overset{\parallel}{}}}{-C-CR_3}$$

3. Stereochemistry,

4. Chemoselectivity,

5. Use of alkenyl, aryl, alkynyl, and hindered alkyl halides.

C-Alkylation Versus O-Alkylation. The C-/O-alkylation ratio is a
function of various factors, of which the following appear to be
some of the most significant. Firstly, as already pointed out in
Sect. 1.6, the oxygen and α-carbon atoms represent the "hard" and
"soft" ends of enolates. Thus the C-/O-alkylation ratio increases as the softness of the alkylating agent increases. For
example, the C-/O-alkylation ratio increases in the order:
ROTs < RCl < RBr < RI. Likewise the C-/O-alkylation ratio generally increases, as the ionicity of the metal-to-enolate bond decreases: N^+R_4 < K^+ < Na^+ < Li^+. Secondly if one compares the
bond energies of C- and O-alkylation products, it is evident
that, in the absence of any special effects, the C-alkylated products are thermodynamically more stable than the O-alkylated products.

85 kcal/mole →

⌐85 kcal/mole

O—R

$-C = C-$

145 kcal/mole

175 kcal/mole →

O   R ←85 kcal/mole

$-C - C-$

85 kcal/mole

If the difference in product stability is strongly reflected in
the transition state, the C-alkylation should predominate.
Thirdly, solvents play an important role in determining the C-/O-
alkylation ratio. If there is strong solvation in the vicinity
of the oxygen atom and a weaker solvation at the carbon atom, relatively greater reactivity at carbon will be observed. One of

the most dramatic manifestations of the solvent effects in enolate alkylation is shown in eq. 4.178 (*4.333*).

(4.178)

Polar aprotic solvents, such as DMF and DMSO, effectively solvate
cations but not anions. The usual mode of phenolate alkylation
involves attack at the oxygen site, and that is what we find. In
CF_3CH_2OH the oxygen end of the phenolate is strongly solvated
through hydrogen bonding. This disfavors O-alkylation.

In general, metal phenolates tend to undergo O-alkylation in
preference to C-alkylation. It is generally difficult to attain
a high C-/O-alkylation ratio. In the alkylation of 1,3-dicar-
bonyl compounds, the C- and O-alkylation processes are often competitive. In most cases, however, their alkylation reactions can
be steered in the desired direction by properly selecting metals,
leaving groups, and solvents. For example, if TlOEt is used as a
base, an exclusive C-methylation of acetylacetone takes place;
see eq. 4.179 (*4.334*).

$$CH_3COCH_2COCH_3 \quad \xrightarrow[\text{2. } CH_3I]{\text{1. TlOEt}} \quad CH_3CH(COCH_3)_2 \quad (4.179)$$

100%

Finally, O-alkylation is usually not competitive in the alkylation of mono carbonyl compounds.

C_α-Alkylation Versus C_α'-Alkylation and Mono-alkylation Versus
Poly-alkylation. One of the most commonly encountered difficulties in the alkylation of ketones is the formation of mixtures of
regioisomers as well as of mono- and poly-alkylated products.
Until recently, this problem had been solved mainly by some indirect methods including (1) malonic ester and acetoacetic ester

syntheses (*4.335*) and related methods for alkylation, for example, eq. 4.180 (*4.336*), (2) ketone syntheses via β-ketosulfones
and β-ketosulfoxides discussed in Sect. 4.3.3.2, and (3) α-alkylation of ketones involving blocking and unblocking of one of the
two α-positions, for example, eq. 4.181 (*4.337*).

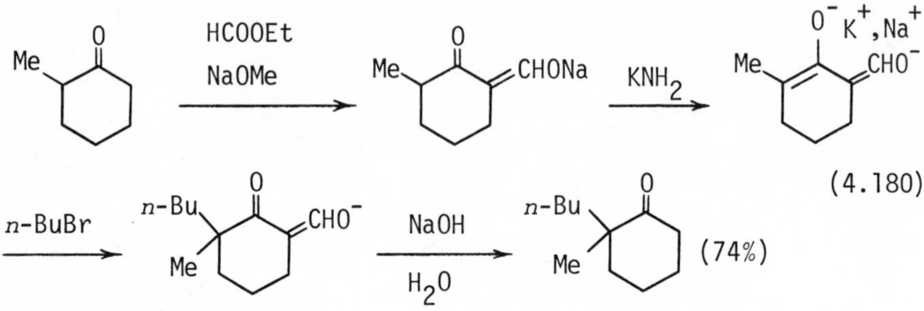

(4.180)

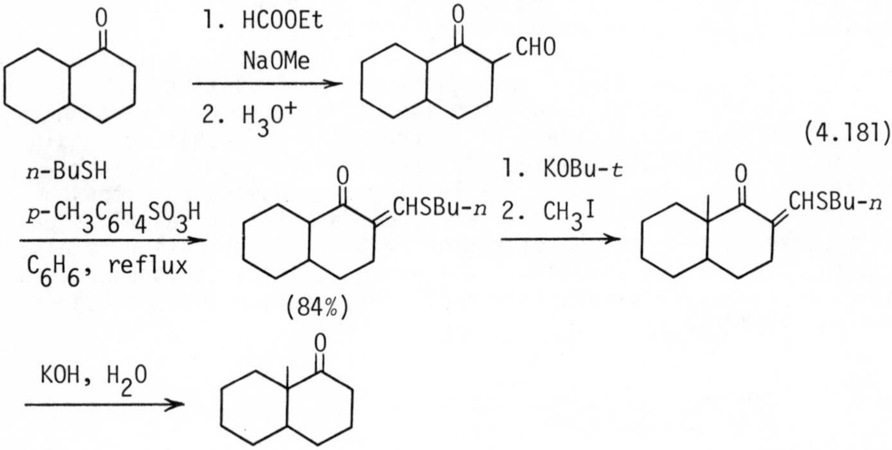

(4.181)

While these multistep methods are generally tedious and are
being replaced gradually by simpler methods, such as those discussed later, some of these conventional methods are still valuable in that (1) they often represent the only practical methods
for certain transformations, (2) regioisomerically pure products
can be obtained, and (3) the regiochemistry of alkylation can be
controlled cleanly through generation of either mono- or dianions, as shown in eq. 4.182 (*4.338*).

$$PhSCH_2\overset{O}{\overset{\|}{C}}CH_3 \quad\begin{cases} \longrightarrow PhS\overset{O}{\overset{\|}{C}}{}^-HCCH_3 \xrightarrow{\ RX\ } PhS\overset{R\ O}{\overset{|\ \ \|}{C}}HCCH_3 \\[2em] \longrightarrow PhS\overset{O}{\overset{\|}{C}}{}^-HCCH_2^- \xrightarrow[2.\ H_3O^+]{1.\ RX} PhSCH_2\overset{O}{\overset{\|}{C}}CH_2R \end{cases} \tag{4.182}$$

In principle, regio-defined enolates, "kinetic" or "thermodynamic", can be regiospecifically alkylated, provided that the
rate of alkylation is higher than that of equilibration. Indeed,
some of the more reactive alkylating agents, such as MeI, allyl
bromide, and benzyl bromide, can participate in highly regiospecific alkylation reactions as shown in eq. 4.169 to 4.171.
Of particular importance is the regiospecific alkylation applicable to the annulation of the cyclohexenone ring (*4.339*) which
represents a significant alternative to the Robinson annulation
(Sect. 4.3.4.4). The key idea is to use those allylic or benzylic halides which are not only highly reactive but capable of
acting as masked carbonyl derivatives. Some of the more promising examples are shown in eq. 4.183 to 4.186.

$$(4.183)$$

$$(4.184)$$

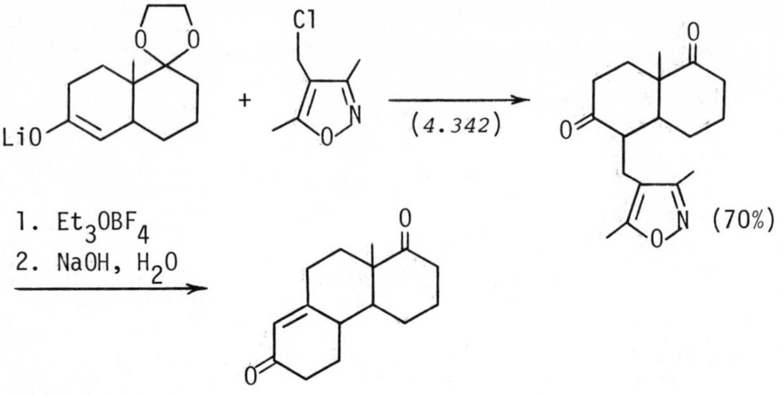
Cl
O
O
LiO
+
N
O
(4.342)
O
O
N
O
(70%)
1. Et3OBF4
2. NaOH, H2O
O
O

(4.185)

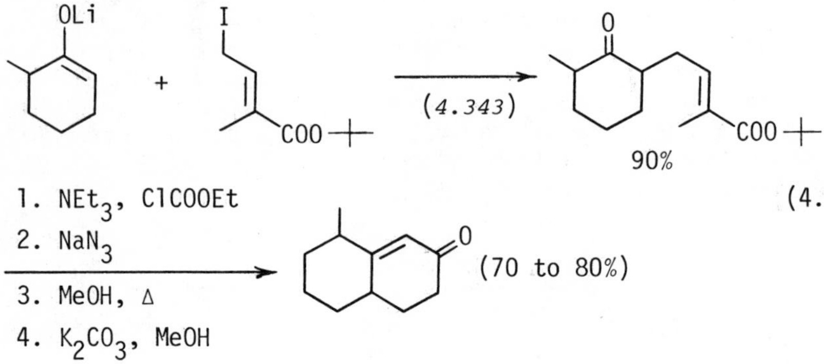
OLi
I
+
COO
(4.343)
O
COO
90%
1. NEt3, ClCOOEt
2. NaN3
3. MeOH, Δ
4. K2CO3, MeOH
O
(70 to 80%)

(4.186)

The alkylation of enolates with less reactive alkylating agents presents the above-mentioned problem of the formation of regioisomers as well as of poly-alkylated products. In principle, this problem can be solved by stopping or slowing down the equilibration of enolates without significantly retarding desired alkylation. One promising approach is to convert regio-defined potassium enolates into the corresponding potassium enoxytriethylborates prior to alkylation; see eqs. 4.187, 4.188 (*4.344*). Under these conditions, neither regiochemical scrambling nor poly-alkylation competes seriously with the desired mono-alkylation. If either lithium enolates or hindered trialkylboranes, such as *s*-Bu_3B, are used, the formation of enoxyborates either does not take place or is incomplete.

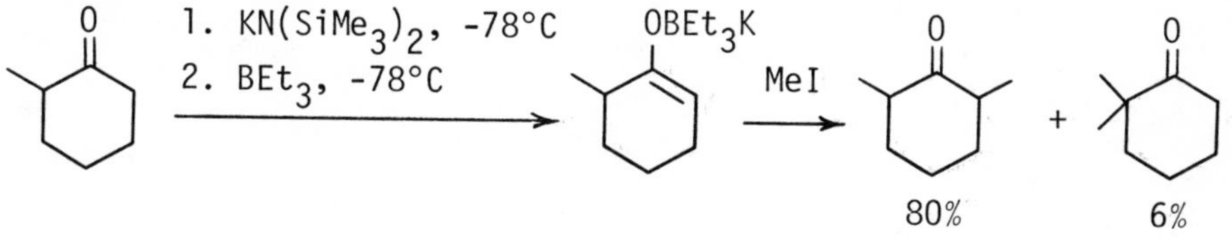

1. KN(SiMe$_3$)$_2$, -78°C
2. BEt$_3$, -78°C

$$80\% \qquad 6\%$$

$$(4.187)$$

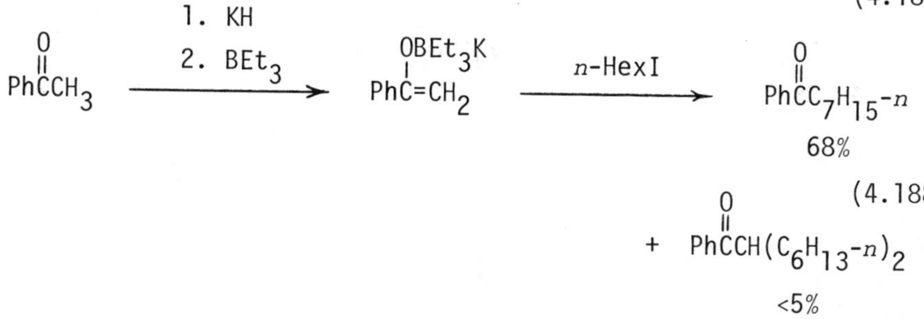

1. KH
2. BEt$_3$

PhCCH$_3$ → OBEt$_3$K PhC=CH$_2$ → n-HexI → PhCC$_7$H$_{15}$-n

$$68\%$$

$$(4.188)$$

$$+ \quad PhCCH(C_6H_{13}\text{-}n)_2$$

$$<5\%$$

Alternatively, the reaction of regio-defined silyl enol ethers with alkyl halides can be promoted by certain fluorides; see eq. 4.189 (4.345), or TiCl$_4$; see eq. 4.190 (4.346). The latter reagent is unique in that it permits the hitherto difficult tertiary alkylation of ketones.

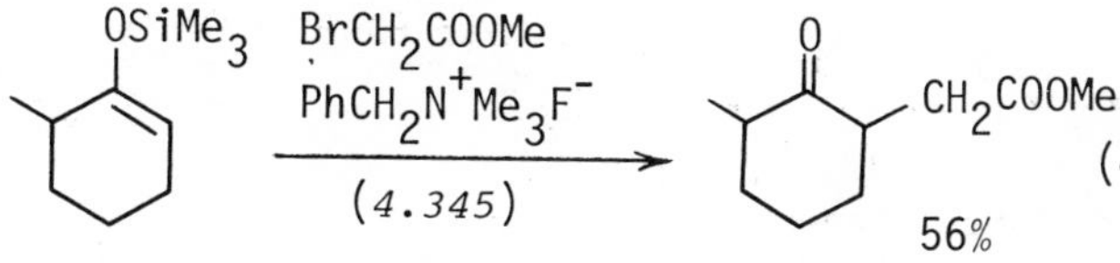

OSiMe$_3$ BrCH$_2$COOMe
PhCH$_2$N$^+$Me$_3$F$^-$
(4.345)
→ CH$_2$COOMe

$$(4.189)$$

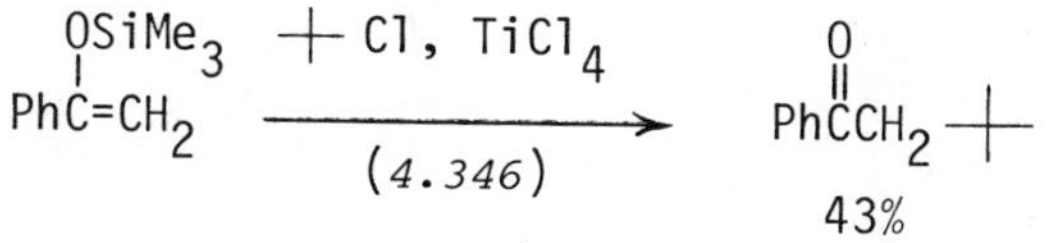

$$56\%$$

OSiMe$_3$
PhC=CH$_2$ + Cl, TiCl$_4$
(4.346)
→ PhCCH$_2$

$$(4.190)$$

$$43\%$$

Stereochemistry of the Alkylation of Enolates. In cases where the steric requirements of the two sides of an enolate anion are considerably different, attack by an alkylating agent from the less hindered side is favored, as expected. The following two examples fall into this category. In analogy with the discussion presented in Sect. 4.3.2.2, the equatorial side of the methylene-cyclohexane system must be the less hindered side of the two.

$$(4.191)$$

$$(4.192)$$

Much less clear is the stereochemistry of the alkylation of endocyclic enolates. The available data suggest that the predominant attack takes place such that the entering alkyl group ends up as an axial substituent, as shown in eq. 4.193 (*4.349*).

$$(4.193)$$

The results have been interpreted in terms of the stereoelectronically favorable arrangement in the axial attack relative to the equatorial attack. On the other hand, an essentially 50:50 mixture of *cis*- and *trans*-isomers is formed in the methylation of 4-*t*-butylcyclohexenolate; see eq. 4.194 (*4.347*).

$$(4.194)$$

Although the significance of a twist-boat conformation was suggested, this explanation does not appear to be totally convincing. It should be noted that the products in eq. 4.194 are epimerizable under the reaction conditions, which is not the case in the three preceding reactions.

α-Arylation and α-Alkenylation. Until recently, the α-arylation and α-alkenylation of enolates had been very difficult. A few methods have recently been developed for these purposes. The most promising by far is Bunnett's photochemical arylation; see eq. 4.195 (*4.350*), and alkenylation; see eq. 4.196 (*4.351*).

$$\underset{\text{MeC=CH}_2}{\overset{\overset{\text{OK}}{|}}{}} \quad \xrightarrow[h\nu]{\text{PhX}} \quad \underset{\text{MeCCH}_2\text{Ph}}{\overset{\overset{\text{O}}{||}}{}} \qquad (4.195)$$

X = Cl (68%), Br (52%), I (71%)

$$\underset{\text{MeC=CH}_2}{\overset{\overset{\text{OK}}{|}}{}} \quad \xrightarrow[h\nu]{\text{PhCH=CHBr}} \quad \text{MeCOCH}_2\text{CH=CHPh} + \text{MeCOCH=CHCH}_2\text{Ph}$$

$$\qquad\qquad\qquad\qquad\qquad 48\% \qquad\qquad\qquad\qquad 34\%$$

$$(4.196)$$

The results have been interpreted in terms of the $S_{RN}1$ mechanism; see eq. 4.197 (*4.350*).

$$\text{ArX} \quad \xrightarrow{\text{e}^-} \quad \text{ArX}^{\bar{\cdot}}$$

$$\text{ArX}^{\bar{\cdot}} \quad \longrightarrow \quad \text{Ar}\cdot + \text{X}^-$$

$$\text{Ar}\cdot + \text{R}^{\bar{\cdot}} \quad \longrightarrow \quad \text{ArR}^{\bar{\cdot}} \qquad (4.197)$$

$$\text{ArR}^{\bar{\cdot}} + \text{ArX} \quad \longrightarrow \quad \text{ArR} + \text{ArX}^{\bar{\cdot}}$$

α-Alkylation of Aldehydes, Esters, Nitriles, and Other Related Compounds. Various aspects of the alkylation of ketones are presented in the above discussions. Some representative alkylation reactions of other carbonyl and related compounds are now briefly presented. The metallated derivatives of esters, aldehydes, and various related compounds are now readily available through appropriately strong but nonnucleophilic bases, as discussed in Sect. 4.3.4.1. In many cases, their alkylation proceeds satisfactorily, as summarized in Table 4.14. In most of the examples listed in Table 4.14, there is no ambiguity regarding the structures of the products. In the alkylation of dienolates, α-alkylation is strongly favored over γ-alkylation, as shown in eqs. 4.198 to 4.200.

$$ZCH_2CH=CHCOOR^1 \quad \xrightarrow{\begin{array}{c}1.\ LDA \\ 2.\ RX\end{array}} \quad ZCH=CHCHCOOR^1 \atop \underset{R}{|} \qquad (4.198)$$

Z = H (*4.353*), SMe (*4.357*), OMe (*4.358*)

$$\xrightarrow[\ (4.376)\ ]{\begin{array}{c}1.\ LDA \\ 2.\ RX\end{array}} \qquad (4.199)$$

$$\xrightarrow[\ (4.366)\ ]{\begin{array}{c}1.\ LDA \\ 2.\ RX\end{array}} \qquad (4.200)$$

In the alkylation of the dianion of 2-butynoic acid, however, γ-alkylation is the predominant path; see eq. 4.201 (*4.370*).

$$LiCH_2C\equiv CCOOLi \qquad \xrightarrow{\hspace{2cm}} \qquad (4.201)$$

The observed α-selectivity in alkylation is in sharp contrast with the γ-selectivity in various carbonyl addition reactions of the corresponding dienolates discussed in the following section. The use of heterocyclic masked enolates is discussed in Sect. 4.3.4.6.

Table 4.14. α-Alkylation of Aldehydes, Esters, Nitriles, and Other Related Compounds

Carbonyl or Related Compound	Base	Alkylating Agent	Yield (%)	Ref.
(Aldehydes)				
R^1R^2CHCHO	KH	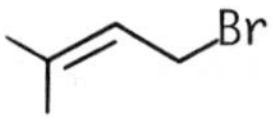 Br	88 to 96	*4.331*
		n-RI	about 50	*4.331*
(Esters and Lactones)				
$CH_3COO{-}\!\!\!\!+$	LICA	RX	42 to 96	*4.329*
n-PrCOOMe	LDA, THF	RX	88 to 98	*4.352*
MeCH=CHCOOEt	LDA, HMPA	RX	90 to 96	*4.353*
${+}\!\!\!\!{-}SCR^1R^2COOMe$	2 Li	MeI or PhCH$_2$Br	72 to 85	*4.354*
	LDA, THF	RX	>90%	*4.355*
$n\text{-}C_6H_{13}$CH=CHCOOMe	LiB(Bu-*s*)$_3$H	RX	50 to 63	*4.356*
MeSCH$_2$CH=CHCOOMe	LDA	RX	63 to 92	*4.357*
MeOCH$_2$CH=CHCOOMe	LDA	RX	42 to 81	*4.358*
HOOCCH$_2$COOEt	LICA	RX	50 to 98	*4.359*
PhCH=NCH$_2$COOEt	LDA	RX	75 to 90	*4.360*
PhCONHCH$_2$COOEt	2 LDA, TMEDA	RX	48 to 71	*4.361*
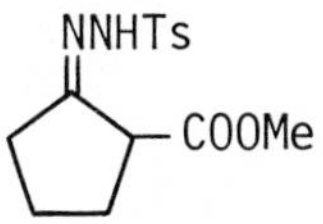	3 LDA	RX	49 to 64	*4.362*
RCH(OH)COOEt	2 LDA	RX	57 to 79	*4.363*
Me$_2$NCH=NCHR^{1}COOMe	LDA	RX	65 to 84	*4.364*
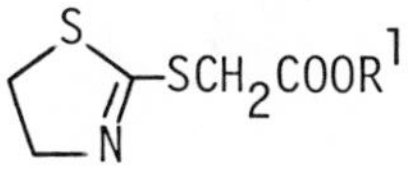	NaH, DMF	RX	34 to 68	*4.365*

195

(continued...)

Table 4.14 continued

Carbonyl or Related Compound	Base	Alkylating Agent	Yield (%)	Ref.
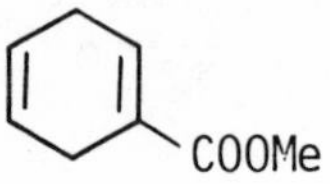 (structure: cyclohexadiene-COOMe)	LDA	RX	52 to 96	*4.366*
(Carboxylic Acids)				
$Me_2CHCOOH$	2 LDA	RCH_2X	46 to 80	*4.367*
Me_3SiCH_2COOH	2 LDA	RX	60 to 95	*4.368a*
$PhSCHR^1COOH$	2 LDA	RX	95 to 98	*4.368b*
$MeSCHR^1COOH$	2 LDA	RX	59 to 80	*4.369*
$MeC{\equiv}CCOOH$	 LiN (structure: 2,2,6,6-tetramethylpiperidide)	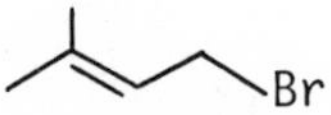 (structure: prenyl bromide) —Br	53 to 59	*4.370*
$R^1CH(COOH)_2$	3 *n*-BuLi	RX	74 to 88	*4.371*
$PhCONHCH_2COOH$	3 LDA	RX	23 to 60	*4.361*
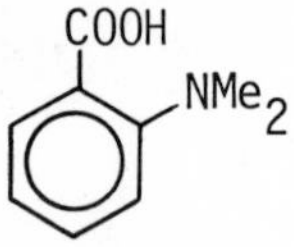 (structure: 2-(dimethylamino)benzoic acid, COOH, NMe$_2$)	Li, NH_3	RX	45 to 80	*4.372*
(Amides)				
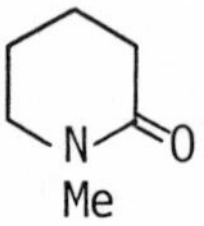 (structure: N-methyl-2-piperidinone)	LDA or $LiNMe_2$	RI	68 to 85	*4.373*
$R^1R^2CHCONMe_2$	$LiNEt_2$, HMPA	RBr	69 to 86	*4.374*
$MeCONMe_2$	LDA	MeI and $PhCH_2Br$	62 and 99	*4.375*

(continued...)

Table 4.14 continued

Carbonyl or Related Compound	Base	Alkylating Agent	Yield (%)	Ref.
(Nitriles)				
$\diagup\!\!\!\!\diagdown\!\!\!\!\diagup$ CN	LDA	RX	72 to 98	*4.376*
PhCHCN $\mid$ OSiMe$_3$	LDA	RX	48 to 96	*4.377*
(Other masked carbonyl compounds)				
NOH $\parallel$ R^{1}CCH$_2$R^2	2 BuLi	RX	37 to 81	*4.378*
=NOMe	LDA	MeI	86 to 94	*4.379*
NNMe$_2$ $\parallel$ CH$_3$CR^1	LDA	RX	83 to 95	*4.380*
R^{1}CH$_2$CH=NNMe$_2$	LiNEt$_2$	RBr	75 to 79	*4.381a*
MeR^{1}C=N $\langle$ $\rangle$	LiNEt$_2$	RX	62 to 80	*4.381b* *4.381c*
Me$_2$CHCH=N $+$	LDA, DME	RBr	59 to 76	*4.382*
R^1(R^{2}CH$_2$)C=NR	LDA	CH$_2$=CCH$_2$Br $\mid$ OMe	43 to 84	*4.383*
—OMe	LDA, HMPA	RI	52 to 60	*4.384*
MeCH=CHCH=N $\langle$ $\rangle$	LDA, HMPA	RX	98 to 100	*4.385*
—CH$_3$	BuLi	RX	70 to 90	*4.386*

(continued...)

Table 4.14   continued

Carbonyl or Related Compound	Base	Alkylating Agent	Yield (%)	Ref.
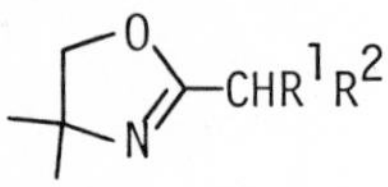 $\diagdown$O$\diagup$—CHR^{1}R^2 (oxazoline)	RLi	RX	60 to 75	*4.387*
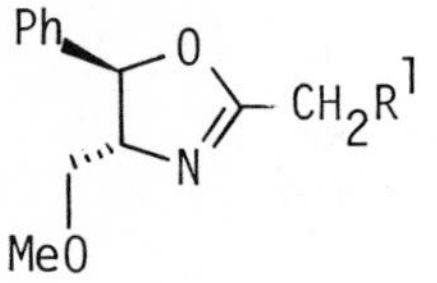 Ph, MeO oxazoline—CH$_2$R^1	BuLi or LDA	RX	30 to 84	*4.388*
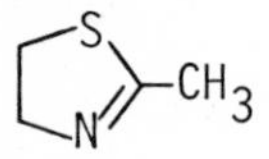 S-thiazoline—CH$_3$	BuLi	RX	80 to 95	*4.389*

4.3.4.3   Reactions of Enolates with Carbonyl Compounds

The reactions of enolates with carbonyl compounds may arbitrarily be divided into the following two types (eq. 4.202).

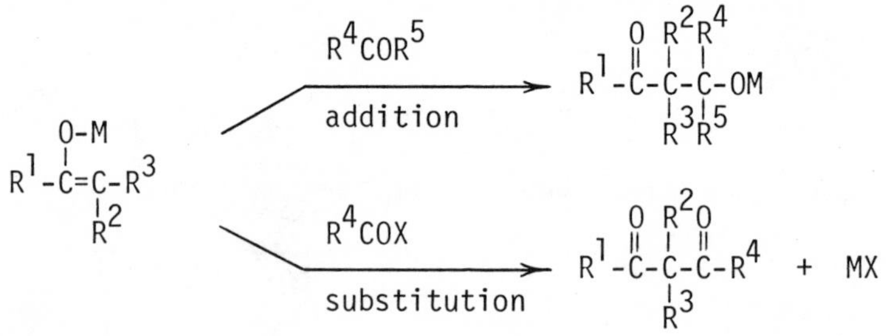

$$
\begin{array}{c}
\text{O-M} \\
R^1\text{-C=C-}R^3 \\
\overset{|}{R^2}
\end{array}
\quad
\begin{array}{l}
\xrightarrow[\text{addition}]{R^4COR^5}
\quad
R^1\text{-}\underset{\overset{|}{R^3}\overset{|}{R^5}}{\overset{O\ R^2R^4}{\overset{\|\ |\ |}{C\text{-}C\text{-}C}}}\text{-OM} \\
\\
\xrightarrow[\text{substitution}]{R^4COX}
\quad
R^1\text{-}\underset{\overset{|}{R^3}}{\overset{O\ R^2\ O}{\overset{\|\ |\ \|}{C\text{-}C\text{-}C}}}\text{-}R^4 \ + \ MX
\end{array}
\qquad (4.202)
$$

The aldol reaction (*4.390*) represents the addition reactions, and the Claisen condensation (*4.391*) represents the substitution reactions. Some representative addition and substitution reactions of enolates with carbonyl compounds is briefly discussed in this section, while the reaction of enolates with α,β-unsaturated carbonyl compounds, that is, the Michael reaction, is discussed in the following section.

The Aldol and Related Reactions.   A wide variety of reactions of enolates with carbonyl compounds fall into the category of the aldol and related reactions.

   (a) The aldol reaction (*4.390*).   The aldol reaction can be carried out under either acidic or basic conditions (*4.390*).   The base-catalyzed aldol condensation is shown in eq. 4.202.   On protonolysis the reaction produces β-hydroxy ketones which can be dehydrated to give α,β-unsaturated carbonyl compounds.   Both β-hydroxy and α,β-unsaturated carbonyl moieties are found among various natural products.   In many cases, their biogenesis is presumed to involve enzyme-catalyzed aldol condensation processes.   The reaction has also played an important role in the laboratory synthesis of many natural products and related substances, especially those containing five- and six-membered rings, as shown in eqs. 4.203 (*4.392*) and 4.204 (*4.393*).

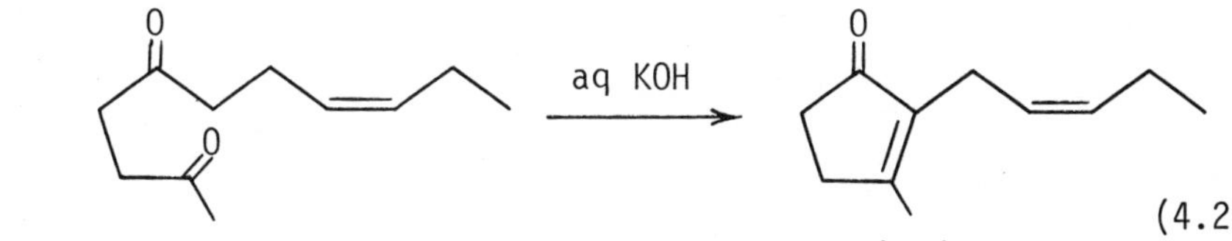

cis-jasmone   (4.203)

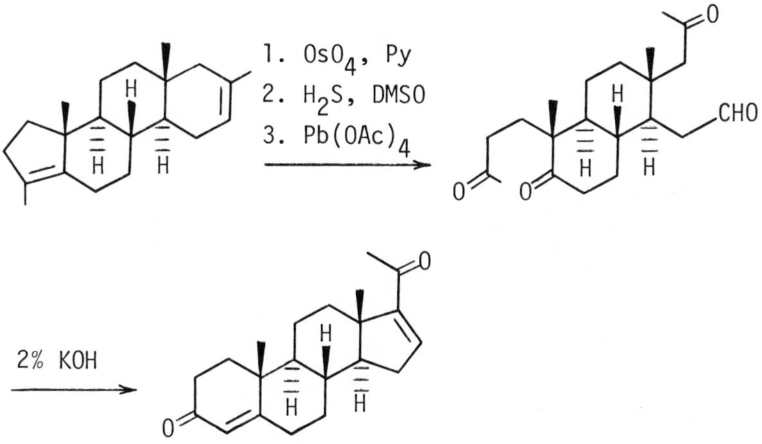

(4.204)

dehydroprogesterone

Despite its obvious significance, however, the aldol reaction has been plagued with various problems.   Some of these problems and their possible solutions are discussed later in detail.

At this point it is useful to note that the following well-known name reactions may be considered variations of the aldol reactions.

(b) The Reformatsky reaction (4.36, 4.37). The Reformatsky reaction may be viewed as the ester version of the aldol reaction.

$$R^1COR^2 + BrZnCH_2COOR \longrightarrow \xrightarrow{H_3O^+} R^1R^2\underset{\underset{HO}{|}}{C}CH_2COOR \xrightarrow{-H_2O} R^1R^2C=CHCOOR$$

(4.205)

(c) The Stobbe reaction (4.394). A variation of the Reformatsky reaction involving the use of potassium salt of ethyl succinate is called the Stobbe reaction.

$$\underset{R^2}{\overset{R^1}{>}}C=0 \quad + \quad \underset{CH_2COOEt}{\overset{CH_2COOEt}{|}} \quad \xrightarrow[\text{HO}+]{\text{KO}+} \quad \underset{R^2}{\overset{R^1}{>}}C\text{----}CHCOOEt$$

(4.206)

$$\longrightarrow \quad \underset{R^2}{\overset{R^1}{>}}\!\!\!\!-COOEt \quad \xrightarrow[\text{2. } H_3O^+]{\text{1. KO}+} \quad \underset{R^2}{\overset{R^1}{>}}C=C\text{-COOEt}$$

(d) The Knoevenagel reaction (4.395). Various reactions of relatively acidic carbonyl compounds with aldehydes and ketones are classified as the Knoevenagel reaction, which does not even require any metal containing reagent.

$$R^1COR^2 + \underset{Z}{\overset{CH_2COY}{|}} \xrightarrow[\text{reflux}]{\overset{\text{amine}}{\text{RCOOH}}} R^1R^2C=\underset{Z}{\overset{CCOY}{|}}$$

(4.207)

Y = R, OR, and so on;   Z = COY, CN, Ph, NO_2, and so on

(e) The Mannich reaction (4.396). The aminomethylation of ketones, aldehydes and other related active hydrogen compounds is called the Mannich reaction. Its similarity to the aldol reaction in the reactant-product relationship is indicated in eq. 4.208.

$$R^1-\overset{\overset{\displaystyle O}{\|}}{C}-\overset{\displaystyle |}{\underset{\displaystyle |}{C}}-H \;+\; CH_2=O \;+\; R_2^2 N^+ H_2 Cl^- \xrightarrow[\text{reflux}]{\substack{\text{HCl} \\ \text{EtOH}}} R^1-\overset{\overset{\displaystyle O}{\|}}{C}-\overset{\displaystyle |}{\underset{\displaystyle |}{C}}-CH_2 NR_2^2 \qquad (4.208)$$

(f) The Claisen-Schmidt reaction (*4.397*). This reaction represents nothing but a special case of the aldol reaction in which one of the carbonyl compounds is an aromatic aldehyde.

$$ArCHO \;+\; H-\overset{\overset{\displaystyle O}{\|}}{\underset{\displaystyle |}{C}}-CH(R) \xrightarrow[\substack{H_2O \\ EtOH}]{NaOH} \underset{H}{\overset{Ar}{>}}C=C\underset{COH(R)}{\overset{H}{<}} \qquad (4.209)$$

trans

(g) The Perkin reaction (*4.398*). It should also be pointed out that various carbonyl olefination reactions, such as the Wadsworth-Emmons reaction and the Peterson reaction discussed in Sect. 4.3.3.3, and epoxidation reactions represented by the Darzens reaction (Sect. 4.3.3.4) are closely related to the aldol reaction.

$$ArCHO \;+\; (CH_3CO)_2O \xrightarrow[\text{2. } H_2O]{\text{1. KOAc}} \underset{H}{\overset{Ar}{>}}C=C\underset{COOH}{\overset{H}{<}} \qquad (4.210)$$

Problems and Solutions in the Aldol Reaction. Of various problems associated with the aldol reaction, the following three are discussed in some detail: (1) cross/homo ratio, (2) regioselectivity, and (3) stereoselectivity.

(a) Cross/homo ratio and regioselectivity. In cases where both of the two carbonyl compounds are enolizable, the aldol reaction can be and has often been accompanied by cross-homo scrambling leading to the formation of all four possible products, as shown in eq. 4.211.

$$R^1CH_2COR^2 \ + \ R^3CH_2COR^4 \ \xrightarrow{\text{base}} \ R^1CH_2\underset{\underset{HO}{|}}{\overset{\overset{R^2}{|}}{C}}-\overset{\overset{R^1}{|}}{C}HCOR^2 \ + \ R^1CH_2\underset{\underset{HO}{|}}{\overset{\overset{R^2}{|}}{C}}-\overset{\overset{R^3}{|}}{C}HCOR^4$$

$$+ \ R^3CH_2\underset{\underset{HO}{|}}{\overset{\overset{R^4}{|}}{C}}-\overset{\overset{R^1}{|}}{C}HCOR^2 \ + \ R^3CH_2\underset{\underset{HO}{|}}{\overset{\overset{R^4}{|}}{C}}-\overset{\overset{R^3}{|}}{C}HCOR^4$$

$$(4.211)$$

It is therefore important to be able to direct the reaction so that only one cross-aldol product is obtained. A number of selective procedures have been developed in recent years, of which the following are worth mentioning here. The first major breakthrough was achieved by Wittig (*4.399*) with preformed lithium salts of imines (eq. 4.212).

$$R^1CH_2COR^2 \ \xrightarrow{RNH_2} \ R^1CH_2\underset{\underset{NR}{\|}}{C}R^2 \ \xrightarrow{LDA} \ R^1CH=\underset{\underset{LiNR}{|}}{C}R^2$$

$$(4.212)$$

An operationally more convenient procedure for cross-aldol condensation has been developed by House (*4.400*), which involves the reaction of preformed lithium enolates with aldehydes in the presence of $ZnCl_2$ or $MgCl_2$, as shown in eq. 4.213.

$$81\% \quad (4.213)$$

The aldol reaction can be further complicated by regiochemical scrambling. It has been reported, however, that, as in alkylation (Sect. 4.3.4.2), even regio-unstable "kinetic" enolates can undergo regio-specific aldol reactions when the desired aldol processes are sufficiently rapid.

$$R = n\text{-}Pr \ (65\%), \ Ph \ (80\%)$$

An indirect but potentially highly useful approach by Kuwajima (4.403) involves the reaction of dianions of β-ketosulfoxides with aldehydes and ketones (eq. 4.216). The products can be further converted into a variety of cross-aldol derivatives.

$$CH_3COCH_2SOPh \quad \xrightarrow[\text{3. } R^1COR^2]{\begin{array}{l}\text{1. NaH}\\\text{2. BuLi}\end{array}} \quad R^1R^2\overset{\overset{\displaystyle OH}{|}}{C}CH_2COCH_2SOPh \quad (4.216)$$

$$65 \text{ to } 91\%$$

A more recent trend in the aldol reaction is to use metal enolates containing highly covalent metal-oxygen bonds, such as B-O and Si-O. Thus Mukaiyama and others (4.404) have developed selective procedures involving the intermediacy of enoxyboranes, as shown in eq. 4.217.

$$CH_3COPh \quad \xrightarrow[\text{3. } H_2O]{\begin{array}{l}\text{1. } n\text{-}Bu_2BOTf\\\quad Et_3N, \ Et_2O\\\text{2. } PhCH_2CH_2CHO\end{array}} \quad PhCH_2CH_2\overset{\overset{\displaystyle HO}{|}}{C}HCH_2COPh \quad (4.217)$$

$$75\%$$

Although not yet fully developed, the following aldol reaction of potassium enoxytrialkylborates discovered in our laboratories

appears to be readily applicable to the hitherto difficult selective ketone-ketone condensation, as shown in eq. 4.218 (*4.405*).

$$\text{(4.218)}$$

Also highly promising are the aldol reaction of silyl enol
ethers, promoted by $TiCl_4$ (Sect. 6.4.2.2) or by trialkylammonium
fluorides (*4.406*), and the copper- and aluminum-moderated reaction of zinc enolates; see eq. 4.219, for example (*4.407*).

$$\text{(4.219)}$$

93%

(b) Stereoselectivity. Aldol condensation can be additionally complicated by the formation of stereoisomeric mixtures.
Consider, for example, the reaction of an enolate derived from a
primary alkyl ketone (R^1CH_2COR) with an aldehyde having an asymmetric carbon atom in the α-position ($R^1R^2R^3CCHO$), which proceeds
cleanly as in eq. 4.220.

$$R^1-\underset{\underset{R^3}{|}}{\overset{\overset{R^2}{|}}{C}}-CHO \;+\; R'HC=\overset{\overset{O^-}{|}}{CR} \longrightarrow R^1-\underset{\underset{\underset{H}{|}}{\underset{R^3}{|}}}{\overset{\overset{R^2}{|}}{C}}\underset{\underset{H}{|}}{\overset{\overset{OH}{|}}{\underset{\beta}{C}}}\underset{\alpha}{\overset{\overset{R'}{|}}{C}}-COR \qquad \text{(4.220)}$$

Since the α- and β-carbon atoms in the product can be either R or
S, there can be four diastereomers even in the event that the γ-
carbon atom is 100% R or S. In such a case, the stereochemical
relationship between the α- and β-carbon atoms, as well as between the β- and γ-carbon atoms, must be considered. The available data suggest that the relationship between the β- and γ-car-
bon atoms may be predicted based on Cram's rule (Sect. 4.3.2.2),
as depicted in eq. 4.221.

Much less clear is the relation between the α- and β-carbon atoms. At least three distinct stereochemical results have been reported. Dubois (4.408) and Heathcock (4.409) have found that aldol condensation is subject to kinetic stereoselection, with (E)- and (Z)-enolates giving exclusively or predominantly the threo and erythro aldols, respectively (eq. 4.222).

$$(4.222)$$

On the other hand, the completely opposite stereospecificity, that is, (E)→erythro and (Z)→threo, has been reported for the reaction of enoxydimethylalanes and acetoaldehyde (4.410). Finally, House (4.400) has reported a nonstereospecific but stereoselective aldol reaction which is carried out under the influence of $ZnCl_2$ or $MgBr_2$. Under such conditions, the threo isomer corresponding to the six-membered metal chelate intermediate with the greater number of equatorial substituents predominates, regardless of the enolate geometry (eq. 4.223).

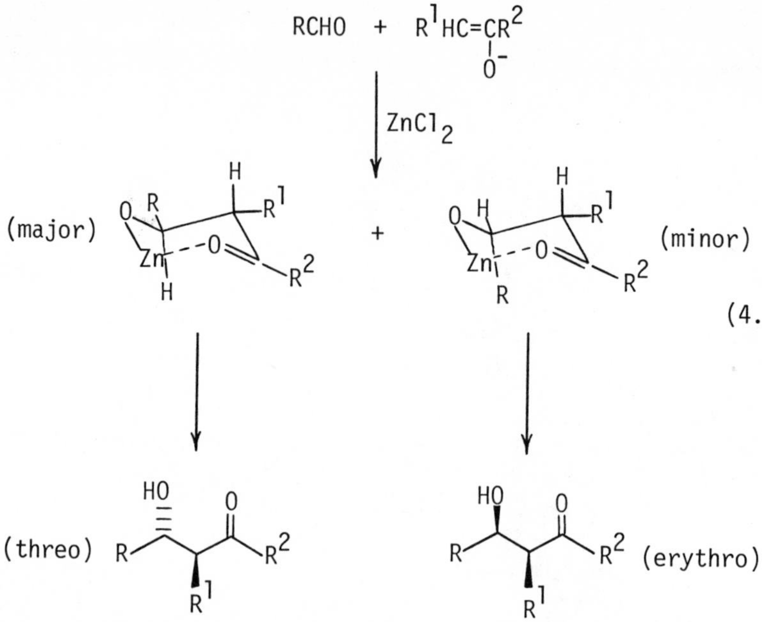

$$RCHO \;+\; R^1HC=CR^2$$
$$\underset{O^-}{|}$$

$$\downarrow ZnCl_2$$

(major) + (minor)

(4.223)

(threo)   (erythro)

The procedure developed by House has recently been applied to the synthesis of lasalocid A (4.23) obtained as the major isomer of a 40:10:7:3 mixture as shown in eq. 4.224 (4.411).

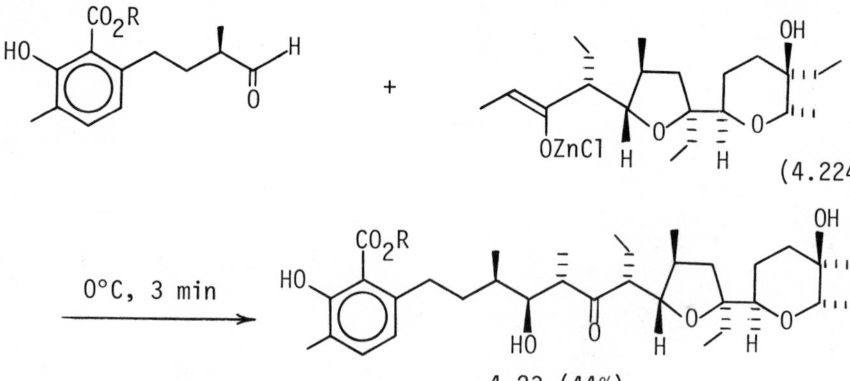

+

(4.224)

0°C, 3 min

4.23 (44%)

Miscellaneous Examples of the Aldol and Related Reactions.   Over the past several years, a number of variations of the aldol reaction have been reported.   Some representative examples are summarized in Table 4.15. In the reaction of dienolates with carbonyl compounds, carbon-carbon bond formation takes place predominantly in the γ-position.

Table 4.15. Miscellaneous Variations of the Aldol Reaction

Enolate Precursor	Base	Carbonyl Compound	Product Yield (%)	Ref.
(Esters and lactones)				
$CH_3COOBu\text{-}t$	LICA	R^1COR^2	high	4.412
Et Me $>$CCOOPr-i Cl	Li	$R^1COH(R^2)$	about 70	4.413
$CH_3COSBu\text{-}t$	LDA	$R^1COH(R^2)$	68 to 80	4.414
$XCH_2CO_2Bu\text{-}t$	LDA	$R^1COH(R^2)$	65 to 85	4.415
(X=OBu-t, NMe$_2$)				
$BrCH_2CO_2Et$	active Zn	$R^1COH(R^2)$	95 to 98	4.416
$C\equiv NCH_2COOR$	BuLi	$R^1COH(R^2)$	55 to 83	4.417
$MeCH=CHCOOEt$	LICA	Me_2CO	78	4.418
(bicyclic lactone structure)	LDA	CH_2O	95	4.419a
(dimethyl lactone structure)	LDA	CO_2	75	4.419b
(Acids)				
$R^1CHCOOH$ Br	Zn	$R^2C\equiv N$	63 to 77	4.420

(continued...)

Table 4.15 continued

Enolate Precursor	Base	Carbonyl Compound	Product Yield (%)	Ref.
$R^1R^2CHCOOH$	$BrMgN(Pr\text{-}i)_2$	$R^3COH(R^4)$	43 to 88	4.421
$Me_2C=CCOOH$ $\quad\ \ \|$ $\quad\ \ Br$	2 BuLi (-100°C)	R^1COR^2	30 to 64	4.422
$MeCH=CHCOOH$	$LiNEt_2$	R^1COR^2	48 to 64	4.423

(Amides)

Enolate Precursor	Base	Carbonyl Compound	Product Yield (%)	Ref.
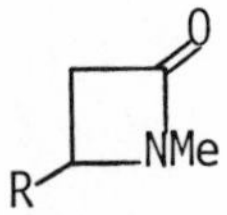	LDA	R^1COR^2	55 to 88	4.424
$R^1R^2CHCONEt_2$	$BrMgN(Pr\text{-}i)_2$	R^3COH	85 to 88	4.421
$R^1R^2CHCONMe_2$	$LiNEt_2$	$R^3COH(R^4)$	39 to 91	4.425
$MeCONMe_2$	LDA	$R^1COH(R^2)$	97 to 99	4.375

(Nitriles)

Enolate Precursor	Base	Carbonyl Compound	Product Yield (%)	Ref.
CH_3CN	$NaOC_8H_{17}\text{-}n$	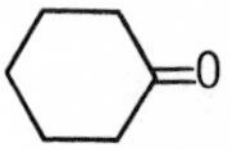 (cyclohexanone)=O	70	4.426
R^1R^2CHCN	$LiNEt_2$	$R^3COH(R^4)$	70 to 96	4.427
$PhCHCN$ $\ \ \|$ $\ \ OSiMe_3$	LDA	$R^1COH(R^2)$	96 to 100	4.428

(Other masked carbonyl compounds)

Enolate Precursor	Base	Carbonyl Compound	Product Yield (%)	Ref.
$CH_2=CHNH$—(cyclohexyl) 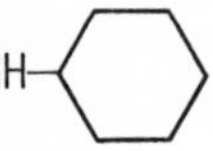	LDA	RCN	71 to 91	4.429
$PhCH_2C=NNHTs$ $\qquad\ \|$ $\qquad\ Me$	MeLi	R^1COR^2	–	4.430
$\qquad NNHTs$ $\qquad\ \|\|$ $R^1CCH_2R^2$	3 BuLi	$R^3COH(R^4)$	48 to 78	4.431

(continued...)

Table 4.15   continued

Enolate Precursor	Base	Carbonyl Compound	Product Yield (%)	Ref.
$R^1 \overset{NOH}{C}CH_2R^2$	2 BuLi	$R^3COH(R^4)$	45 to 55	*4.378b*
$R^1R^2CHCH=NNMe_2$	LiNEt$_2$	R^3COR^4	87 to 90	*4.432*
$Me_3SiCHR^1CH=NX$	LDA	R^2CHO	88 to 94	*4.433*
(X=Bu-t, NMe$_2$)				
Ph, O / N—CH$_3$ (MeO)	BuLi	RCHO	–	*4.434*
S / N—CH$_3$	n-BuLi	R^1COR^2	–	*4.435*

The Claisen Condensation and Related Reactions.   Since the results of earlier investigations of the Claisen, eq. 4.225 (*4.391*), Dieckmann, eq. 4.226 (*4.436*), and Thope-Ziegler, eq. 4.227 (*4.436*), reactions have been extensively reviewed, no detailed discussion of these relatively well-established reactions is presented.   The following comments may be of use, however, in applying these reactions to organic synthesis.

$$\text{(cyclohexanone)} + HCOOEt \xrightarrow{\text{NaH, EtOH}} \text{(2-formylcyclohexanone, CHO)} \quad 74\% \quad (4.225)$$

$$EtOC(CH_2)_4COEt \xrightarrow{\text{Na, EtOH}} \text{(ONa enol, COOEt)} \xrightarrow[H_2O]{\text{HOAc}} \text{(cyclopentanone, COOEt)} \quad 80\% \quad (4.226)$$

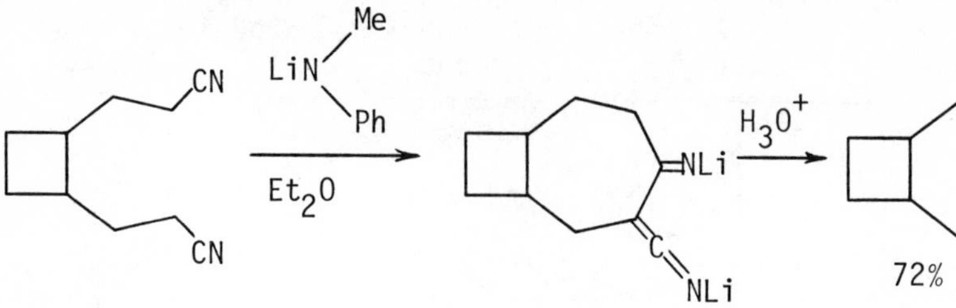

(4.227)

The Claisen condensation is an equilibrium reaction that tends to reverse itself in many cases. This difficulty can, in principle, be overcome by shifting the equilibrium through metallation of the 1,3-dicarbonyl product.

Under equilibrating conditions, the Claisen condensation may give mixtures of cross- and homo-condensation products. On the other hand, the regioselectivity of the Claisen condensation carried out under these conditions can often be distinctly higher than that in the alkylation of the same enolate, as shown in eq. 4.228 (4.336).

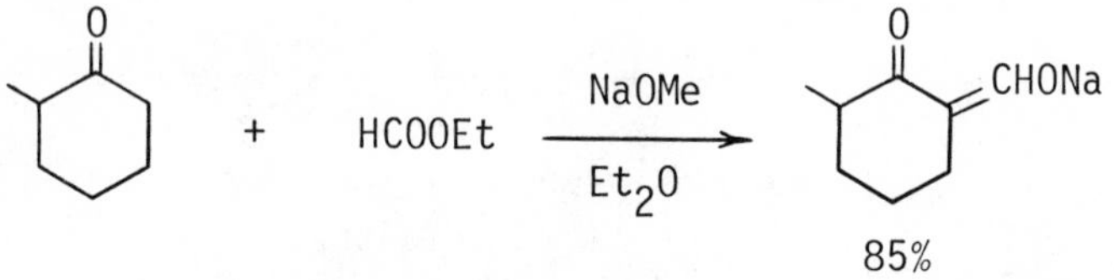

(4.228)

In contrast to the corresponding alkylation, the Claisen condensation tends to take place exclusively or predominantly in the less hindered α-position which has been interpreted in terms of the relative stability of the enolates of two possible 1,3-dicarbonyl compounds.

A recent trend in the Claisen and related reactions is to use preformed metal enolates and reactive acylating agents, such as acyl halides, as shown in eq. 4.229 to 4.231.

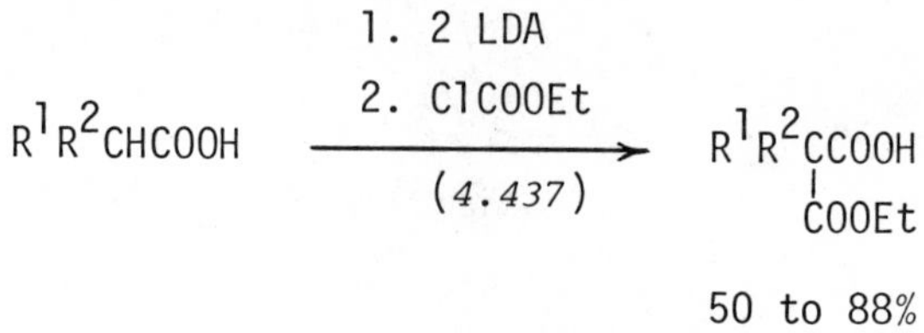

$$R^1R^2CHCOOH \xrightarrow[\text{(4.437)}]{\begin{array}{l}\text{1. 2 LDA}\\\text{2. ClCOOEt}\end{array}} R^1R^2\underset{\overset{|}{COOEt}}{CCOOH}$$

(4.229)

50 to 88%

$$R^1R^2CHCOOH \quad \xrightarrow[\substack{(4.438)}]{\substack{\text{1. 2 LDA} \\ \text{2. } R^3COCl \\ \text{3. aq HCl} \\ \text{4. } 150\text{-}200°C}} \quad R^1R^2CHCOR^3 \qquad (4.230)$$

30 to 70%

$$R^1R^2CHCOR^3 \quad \xrightarrow[\substack{(4.439)}]{\substack{\text{1. LDA} \\ \text{2. } RCO_2CO_2Et}} \quad \underset{R^2}{\overset{R^1}{RCOCCOR^3}} \qquad (4.231)$$

49 to 76%

It is noteworthy that, under these kinetic conditions, even race-mizable chiral centers can retain much of their chirality, as shown in eq. 4.232 (*4.440*).

$$CH_3CO-C\overset{Me}{\underset{Et}{\diagdown_{''''H}}} \quad \xrightarrow[\substack{\text{2. } +COCl}]{\substack{\text{1. LDA}}} \quad +COCH_2CO-C\overset{Me}{\underset{Et}{\diagdown_{''''H}}}$$

retention of configuration

(4.232)

4.3.4.4   The Michael Reaction

The Michael Reaction and the Robinson Annulation Reaction.   The conjugate addition of enolate anions to α,β-unsaturated carbonyl and related compounds is known as the Michael reaction; see eq. 4.233 (*4.441*).

$$\underset{|}{\overset{O-M}{\underset{|}{-C=C-}}} + \underset{|\quad|}{\overset{O}{\overset{\|}{-C=C-C-}}} \rightleftharpoons \underset{|\quad|\quad|}{\overset{O\quad\quad OM}{\overset{\|}{-C-C-C-C=C-}}} \qquad (4.233)$$

As is the case in any other conjugate addition reaction of or-ganometallics, the products of the Michael reaction are metal en-olates that can participate in various reactions that typical en-olates undergo.   Unlike the reaction of "ordinary" proximally nonfunctional organoalkali and organoalkaline earth metals with

α,β-unsaturated carbonyl compounds, in which the 1,2-addition
path tends to predominate, the corresponding reaction of metal
enolates usually undergoes the 1,4-addition. Although it is not
completely clear, the above-mentioned difference, in part, must
be due to the charge delocalized or "soft" nature of enolates,
the effects of which are at least twofold. Firstly, the softness
of enolates should favor attack at the soft olefinic site. It is
instructive to note that the dianions of free carboxylic acids
incapable of effective charge delocalization tend to undergo 1,2-
addition, as shown in eq. 4.234 (*4.442*).

$$1. \ PhSC^-HCOO^-$$
$$2. \ CH_2N_2$$

OH
CH(SPh)COOMe

(4.234)

$$PhSC^-HCOOMe$$

PhSCHCOOMe

Secondly, charge delocalization makes enolate anions relatively
weak bases in the Brønsted sense, which should promote the reversal of both 1,2- and 1,4-addition processes. Under equilibrating
conditions, the formation of the 1,4-adduct should be favored, as
discussed in Sect. 4.3.2.4.
    The Michael reaction represents one of the most widely used
carbon-carbon bond-forming reactions. Particularly attractive
from the viewpoint of natural product synthesis is the combination of the Michael reaction with the aldol reaction, known as
the Robinson annulation (*4.443*). In favorable cases, the Michael
reaction and the Robinson annulation proceed quite satisfactorily. Thus, for example, relatively acidic carbonyl compounds,
such as 2-methyl-1,3-cyclohexanedione, can be converted into
Michael adducts and annulated products in high yields, as shown
in eq. 4.235 (*4.444*).

KOH
MeOH

64%

(4.235)

With less acidic carbonyl compounds the Michael reaction
tends to be complicated by various problems including (1) polymerization of α,β-unsaturated carbonyl compounds, (2) regiochemical scrambling, (3) the reverse Michael reaction, and (4) competition arising from the aldol, Claisen, and other reactions of
enolates. Various procedures have been developed to circumvent
these difficulties. An early trend was to use the mildest possible bases, such are amines and hydroxides, and to run the reaction in protic solvents in order to minimize certain side reactions. Other early modifications include (1) the use of the
Mannich bases (4.445) and β-halo ketones as precursors of enones
and (2) the use of enamines (4.446) in place of metal enolates.
For extensive discussions of these older techniques, the reader
is referred to reviews by Bergamann (4.441), House (4.316), and
Jung (4.339a).
     None of the above modifications, however, permits clean re-
giospecific trapping of "kinetic" enolates by α,β-unsaturated
carbonyl compounds, which would require generation of "kinetic"
enolates under aprotic conditions and their consumption via the
desired conjugate addition at rates faster than undesirable side
reactions. This was first achieved by Stork (4.447) with α-
silylated enones, such as methyl α-trimethylsilylvinyl ketone, as
shown in eq. 4.236.

(60%)   (4.236)

     The trimethylsilyl group presumably accelerates the Michael
reaction through its anion-stabilizing effect and, perhaps more
importantly, retards the polymerization of the enone through its
steric hindrance. The results shown in eq. 4.236 indicate that
the desired Michael reaction is faster than the regio-isomeriza-

tion of the "kinetic" enolate intermediate.

Another recent trend in the area of annulation is to use the alkylation of enolates as a more flexible substitute for the Michael reaction, as discussed in Sect. 4.3.4.2. Taking advantage of the fact that the product of the Michael reaction is an enolate, the following efficient syntheses of bicyclic; see eq. 4.237 (*4.448*), and tricyclic; see eq. 4.238 (*4.449*), compounds involving the double Michael annulation have been achieved.

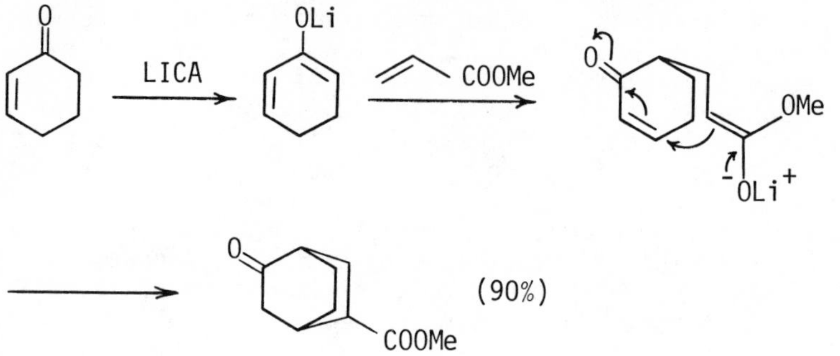

(4.237)

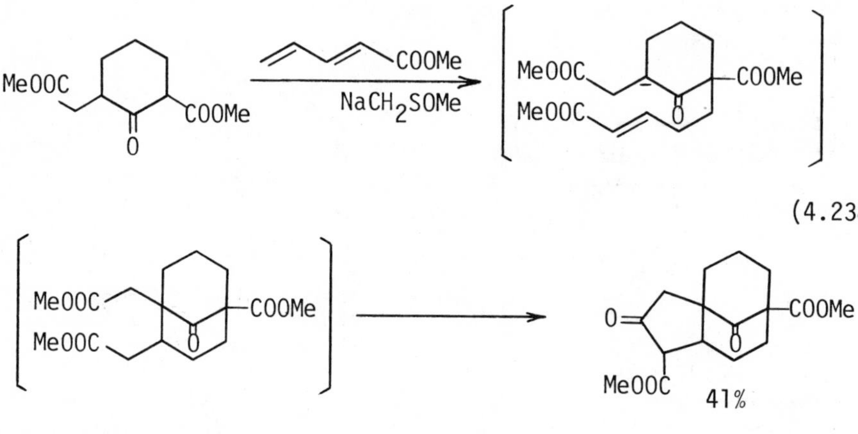

(90%)

(4.238)

41%

Miscellaneous Variations of the Michael Reaction. Various α-hetero-substituted enolates have successfully been used in the Michael reaction. The structures of some α-hetero-substituted enolates are shown in Table 4.16.

Table 4.16.   The Use of α-Hetero-Substituted Enolates in the Michael Reaction

Enolate	Acceptor	Yield (%)	Ref.
ONa $\text{MeO}\overset{\mid}{\text{C}}=\text{C(SEt)}_2$	$\text{H}_2\text{C}=\text{CHCOX}$ $(\text{X} = \text{Me, OMe, OBu-}t)$	92 to 95	*4.211*
LiO, EtO $\text{C}=\text{C}$ (dithiolane ring)	$\text{MeC}=\text{CHCOOEt}$ $\overset{\mid}{\text{Cl}}$	80	*4.204*
$\text{MN}=\text{C}$(allyl, OR)	cyclohexenone	90	*4.450*
LiO, EtO $\text{C}=\text{CHN}=\text{CHPh}$	cyclohexenone	90	*4.451*
$\text{LiN}=\text{C}$(C_6H_{13}, SePh)	cyclohexenone	91	*4.452*

In addition to the α-silyl enones, α-sulfur-substituted α,β-unsaturated carbonyl compounds, such as 4.24 (*4.453*), have also been used as Michael acceptors. Other noteworthy additions to the list of acceptors include an iron π-complex 4.25 (*4.454*), cyclopropyl ketones and esters, such as 4.26 (*4.455*), 2-vinyl-6-methylpyridine, such as 4.27 (*4.456*), and a vinyl sulfide sulfoxide 4.28 (*4.457*).

SMe
$\text{CH}_2=\overset{\mid}{\text{C}}\text{COOMe}$ $\text{CH}_2=\text{CHCOMe}$ COOMe (cyclopropyl, $^+\text{PPh}_3\text{Br}^-$)   (pyridine vinyl) SMe
                              $\eta^5\text{-C}_p\text{Fe(CO)}_2$ $\text{CH}_2=\overset{\mid}{\text{C}}\text{SOMe}$

4.24   4.25   4.26   4.27   4.28

The 2-vinylpyridine derivative (4.27) has been developed as an efficient bis-annulating agent; see eq. 4.239 (*4.456*).

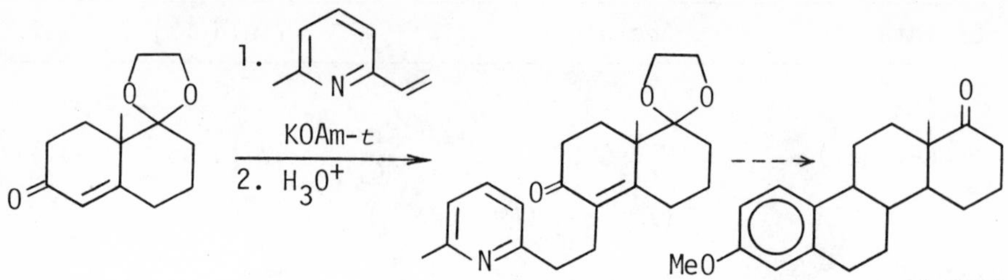

(4.239)

80%

A vinyl sulfide sulfoxide 4.29 has been used in an efficient synthesis of 4.30; see eq. 4.240 (*4.458*).

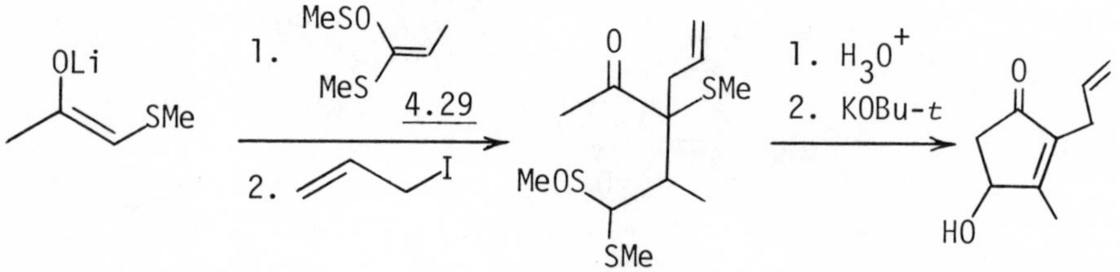

99%

4.30 (80%)

(4.240)

4.3.4.5 Reactions of Metal Enolates and Related Compounds with Epoxides

As shown in Scheme 4.6, the reaction of enolates with epoxides
produces γ-hydroxy carbonyl compounds. A few noteworthy examples
follow (eqs. 4.241 to 4.244).

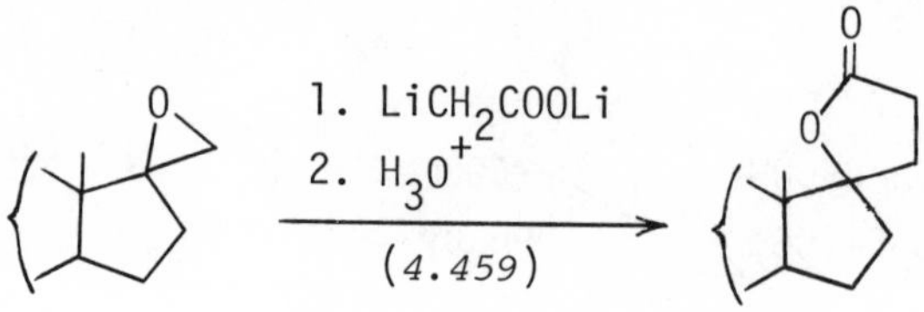

(55%)   (4.241)

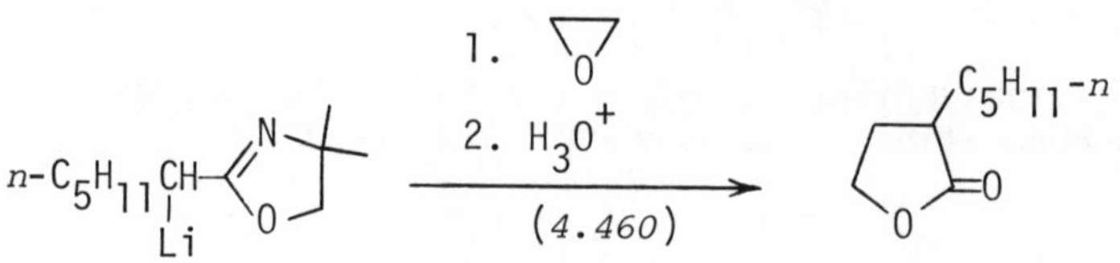

(72%)   (4.242)

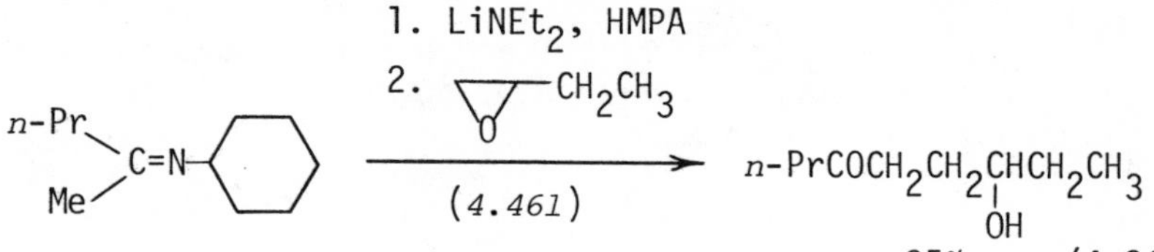
n-Pr
C=N
Me
1. LiNEt2, HMPA
2. CH2CH3
O
(4.461)
n-PrCOCH2CH2CHCH2CH3
OH

95% (4.243)

Note that the reaction shown in eq. 4.243 is readily applicable to the synthesis of 1,4-diketones.

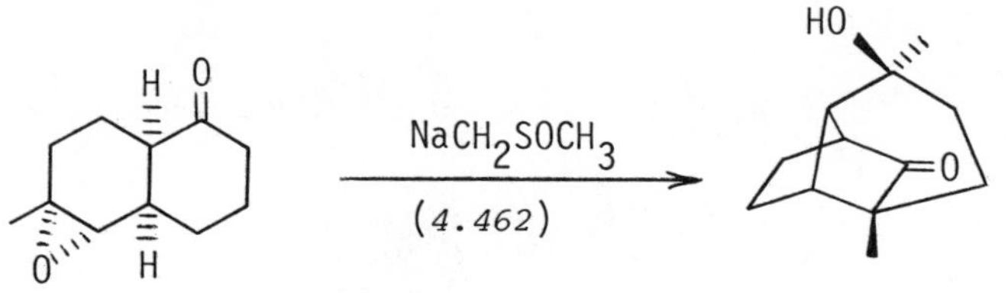
H
O
O
H
NaCH2SOCH3
(4.462)
HO
O

(93%)

(4.244)

Very interesting results have been observed in the epoxynitrile cyclization, which can be explained in terms of stereoelectronic constraints (*4.463*). For example, the four-membered ring formation is favored over the usually favored five-membered ring formation in the reaction shown in eq. 4.245 (*4.463b*).

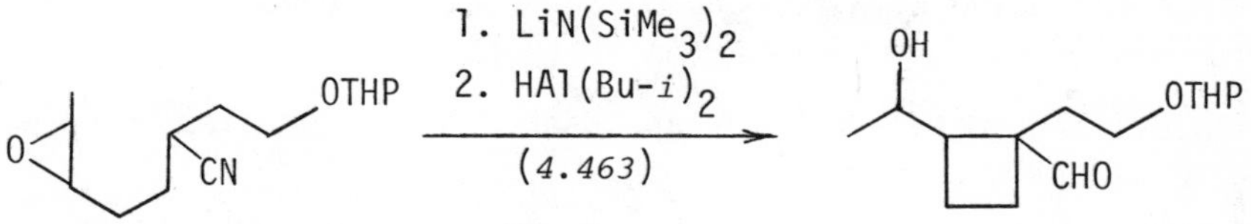
O
OTHP
CN
1. LiN(SiMe3)2
2. HAl(Bu-i)2
(4.463)
OH
OTHP
CHO

52% (4.245)

The back-side attack at the tertiary carbon atom of the epoxide ring leading to the five-membered ring formation is not competitive, presumably because the carbanionic center cannot readily reach around and attack the tertiary carbon from the side opposite the oxygen atom, as depicted in eq. 4.246.

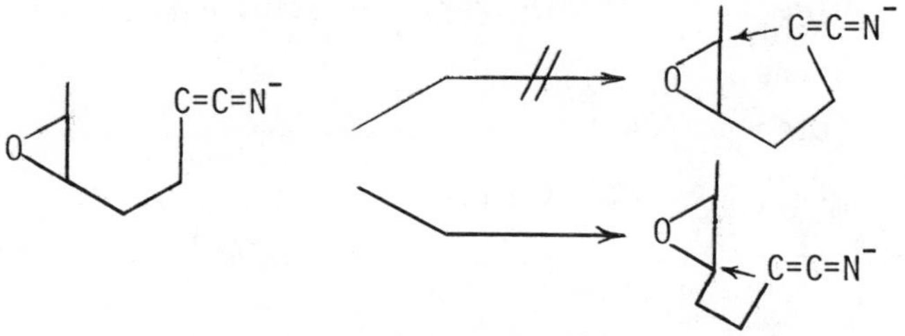
O
C=C=N⁻
O
C=C=N⁻
O
C=C=N⁻

(4.246)

As expected, the three-membered ring formation is favored over the four-membered ring formation, and partition between the five- and six-membered rings appears to be governed by various factors including steric hindrance.

4.3.4.6   Carbon-Carbon Bond Formation via Masked Enolates

The chemistry of metal enolates can, in principle, be expanded by changing various structural parameters of the metal enolates. A priori, any metal-containing species that may be classified either as metal enolates or their hetero analogues would be represented by either 4.31 or 4.32.

$$
\begin{array}{cc}
\text{M-X-C=C-} & \text{M-X=C=C-} \\
\quad\;\; | \;\; | & \qquad\quad | \\
\quad\;\; Y & \\
\end{array}
$$

4.31   4.32

In reality, metallated nitriles appear to be the only readily accessible species represented by 4.32. Similarly, only O, N, and S can be X in 4.31. Interestingly, few species represented by 4.31 in which X is S have been developed as useful reagents or intermediates in organic synthesis. We therefore restrict our discussion to those species represented by 4.31 in which X is either O (4.31a) or N (4.31b). On the other hand, various elements including C, N, O, and S can be Y, as shown in Table 4.17.

Table 4.17.   Metal Enolates and Masked Metal Enolates

Structure	Organic Precursors	Structure	Organic Precursors

M-O-C=C-
   | |
  -C-
    |
	aldehydes,
	ketones
		M-N-C=C-
		   | | |
		  -C-
		    |
			imines, oximes,
			hydrazones, and so on

M-O-C=C-
   | |
  -O
	acids,
	esters,
	lactones
		M-N-C=C-
		   | | |
		  -O
			-N=C-CH-
			   | |
			  -O

M-O-C=C-
   | |
   N
  ⁄ ⧹
	amides,
	lactam
		M-N-C=C-
		   | | |
		   N
		  ⁄ ⧹
			-N=C-CH-
			   | |
			   N
			  ⁄ ⧹

M-O-C=C-   thiol-ester   M-N-C=C-   -N=C-CH-
   | | | | | | |
  -S   -S   -S

If the metallated thiol-esters are excluded, the chemistry of 4.31a has now been reasonably well established, as discussed in several preceding sections. On the other hand, active and systematic investigations of the chemistry of 4.31b are largely of recent origin. For the sake of convenience, we arbitrarily term the metallated nitriles and species represented by 4.31b masked metal enolates. Many examples of these species are listed in Tables 4.14 and 4.15, and a few of their reactions have already been discussed in Sects. 4.3.4.2 to 4.3.4.5. In this section, we focus our attention on just one well-investigated case, that is, the chemistry of oxazolines and dihydro-1,3-oxazines, which seems to represent new directions in the chemistry of enolates. Although not discussed here, it may be pointed out that investigations of metal enolates and their masked analogues containing various metals other than Group IA and Group II elements, such as B, Al, Si, Sn, and transition metals, also appear to be a promising area.

Carbon-Carbon Bond Formation via Oxazolines and Dihydro-1,3-oxazines. Over the past decade Meyers and his associates have developed oxazolines, 4.33 and 4.34, for example, and dihydro-1,3-oxazines, 4.35, for example, as useful masked enolates. Several reviews on the subject are available (4.464, 4.465).

4.33   4.34   4.35

The oxazoline and dihydro-1,3-oxazine rings are relatively stable to bases. Although organolithiums can attack the C=N bond of these heterocycles (eq. 4.247), it is generally inert to Grignard reagents, unless it is converted to the corresponding iminium salt. Thus their reactivity as masked carbonyl derivatives is generally lower than that of cyanides. On the other hand, these heterocycles are relatively sensitive to acids and can readily be converted into the corresponding carboxylic acids and esters on acidic solvolysis, as shown in eq. 4.248.

$$\text{(4.247)}$$

$$\text{(4.248)}$$

Alternatively, dihydro-1,3-oxazines can be converted into aldehydes via reduction with $NaBH_4$ followed by hydrolysis. Furthermore, the α-hydrogen atom of an alkyl substituent in the 2-posi-
tion can readily be abstracted with strong bases, such as n-BuLi
and LDA. These properties can now be exploited in developing
various procedures for the syntheses of aldehydes, as shown in
eq. 4.249 (*4.386*), ketones, as shown in eq. 4.247 (*4.387a*), carboxylic acids and esters, as shown in eq. 4.250 (*4.387b*), and lactones, as shown in eq. 4.242 (*4.460*).

$$\text{(4.249)}$$

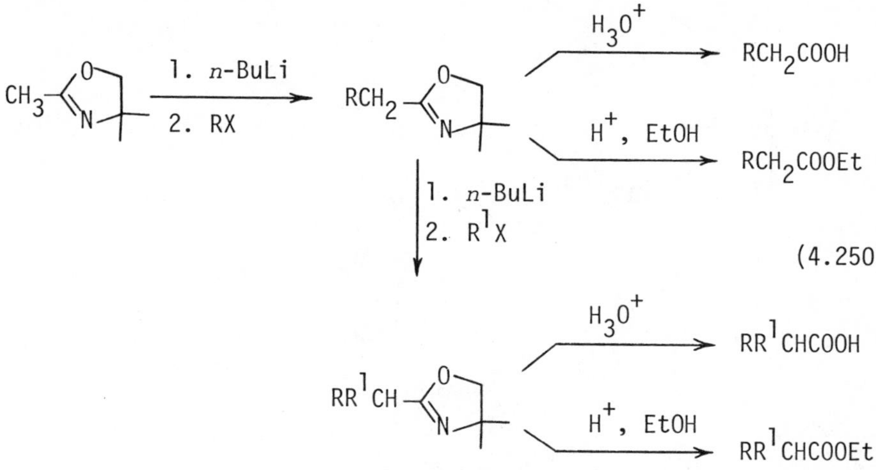

$$(4.250)$$

The results shown above indicate that the whole of enolate chemistry could be rewritten using 1,3-oxazolines and dihydro-1,3-oxazines as masked carbonyl compounds. One critical question that might be raised here is whether or not it is necessary or desirable to attempt to do so. This author believes that the development of new procedures for carbon-carbon bond formation is still a viable research project so long as the procedures developed are satisfactory and offer some unique synthetic utilities.

Recent investigations by Meyers have established that certain readily available chiral oxazolines, for example, 4.34, are highly promising reagents for asymmetric carbon-carbon bond formation. For a highly stimulating discussion, the reader is referred to a review by Meyers (4.464c). The following examples represent a few of the as yet very small number of asymmetric carbon-carbon bond-forming reactions that may be considered useful (ee > 70 to 80%).

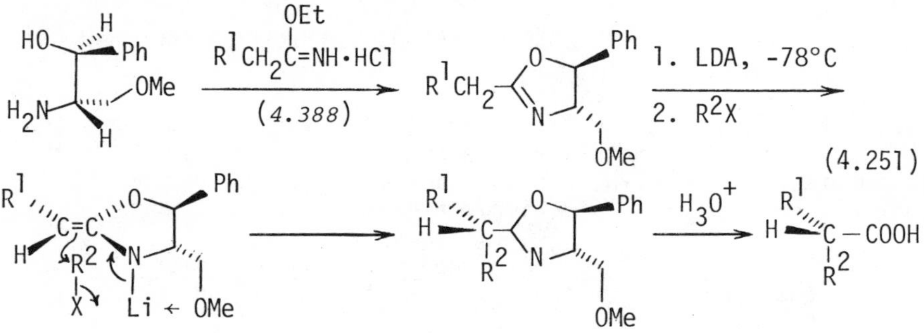

$$(4.251)$$

(4.466)

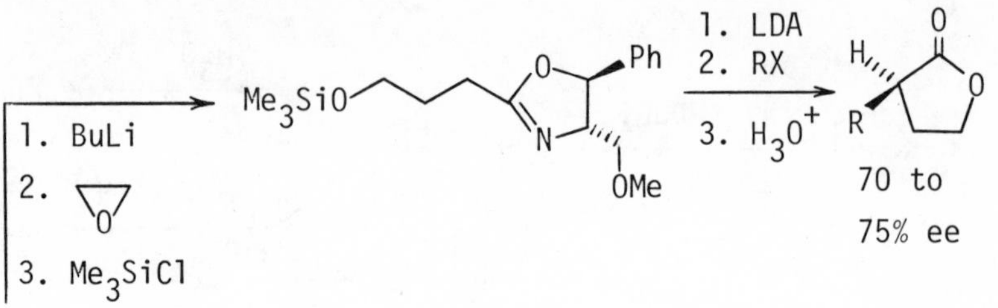

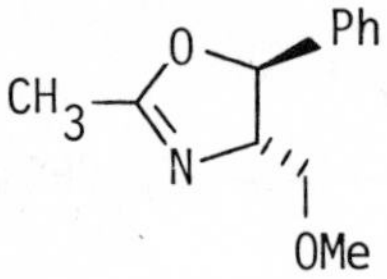

(4.252)

$(4.464c)$

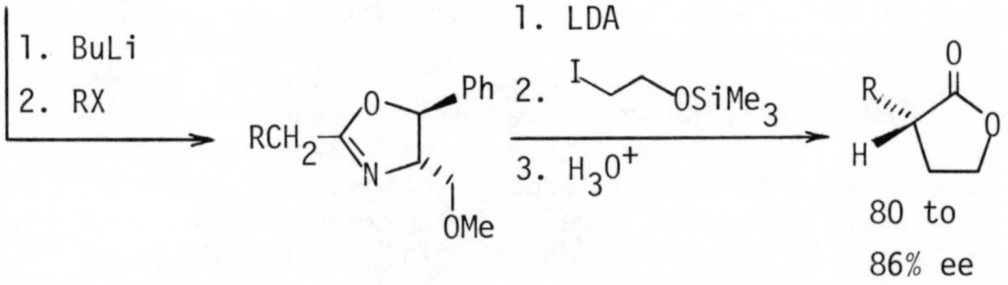

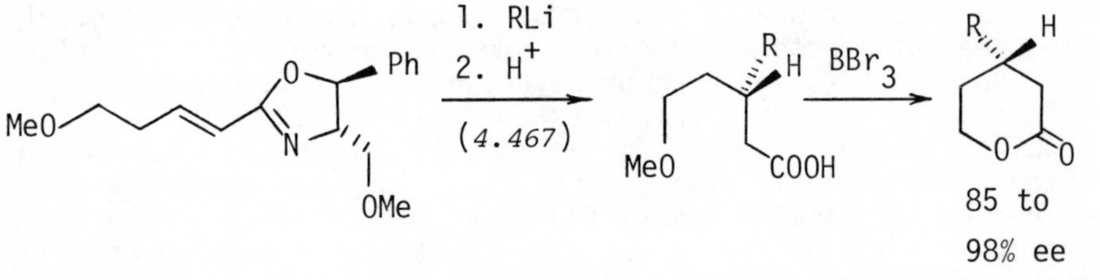

(4.467)

4.4   CARBON-HETERO ATOM BOND FORMATION VIA ORGANOALKALI AND OR-GANOALKALINE EARTH METALS

The metal atom of the metal-carbon bond of organoalkali and or-ganoalkaline earth metals can be replaced with either nonmetallic elements, such as H, N, O, P, S, and halogens, or various metal-lic elements (transmetallation).

4.4.1   Formation of Carbon-Hydrogen Bonds -- Reduction of Organic Compounds via Organometallics of Group IA and Group II Elements

Conversion of the metal-carbon bond of organometallics of Group
IA and Group II metals can be achieved either via protonolysis
or via hydrogenolysis (eq. 4.254). Deuterium labeling can be
achieved through the use of the corresponding deuterium compounds.

$$
RM
\begin{array}{l}
\xrightarrow{\ HX(DX)\ } RH(D)\ +\ MX \\[2ex]
\xrightarrow[H_2(D_2)]{} RH(D)\ +\ MH(D)
\end{array}
\qquad (4.254)
$$

Except for organomercury compounds, organometallics of Group IA
and Group II metals are readily protonolyzed by water (hydrolysis), alcohols, and various other proton acids. Many of these
reactions take place cleanly and quantitatively. Since the protonation is operationally simpler and generally cleaner than the
corresponding hydrogenation, the latter reaction is seldom used
in organic synthesis. In cases where the metal-bearing carbon
atom is configurationally stable, the hydrolysis of organoalkali
and organoalkaline earth metals appears to proceed generally with
retention of configuration.

$$
\begin{array}{c}
R^1 \diagdown \quad \diagup H \\
\qquad C=C \\
R^2 \diagup \quad \diagdown Li
\end{array}
\xrightarrow{\ D_2O\ }
\begin{array}{c}
R^1 \diagdown \quad \diagup H \\
\qquad C=C \\
R^2 \diagup \quad \diagdown D
\end{array}
\qquad (4.255)
$$

$$
\xrightarrow[\ (4.468)\ ]{\ D_2O\ }
\qquad (4.256)
$$

Although protonolysis is one of the most significant and
widely used methods for the conversion of organometallics of
Group IA and Group II metals into organic products, the successful use of simple protonation reactions does not require extensive discussion here. In this section, only those protonation
reactions that pertain to the reduction of organic compounds are
briefly discussed.

4.4.1.1 Reduction with Dissolving Metals -- General Discussions

Addition of electrons to organic substances usually takes place
either by one-electron or two-electron transfer. Dissolving

metal reduction generally involves a series of one-electron transfer steps, while reduction with metal hydrides, including catalytic hydrogenation, usually, but not always, involves two-electron transfer reactions.

Metals. The metals used commonly in the dissolving metal reduction include Li, Na, K, Mg, Ca, Zn, Al, Sn, Ti, and Fe. The versatility of the alkali and alkaline earth metals in the dissolving metal reduction stems from the fact that they are readily oxidized to the corresponding cations, as indicated by their low ionization energies (Appendix IV) and by their high oxidation potentials (Table 4.18).

Table 4.18. Oxidation Potentials[a] (in liquid NH_3 at 25°C)

Reaction	ε^0 (V)
$Li \longrightarrow Li^+ + e^-$	+2.34
$Na \longrightarrow Na^+ + e^-$	+1.89
$K \longrightarrow K^+ + e^-$	+2.04
$Mg \longrightarrow Mg^{+2} + 2e^-$	+1.74
$Ca \longrightarrow Ca^{+2} + 2e^-$	+2.17
$Zn \longrightarrow Zn^{+2} + 2e^-$	+0.54

[a]The oxidation potentials in the table are taken from Jolly, W. L., *J. Chem. Educ.*, **33**, 512 (1956).

General Scheme and Selection of Proton Donors. Both multiple bonds and single bonds have been reduced with dissolving metals. Although the exact course of reduction depends on a number of factors, including metals, proton donors, and solvents, the general scheme for the dissolving metal reduction may be represented by Scheme 4.9. For the sake of simplicity, only the scheme for the double bond reduction is shown. The scheme for the reduction of single or triple bonds may be obtained by deleting or adding one X-Y bond, respectively.

Scheme 4.9

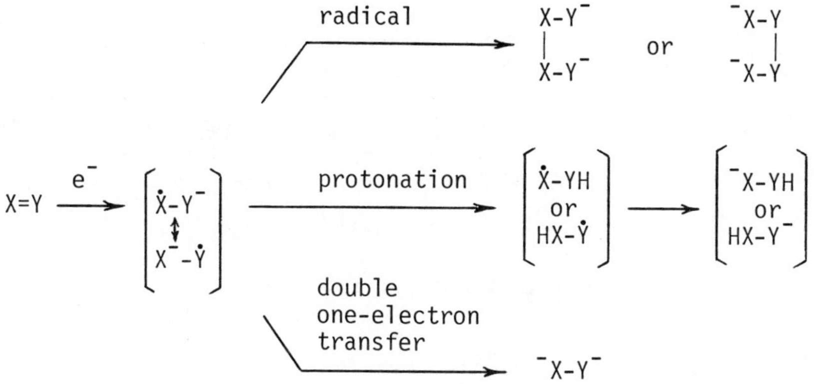

Of the three paths shown in Scheme 4.9, the double one-electron transfer process is relatively uncommon and appears to be practically limited to cases where the electrostatic repulsion between the two negatively charged centers can be minimized (1) through delocalization, (2) by the formation of tight ion pairs or highly covalent species, or (3) by the use of solvents of low dielectric constant. A typical example is shown in eq. 4.257 (4.469).

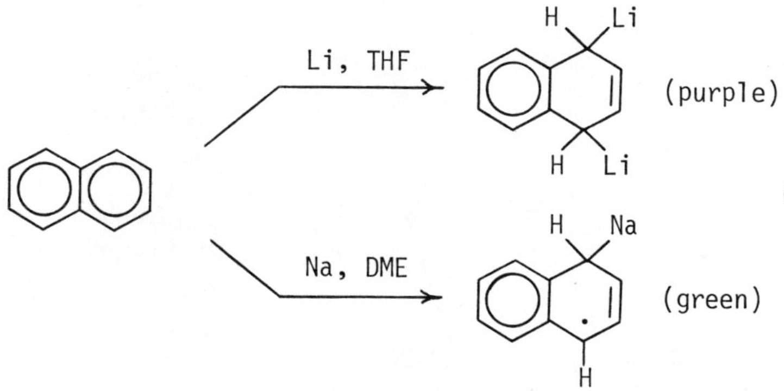

(4.257)

The relative rates of radical coupling and protonation reactions can be controlled most readily by controlling the rate of protonation which, in turn, can be achieved by choosing an appropriate proton donor. In the absence of a substance that can act as an efficient proton donor, radical coupling becomes the predominant reaction path. Even if a proton donor is used, radical coupling can still be a serious side reaction. Effective suppression of radical coupling can, in principle, be achieved by promoting protonation through the use of strong acids. Unfortunately, however, metals of high oxidation potentials are readily decomposed by strong acids, which severely restricts the selection of proton donors. From the practical viewpoint, the dissolving metal reductions may be classified into the following four categories according to the proton donor used:

1. Reduction with metals and protic acids ($pK_a < 10$).

2. Reduction with metals and neutral protic compounds (pK_a 15 to 30),

3. Reduction with metals and protic bases ($pK_a > 30$),

4. Reduction with metals in aprotic solvents.

Reduction Potential of the Organic Substrates. It is essential to select a proper reagent combination that best suits the desired transformation. In addition to the parameters discussed above, the reduction potential of the organic substrate, which can be quantitatively expressed in terms of the polarographic half-wave reduction potential defined as shown in Fig. 4.1, is an important criterion in this respect.

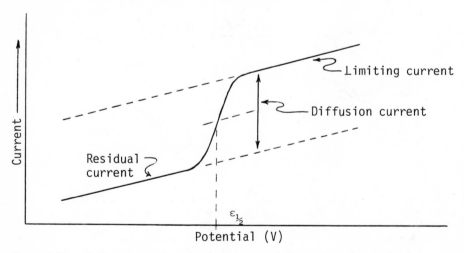

Fig. 4.1. Polarographic half-wave reduction potential ($\varepsilon_{\frac{1}{2}}$)

When the electrode potential is low, only the residual current
is observed. As it approaches the reduction potential of the
substrate, a rapid reduction, which is reflected by a sudden increase in current, is observed. It reaches the limiting value,
however, as the reduction becomes of diffusion control. The
electrode potential corresponding to the half-way point of the
diffusion current is defined as the polarographic half-wave pot-
ential ($\varepsilon_{\frac{1}{2}}$). The reduction potential is a measure of the electron affinity, that is, the energy liberated when an electron is
added to the substrate. Thus it should be roughly proportional
to the energy of the lowest unoccupied molecular orbital (LUMO).
     Table 4.19 lists the half-wave reduction potentials of some
organic substances versus a standard calomel electrode (SCE).
The smaller the absolute value is, the easier the reduction is.

Table 4.19. The Half-Wave Reduction Potentials of Some Organic
             Compounds[a]

Compound	$\varepsilon_{\frac{1}{2}}$ versus SCE	Compound	$\varepsilon_{\frac{1}{2}}$ versus SCE
(C₆H₅)₃CH	-1.05	$CH_3CH=CHCO_2Me$	-2.33[b]
anthracene-CHO	-1.21	C₆H₅-CH=CH₂	-2.37
C₆H₅-CH=CHCHO	-1.34	phenanthrene	-2.46
C₆H₅-CHO	-1.59	t-Bu-cyclohexene-CO₂Me	-2.50[b]
C₆H₅-CH=CHCOCH₃	-1.64[b]	t-Bu-cyclohexene-CN	-2.55[b]

(anthracene structure)	-1.96	$CH_2=CHCH=CH_2$	-2.63
(stilbene structure) —CH=CH—	-2.16	(biphenyl structure)	-2.70

[a]Unless otherwise noted, the values are taken from Streitwieser, A., Jr., *Molecular Orbital Theory for Organic Chemists*, John Wiley & Sons, New York, 1961.

[b]House, H. O., *Acc. Chem. Res.*, **9**, 59 (1976).

The following discussion of specific dissolving metal reductions and related reactions is very brief. For further details, the reader is referred to a textbook by House (*4.316*) and references cited in this section.

4.4.1.2 Reduction of Carbonyl Compounds with Alkali and Alkaline Earth Metals

The Bouveault-Blanc and Related Reductions. Aldehydes, ketones, and esters can be reduced to alcohols with alkali metals in the presence of alcohols, such as EtOH, used as proton donors. The reduction of esters by this method is called the Bouveault-Blanc reduction which most likely proceeds by the following mechanism (eq. 4.258).

$$RCOOEt \xrightarrow{Na} R\overset{\displaystyle\cdot}{\underset{\displaystyle ONa}{C}}OEt \xrightarrow{EtOH} R\overset{\displaystyle\cdot}{\underset{\displaystyle OH}{C}}OEt \xrightarrow{Na} R\overset{\displaystyle Na}{\underset{\displaystyle OH}{C}}OEt$$

$$\xrightarrow{EtOH} R\overset{\displaystyle H}{\underset{\displaystyle OH}{C}}OEt \longrightarrow \underset{\displaystyle O}{RCH} \xrightarrow{Na} \underset{\displaystyle ONa}{R\overset{\displaystyle\cdot}{C}H} \xrightarrow{EtOH} \underset{\displaystyle OH}{RCH} \quad (4.258)$$

$$\xrightarrow{Na} \underset{\displaystyle OH}{R\overset{\displaystyle Na}{C}H} \xrightarrow{EtOH} RCH_2OH$$

Similar mechanisms may be written for the reduction of aldehydes and ketones. The dissolving metal reduction of ketones tends to produce thermodynamically more stable products, as shown in eq. 4.259 (4.470).

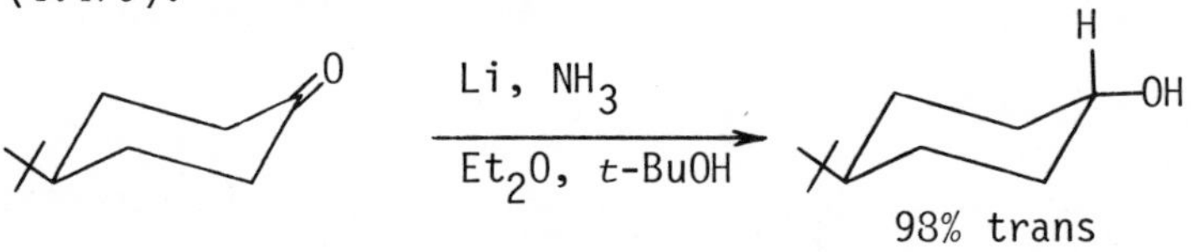

$$\text{(4.259)}$$

98% trans

In this respect, the dissolving metal reduction is often complementary with the metal hydride reduction, which tends to give the thermodynamically less stable isomers (Sect. 5.2.3.6).

The Clemmensen Reduction. In the presence of efficient proton donors, such as HCl, the reduction of ketones and aldehydes with zinc amalgam can produce the corresponding methylene and methyl compounds, respectively. The Clemmensen reduction has been reviewed in a comprehensive manner by Martin and Vedejs (4.471). The reaction is commonly run in a three-phase system consisting of zinc amalgam, aqueous HCl, and a nonpolar organic solvent, such as toluene, to minimize unwanted radical coupling. Amalgamation of zinc raises its hydrogen overvoltage to the point where it can resist the attack by HCl. Hydrogen halides are about the only strong acids whose anions are not reduced by zinc amalgam. It should be noted that the potential required for reduction of ketones in strong acidic media is substantially less (ca. -1.0 V) than that in neutral media (-2.0 to -2.5 V).

The mechanism of the Clemmensen reduction is unclear, since alcohols are not reduced under the Clemmensen conditions, they cannot be intermediates. The reaction is often complicated by skeletal rearrangements. The following scheme is based on the one that was proposed in the literature; see eq. 4.260 (4.472).

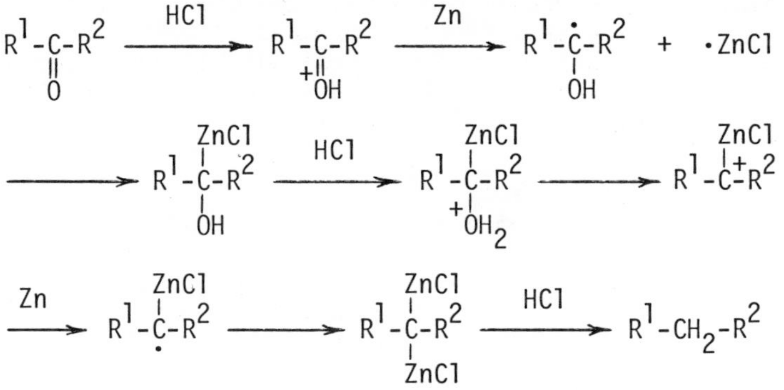

$$\text{(4.260)}$$

The intermediacy of a carbocationic intermediate in the above mechanism is consistent with frequently observed skeletal rearrangements. Any of the organozinc intermediates shown in eq. 4.260 can be protonolyzed under the reaction conditions. Thus various minor modifications of the above mechanism may also be considered as plausible paths.

It is useful to know that the deoxygenation of ketones can also be achieved by various other methods including (1) the Wolff-Kishner reduction, (2) treatment of tosylhydrazones with $NaBH_3CN$ (Sect. 5.2.3.6), (3) reduction of ketones to alcohols followed by tosylation and reduction with $LiAlH_4$ or $LiBR_3N$ (Sect. 5.2.3.6), and (4) conversion of ketones into thioketals followed by hydrogenation over Raney nickel (Sect. 10.1.3).

The reduction of α-halo, α-hydroxy, and α-acetoxy ketones with zinc and HOAc follows a different reaction path, the overall change being the removal of the α-hetero substituents (eq. 4.261).

$$R^1-\overset{\overset{O}{\|}}{C}-\overset{\overset{X}{|}}{C}R^2R^3 \xrightarrow{\text{Zn, HOAc}} R^1-\overset{\overset{O}{\|}}{C}-\overset{\overset{H}{|}}{C}R^2R^3 \qquad (4.261)$$

X = halogen, OH, OAc, and so on

The reaction can be used in tandem with the acyloin reaction for the conversion of dicarboxylic acid esters into cyclic ketones, as shown in eq. 4.262 (*4.473*).

$$EtOOC(CH_2)_8COOEt \xrightarrow[\text{2. }H_3O^+]{\text{1. Na}} (CH_2)_8\begin{array}{c}C=O\\ \\ CHOH\end{array} \xrightarrow[\substack{HCl\\HOAc}]{\text{Zn}} (CH_2)_8\begin{array}{c}C=O\\ \\ CH_2\end{array} \qquad (4.262)$$

The mechanism of the reaction has not been clarified. It may involve the initial attack by zinc either at the carbonyl group, as in the Clemmensen reduction, or at the C-X bond. Alternatively, zinc may interact simultaneously with both the carbonyl and the C-X bond, as shown in eq. 4.263.

$$R^1-\overset{\overset{O}{\|}}{C}-\overset{\overset{X}{|}}{C}R^2R^3 \xrightarrow{\text{Zn}} R^1-\overset{\overset{O}{\|}}{C}\overset{Zn}{\curvearrowright}\overset{X}{C}R^2R^3 \longrightarrow R^1-\overset{\overset{OZnX}{|}}{C}=CR^2R^3 \qquad (4.263)$$

The Acyloin Reaction and Related Reactions. In the absence of efficient proton donors, the radical coupling process predominates. The reductive coupling of esters with sodium gives acyloins (4.474). Unlike essentially all other known cyclization reactions of chain molecules, the acyloin reaction can readily produce medium ring carbocycles in high yields. The synthesis of large rings by ordinary cyclization reactions requires the use of either the high dilution technique or other special reaction conditions. On the other hand, the acyloin reaction does not require such special reaction conditions.

The reaction presumably proceeds by the mechanism shown in eq. 4.264.

$$2RCOOR \xrightarrow[\text{xylene}]{Na} 2R\overset{\bullet}{C}\underset{\underset{ONa}{|}}{OR'} \longrightarrow \begin{array}{c} \overset{ONa}{|} \\ RC-OR' \\ | \\ RC-OR' \\ | \\ ONa \end{array} \quad (4.264)$$

$$\longrightarrow \begin{array}{c} RC=O \\ | \\ RC=O \end{array} \xrightarrow{2Na} \begin{array}{c} RC-ONa \\ || \\ RC-ONa \end{array} \xrightarrow{H_3O^+} \begin{array}{c} RC=O \\ | \\ RCHOH \end{array}$$

Although small and common rings (four- to seven- membered) can also be prepared by the acyloin reaction, the Dieckmann cyclization reaction (4.475) induced by the endiolate intermediate tends to be competitive in these cases. This problem, however, can be alleviated by trapping the endiolate intermediate with Me_3SiCl, as shown in eq. 4.265 (4.476).

$$EtOOC(CH_2)_5COOEt \xrightarrow[\text{Me}_3\text{SiCl}]{\text{Na, toluene}} \begin{array}{c} -OSiMe_3 \\ -OSiMe_3 \end{array} \quad (4.265)$$

$$75\%$$

Reduction of ketones with alkali and alkaline earth metals in the absence of proton donors produces pinacols. In the past, Mg, Mg(Hg), and Al(Hg) have been used extensively. More recently, however, the combined use of Mg(Hg) and $TiCl_4$ has been recommented for this transformation; see eq. 4.266 (4.477).

$$(4.266)$$

4.4.1.3 Reduction of Alkenes, Alkynes, and Arenes with Alkali and Alkaline Earth Metals

Alkenes. Simple alkenes are not readily reduced with dissolving metals. They can, however, be reduced with sodium in a mixture of t-BuOH and HMPA at 25°C (4.478). A wide variety of olefins, such as 1-hexene, methylenecyclohexane, 3-hexene, cyclohexene, norbornene, and $\Delta^{9,10}$-octalin, have been reduced to the corresponding alkanes and cycloalkanes.

Alkenyl groups that are conjugated with carbonyl, alkenyl, or aryl groups are readily reduced with dissolving alkali metals. The reduction of α,β-unsaturated carbonyl compounds with alkali metals provides a regiospecific route to the corresponding metal enolates, as discussed in Sect. 4.3.4.1.

In contrast with catalytic hydrogenation (Sect. 10.1.3), the dissolving metal reduction of α,β-unsaturated ketones generally produces thermodynamically more stable stereoisomers, as shown in eq. 4.267 (4.327).

$$(4.267)$$

Conjugated dienes can be readily reduced with alkali metals, as exemplified by the reaction shown in eq. 4.268 (4.479).

$$CH_2=CHCH=CH_2 \xrightarrow[\text{liquid } NH_3]{Na} CH_3CH=CHCH_3 \qquad (4.268)$$

(E)- and (Z)-isomers

Styrene and its derivatives can also be reduced with alkali metals. This reaction provides an important procedure that has been used for the synthesis of 19-norsteroids; see eq. 4.269

(*4.480*), including a number of oral contraceptives.

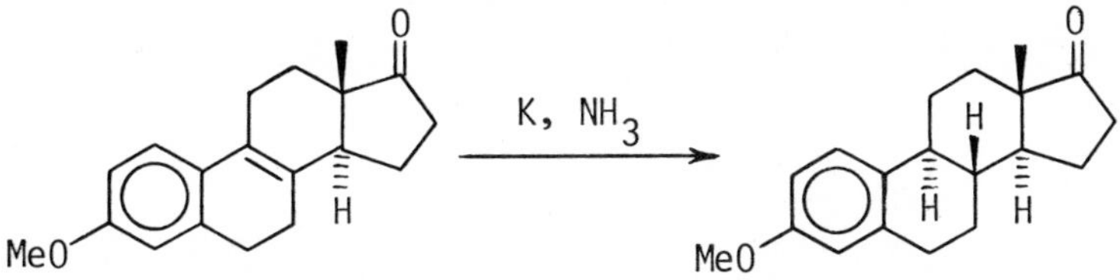

(4.269)

The more stable trans isomer is formed selectively.

Acetylenes. Internal acetylenes are readily reduced to trans-olefins with alkali metals in liquid NH_3. The reaction most probably proceeds by the mechanism shown in eq. 4.270.

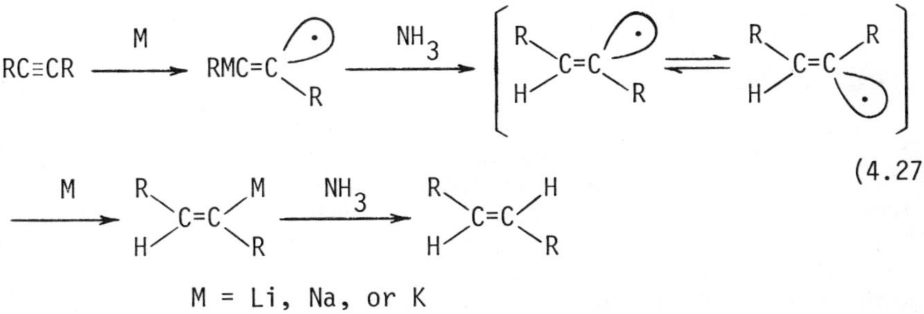

(4.270)

M = Li, Na, or K

Trans stereochemistry may be explained based on reasonable as-sumptions that the *trans*-alkenyl radical is the more stable ster-eoisomer and that it is this isomer that is converted into the configurationally rigid alkenylmetal intermediate with retention. With respect to stereoselectivity, the reaction is complementary with the catalytic hydrogenation over Lindlar's catalyst (Sect. 10.1.3).

In the absence of an efficient proton donor, terminal ace-tylenes are converted to metal acetylides which are resistant to the attack by alkali metals in liquid NH_3. In the presence of NH_4Cl, however, the formation of metal acetylides can be prevent-ed, and the terminal acetylenes can also be converted into ole-fins (eq. 4.271).

It has recently been reported that internal acetylenes can be reduced to *cis*-olefins with zinc powder in 50% aq 1-propanol (*4.481*). *Cis* stereochemistry can be explained in terms of either the metallocyclic intermediate 4.36 or a heterogeneous surface reaction mechanism 4.37.

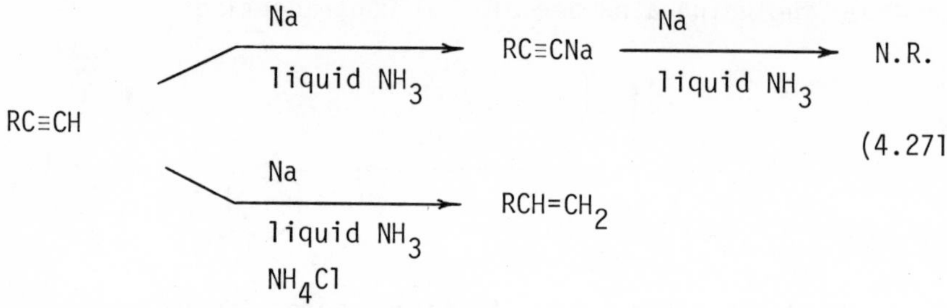

$$RC\equiv CH \quad \xrightarrow[\text{liquid } NH_3]{Na} \quad RC\equiv CNa \quad \xrightarrow[\text{liquid } NH_3]{Na} \quad N.R.$$

$$RC\equiv CH \quad \xrightarrow[\substack{\text{liquid } NH_3 \\ NH_4Cl}]{Na} \quad RCH=CH_2$$

(4.271)

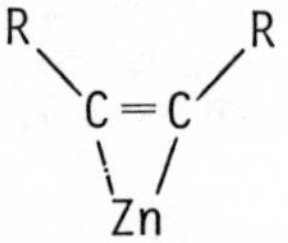

4.36

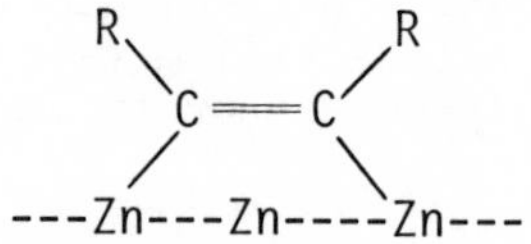

4.37

The Birch Reduction. The reduction of arenes with alkali metals
is called the Birch reduction (4.482). The reaction provides a
unique and synthetically highly useful route to dihydroaromatic
derivatives. The reaction typically involves treatment of arenes
with lithium or sodium in liquid NH_3 either in the presence or in
the absence of an alcohol, such as EtOH and t-BuOH. Lithium has
both a higher molar solubility and a higher redox potential in
NH_3 than sodium or potassium, and reactions with lithium are
often cleaner than those with the heavier alkali metals.
    Ease of reduction can be roughly correlated with the reduc-
tion potential of the substrate and decreases in the order:
anthracene > phenanthrene > naphthalene > biphenyl > benzene.
Benzene itself cannot readily be reduced with alkali metals in
liquid NH_3, and its reduction to 1,4-cyclohexadiene requires the
use of a more efficient proton donor, such as EtOH. The results
can be accommodated by the following scheme (eq. 4.272).

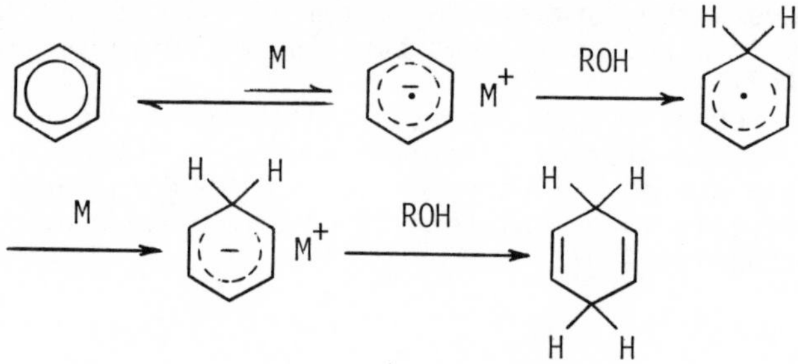

(4.272)

For the reduction of biphenyl and fused polycyclic aromatics, liquid NH_3 is a sufficiently strong acid.

The regiochemistry of the Birch reduction presents an interesting puzzle. The major products are skipped dienes rather than conjugated dienes or styrene derivatives. An early explanation was based on electron repulsion between the two carbanionic centers of a dianion intermediate. The available data, however, indicate that in the great majority of cases dianions are not formed. Although there is no intuitively obvious explanation, the most satisfactory explanation is based on the MO theory. It is reasonable to predict that protonation takes place on the carbon atom of the highest electron density, as determined by the coefficients of the LUMO. Thus, for example, the first protonation in the reduction of naphthalene may be predicted to take place in the α-position based on the coefficients of the LUMO indicated in 4.38.

4.38

(4.273)

4.39

Simple HMO theory fails to correctly predict the regiochemistry of the second protonation, since it gives the same coefficient of 0.50 to both α- and γ-positions of the allylic moiety in 4.39. When the electron-donating and electron-attracting effects of the benzyl and the aryl groups, respectively, are taken into consideration, we can predict that the second protonation will also take place in the α-position.

The contrasting regioselectivities exhibited by electron-donating and electron-attracting substituents can also be explained in a similar manner. The available information is summarized in Scheme 4.10. Typical examples follow. Arenes containing electron-donating substituents tend to give the most highly substituted 1,4-dienes, as shown in eq. 4.276.

Scheme 4.10

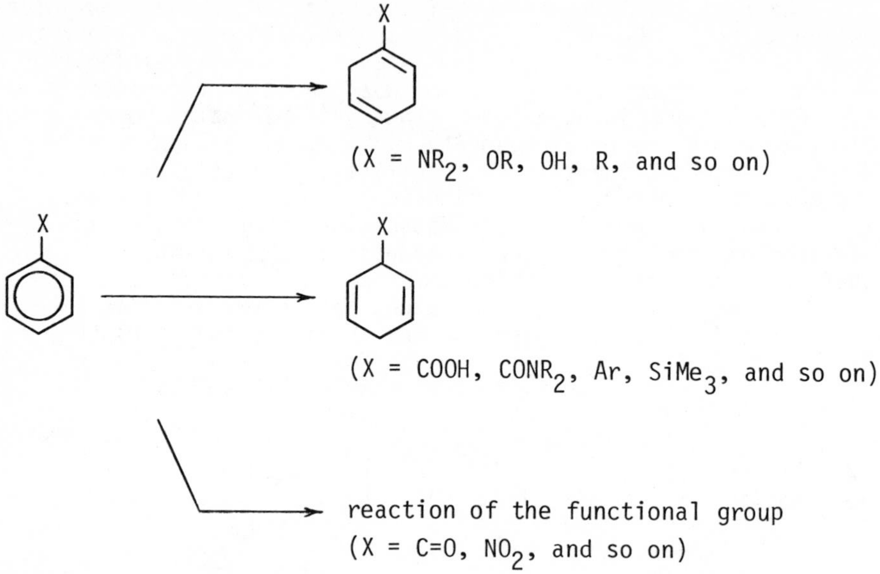
X
(X = NR2, OR, OH, R, and so on)
(X = COOH, CONR2, Ar, SiMe3, and so on)
reaction of the functional group
(X = C=O, NO2, and so on)

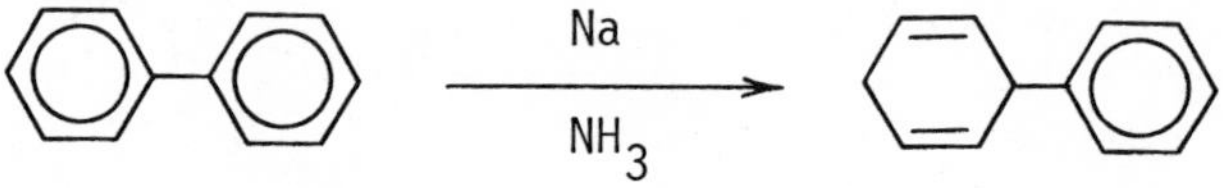
Na
NH3

(4.274)

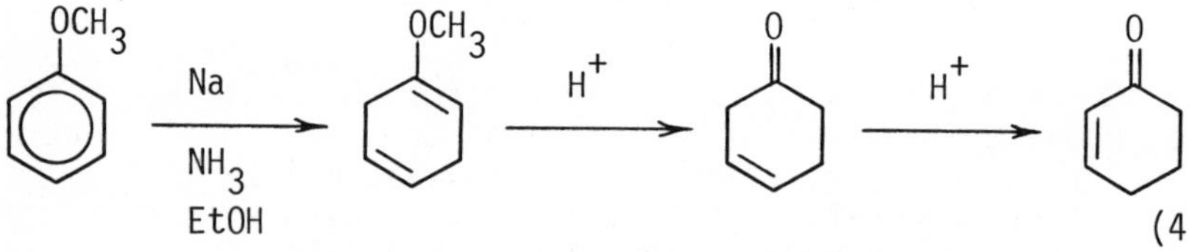
OCH3
Na
NH3
EtOH
H+
O

(4.275)

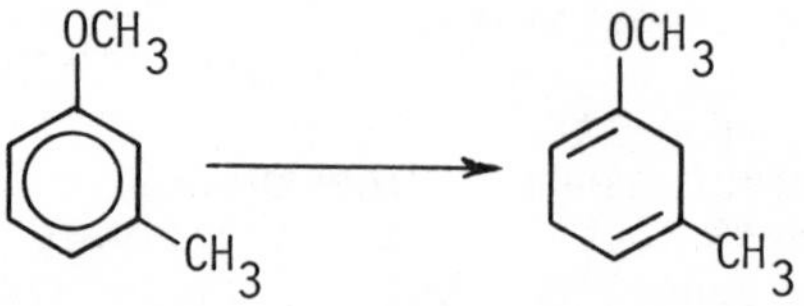
OCH3
CH3

(4.276)

As might be readily expected, electron-attracting groups activate the aromatic ring, while electron-donating groups deactivate it, as indicated for the cases of naphthalene derivatives in Scheme 4.11.

Scheme 4.11

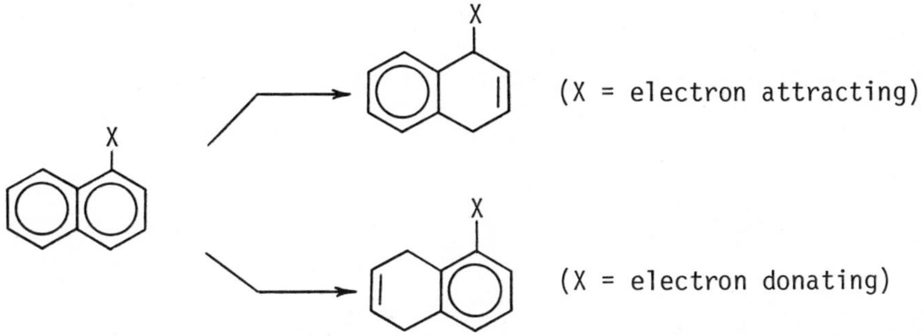
X
X
X
(X = electron attracting)
(X = electron donating)

The course of the Birch reduction can be further controlled in a delicate manner either by varying proton donors or by using suitable additives, as exemplified by the results shown in eq. 4.277 (*4.482c*).

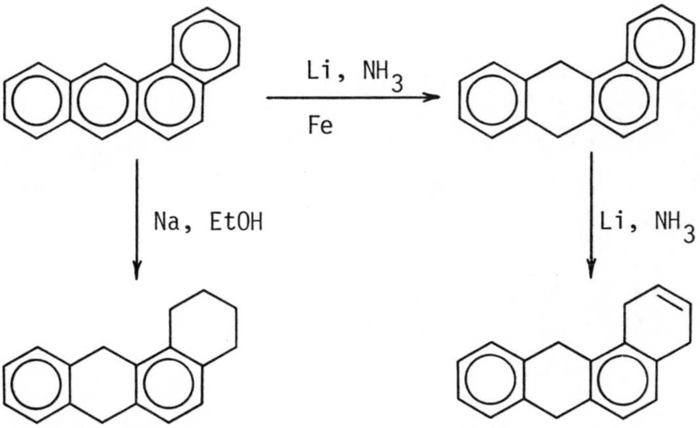
Li, NH_3
Fe
Na, EtOH
Li, NH_3
(4.277)

Iron catalyzes the reaction of alkali metals with NH_3 to form metal amides, thereby preventing or minimizing overreduction by excess alkali metals.

4.4.1.4 Reduction of Carbon-Hetero Atom Single Bonds with Alkali and Alkaline Earth Metals

<u>C-X(halogen) Bonds</u>. The two-step procedure for the reduction of carbon-halogen bonds is discussed in Sect. 4.4.1. β-Chloro enones can be converted into dehalogenated enones with Zn-Ag in MeOH (*4.483*).

<u>C-O Bonds</u>. Various types of activated C-O bonds can be cleaved with alkali metals. The C-O bonds between benzylic or allylic groups and the oxygen atom as well as those of epoxides are sufficiently reactive so that they do not require an additional activating group, as indicated by eqs. 4.278 to 4.281.

$$PhCH_2\text{-}OR \xrightarrow{\text{Na, } NH_3} PhCH_3 \ + \ HOR \qquad (4.278)$$

$$\underset{\underset{R^2}{|}}{\overset{\overset{R^1}{|}}{Ar\text{-}C\text{-}OH}} \xrightarrow[\text{(4.484)}]{\text{Li, } NH_3, \ NH_4Cl} \underset{\underset{R^2}{|}}{\overset{\overset{R^1}{|}}{Ar\text{-}C\text{-}H}} \qquad (4.279)$$

The reaction shown in eq. 4.279 can be used in tandem with the carbonyl addition reactions of organolithiums or Grignard reagents to produce a variety of alkyl-substituted arenes, as shown in eq. 4.280.

$$(4.280)$$

$$90\%$$

It is noteworthy that neither regiochemical nor stereochemical scrambling takes place to any significant extent.

1. Li, $NH_2CH_2CH_2NH_2$
2. H_2O

$$(4.485)$$

$$(4.281)$$

$$93\%$$

The reaction involves predominant cleavage at the less hindered epoxide carbon atom.

Cleavage of other types of C-O bonds generally requires the presence of an activating group. Phosphoryl groups are some of the most commonly used activating groups, as shown in the following examples.

$$R_2CHOH \quad \xrightarrow[\text{(4.486)}]{\begin{array}{l}1.\ ClPO(NMe_2)_2\\ 2.\ Li,\ EtNH_2\end{array}} \quad R_2CH_2 \qquad (4.282)$$

$$\underset{\substack{|\quad|}}{-C{=}C{-}OLi} \quad \xrightarrow[\text{(4.486)}]{\begin{array}{l}1.\ ClPO(NMe_2)_2\\ 2.\ Li,\ EtNH_2\end{array}} \quad \begin{array}{c}\diagdown\\ \diagup\end{array}C{=}C\begin{array}{c}H\\ \diagup\end{array} \qquad (4.283)$$

$$R{-}\bigcirc{-}OH \quad \xrightarrow[\text{(4.487)}]{\begin{array}{l}1.\ ClPO(OEt)_2,\ NaOH\\ 2.\ Li,\ Na,\ or\ K,\ NH_3\end{array}} \quad R{-}\bigcirc{-}H \qquad (4.284)$$

(4.285)

C-S Bonds. Reductive cleavage of the C-S bond with alkali metals is facile and general with respect to the organic group bonded to sulfur. Its scope may be indicated by the following examples.

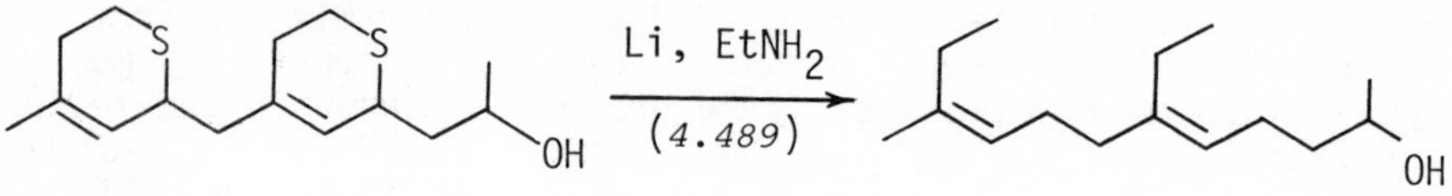

(4.286)

It should be noted that the cleavage of the allyl-S bonds with lithium in $EtNH_2$ proceeds in a regio- and stereoselective manner, as mentioned in the discussion of the Biellmann 1,5-diene synthesis (Sect. 4.3.2.1). Even secondary alkyl-S bonds can be cleaved with dissolving metals; see eq. 4.287 (4.490).

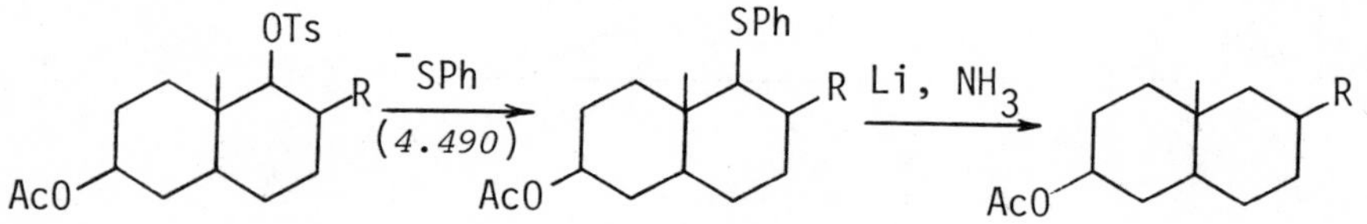

(4.287)

C-N and C-P Bonds. Although the C-N and C-P bonds of amines and phosphines are highly resistant to the action of dissolving metals, certain activated ammonium and phosphonium salts can be reduced with dissolving metals, as shown in eqs. 4.288 and 4.289.

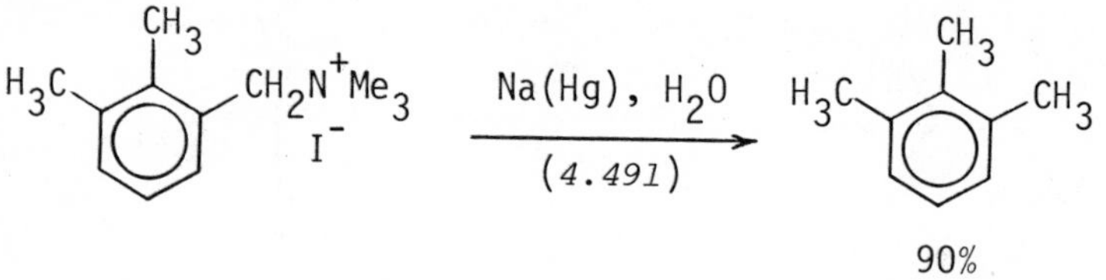

(4.288)

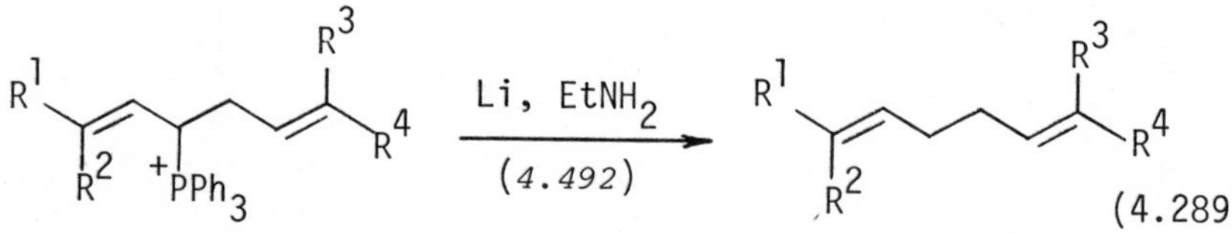

90%

(4.289)

4.4.2 Formation of Other Carbon-Hetero Atom Bonds Containing Nonmetallic Elements

The metal-carbon bond of organometallics of Group IA and Group II metals can be converted into various other carbon-hetero atom bonds including C-X(halogen), C-O, C-S, C-Se, C-N, and C-P bonds (Scheme 4.12).

Scheme 4.12

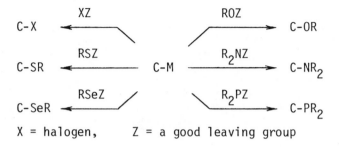

X = halogen,   Z = a good leaving group

These reactions seem to require reagents, represented by Y-Z, which are either highly polarizable or polarized such that the Y and Z groups are partially positively and partially negatively charged, respectively (eq. 4.290). The mechanistic details of these reactions, however, are not very clear. There are indications that some of these reactions proceed by one-electron processes.

$$C^{\delta-}-M^{\delta+} \ + \ Y^{\delta+}-Z^{\delta-} \longrightarrow C-Y \ + \ M-Z \qquad (4.290)$$

Since the Group V, Group VI, and Group VII elements of our interest, that is, halogens, O, S, Se, N, and P, are generally highly electronegative, the availability of suitable leaving groups Z, which can induce the desired bond polarization, is very limited.

4.4.2.1 Halogenolysis

Conversion of the carbon-metal bond of organometallics of Group IA and Group II metals into carbon-iodine and carbon-bromine bonds is generally a facile process of high synthetic significance. Although less facile, formation of the carbon-chlorine bond is also a synthetically useful reaction. On the other hand, the corresponding fluorination does not appear to be of much synthetic use.

Preferred reagents are molecular halogens, their complexes with phosphines and amines, for example, $Br_2 \cdot PPh_3$ and $Br_2 \cdot$ pyridine, and N-halosuccinimides. The use of complex reagents and N-halosuccinimides is desirable in cases where organometallic intermediates contain some functional groups that are sensitive to molecular halogens, such as olefins.

Halogenolysis of the metal-alkenyl carbon bond can often proceed with retention of configuration. On the other hand, the stereochemical details of the halogenolysis of the metal-alkyl carbon bonds are not very clear. Some representative examples of halogenolysis follow.

$$n\text{-BuC}{\equiv}\text{CLi} \quad \xrightarrow[\text{THF}]{\text{I}_2} \quad n\text{-BuC}{\equiv}\text{CI} \qquad (4.291)$$

$$\text{Ph}_2\text{C=CClLi} \quad \xrightarrow[(4.493)]{\text{Br}_2} \quad \text{Ph}_2\text{C=CClBr} \;(94\%) \quad (4.292)$$

$$(64\%) \quad (4.293)$$

$$(85\%) \qquad (4.294)$$

$$(85\%)$$
$$(4.295)$$

4.4.2.2 Formation of C-O, C-S, C-Se, C-N, and C-P Bonds

Formation of C-O Bonds. Autoxidation of organoalkali and organo-alkaline earth metals can give a variety of products, the major products obtained after hydrolysis being alcohols and coupled hydrocarbons. Autoxidation of Grignard reagents generally gives better yields of alcohols than the corresponding reaction of organolithiums (4.497). Although high yields of alcohols have been obtained in some favorable cases, for example, eq. 4.296 (4.498), these reactions have not yet been developed into synthetically useful procedures.

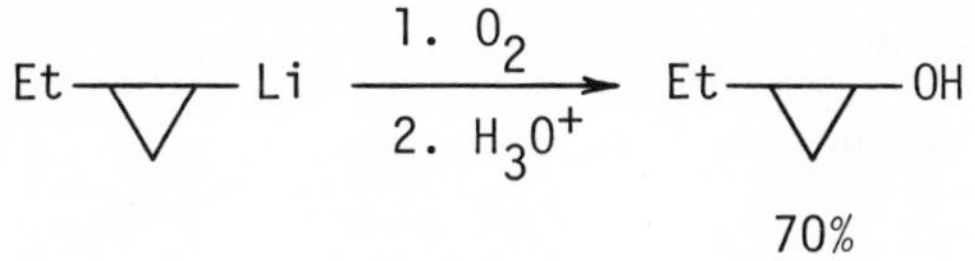

$$\text{Et} \triangledown \text{Li} \xrightarrow[\text{2. } H_3O^+]{\text{1. } O_2} \text{Et} \triangledown \text{OH} \qquad (4.296)$$

$$70\%$$

An indirect but more dependable method for converting organometallics of Group IA and Group II metals into alcohols involves their conversion into organodimethoxyboranes followed by oxidation with alkaline hydrogen peroxide; see eq. 4.297 (*4.499*).

$$\diagup\!\!\!\!\diagdown\!\!\!\!\diagup\text{K} \xrightarrow{FB(OMe)_2} \diagup\!\!\!\!\diagdown\!\!\!\!\diagup B(OMe)_2 \xrightarrow[H_2O_2]{NaOH} \diagup\!\!\!\!\diagdown\!\!\!\!\diagup OH$$
$$(4.297)$$

The following reactions appear to be satisfactory for the hydroxylation of metal enolates.

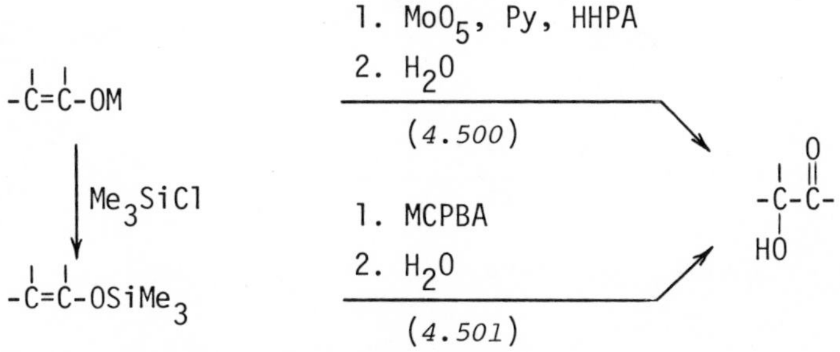

$$-\overset{|}{C}=\overset{|}{C}-OM \xrightarrow[\quad(4.500)\quad]{\begin{array}{l}\text{1. } MoO_5, \text{ Py, HHPA}\\\text{2. } H_2O\end{array}} $$

$$\Big\downarrow Me_3SiCl$$

$$-\overset{|}{C}=\overset{|}{C}-OSiMe_3 \xrightarrow[\quad(4.501)\quad]{\begin{array}{l}\text{1. MCPBA}\\\text{2. } H_2O\end{array}} \overset{O}{\underset{HO}{-\overset{|}{C}-\overset{\|}{C}-}} \qquad (4.298)$$

<u>Formation of C-S and C-Se Bonds.</u> Organometallics of Group IA and Group II metals can react with disulfides (e.g., MeSSMe, and PhSSPh), sulfenyl halides (e.g., MeSCl, and PhSCl), diselenides (e.g., PhSeSePh), and selenenyl halides (e.g., PhSeCl) to form the corresponding sulfides and selenides, respectively.

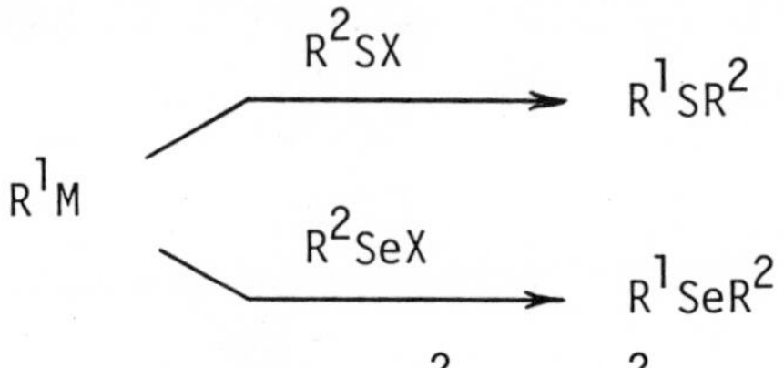

$$R^1M \quad \begin{array}{c} \xrightarrow{R^2SX} R^1SR^2 \\ \\ \xrightarrow{R^2SeX} R^1SeR^2 \end{array} \qquad (4.299)$$

$$X = \text{halogen, } SR^2 \text{ or } SeR^2$$

Some typical examples follow.

$$\text{(4.502)} \qquad \text{(4.300)}$$

$$\text{(4.503)} \qquad \text{(4.301)}$$

$$-\overset{|}{\underset{|}{C}}=\overset{|}{\underset{|}{C}}-OLi \xrightarrow[\text{(4.504)}]{\text{RSCl}} -\overset{|}{\underset{\underset{RS}{|}}{C}}-\overset{O}{\overset{||}{C}}- \qquad \text{(4.302)}$$

$$R = Me, Ph, CSNMe_2 \qquad\qquad 60 \text{ to } 90\%$$

$$\text{(4.505)} \qquad \text{(4.303)}$$

$$\text{(4.506)} \qquad \text{(4.304)}$$

$$\text{(4.305)}$$

$$\text{(4.507)}$$

85%   (4.306)

While sulfenyl and selenenyl halides are more reactive than the corresponding disulfides and diselenides, the latter reagents are more convenient. In the sulfenylation, ester or lactone enolates react readily with both MeSSMe and PhSSPh, ketone enolates react only with the more reactive PhSSPh in THF. The use of HMPA as a cosolvent permits the use of MeSSMe, however. The relative reactivities of ketone and ester enolates are indicated by eq. 4.307 (4.312).

(4.307)

The corresponding selenenylation appears to be much more facile than sulfenylation and can be achieved with PhSeSePh, PhSeCl, or PhSeBr. Although the method does not appear to have been used with aldehyde enolates, it works well with lithium enolates derived from esters (4.509), lactones (4.506, 4.508), nitriles (4.507), and lactams (4.510). Ketone enolates have not successfully been selenenylated with PhSeSePh. Selenenyl halides, such as PhSeCl and PhSeBr, however, are satisfactory (4.506).

One of the major synthetic applications of sulfides and selenides is based on the fact that they can readily be converted into olefins via oxidation-elimination. Conversion of sulfides and selenides into the corresponding oxides is achieved with various oxidizing agents, such as $NaIO_4$ and H_2O_2. The elimination reaction of β-carbonyl-substituted sulfoxides takes place at 25 to 80°C with aryl sulfides and at 110 to 130°C with alkyl sulfides. The corresponding elimination reaction of selenoxides is much more facile and usually takes place at or below room temperature. The regiochemistry of the 2,3-shift of sulfoxides in acyclic systems generally, but not always, follows the order: C=C-CH$_2$ ~ C C-CH$_2$ > ArCH$_2$ ~ CH$_3$ > $\rangle$C-CH$_2$ >> $\rangle$C-CHR. The regiochemistry in cyclic systems is controlled to some degree by the cis, syn nature of the 2,3-shift process. The following examples indicate the synthetic utility of the olefination sequence.

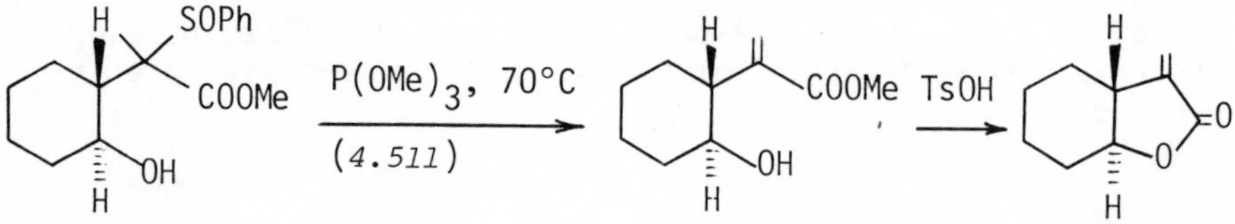

(4.308)

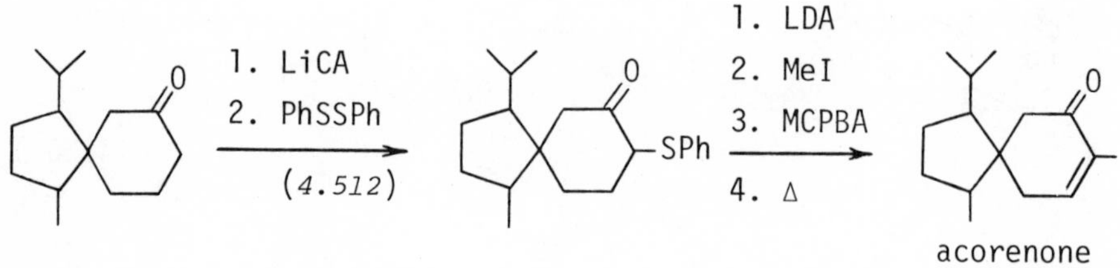

acorenone

(4.309)

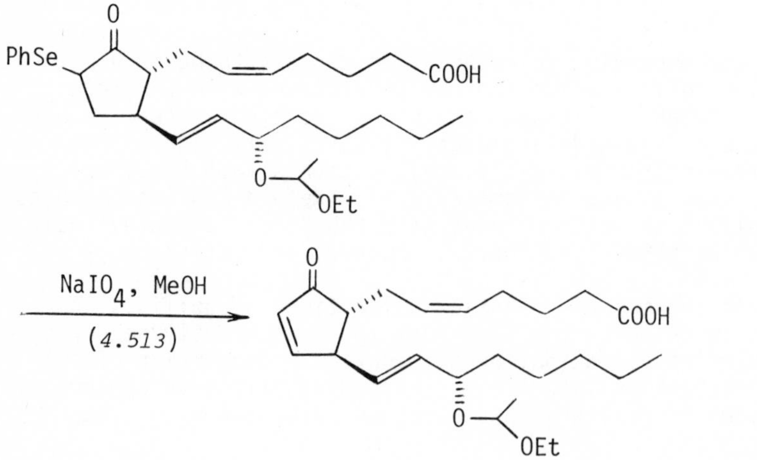

(4.310)

Introduction of two sulfur substituents in the α-position of carbonyl compounds can readily be achieved, as shown in eq. 4.311 (4.514). This bissulfenylation reaction can be adapted to achieve a 1,2-carbonyl transposition, as shown in eq. 4.312 (4.515).

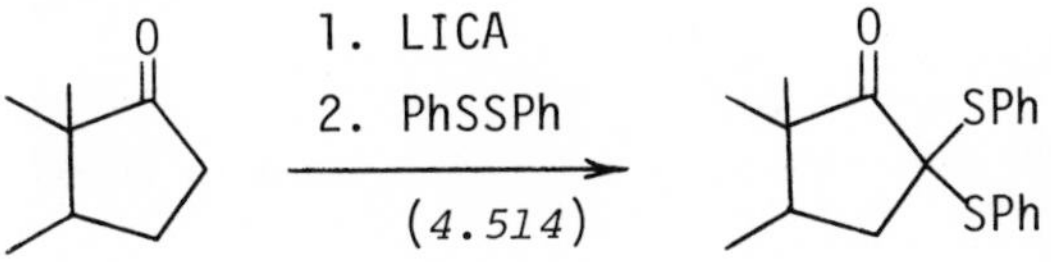

(90%) (4.311)

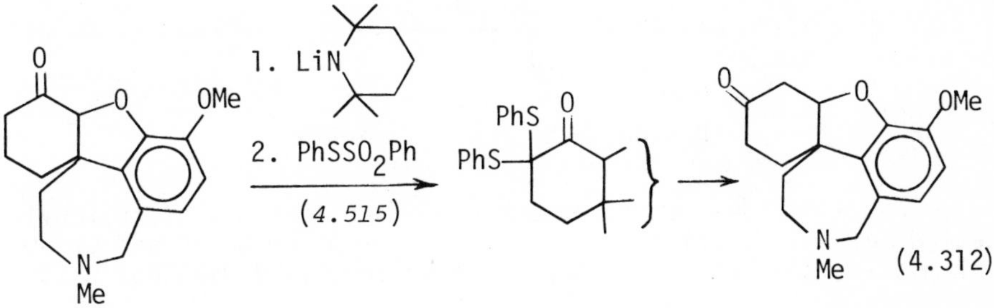

(4.312)

Another useful application of α-sulfenylation of carbonyl compounds is the following oxidative decarboxylation; see eq. 4.313 (4.516).

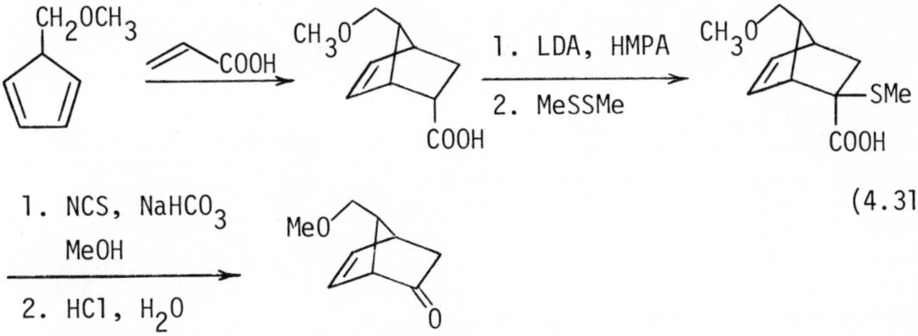

(4.313)

Various other sulfur compounds react with organometallics of Group IA and Group II metals. A few examples are shown below. It is interesting to note that the reaction of sulfonates gives sulfones, whereas that of thiosulfonates gives sulfides.

$$ArSO_2OR \xrightarrow[\substack{(4.517)}]{R^1Li} ArSO_2R^1 \qquad (4.314)$$

$$R = Ph,\ n\text{-alkyl}; \qquad R^1 = \text{alkyl, aryl}$$

$$ArSO_2SPh \xrightarrow[\substack{(4.515)}]{R^1Li} PhSR^1 \qquad (4.315)$$

$$RLi \xrightarrow[\substack{(4.518)}]{SO_2Cl_2\ -85\ \text{to}\ -20°C} RSO_2Cl \qquad (4.316)$$

$$40\ \text{to}\ 60\%$$

$$R = Bu,\ s\text{-Bu},\ i\text{-Bu, neo-Pent}$$

For additional examples and other applications of sulfenyl-ation and selenylation of organometallics of Group IA and Group II metals, see recent reviews by Trost (*4.312, 4.519*) and Clive (*4.290*).

Formation of C–N and C–P Bonds. Although conversion of organo-metallics of Group IA and Group II metals into amines has not been investigated extensively, the following procedure appear promising.

$$RMgX\ +\ Me_2NOTs \xrightarrow[\substack{(4.520)}]{} RNMe_2 \qquad (4.317)$$

$$\underset{\substack{| \\ Li}}{RCHCOOLi}\ +\ NH_2OCH_3 \xrightarrow[\substack{(4.521)}]{} \underset{\substack{| \\ NH_2}}{RCHCOOH} \qquad (4.318)$$

Far more general and straightforward is conversion of organomet-allics of Group IA and Group II metals into phosphines. A few representative examples are shown below.

$$+C{\equiv}CLi\ +\ PCl_3 \xrightarrow[\substack{(4.522)}]{} (+C{\equiv}C-)_3P \qquad (4.319)$$

$$98\%$$

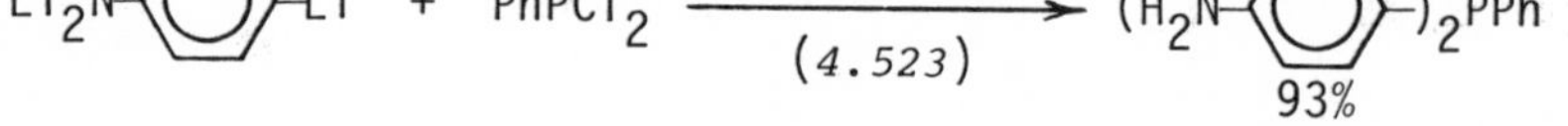

$$Li_2N{-}\langle\bigcirc\rangle{-}Li\ +\ PhPCl_2 \xrightarrow[\substack{(4.523)}]{} (H_2N{-}\langle\bigcirc\rangle{-})_2PPh \qquad (4.320)$$

$$93\%$$

$$\text{Li-}\bigcirc\text{-Li} \;+\; Ph_2PCl \xrightarrow[(4.524)]{} Ph_2P\text{-}\bigcirc\text{-}PPh_2 \quad (4.321)$$

$$68\%$$

4.4.3 Transmetallation Reactions of Organometallics of Group IA and Group II Metals

As pointed out in Sect. 2.11, the ready availability of organolithiums and Grignard reagents and the high electropositivity of lithium and magnesium make these organometallics highly desirable reagents for the preparation of organometallics containing various other metals. The reaction is also discussed in almost all of the following chapters. For extensive discussions of the reaction, the reader is referred to monographs by Wakefield (4.1), Coates (0.6), and Nesmeyanov (0.8). In this section, a brief survey of the reaction and a few specific examples which are not discussed in other chapters will be presented.

Organolithiums can be and have been converted into organometallics containing essentially all metals except some of Group IA and Group IIA metals (4.1). Although it may seem contradictory to the generalization presented in Sect. 2.11, organolithiums have been used even for the preparation of organosodiums and organopotassiums by making use of the fact that organosodiums and organopotassiums tend to be insoluble in hydrocarbons; see eq. 4.322 (4.525).

$$RLi \;+\; MOBu\text{-}t \longrightarrow RM\downarrow \;+\; LiOBu\text{-}t \quad (4.322)$$

$$M = Na \text{ or } K$$

While organolithiums do not appear to have been used in preparing organometallics containing Ca, Sr, and Ba, organoberylliums and organomagnesiums can be prepared from organolithiums (eq. 4.323).

$$RLi \xrightarrow{MgCl_2} RMgCl \xrightarrow{RLi} R_2Mg \quad (4.323)$$

Although the transmetallation reaction of organolithiums and Grignard reagents is of wide applicability and often proceeds cleanly, several side reactions can complicate the desired reaction. Two of the most common side reactions are "ate" complexation and disproportionation. "Ate" complexes are often formed as intermediates in the transmetallation reaction. When they contain a good leaving group, they can decompose under the reac-

tion conditions to form the desired transmetallated products. If
not, they are obtained as unwanted products. A general reaction
scheme for the reaction between an organolithium and a chloride
of a divalent metal, for example, is as follows.

$$MCl_2 \xrightarrow{\text{LiR}} LiMRCl_2 \xrightarrow[-LiCl]{} MRCl$$

$$\text{"ate" complex} \tag{4.324}$$

$$\xrightarrow{\text{LiR}} LiMR_2Cl \xrightarrow[-LiCl]{} MR_2 \xrightarrow{\text{LiR}} LiMR_3$$

$$\text{"ate" complex} \qquad\qquad \text{"ate" complex}$$

In the reaction shown in eq. 4.324, it might appear that the use
of MCl_2 and LiR in the 1:2 ratio would exclusively produce MR_2.
In reality, however, the formation of $LiMR_3$ is often competitive.
    Another very commonly encountered reaction complicating the
transmetallation reaction is disproportionation. Except in cases
where the overall structure of organometallics is important as in
the carbonylation of organoboranes (Sect. 5.2.3.3), for example,
the disproportionation reaction does not usually affect the
course of an organic synthesis in a serious manner. Thus chemists can successfully use Grignard reagents without knowing the
exact Schlenk equilibrium constants.

4.5   OTHER SYNTHETIC APPLICATIONS OF ORGANOMETALLICS OF GROUP IA AND GROUP II METALS

4.5.1   Other Synthetic Applications as Intermediates

So far we have focused our attention mainly on those synthetic
applications of organoalkali and organoalkaline earth metals that
involve only the formation and scission of metal-carbon bonds
(Type A process in Scheme 3.1). Although much less commonly encountered, those that additionally involve organometallic interconversions (Type B process) can also be synthetically useful.
Organometallic interconversion reactions are classified into
three types in Sect. 3.2.2.1. Of the three, only those that involve organic ligand transformations accompanied by both formation and cleavage of the metal-carbon bonds (Type III) are worth
mentioning here. They include metallotropy, disproportionation

ligand displacement, carbometallation, and ligand migration (Sect. 3.2.2). The synthetic utility of the carbometallation of organoalkali and organoalkaline earth metals is discussed in Sect. 4.3.2.5. Of particular interest here are some metallotropic reactions of organometallics of Group IA and Group II metals.

4.5.1.1 Allylic Rearrangement and Acetylene "Zipper"

As pointed out in Sect. 1.6, the 1,3-shift or allylic rearrangement of organometallics is often a very facile process. Although it usually represents a nuisance, it can provide useful synthetic procedures in favorable instances. A few recent discussions of the allylic rearrangement of organomagnesiums are available (*4.526* to *4.528*).

One of the most interesting and useful functional modification reactions involving organoalkali metals is the acetylene "zipper" reaction which presumably proceeds via a series of allylic (propargyl-allenyl) rearrangements. The reaction involves treatment of internal acetylenes with potassium 3-aminopropylamide (KAPA) in 1,3-diaminopropane (APA), which induces an exceedingly facile multipositional isomerization of the internal acetylenes into terminal acetylenes; see eq. 4.325 (*4.325*).

$$R^1C \equiv C(CH_2)_nCH_3 \xrightarrow[NH_2(CH_2)_3NH_2]{KNH(CH_2)_3NH_2} R^1(CH_2)_{n\ +\ 1}C \equiv CK$$

$$(4.325)$$

Although many related acetylene isomerization reactions had been reported previously (*4.530*), the new procedure appears to be far more facile and generally applicable than any of the previously known procedures. Typically, the isomerization of 7-tetradecyne into 1-tetradecyne is over in a few minutes at 15 to 20°C (eq. 4.326).

$$(4.326)$$

$$\xrightarrow[\substack{NH_2(CH_2)_3NH_2 \\ 15\ to\ 20°C \\ 1\ to\ 2\ min}]{KNH(CH_2)_3NH_2}$$

The product yields are often quantitative in the isomerization of simple acetylenes.

Although the exact course of the reaction is unclear, it can be rationalized in terms of a series of deprotonation-allylic rearrangement-protonation processes, which may be either stepwise (eq. 4.327) or synchronous (eq. 4.328).

$$R^1-C{\equiv}C-\underset{\underset{H}{|}}{\overset{\overset{H}{|}}{C}}-R^2 \xrightarrow{\;KHN\frown NH_2\;} R^1-C{\equiv}C-\underset{\underset{H}{|}}{\overset{\overset{K}{|}}{C}}-R^2 \longrightarrow R^1-\underset{\underset{H}{|}}{\overset{\overset{K}{|}}{C}}{=}C{=}C-R^2$$

$$\xrightarrow{\;H_2N\;} R^1-\underset{\underset{H}{|}}{C}{=}C{=}C-R^2 \xrightarrow{\;KHN\frown NH_2\;} R^1-\underset{\underset{K}{|}}{\overset{\overset{H}{|}}{C}}{=}C{=}C-R^2 \qquad (4.327)$$

$$\longrightarrow R^1-\underset{\underset{K}{|}}{\overset{\overset{H}{|}}{C}}-C{\equiv}C-R^2 \xrightarrow{\;H_2N\frown NH_2\;} R^1-\underset{\underset{H}{|}}{\overset{\overset{H}{|}}{C}}-C{\equiv}C-R^2$$

$$R^1-C{\equiv}C-\underset{\underset{H}{|}}{\overset{\overset{H}{|}}{C}}-R^2 \xrightarrow{\;H_2N\frown NHK\;} R^1-C{\equiv}C{\cdots}\underset{\underset{H}{|}}{C}-R^2$$

$$\qquad (4.328)$$

$$\longrightarrow R^1-C{=}C{\cdots}C-R^2 \longrightarrow R^1-\underset{\underset{H}{|}}{\overset{\overset{H}{|}}{C}}-C{\equiv}C-R^2$$

Although both schemes are consistent with the fact that isomerization does not proceed past a tertiary carbon center, the latter scheme, proposed by Brown (4.529), appears to be more plausible than the former in that the latter can readily explain the high efficiency of KAPA. Clearly, the major driving force of

the reaction is provided by the high stability of potassium acetylides formed as the final isomerization product.

Hetero-functional groups tend to interfere with the acetylene "zipper" reaction. The hydroxyl (4.531) and diorganoboryl
(4.532) groups which can be readily converted into negatively
charged groups, however, are tolerated. As might be expected on
the basis of electrostatic repulsion between the carbanionic centers and the negatively charged functional groups, the acetylene
isomerization proceeds away from the negatively charged functional groups. A few synthetically useful examples are shown below.

$$HO(CH_2)_2C{\equiv}C(CH_2)_7CH_3 \xrightarrow[\substack{(4.531a)}]{\substack{1.\ KAPA-APA \\ 2.\ H_2O}} HO(CH_2)_{10}C{\equiv}CH \quad (4.329)$$

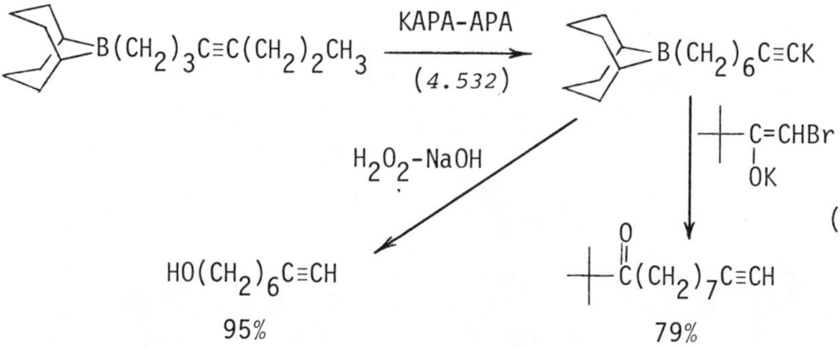

90%

(4.330)

HO(CH$_2$)$_6$C≡CH

95%   79%

4.5.1.2 Cationotropic 1,2-Shift Reactions

Although cationotropic 1,2-shift reactions are relatively rare,
certain types of organoalkali metals undergo such 1,2-shift reactions. The Wittig rearrangement (4.533), for example, eq. 4.331,
represents a classic example of this class of reactions.

$$CH_3OCH_2Ph \xrightarrow{PhLi} CH_3O\overset{\overset{\displaystyle Li}{|}}{C}HPh \longrightarrow LiO\overset{\overset{\displaystyle CH_3}{|}}{C}HPh \quad (4.331)$$

In the Grovenstein-Zimmerman rearrangement (4.534), an aryl
group migrates from a carbon atom to an adjacent carbanionic center as shown in eq. 4.332.

$$Ph_3CCH_2Cl \xrightarrow[\text{-65 to -30°C}]{\text{2 Li}} Ph_3CCH_2Li \xrightarrow{0°C} LiCPh_2CH_2Ph \quad (4.332)$$

Although no 1,3-shift reaction of this type appears to have been reported, the corresponding 1,4- and 1,5-shift reactions are known. A typical example of the 1,4-migration reaction is shown in eq. 4.333 (4.535).

$$Ph_2C(CH_2)_3K \longrightarrow \longrightarrow Ph_2C(CH_2)_2CH_2 \quad (4.333)$$

While these organometallic rearrangement reactions are interesting, their synthetic usefulness appears to be rather limited.

4.5.2 Synthetic Utilities of Organometallics of Group IA and Group II Metals as Reagents

One of the most important applications of organometallics of Group IA and Group II metals is their use as strong Brønsted bases. It does not, however, require a detailed discussion here. Although organometallics of Group IA and Group II metals that contain β-hydrogens can act as hydride donors, for example, eq. 3.65, it does not usually represent a synthetically useful application.

In the great majority of cases, conversion of organic compounds into the corresponding organoalkali and organoalkaline earth metals represents activation of the organic compounds. In some cases, however, it can be used to deactivate or protect certain functional groups. For example, conversion of terminal acetylenes into the acetylides of alkali or alkaline earth metals makes the terminal alkynyl group resistant to hydrogenation and dissolving metal reduction (eq. 4.334).

$$CH_3(CH_2)_2C{\equiv}C(CH_2)_4C{\equiv}CH \xrightarrow[\text{liquid } NH_3]{NaNH_2} CH_3(CH_2)_2C{\equiv}C(CH_2)_4C{\equiv}CNa$$

(4.334)

$$\xrightarrow[\text{2. } H_3O^+]{\text{1. Na, } NH_3}$$

$$CH_3(CH_2)_2 \diagdown C{=}C \diagup^H \diagdown_{(CH_2)_4C{\equiv}CH}$$

REFERENCES

4.1   Wakefield, B. J., *The Chemistry of Organolithium Compounds*, Pergamon Press, New York, 1974, 335 pp.

4.2   Kharasch, M. S., and Reinmuth, O., *Grignard Reactions of Nonmetallic Substances*, Prentice-Hall, New York, 1954, 1384 pp.

4.3   (a) Brown, T. L., *Adv. Organometal. Chem.*, **3**, 365 (1965); (b) Brown, T. L., *Acc. Chem. Res.*, **1**, 23 (1968); (c) Brown, T. L., *Pure Appl. Chem.*, **23**, 447 (1970).

4.4   Brown, T. L., Seitz, L. M., and Kimura, B. Y., *J. Am. Chem. Soc.*, **90**, 3245 (1968).

4.5   West, P., Purmort, J. I., and McKinley, S. V., *J. Am. Chem. Soc.*, **90**, 797 (1968).

4.6   For a review, see Ashby, E. C., Laemmle, J., and Neumann, H. M., *Acc. Chem. Res.*, **7**, 272 (1974).

4.7   Stucky, G. D., and Rundle, R. E., *J. Am. Chem. Soc.*, **85**, 1002 (1963); **86**, 4825 (1964).

4.8   Guggenberger, L. J., and Runcle, R. E., *J. Am. Chem. Soc.*, **86**, 5344 (1964).

4.9   (a) Nordlander, J. E., and Roberts, J. D., *J. Am. Chem. Soc.*, **81**, 1769 (1959); (b) Nordlander, J. E., Young, W. G., and Roberts, J. D., *J. Am. Chem. Soc.*, **83**, 494 (1961); (c) Whitesides, G. M., Nordlander, J. E., and Roberts, J. D., *J. Am. Chem. Soc.*, **84**, 2010 (1962).

4.10 Witanowski, M., and Roberts, J. D., *J. Am. Chem. Soc.*, **88**, 737 (1966).

4.11 Ziegler, K., and Colonius, H., *Ann. Chem.*, **479**, 135 (1930).

4.12 Wittig, G., and Leo, M., *Chem. Ber.*, **64**, 2395 (1931).

4.13 Gilman, H., Zoellner, E. A., and Selby, W. M., *J. Am. Chem. Soc.*, **54**, 1957 (1932).

4.14   Linstrumelle, G., and Michelot, D., *J. C. S. Chem. Comm.*, 561 (1975).

4.15   Kitatani, K., Hiyama, T., and Nozaki, H., *J. Am. Chem. Soc.*, 97, 949 (1975).

4.16   (a) Parham, W. E., Jones, L. D., and Sayed, Y., *J. Org. Chem.*, 40, 2394 (1975); 41, 1184, 1187 (1976); (b) Parham, W. E., and Jones, L. D., *J. Org. Chem.*, 41, 2704 (1976); (c) *J. Org. Chem.*, 42, 257 (1977).

4.17   Gilman, H., and Morton, J. W., Jr., *Org. React.*, 8, 258 (1954).

4.18   Midland, M. M., *J. Org. Chem.*, 40, 2250 (1975).

4.19   Bhanu, S., and Scheinmann, F., *J. C. S. Chem. Comm.*, 817 (1975).

4.20   Seebach, D., and Geiss, K.-H., *J. Organometal. Chem. Libr.*, 1, 1 (1976).

4.21   Juenge, E. C., and Seyferth, D., *J. Org. Chem.*, 26, 563 (1961).

4.22   (a) Seebach, D., *Chem. Ber.*, 105, 487 (1972); (b) Seebach, D., Geiss, K.-H., Beck, A. K., Graf, B., and Daum, H., *Chem. Ber.*, 105, 3280 (1972).

4.23   Russell, G. A., and Ochrymowycz, L. A., *J. Org. Chem.*, 35, 764 (1970).

4.24   Coates, R. M., Pigott, H. D., and Ollinger, J., *Tetrahedron Lett.*, 3955 (1974).

4.25   (a) Seebach, D., and Peleties, N., *Angew. Chem. Int. Ed. Engl.*, 8, 450 (1969); (b) *Chem. Ber.*, 105, 511 (1972); (c) Seebach, D., and Beck, A. K., *Angew. Chem. Int. Ed. Engl.*, 13, 806 (1974).

4.26   Dumont, W., Gayet, P., and Krief, A., *Angew. Chem. Int. Ed. Engl.*, 13, 804 (1974); 14, 350 (1975).

4.27   Seebach, D., and Beck, A. K., *Chem. Ber.*, 108, 314 (1974).

4.28   (a) Gilman, H., and McNinch, H. A., *J. Org. Chem.*, 26, 3723 (1961); (b) Gilman, H., and Schwebke, G. L., *J. Org. Chem.*, 27, 4259 (1962); (c) Eisch, J. J., and Jacobs, A. M., *J. Org. Chem.*, 28, 2145 (1963).

4.29   Jung, M. E., and Blum, R. B., *Tetrahedron Lett.*, 3791 (1977).

4.30   Costa, L. C., Whitesides, G. M., *J. Am. Chem. Soc.*, 99, 2390 (1977).

4.31   Normant, H., *Compt. Rend.*, <u>239</u>, 1510 (1954).

4.32   Sheverdina, N. I., and Kocheshkov, K. A., *The Organic Compounds of Zinc and Cadmium*, North-Holland Publishing, Amsterdam, 1967, 252 pp.

4.33   Frankland, E., *Am. Chem.*, <u>71</u>, 171 (1849).

4.34   (a) Gaudemar, M., *Ann. Chim. (France)*, 161 (1956); (b) *Compt. Rend.*, <u>245</u>, 2054 (1957).

4.35   Emschwiller, G., *Compt. Rend.*, <u>183</u>, 665 (1926); <u>188</u>, 1555 (1929).

4.36   Shriner, R. L., *Org. React.*, <u>1</u>, 1 (1942).

4.37   Rathke, M. W., *Org. React.*, <u>22</u>, 423 (1975).

4.38   Chenault, J., and Tatibouet, F., *C. R. Acad. Sci.*, *Ser. C*, <u>264</u>, 213 (1967).

4.39   Gaudemar, M., *C. R. Acad. Sci.*, *Ser. C*, <u>268</u>, 1439 (1969).

4.40   Mathieu, J., and Weill-Raynal, J., *Formation of C-C Bonds*, Vol. 2, Thieme, Stuttgart, 1975.

4.41   Beletskaya, I. P., Artamkina, G. A., and Reutov, O. A., *Russ. Chem. Rev.*, <u>45</u>, 330 (1976).

4.42   Letsinger, R. L., *J. Am. Chem. Soc.*, <u>72</u>, 4842 (1950).

4.43   Zook, H., and Goldey, R. N., *J. Am. Chem. Soc.*, <u>75</u>, 3976 (1953).

4.44   Letsinger, R. L., *Angew. Chem.*, <u>70</u>, 151 (1958).

4.45   Sauer, S., and Braig, W., *Tetrahedron Lett.*, 4275 (1969).

4.46   Sommer, L. H., and Korte, W. D., *J. Org. Chem.*, <u>35</u>, 22 (1970).

4.47   Korte, W. D., and Kenner, J., *Tetrahedron Lett.*, 603 (1970).

4.48   Letsinger, R. L., and Traynham, J. G., *J. Am. Chem. Soc.*, <u>72</u>, 849 (1950).

4.49   Ward, H. R., *J. Am. Chem. Soc.*, <u>89</u>, 5517 (1967).

4.50   Ward, H. R., and Lawler, R. G., *J. Am. Chem. Soc.*, <u>89</u>, 5518 (1967).

4.51   Ward, H. R., Lawler, R. G., and Cooper, R. A., *J. Am. Chem. Soc.*, <u>91</u>, 746 (1969).

4.52   Garst, J. F., and Cox, R. H., *J. Am. Chem. Soc.*, <u>92</u>, 6389 (1970).

4.53   Ward, H. R., Lawler, R. G., and Narzilli, T. A., *Tetrahedron Lett.*, 521 (1970).

4.54   Russell, G. A., and Lampson, D. W., *J. Am. Chem. Soc.*, 91, 3967 (1969).

4.55   Osawa, E., Majerski, Z., and Schleyer, P. von R., *J. Org. Chem.*, 36, 205 (1971).

4.56   Korte, K. D., Cripe, K., and Cooke, R., *J. Org. Chem.*, 39, 1168 (1974).

4.57   Corey, E. J., and Kirst, H. A., *Tetrahedron Lett.*, 5041 (1968).

4.58   Ireland, R. E., Dawson, M. I., and Lipinski, C. A., *Tetrahedron Lett.*, 2247 (1970).

4.59   Corey, E. J., Kirst, H. A., and Katzenellenbogen, J. A., *J. Am. Chem. Soc.*, 92, 6314 (1970).

4.60   Hoeg, D. F., and Lusk, D. I., *J. Am. Chem. Soc.*, 86, 928 (1964).

4.61   (a) Biellmann, J. F., and Ducep, J. B., *Tetrahedron Lett.*, 3707 (1969); *Tetrahedron*, 27, 5861 (1971); (b) Altman, L. J., Ash, L., and Marson, S., *Synthesis*, 129 (1974).

4.62   Grieco, P. A., and Masaki, Y., *J. Org. Chem.*, 39, 2135 (1974).

4.63   (a) Akiyama, S., and Hooz, J., *Tetrahedron Lett.*, 4115 (1973); (b) Cardillo, G., Contento, M., and Sandri, S., *Tetrahedron Lett.*, 2215 (1974).

4.64   van Tamelen, E. E., Grieder, A., and Lees, R. G., *J. Am. Chem. Soc.*, 96, 2253 (1974).

4.65   Mori, K., Tominaga, M., and Matsui, M., *Tetrahedron*, 31, 1846 (1975).

4.66   Merrill, R. E., and Negishi, E., *J. Org. Chem.*, 39, 3452 (1974).

4.67   (a) Linstrumelle, G., *Tetrahedron Lett.*, 3809 (1974); (b) Millon, J., Lorne, R., and Linstrumelle, G., *Synthesis*, 434 (1975).

4.68   (a) Hall, S. S., and Lipsky, S. D., *J. Org. Chem.*, 38, 1735 (1973); (b) Hall, S. S., and McEnroe, F. J., *J. Org. Chem.*, 40, 271 (1975).

4.69   Shirley, D. A., *Org. React.*, 8, 28 (1954).

4.70   Jorgenson, M. J., *Org. React.*, 18, 1 (1970).

4.71   Mukaiyama, T., Araki, M., and Takei, H., *J. Am. Chem. Soc.*, 95, 4763 (1973).

4.72   Abe, K., Sato, T., Nakamura, N., and Sakan, T., *Chem. Lett.*, 645 (1977).

4.73   Borch, R. F., Levitan, S. R., and Van Catledge, F. A., *J. Org. Chem.*, 37, 726 (1972).

4.74   (a) Fauvarque, J., Ducom, J., and Fauvarque, J.-F., *Compt. Rend.*, 275, 511 (1972); (b) Huet, F., Emptoz, G., and Jubier, A., *Tetrahedron*, 29, 479 (1973).

4.75   Whitmore, F. C., and George, R. S., *J. Am. Chem. Soc.*, 64, 1239 (1942).

4.76   (a) Bartlett, P. D., and Lefferts, E. B., *J. Am. Chem. Soc.*, 77, 2804 (1955); (b) Petrov, A. D., Sokolova, E. G., and Chin-Lan, G., *Izv. Akad. Nauk SSSR, Ser. Khim.*, 871 (1957).

4.77   Negishi, E., and Brown, H. C., unpublished results. Also see Buhler, J. D., *J. Org. Chem.*, 38, 904 (1973).

4.78   (a) Cram, D. J., and Kopecky, K. R., *J. Am. Chem. Soc.*, 81, 2748 (1959); (b) Cram, D. J., and Abd Elhafez, F. A., *J. Am. Chem. Soc.*, 74, 5828 (1952).

4.79   (a) Cram, D. J., and Wilson, D. R., *J. Am. Chem. Soc.*, 85, 1245 (1963); (b) Cornforth, J. W., Cornforth, R. H., and Mathew, K. J., *J. Chem. Soc.*, 112 (1959).

4.80   For an extensive review, see Ashby, E. C., and Laemmle, J. T., *Chem. Rev.*, 75, 521 (1975).

4.81   Macdonald, T. L., and Still, W. C., *J. Am. Chem. Soc.*, 97, 5280 (1975).

4.82   Smith, S. G., Charbonneau, L. F., Novak, D. P., and Brown, T. L., *J. Am. Chem. Soc.*, 94, 7059 (1972).

4.83   Holm, T., and Crossland, I., *Acta Chem. Scand.*, 25, 59 (1971).

4.84   Ashby, E. C., Laemmle, J. T., and Neumann, H. M., *Acc. Chem. Res.*, 7, 272 (1974).

4.85   Gaylord, N. G., and Becker, E. I., *Chem. Rev.*, 49, 413 (1951).

4.86   Coke, J. K., and Richon, A. B., *J. Org. Chem.*, 41, 3516 (1976).

4.87   Cristol, S. J., Douglass, J. R., and Meek, J. S., *J. Am. Chem. Soc.*, 78, 816 (1951).

4.88   Crandall, J. K., Arrington, J. P., and Hen, J., *J. Am. Chem. Soc.*, 89, 6208 (1967).

4.89   Gilman, H., and Kirby, R. H., *J. Am. Chem. Soc.*, 63, 2046 (1941).

4.90   Whitmore, F. C., and Pedlow, G. W., Jr., *J. Am. Chem. Soc.*, 63, 758 (1941).

4.91   Munch-Peterson, J., *Org. Synth.*, 41, 60 (1961).

4.92   Cooks, M. P., Jr., and Goswami, R., *J. Am. Chem. Soc.*, 99, 642 (1977).

4.93   Schöllkopf, U., and Meyer, R., *Ann. Chem.*, 1174 (1977).

4.94   (a) Meyers, A. I., and Whitten, C. E., *J. Am. Chem. Soc.*, 97, 6266 (1975); (b) Meyers, A. I., and Whitten, C. E., *Heterocycles*, 4, 1687 (1976); (c) Meyers, A. I., and Whitten, C. E., *Tetrahedron Lett.*, 1947 (1976).

4.95   Liu, S.-H., *J. Org. Chem.*, 42, 3209 (1977).

4.96   Ziegler, K., and Gellert, H. G., *Ann. Chem.*, 567, 195 (1950).

4.97   (a) Bartlett, P. D., Friedman, S., and Stiles, M., *J. Am. Chem. Soc.*, 75, 1771 (1953); (b) Bartlett, P. D., Tauber, S. J., and Weber, W. P., *J. Am. Chem. Soc.*, 91, 6362 (1969).

4.98   (a) Lehmkuhl, H., and Olbrysch, O., *Ann. Chem.*, 1162 (1975); (b) Auger, J., Cortois, G., and Miginiac, L., *J. Organometal. Chem.*, 133, 285 (1977).

4.99   Welch, J. G., and Magid, R. M., *J. Am. Chem. Soc.*, 89, 5300 (1967).

4.100   Mulvaney, J. E., and Gardlund, Z. G., *J. Org. Chem.*, 30, 917 (1965).

4.101   Drozd, V. N., Ustunyuk, Y. A., Tseleva, M. A., and Dmitriev, L. B., *Zhur. Obshchei Knim.*, 38, 2114 (1968); 39, 1991 (1969).

4.102   Maercker, A., and Roberts, J. D., *J. Am. Chem. Soc.*, 88, 1742 (1966).

4.102a For a review on the corresponding reaction of Grignard reagents, see Hill, E. A., *Adv. Organometal. Chem.*, 16, 131 (1977).

4.103   (a) Felkin, H., Swierczewski, G., and Tambuté, A., *Tetra-hedron Lett.*, 707 (1969); (b) Crandall, J. K., and Clark, A. C., *J. Org. Chem.*, 37, 4237 (1972).

4.104   Mounet, R., and Gouin, L., *Bull. Soc. Chim. France*, 737 (1977).

4.105   Chan, T. H., and Chang, E., *J. Org. Chem.*, 39, 3264 (1974).

4.106   Kauffmann, T., Ahlers, H., Tilhard, H.-J., and Woltermann, A., *Angew. Chem. Int. Ed. Engl.*, 16, 710 (1977).

4.107   (a) Bernadou, F., and Miginiac, L., *Compt. Rend. C*, 280, 1473 (1975); (b) Frangin, Y., and Gaudemar, M., *Bull. Soc. Chim. France*, 1173 (1976).

4.108   Lehmkuhl, H., Rienehr, D., Schomburg, G., Henneberg, D., Damen, H., and Schroth, G., *Ann. Chem.*, 103, 119, 1176 (1975).

4.109   Hsieh, H. L., and Tobolsky, A. V., *J. Polymer Sci.*, 25, 245 (1957).

4.110   Stearns, R. S., and Forman, L. E., *J. Polymer Sci.*, 41, 381 (1959).

4.111   (a) Bawn, C. E. H., and Ledwith, A., *Quart. Rev.*, 16, 361 (1962); (b) Bywater, S., *Adv. Polymer Sci.*, 4, 66 (1965); (c) Kamienski, C. W., *Ind. Eng. Chem.*, No. 1, 38 (1965); (d) Hsieh, H. L., and Glaze, W. H., *Rubber Chem. Technol.*, 43, 22 (1970).

4.112   Takabe, K., Katagiri, T., and Tanaka, J., *Chem. Lett.*, 1025 (1977).

4.113   Dixon, J. A., Fishman, D. H., and Dudinyak, R. S., *Tetra-hedron Lett.*, 613 (1964).

4.114   Köbrich, G., and Fischer, R. H., *Tetrahedron*, 24, 4343 (1968).

4.115   Köbrich, G., Flory, K., and Drischel, W., *Angew. Chem. Int. Ed. Engl.*, 3, 513 (1964).

4.116   Taguchi, H., Yamamoto, H., and Nozaki, H., *J. Am. Chem. Soc.*, 96, 3010 (1974).

4.117   Taguchi, H., Tanaka, S., Yamamoto, H., and Nozaki, H., *Tetrahedron Lett.*, 2465 (1973).

4.118   Bacquet, C., Villieras, J., and Normant, J. F., *Compt. Rend.*, 278, 929 (1974).

4.119   (a) Hine, J., *J. Am. Chem. Soc.*, 72, 2438 (1950); (b) Hine, J., and Dowell, A. M., *J. Am. Chem. Soc.*, 76, 2688

(1954).

4.120 Köbrich, G., et al., *Chem. Ber.*, 99, 670, 680, 689, 1773, 1782, 1793 (1966); 100, 961 (1967).

4.121 Cainelli, G., Tangari, N., and Umani-Ronchi, A., *Tetrahedron*, 28, 3009 (1972).

4.122 Villieras, J., Bacquet, C., Masure, D., and Normant, J. F., *J. Organometal. Chem.*, 50, C7 (1973).

4.123 Fischer, R. H., and Kobrich, G., *Chem. Ber.*, 101, 3230 (1968).

4.124 Köbrich, G., and Heinemann, H., *Angew. Chem. Int. Ed. Engl.*, 4, 594 (1965).

4.125 Braun, M., Dammann, R., and Seebach, D., *Chem. Ber.*, 108, 2368 (1975).

4.126 Hasselmann, D., *Chem. Ber.*, 107, 3486 (1974).

4.127 Simmons, H. E., Cairns, T. L., Vladuchick, S. A., and Hoiness, C. M., *Org. React.*, 20, 1 (1973).

4.128 Seyferth, D., and Lambert, R. L., *J. Organometal. Chem.*, 54, 123 (1973).

4.129 Olofson, R. A., Lotts, K. D., and Barker, G. N., *Tetrahedron Lett.*, 3379 (1976).

4.130 Tavares, D. F., and Estep, R. E., *Tetrahedron Lett.*, 1229 (1973).

4.131 (a) Corey, E. J., and Durst, T., *J. Am. Chem. Soc.*, 88, 5656 (1966); 90, 5548 (1968); (b) Durst, T., *J. Am. Chem. Soc.*, 91, 1034 (1967); *Tetrahedron Lett.*, 2369 (1970).

4.132 Reutrakul, V., and Kanghae, W., *Tetrahedron Lett.*, 1225 (1977).

4.133 (a) Truce, W. E., and Christensen, L. W., *Tetrahedron Lett.*, 3075 (1969); (b) Christensen, L. W., Seaman, J. M., and Truce, W. E., *J. Org. Chem.*, 38, 2243 (1973).

4.134 Stetter, H., and Seinbeck, K., *Ann. Chem.*, 766, 89 (1972).

4.135 Wittig, G., and Schlosser, M., *Chem. Ber.*, 94, 1373 (1961).

4.136 Savignac, P., Petrova, J., Dreux, M., and Coutrot, P., *Synthesis*, 535 (1975).

4.137 Savignac, P., and Coutrot, P., *Synthesis*, 197 (1976).

4.138   (a) Cook, F., and Magnus, P., *J. C. S. Chem. Comm.*, 513
        (1977); (b) Magnus, P., and Roy, G., *J. C. S. Chem. Comm.*,
        297 (1978).

4.139   Seyferth, D., Lefferts, J. L., and Lambert, R. L., Jr.,
        J. Organometal. Chem., 142, 39 (1977).

4.140   For reviews, see (a) Newman, M. S., and Magerlein, B. J.,
        Org. React., 5, 413 (1949); (b) Ballester, M., *Chem. Rev.*,
        55, 283 (1955).

4.141   Borch, R. F., *Tetrahedron Lett.*, 3761 (1972).

4.142   Ballester, M., and Bartlett, P. D., *J. Am. Chem. Soc.*,
        75, 2042 (1953).

4.143   Deyrup, J. A., *J. Org. Chem.*, 32, 3489 (1967).

4.144   Jonezyk, A., Fedorynski, M., and Makosza, M., *Tetrahedron
        Lett.*, 2395 (1972).

4.145   Koppel, G. A., *Tetrahedron Lett.*, 1507 (1972).

4.146   Dilling, W. L., Hickner, R. A., and Farker, H. A., *J. Org.
        Chem.*, 32, 3489 (1967).

4.147   Schöllkopf, U., and Hanssle, P., *Ann. Chem.*, 763, 208
        (1972).

4.148   Baldwin, J. E., Hofle, G. A., and Lever, O. W., Jr., *J.
        Am. Chem. Soc.*, 96, 7125 (1974).

4.149   Deuchert, K., Hertenstein, U., and Hünig, S., *Synthesis*,
        777 (1973).

4.150   Kalir, A., and Balderman, D., *Synthesis*, 358 (1973).

4.151   Beak, P., and McKinnie, B. G., *J. Am. Chem. Soc.*, 99,
        5213 (1977).

4.152   Still, W. C., *J. Am. Chem. Soc.*, 100, 1481 (1978).

4.153   Still, W. C., and MacDonald, T. L., *J. Am. Chem. Soc.*, 96,
        5561 (1974).

4.154   Still, W. C., and MacDonald, T. L., *J. Org. Chem.*, 41,
        3620 (1976).

4.155   Evans, D. A., Andrews, G. C., and Buckwalter, B., *J. Am.
        Chem. Soc.*, 96, 5560 (1974).

4.156   Carlson, R. M., Jones, R. W., and Hatcher, A. S., *Tetra-
        hedron Lett.*, 1741 (1975).

4.157   Schank, K., Hasenfrantz, H., and Weber, A., *Chem. Ber.*,
        106, 1107 (1973).

4.158 Fuji, K., Ueda, M., and Fujita, E., *J. C. S. Chem. Comm.*, 814 (1977).

4.159 Peterson, D. J., *J. Org. Chem.*, **32**, 1717 (1967).

4.160 (a) Corey, E. J., and Seebach, D., *J. Org. Chem.*, **31**, 4097 (1966); (b) Corey, E. J., and Jautelat, M., *Tetrahedron Lett.*, 5787 (1968).

4.161 Shanklin, J. R., Johnson, C. R., Ollinger, J., and Coates, R. M., *J. Am. Chem. Soc.*, **95**, 3430 (1973).

4.162 Anciaux, A., Eman, A., Dumont, W., and Krief, A., *Tetrahedron Lett.*, 1617 (1975).

4.163 Attani, P. M., Biellmann, J. F., Dube, S., and V cens, J. J., *Tetrahedron Lett.*, 2665 (1974).

4.164 Seebach, D., Geiss, K.-H., and Pohmakotr, M., *Angew. Chem. Int. Ed. Engl.*, **15**, 437 (1976).

4.165 Seebach, D., and Geiss, K.-H., *Angew. Chem. Int. Ed. Engl.*, **13**, 202 (1974).

4.166 (a) Trost, B. M., Keeley, D. E., and Bogdanowicz, M. J., *J. Am. Chem. Soc.*, **95**, 3068 (1973); (b) Trost, B. M., and Keeley, D. E., *J. Am. Chem. Soc.*, **96**, 1254 (1974); (c) Trost, B. M., Bogdanowicz, M. J., and Kern, J., *J. Am. Chem. Soc.*, **97**, 2218 (1975).

4.167 Oshima, K., Shimoji, K., Takahashi, H., Yamamoto, H., and Nozaki, H., *J. Am. Chem. Soc.*, **95**, 2694 (1973).

4.168 Vlattas, I., Vecchia, L. D., and Lee, A. O., *J. Am. Chem. Soc.*, **98**, 2008 (1976).

4.169 Hirai, K., and Kishida, Y., *Tetrahedron Lett.*, 2743 (1972).

4.170 Johnson, C. R., Nakanishi, A., Nakanishi, N., and Tanaka, K., *Tetrahedron Lett.*, 2865 (1975).

4.171 Hirai, K., and Kishida, Y., *Tetrahedron Lett.*, 2117 (1972).

4.172 Oshima, K., Takahashi, H., Yamamoto, H., and Nozaki, H., *J. Am. Chem. Soc.*, **95**, 2693 (1973).

4.173 Oshima, K., Yamamoto, H., and Nozaki, H., *J. Am. Chem. Soc.*, **95**, 4446 (1973).

4.174 Hayashi, T., and Baba, H., *J. Am. Chem. Soc.*, **97**, 1608 (1975).

4.175   Takahashi, H., Oshima, K., Yamamoto, H., and Nozaki, H., *J. Am. Chem. Soc.*, 95, 5803 (1973).

4.176   Nakai, T., Shiono, H., and Okawara, M., *Tetrahedron Lett.*, 3625 (1974); 4027 (1975).

4.177   (a) Corey, E. J., and Chaykovsky, M., *J. Am. Chem. Soc.*, 87, 1345 (1965); (b) Russell, G. A., and Becker, H. D., *J. Am. Chem. Soc.*, 85, 3406 (1963).

4.178   Tsuchihashi, G., Iriuchijima, S., and Maniwa, K., *Tetrahedron Lett.*, 3389 (1973).

4.179   Trost, B. M., and Bridges, A. J., *J. Org. Chem.*, 40, 2014 (1975).

4.180   Corey, E. J., and Durst, T., *J. Am. Chem. Soc.*, 88, 5656 (1966); 90, 5548, 5553 (1968).

4.181   Jung, F., Sharma, N. K., and Dart, T., *J. Am. Chem. Soc.*, 95, 3420 (1973).

4.182   Evans, D. A., Andrews, G. C., Fujimoto, T. T., and Wells, D., *Tetrahedron Lett.*, 1385 (1973).

4.183   Evans, D. A., Crawford, T. C., Fujimoto, T. T., and Thomas, R. C., *J. Org. Chem.*, 39, 3178 (1974).

4.184   Becker, H. D., and Russell, G. A., *J. Org. Chem.*, 28, 1896 (1963).

4.185   McFarland, J. W., and Buchanan, D. N., *J. Org. Chem.*, 30, 2003 (1965).

4.186   (a) Kondo, K., and Tunemoto, D., *Tetrahedron Lett.*, 1007 (1975); (b) 1397 (1975); (c) Kondo, K., Saito, E., and Tunemoto, D., *Tetrahedron Lett.*, 2275 (1975).

4.187   Grieco, P. A., and Masaki, Y., *J. Org. Chem.*, 39, 2135 (1974).

4.188   Yoshimoto, M., Ishida, N., and Kishida, Y., *Chem. Pharm. Bull.*, 20, 2137 (1972).

4.189   Kaiser, E. M., and Knutson, P. L. A., *J. Org. Chem.*, 40, 1342 (1975).

4.190   Johnson, C. R., Shanklin, J. R., and Kirchhoff, R. A., *J. Am. Chem. Soc.*, 95, 6462 (1973).

4.191   Johnson, C. R., Kirchhoff, R. A., Reischer, R. J., and Katekar, G. F., *J. Am. Chem. Soc.*, 95, 4287 (1973).

4.192   Kamata, S., Uyeo, S., Haga, N., and Nagata, W., *Synth. Commun.*, 3, 265 (1973).

4.193   Inomata, K., Aoyama, S., and Kotake, H., *Bull. Chem. Soc. Japan*, <u>51</u>, 930 (1978).

4.194   Tanaka, K., Yamagishi, N., Tanikaga, R., and Kaji, A., *Chem. Lett.*, 471 (1977).

4.195   Tanaka, K., Yamagishi, N., Uneme, H., Tanikaga, R., and Kaji, A., *Chem. Lett.*, 197 (1978).

4.196   Yamagiwa, S., Hoshi, N., Sato, H., Kosugi, H., and Uda, H., *J. Chem. Soc. Perkin I*, 214 (1978).

4.197   Gassman, P. G., and Richmond, G. D., *J. Org. Chem.*, <u>31</u>, 2355 (1966).

4.198   Russell, G. A., and Ochrymowycz, L. A., *J. Org. Chem.*, <u>34</u>, 3618, 3624 (1969); Russell, G. A., and Hamprecht, G., *J. Org. Chem.*, <u>35</u>, 3007 (1970).

4.199   Grieco, P. A., Boxler, D., and Pogonowski, C. S., *J. C. S. Chem. Comm.*, 497 (1974).

4.200   (a) Koutek, B., Pavlickova, L., and Soucek, M., *Coll. Czech. Chem. Comm.*, <u>38</u>, 3872 (1973); (b) <u>39</u>, 192 (1974).

4.201   (a) Schöllkopf, U., and Schröder, R., *Angew. Chem. Int. Ed. Engl.*, <u>11</u>, 311 (1972); (b) <u>12</u>, 407 (1973).

4.202   (a) Corey, E. J., and Seebach, D., *Angew. Chem. Int. Ed. Engl.*, <u>4</u>, 1075, 1077 (1965); (b) Seebach, D., and Corey, E. J., *J. Org. Chem.*, <u>40</u>, 231 (1975).

4.203   (a) Seebach, D., *Synthesis*, 17 (1969); (b) Seebach, D., *Synthesis*, 357 (1977).

4.204   Herrmann, J. L., Richman, J. E., and Schlessinger, R. H., *Tetrahedron Lett.*, 2599 (1973).

4.205   (a) Seebach, D., Kolb, M., and Gröbel, B. T., *Tetrahedron Lett.*, 3171 (1974); (b) Corey, E. J., and Koziekowski, A. P., *Tetrahedron Lett.*, 925 (1975).

4.206   Eliel, E. L., and Hartmann, A. A., *J. Org. Chem.*, <u>37</u>, 505 (1972).

4.207   Seebach, D., Corey, E. J., and Beck, A. K., *Chem. Ber.*, <u>107</u>, 367 (1974).

4.208   Mcube, S., Pelter, A., Smith, K., Blatcher, P., and Warren, S., *Tetrahedron Lett.*, 2345 (1978).

4.209   Mori, K., Hashimoto, H., Takenaka, Y., and Takigawa, T., *Synthesis*, 720 (1975).

4.210   Schill, G., and Merkel, C., *Synthesis*, 387 (1975).

4.211   Cregge, R. J., Herrmann, J. L., Richman, J. E., Romanet, R. F., and Schlessinger, R. H., *Tetrahedron Lett.*, 2595 (1973).

4.212   Hori, I., Hayashi, T., and Midorikawa, H., *Synthesis*, 705 (1974).

4.213   (a) Ogura, K., and Tsuchihashi, G., *Tetrahedron Lett.*, 3151 (1971); 1383, 2681 (1972); (b) Ogura, K., Yamashita, M., Suzuki, M., Tsuchihashi, G., *Tetrahedron Lett.*, 3653 (1974); 2767 (1975).

4.214   (a) Richman, J. E., Herrmann, J. L., and Schlessinger, R. H., *Tetrahedron Lett.*, 3267 (1973); (b) Herrmann, J. L., Richman, J. E., Wepplo, P. J., and Schlessinger, R. H., *Tetrahedron Lett.*, 4707 (1973).

4.215   Seebach, D., *Chem. Ber.*, 105, 487 (1972).

4.216   Manas, A. R. B., and Smith, R. A. J., *J. C. S. Chem. Comm.*, 216 (1975).

4.217   Corey, E. J., and Boger, D. L., *Tetrahedron Lett.*, 5, 9, 13 (1978).

4.218   (a) Dumont, W., Bayet, P., and Krief, A., *Angew. Chem. Int. Ed. Engl.*, 13, 805 (1974); (b) Seebach, D., and Beck, A. K., *Angew. Chem. Int. Ed. Engl.*, 13, 806 (1974); (c) Van Ende, D., Dumont, W., and Krief, A., *Angew. Chem. Int. Ed. Engl.*, 14, 700 (1975).

4.219   (a) Reich, H. J., and Shah, S. K., *J. Am. Chem. Soc.*, 97, 3250 (1975); (b) Reich, H. J., and Chow, F., *J. C. S. Chem. Comm.*, 790 (1975).

4.220   Reich, H. J., *J. Org. Chem.*, 40, 2570 (1975).

4.221   Reich, H. J., and Shah, S. K., *J. Am. Chem. Soc.*, 99, 263 (1977).

4.222   Sachdev, K., and Sachdev, H. S., *Tetrahedron Lett.*, 4223 (1976).

4.223   (a) Seebach, D., and Enders, D., *Angew. Chem. Int. Ed. Engl.*, 11, 1101 (1972); *Chem. Ber.*, 108, 1293 (1975); (b) For a review, see Seebach, D., and Enders, D., *Angew. Chem. Int. Ed. Engl.*, 14, 15 (1975).

4.224   Savignac, P., Leroux, Y., and Normant, H., *Tetrahedron*, 31, 877 (1975).

4.225   Ahlbrecht, H., and Eichler, J., *Synthesis*, 672 (1974).

4.226   Martin, S. F., and DuPriest, M. T., *Tetrahedron Lett.*, 3925 (1977).

4.227   For a review, see McMurry, J. E., *Acc. Chem. Res.*, 7, 281 (1974).

4.228   (a) Pubs, P., and Schenk, H. P., *Helv. Chim. Acta*, 61, 984 (1978); (b) Dubs, P., and Stüssi, R., *Helv. Chim. Acta*, 61, 990 (1978).

4.229   Meyer, P., Schöllkopf, U., and Böhme, P., *Ann. Chem.*, 1183 (1977).

4.230   Schöllkopf, U., Stafforst, D., and Jentsch, R., *Ann. Chem.*, 1167 (1977).

4.231   Nizuik, G. E., Morrison, W. H., III, and Walborsky, H. M., *J. Org. Chem.*, 39, 600 (1974).

4.232   Fraser, R. R., and Hubert, P. R., *Can. J. Chem.*, 52, 185 (1974).

4.233   (a) Wadsworth, W. S., Jr., and Emmons, W. D., *Org. Synth.*, 45, 44 (1965); (b) House, H. O., Jones, V. K., and Frank, G., *J. Org. Chem.*, 29, 3327 (1964); (c) Wadsworth, D. H., Schupp., O. E., Seus, E. J., and Ford, J. A., Jr., *J. Org. Chem.*, 30, 680 (1965); (d) Kinstle, T. H., and Mandanas, B. Y., *J. C. S. Chem. Comm.*, 1699 (1968).

4.234   Jones, G., and Maisey, R. F., *J. C. S. Chem. Comm.*, 543 (1968).

4.235   Kondo, K., Negishi, A., and Tunemoto, D., *Angew. Chem. Int. Ed. Engl.*, 13, 407 (1974).

4.236   Schweizer, E. E., and Bach, R. D., *Org. Synth.*, 48, 129 (1968).

4.237   Mikolajczyk, M., Grzejszczak, S., and Zatorski, A., *J. Org. Chem.*, 40, 1979 (1975).

4.238   Posner, G. H., and Brunelle, D. J., *Tetrahedron Lett.*, 935 (1973).

4.239   Comasseto, J., and Petragnani, N., *J. Organometal. Chem.*, 152, 295 (1978).

4.240   Peterson, D. J., *J. Org. Chem.*, 33, 780 (1968).

4.241   Dumont, W., and Krief, A., *Angew. Chem. Int. Ed. Engl.*, 15, 161 (1976).

4.242   Chan, T. H., and Chang, E., *J. Org. Chem.*, 39, 3264
        (1974).

4.243   Ayalon-Chass, D., Ehlinger, E., and Magnus, P., *J. C. S.
        Chem. Comm.*, 772 (1977).

4.244   Gröbel, B. T., and Seebach, D., *Chem. Ber.*, 110, 852,
        867 (1977).

4.245   (a) Shimoji, K., Taguchi, H., Oshima, K., Yamamoto, H.,
        and Nozaki, H., *J. Am. Chem. Soc.*, 96, 1620 (1974); *Bull.
        Chem. Soc. Japan*, 47, 2529 (1974); (b) Hartzell, S. L.,
        Sullivan, D. F., and Rathke, M. W., *Tetrahedron Lett.*,
        1403 (1974).

4.246   Grieco, P. A., Wang, C.-L. J., and Burke, S. D., *J. C. S.
        Chem. Comm.*, 537 (1975).

4.247   Ojima, I., and Kumagai, M., *Tetrahedron Lett.*, 4005
        (1974).

4.248   Rathke, M. W., and Woodbury, R., *J. Org. Chem.*, 43, 1947
        (1978).

4.249   Corey, E. J., Enders, D., and Bock, M. G., *Tetrahedron
        Lett.*, 7 (1976).

4.250   (a) Cook, M. A., Eaborn, C., Jukes, A. E., and Walton, D.
        M. R., *J. Organometal. Chem.*, 24, 529 (1970); (b) Sakurai,
        H., Nishiwaki, K.-I., and Kira, M., *Tetrahedron Lett.*,
        4193 (1973).

4.251   Gröbel, B. T., and Seebach, D., *Angew. Chem. Int. Ed.
        Engl.*, 13, 83 (1974).

4.252   Carey, F. A., and Court, A. S., *J. Org. Chem.*, 37, 939
        (1972).

4.253   Carey, F. A., and Hernandez. O., *J. Org. Chem.*, 38, 2670
        (1973).

4.254   (a) Seebach, D., Gröbel, B. T., Beck, A. K., Braun, M.,
        and Geiss, K. H., *Angew. Chem. Int. Ed. Engl.*, 11, 443
        (1972); (b) Seebach, D., Kolb, M., and Gröbel, B. T.,
        Chem. Ber., 106, 2277 (1973); (c) Seebach, D., and
        Bürstinghaus, R., *Angew. Chem. Int. Ed. Engl.*, 14, 57
        (1975).

4.255   Jones, P. F., and Lappert, M. F., *J. C. S. Chem. Comm.*,
        526 (1972); *J. C. S. Perkin I*, 2272 (1973).

4.256   Seebach, D., *Angew. Chem. Int. Ed. Engl.*, 9, 639 (1969).

4.257   Gröbel, B. T., and Seebach, D., *Synthesis*, 357 (1977).

4.258   Lever, O. W., Jr., *Tetrahedron*, 32, 1943 (1976).

4.259   (a) Starks, C. M., and Owens, R. M., *J. Am. Chem. Soc.*, 95, 3613 (1973); (b) Zubrick, J. W., Dunbar, B. I., and Durst, H. D., *Tetrahedron Lett.*, 71 (1975).

4.260   Evans, D. A., Truedale, L. K., and Carroll, G. L., *J. C. S. Chem. Comm.*, 55 (1973).

4.261   For a review, see Noland, W. E., *Chem. Rev.*, 55, 137 (1955).

4.262   Banhidai, B., and Schöllkopf, U., *Angew. Chem. Int. Ed. Engl.*, 12, 836 (1973).

4.263   Seebach, D., Lubosch, W., and Enders, D., *Chem. Ber.*, 109, 1309 (1976).

4.264   Jutzi, P., and Schroder, F. W., *Angew. Chem. Int. Ed. Engl.*, 10, 339 (1971).

4.265   (a) Corey, E. J., and Crouse, D., *J. Org. Chem.*, 33, 298 (1968); (b) Vedejs, E., and Fuchs, P. L., *J. Org. Chem.*, 36, 366 (1971).

4.266   Mukaiyama, T., Narasaka, K., and Furusato, M., *J. Am. Chem. Soc.*, 94, 8641 (1972).

4.267   Seebach, D., and Neumann, H., *Chem. Ber.*, 107, 847 (1974).

4.268   Ho, T. L., and Wong, C. M., *Can. J. Chem.*, 50, 3740 (1972).

4.269   Ho, T. L., Ho, H. C., and Wong, C. M., *J. C. S. Chem. Comm.*, 791 (1972).

4.270   Huurdeman, W. F. J., Wynberg, H., and Emerson, D. W., *Synth. Comm.*, 2, 7 (1972).

4.271   Tamura, Y., Sumoto, K., Fujii, S., Satoh, H., and Ikeda, M., *Synthesis*, 312 (1973).

4.272   Corey, E. J., and Erickson, B. W., *J. Org. Chem.*, 36, 3553 (1971).

4.273   (a) Chang, H. W., *Tetrahedron Lett.*, 1989 (1972); (b) Fetizon, M., and Jurion, M., *J. C. S. Chem. Comm.*, 382 (1972); (c) Oishi, T., Kamemoto, K., and Ban, Y., *Tetrahedron Lett.*, 1085 (1972).

4.274   Seebach, D., and Bürstinghaus, R., *Angew. Chem. Int. Ed. Engl.*, 14, 57 (1975).

4.275   For a review, see Pettit, G. R., and van Tamelem, E. E., *Org. React.*, 12, 356 (1962).

4.276   Mukaiyama, T., Narasaka, K., Maekawa, K., and Furusato, M., *Bull. Chem. Soc. Japan*, 44, 2285 (1971).

4.277   Mukaiyama, T., Hayashi, M., and Narasaka, K., *Chem. Lett.*, 291 (1973).

4.278   Baarshers, W. H., and Loh, T. L., *Tetrahedron Lett.*, 3483 (1971).

4.279   Newman, B. C., and Eliel, E. L., *J. Org. Chem.*, 35, 3641 (1970).

4.280   (a) Herrmann, J. L., Richman, J. E., and Schlessinger, R. H., *Tetrahedron Lett.*, 3271, 3275 (1973); (b) Ogura, K., Yamashita, M., and Tsuchihashi, G., *Tetrahedron Lett.*, 1303 (1978).

4.281   Evans, D. A., and Andrews, G. C., *Acc. Chem. Res.*, 7, 147 (1974).

4.282   Still, W. C., and Mitra, A., *J. Am. Chem. Soc.*, 100, 1927 (1978). Also see Still, W. C., *J. Am. Chem. Soc.*, 100, 1481 (1978).

4.283   Cooke, F., and Magnus, P., *J. C. S. Chem. Comm.*, 519 (1976).

4.284   Kato, T., Takayanagi, H., Uyehara, T., and Kitahara, Y., *Chem. Lett.*, 1009 (1977).

4.285   Atlani, P. M., Biellmann, J. F., Dube, S., Vicens, J. J., *Tetrahedron Lett.*, 2665 (1974).

4.286   Maercker, A., *Org. React.*, 14, 270 (1965).

4.287   Schlosser, M., *Top. Stereochem.*, 5, 1 (1970).

4.288   Johnson, A. W., *Ylid Chemistry*, Academic Press, New York, 1966, pp. 5 to 247.

4.289   Corey, E. J., and Shulman, J. I., *J. Org. Chem.*, 35, 777 (1970). Also see Ref. *4.236*.

4.290   Clive, D. L. J., *Tetrahedron*, 34, 1049 (1978).

4.291   Trost, B. M., *Chem. Rev.*, 78, 363 (1978).

4.292   For a review, see Paquette, L. A., *Org. React.*, 25, 1 (1977).

4.293   Buchi, G., and Freidinger, R. M., *J. Am. Chem. Soc.*, 96, 3332 (1974).

4.294   (a) Newman, M. S., and Magerlein, B. J., *Org. React.*, 5, 413 (1949); (b) Ballester, M., *Chem. Rev.*, 55, 283 (1955).

4.295   Trost, B. M., and Melvin, L. S., Jr., *Sulfur Ylides: Emerging Synthetic Intermediates*, Academic Press, New York, 1975, 344 pp.

4.296   Skell, P. S., and Woodworth, R. C., *J. Am. Chem. Soc.*, <u>78</u>, 4496 (1956).

4.297   Weyerstahl, P., Mathias, R., and Blume, G., *Tetrahedron Lett.*, 611 (1973).

4.298   Fedorynski, F., *Synthesis*, 783 (1977).

4.299   Kostikov, R. R., Molchanov, A. P., and Bespalov, A. Y., *J. Org. Chem. USSR*, <u>10</u>, 8 (1974).

4.300   For a review, see Köbrich, G., *Angew. Chem. Int. Ed. Engl.*, <u>11</u>, 473 (1972).

4.301   Martel, B., and Hiriart, J. M., *Synthesis*, 201 (1972).

4.302   Sakan, F., Sugiura, E., Matsumoto, T., and Shirahama, H., *Bull. Soc. Chem. Japan*, <u>47</u>, 1037 (1974).

4.303   Boche, G., and Schneider, D. R., *Tetrahedron Lett.*, 4247 (1975).

4.304   Olofson, R. A., and Dougherty, C. M., *J. Am. Chem. Soc.*, <u>95</u>, 581 (1973).

4.305   Arora, S., Binger, P., and Köster, R., *Synthesis*, 146 (1973).

4.306   Simmons, H. E., and Smith, R. D., *J. Am. Chem. Soc.*, <u>80</u>, 5323 (1958); <u>81</u>, 4256 (1959).

4.307   (a) Blanchard, E. P., and Simmons, H. E., *J. Am. Chem. Soc.*, <u>86</u>, 1337 (1964); (b) Simmons, H. E., Blanchard, E. P., and Smith, R. D., *J. Am. Chem. Soc.*, <u>86</u>, 1347 (1964).

4.308   Dauben, W. G., and Ashcraft, A. C., *J. Am. Chem. Soc.*, <u>85</u>, 3673 (1963).

4.309   Corey, E. J., Yamamoto, H., Herron, D. K., and Achiwa, K., *J. Am. Chem. Soc.*, <u>92</u>, 6635 (1970).

4.310   (a) For a review, see Conia, J. M., *Pure Appl. Chem.*, <u>43</u>, 317 (1975); (b) Le Goaller, R., and Pierre, J.-L., *Bull. Soc. Chim. France*, 1531 (1973); (c) Rubottom, G. H., and Lopez, M. I., *J. Org. Chem.*, <u>38</u>, 2097 (1973); (d) Murai, S., Aya, T., and Sonoda, N., *J. Org. Chem.*, <u>38</u>, 4354 (1973); (e) Murai, S., Aya, T., Renge, T., Ryu, I., and Sonoda, N., *J. Org. Chem.*, <u>39</u>, 858 (1974); (f) Girard, G., and Conia, H. C. J. M., *Tetrahedron Lett.*, 3327, 3329, 3333 (1974).

4.311   (a) Furukawa, J., Kawabata, N., and Nishimura, J., *Tetrahedron Lett.*, 3353 (1966); (b) For a review, see Furukawa, J., and Kawabata, N., *Adv. Organometal. Chem.*, **12**, 83 (1974).

4.312   Trost, B. M., *Pure Appl. Chem.*, **43**, 563 (1975), and pertinent references cited therein.

4.313   Trost, B. M., and Keeley, D. E., *J. Am. Chem. Soc.*, **98**, 248 (1976).

4.314   Trost, B. M., Keeley, D. E., Arndt, H. C., and Bogdanowicz, M. J., *J. Am. Chem. Soc.*, **99**, 3088 (1977).

4.315   Trost, B. M., and Rigby, J. H., *J. Org. Chem.*, **41**, 3217 (1976).

4.316   House, H. O., *Modern Synthetic Reactions*, 2nd ed., W. A. Benjamin, Menlo Park, Calif., 1972, 856 pp.

4.317   Jackman, L. M., and Lange, B. C., *Tetrahedron*, **33**, 2727 (1977).

4.318   d'Angelo, J., *Tetrahedron*, **32**, 2979 (1976).

4.319   For a review, see Nesmeyanov, A. N., *J. Organometal. Chem.*, **100**, 161 (1975).

4.320   House, H. O., Auerback, R. A., Gall, M., and Peet, N. P., *J. Org. Chem.*, **38**, 514 (1973).

4.321   Jackman, L. M., and Hadden, R. C., *J. Am. Chem. Soc.*, **95**, 3687 (1973).

4.322   Jackman, L. M., and Szeverenji, N., *J. Am. Chem. Soc.*, **99**, 4954 (1977).

4.323   Posner, G. H., and Sterling, J. J., *J. Am. Chem. Soc.*, **95**, 3076 (1973).

4.324   Stork, G., and Hudrlik, P. F., *J. Am. Chem. Soc.*, **90**, 4464 (1968).

4.325   Näf, F., Decorzant, R., and Thommen, W., *Helv. Chim. Acta*, **58**, 1808 (1975).

4.326   Fortunato, J. M., and Ganem, B., *J. Org. Chem.*, **41**, 2194 (1976).

4.327   Stork, G., Rosen, P., and Goldman, N. L., *J. Am. Chem. Soc.*, **83**, 2965 (1961).

4.328   Stork, G., Uyeo, S., Wakamatsu, T., Grieco, P., and Labowitz, J., *J. Am. Chem. Soc.*, **93**, 4945 (1971).

4.329   Rathke, M. W., and Lindert, A., *J. Am. Chem. Soc.*, <u>93</u>, 2318 (1971).

4.330   Brown, C. A., *J. Org. Chem.*, <u>34</u>, 2324 (1969).

4.331   Groenewegen, P., Kallenberg, H., and van der Gen, A., *Tetrahedron Lett.*, 491 (1978). Also see de Graaf, S. A. G., Oosterhoff, P. E. R., and van der Gen, A., *Tetrahedron Lett.*, 1653 (1974).

4.332   Brown, C. A., *Synthesis*, 326 (1975).

4.333   Kornblum, N., Berrigan, P. J., and LeNoble, W. J., *J. Am. Chem. Soc.*, <u>85</u>, 1141 (1963).

4.334   Taylor, E. C., Hawkes, G. H., and McKillop, A., *J. Am. Chem. Soc.*, <u>90</u>, 2421 (1968).

4.335   Hauser, C. R., and Hudson, B. E. Jr., *Org. React.*, <u>1</u>, 266 (1942).

4.336   Boatman, S., Harris, T. M., and Hauser, C. R., *Org. Synth.*, <u>48</u>, 40 (1968).

4.337   Ireland, R. E., and Marshall, J. A., *J. Org. Chem.*, <u>27</u>, 1615, 1620 (1962).

4.338   (a) Grieco, P. A., and Pogonowski, C. S., *J. Org. Chem.*, <u>39</u>, 732 (1974); (b) *J. C. S. Chem. Comm.*, 72 (1975).

4.339   For extensive reviews, see (a) Jung, M. E., *Tetrahedron*, <u>32</u>, 3 (1976); (b) Gawley, R. E., *Synthesis*, 777 (1976).

4.340   (a) Stork, G., and Jung, M. E., *J. Am. Chem. Soc.*, <u>94</u>, 3682 (1974); (b) Stork, G., Jung, M. E., Colvin, E., and Noel, Y., *J. Am. Chem. Soc.*, <u>94</u>, 3684 (1974).

4.341   Wichterle, O., Prochaska, J., and Hoffman, J., *Coll. Czech, Chem. Comm.*, <u>13</u>, 300 (1948).

4.342   (a) Stork, G., Danishefsky, S., and Ohashi, M., *J. Am. Chem. Soc.*, <u>89</u>, 5459 (1967); (b) Stork, G., and McMurry, J. E., *J. Am. Chem. Soc.*, <u>89</u>, 5463, 5464 (1967).

4.343   Stotter, P. L., and Hill, K. A., *J. Am. Chem. Soc.*, <u>96</u>, 6524 (1974).

4.344   Negishi, E., Idacavage, M. J., DiPasquale, F., and Silveira, A., Jr., *Tetrahedron Lett.*, 845 (1979).

4.345   Kuwajima, I., and Nakamura, E., *J. Am. Chem. Soc.*, <u>97</u>, 3257 (1975).

4.346   Chan, T. H., Paterson, I., and Pinsonnault, J., *Tetrahedron Lett.*, 4183 (1977).

4.347   House, H. O., Tefertiller, B. A., and Olmstead, H. D., *J. Org. Chem.*, 33, 935 (1968).

4.348   House, H. O., and Trost, B. M., *J. Org. Chem.*, 30, 2502 (1965).

4.349   House, H. O., and Umen, M. J., *J. Org. Chem.*, 38, 1000 (1973).

4.350   (a) Rossi, R. A., and Bunnett, J. F., *J. Am. Chem. Soc.*, 94, 638 (1972); (b) Bunnett, J. F., and Sundberg, J. E., *Chem. Pharm. Bull.*, 23, 2621 (1975).

4.351   Bunnett, J. F., Creary, X., Sundberg, J. E., *J. Org. Chem.*, 41, 1707 (1976).

4.352   Gregge, R. J., Herrmann, J. L., Lee, C. S., Richman, J. E., and Schlessinger, R. H., *Tetrahedron Lett.*, 2425 (1973).

4.353   Herrmann, J. L., Kieczykowski, G. R., and Schlessinger, R. H., *Tetrahedron Lett.*, 2433 (1973).

4.354   Kamata, S., Uyeo, S., Haga, N., and Nagata, W., *Synth. Commun.*, 3, 265 (1973).

4.355   Herrmann, J. L., and Schlessinger, R. H., *J. C. S. Chem. Comm.*, 711 (1973).

4.356   Ganem, B., and Fortunato, J. M., *J. Org. Chem.*, 40, 2846 (1975).

4.357   Kende, A. S., Constantinides, D., Lee, S. J., and Liebeskind, L., *Tetrahedron Lett.*, 405 (1975).

4.358   Zimmerman, M. P., *Synth. Commun.*, 7, 189 (1977).

4.359   McMurry, J. E., and Musser, J. H., *J. Org. Chem.*, 40, 2556 (1975).

4.360   (a) Stork, G., Leong, A. Y. W., and Touzin, A. M., *J. Org. Chem.*, 41, 3491 (1976); (b) Bey, P., and Vevert, J. P., *Tetrahedron Lett.*, 1455 (1977).

4.361   Krapcho, A. P., and Dundulis, E. A., *Tetrahedron Lett.*, 2205 (1976).

4.362   Bunnell, C. A., and Fuchs, P. L., *J. Am. Chem. Soc.*, 99, 5184 (1977).

4.363   Ciochetto, L. J., Bergbreiter, D. E., and Newcomb, M., *J. Org. Chem.*, 42, 2948 (1977).

4.364   Fitt, J. J., and Gschwend, H. W., *J. Org. Chem.*, 42, 2639 (1977).

4.365   Hirai, K., Iwano, Y., and Kishida, Y., *Tetrahedron Lett.*, 2677 (1977).

4.366   Boeckman, R. K., Jr., Ramaiah, M., and Medwid, J. B., *Tetrahedron Lett.*, 4485 (1977).

4.367   Creger, P. L., *J. Am. Chem. Soc.*, 89, 2500 (1967); 92, 1396, 1397 (1970).

4.368   (a) Grieco, P. A., Wang, C. L. J., and Burke, S. D., *J. C. S. Chem. Comm.*, 537 (1975); (b) Grieco, P. A., and Wang, C. L. J., *J. C. S. Chem. Comm.*, 714 (1975).

4.369   Trost, B. M., and Tamaru, Y., *Tetrahedron Lett.*, 3797 (1975).

4.370   Pitzele, B. S., Baran, J. S., and Steinman, D. H., *J. Org. Chem.*, 40, 269 (1975).

4.371   Krapcho, A. P., and Kashdan, D. S., *Tetrahedron Lett.*, 707 (1975).

4.372   Birch, A. J., and Slobbe, J., *Aust. J. Chem.*, 30, 1045 (1977).

4.373   (a) Trost, B. M., Kunz, R. A., *J. Org. Chem.*, 39, 2475 (1974); (b) Deslonchamps, P., Cheriyan, U. O., and Patterson, D. R., *Can. J. Chem.*, 53, 1682 (1975).

4.374   Hullot, P., Cuvigny, T., Larcheveque, M., and Normant, H., *Can. J. Chem.*, 54, 1098 (1976).

4.375   Woodbury, R. P., and Rathke, M. W., *J. Org. Chem.*, 42, 1688 (1977).

4.376   Kieczykowski, G. R., Schlessinger, R. H., and Sulsky, R. B., *Tetrahedron Lett.*, 4647 (1975).

4.377   Hünig, S., and Wehner, G., *Synthesis*, 180 (1975).

4.378   (a) Jung, M. E., Blair, P. A., and Lowe, J. A., *Tetrahedron Lett.*, 1439 (1976); (b) Kofron, W. G., and Yeh, M. K., *J. Org. Chem.*, 41, 439 (1976).

4.379   Fraser, R. R., and Dhawau, K. L., *J. C. S. Chem. Comm.*, 674 (1976).

4.380   Corey, E. J., and Enders, D., *Tetrahedron Lett.*, 3 (1976).

4.381   (a) Cuvigny, T., LeBorgne, J. F., Larcheveque, M., and Normant, H., *Synthesis*, 237, 238 (1976); (b) Cuvigny, T., Larcheveque, M., and Normant, H., *Compt. Rend.*, 277, 511 (1973); *Ann. Chem.*, 719 (1975); (c) LeBorgne, J. F., *J. Organometal. Chem.*, 122, 123, 129 (1977).

4.382   House, H. O., Liang, W. D., and Weeks, P. D., *J. Org. Chem.*, 39, 3102 (1974).

4.383   Jacobson, R. M., Raths, R. A., and McDonald, J. H., III, *J. Org. Chem.*, 42, 2545 (1977).

4.384   Trost, B. M., and Kunz, R. A., *J. Am. Chem. Soc.*, 97, 7152 (1975).

4.385   Kieczykowski, G. R., Schlessinger, R. H., and Sulsky, R. B., *Tetrahedron Lett.*, 597 (1976).

4.386   (a) Meyers, A. I., et al., *J. Org. Chem.*, 38, 36 (1973); (b) Meyers, A. I., and Nazarenko, N., *J. Org. Chem.*, 38, 175 (1973); (c) Meyers, A. I., Smith, E. M., and Ao, M. S., *J. Org. Chem.*, 38, 2129 (1973).

4.387   (a) Lion, C., and Dubois, J. E., *Tetrahedron*, 29, 3417 (1973); (b) Meyers, A. I., Temple, D. L., Nolen, R. L., and Nihelich, E. D., *J. Org. Chem.*, 39, 2778 (1974).

4.388   (a) Meyers, A. I., Knaus, G., Kamata, K., and Ford, M. E., *J. Am. Chem. Soc.*, 98, 567 (1976); (b) Meyers, A. I., and Kamata, K., *J. Am. Chem. Soc.*, 98, 2290 (1976).

4.389   Meyers, A. I., and Durandetta, J. L., *J. Org. Chem.*, 40, 2021 (1975).

4.390   Nielsen, A. T., and Houlihan, W. J., *Org. React.*, 16, 1 (1968).

4.391   Hauser, C. R., Swamer, F. W., and Adams, J. T., *Org. React.*, 8, 59 (1954).

4.392   For a review, see Ho, T. L., *Synth. Commun.*, 4, 265 (1974).

4.393   Johnson, W. S., Semmelhack, M. F., Sultanbawa, M. U. S., and Dolak, L. A., *J. Am. Chem. Soc.*, 90, 2994 (1968).

4.394   Johnson, W. S., and Daub, G. H., *Org. React.*, 6, 1 (1951).

4.395   Johnson, J. R., *Org. React.*, 1, 210 (1942).

4.396   Blicke, F. F., *Org. React.*, 1, 303 (1942).

4.397   Ref. *4.316*, p. 633.

4.398   Ref. *4.316*, p. 660.

4.399   For reviews, see (a) Wittig, G., *Rec. Chem. Progr.*, 28, 45 (1967); (b) Wittig, G., *Fortschr. Chem. Forsch.*, 67, 1 (1976).

4.400   House, H. O., Crumrine, D. S., Teranishi, A. Y., and Olmstead, H. D., *J. Am. Chem. Soc.*, 95, 3310 (1973).

4.401   Stork, G., Kraus, G. A., and Garcia, G. A., *J. Org. Chem.*, 39, 3459 (1974).

4.402   Gaudemar, M., *Compt. Rend.*, 278, 533 (1974); 279, 961 (1974).

4.403   Kuwajima, I., and Iwasawa, H., *Tetrahedron Lett.*, 107 (1974). For a related procedure using acetoacetic esters, see Huckin, S. N., and Weiler, L., *Can. J. Chem.*, 52, 2157 (1974).

4.404   (a) Mukaiyama, T., and Inoue, T., *Chem. Lett.*, 559 (1976); (b) Mukaiyama, T., Inomata, K., and Muraki, M., *J. Am. Chem. Soc.*, 95, 967 (1973); (c) Inomata, K., Muraki, M., and Mukaiyama, T., *Bull. Chem. Soc. Japan*, 46, 1807 (1973).

4.405   Negishi, E., Idacavage, M. J., unpublished results.

4.406   Noyori, R., Yokoyama, K., Sakata, J., Kuwajima, I., Nakamura, E., and Shimizu, M., *J. Am. Chem. Soc.*, 99, 1265 (1977).

4.407   Maruoka, K., Hashimoto, S., Kitagawa, Y., Yamamoto, H., and Nozaki, H., *J. Am. Chem. Soc.*, 99, 7705 (1977).

4.408   (a) Dubois, J. E., and Dubois, M., *Tetrahedron Lett.*, 4215 (1967); (b) Dubois, J. E., and Fellmann, P., *C. R. Acad. Sci.*, 274, 1307 (1972); *Tetrahedron Lett.*, 1225 (1975).

4.409   (a) Kleschick, W. A., Buse, C. T., and Heathcock, C. H., *J. Am. Chem. Soc.*, 99, 247 (1977); (b) Buse, C. T., and Heathcock, C. H., *J. Am. Chem. Soc.*, 99, 8109 (1977).

4.410   Jeffery, E. A., Meisters, A., and Mole, T., *J. Organometal. Chem.*, 74, 365, 373 (1974).

4.411   Nakata, T., Schmid, G., Vranesic, B., Okigawa, M., Smith-Palmer, T., and Kishi, Y., *J. Am. Chem. Soc.*, 100, 2933 (1978).

4.412   Rathke, M. W., and Sullivan, D. F., *J. Am. Chem. Soc.*, 95, 3050 (1973).

4.413   Villieras, J., Perriot, P., Bourgain, M., and Normant, J. F., *J. Organometal. Chem.*, 102, 129 (1975).

4.414   Wemple, J., *Tetrahedron Lett.*, 3255 (1975).

4.415   Touzin, A. M., *Tetrahedron Lett.*, 1477 (1975).

4.416   Rieke, R. D., and Uhm, S. J., *Synthesis*, 452 (1975).

4.417 Schöllkopf, U., Gerhart, F., Schroder, R., and Hoppe, D., *Ann. Chem.*, 766, 116 (1972).

4.418 Rathke, M. W., and Sullivan, D., *Tetrahedron Lett.*, 4249 (1972).

4.419 (a) Grieco, P. A., and Hiroi, K., *J. C. S. Chem. Comm.*, 1317 (1972); (b) *J. C. S. Chem. Comm.*, 500 (1973).

4.420 Bellassoued, M., and Gaudemar, M., *J. Organometal. Chem.*, 81, 139 (1974).

4.421 Bellassoued, M., Dardoize, F., Gaudemar, M., and Goasdoue, N., *Compt. Rend.*, 281, 893 (1975).

4.422 Parham, W. E., and Boykin, D. W., *J. Org. Chem.*, 42, 260 (1977).

4.423 Watanabe, S., Suga, K., Fujita, T., and Fujiyoshi, K., *Chem. Ind.*, 80 (1972).

4.424 (a) Durst, T., and LeBelle, M. J., *Can. J. Chem.*, 50, 3196 (1972); (b) Durst, T., Elzen, R. V. D., Legault, R., *Can. J. Chem.*, 52, 3206 (1974).

4.425 Hullot, P., Cuvigny, T., Larcheveque, M., and Normant, H., *Can. J. Chem.*, 55, 266 (1977).

4.426 Arpe, H. J., and Leupold, I., *Angew. Chem. Int. Ed. Engl.*, 11, 722 (1972). Also see Gokel, G. W., DiBiase, S. A., and Lipisko, B. A., *Tetrahedron Lett.*, 3495 (1976).

4.427 Cuvigny, T., Hullot, P., and Larcheveque, M., *J. Organometal. Chem.*, 57, C36 (1973).

4.428 Hünig, S., and Wehner, G., *Synthesis*, 391 (1975).

4.429 Wittig, G., Fischer, S., and Tanaka, M., *Ann. Chem.*, 1075 (1973).

4.430 Shapiro, R. H., Lipton, M. F., Kolonko, K. J., Buswell, R. L., and Capnano, L. A., *Tetrahedron Lett.*, 1811 (1975).

4.431 Stemke, J. E., Chamberlin, A. R., and Bond, F. T., *Tetrahedron Lett.*, 2947 (1976).

4.432 LeBorgne, J. F., Cuvigny, T., Larcheveque, M., and Normant, H., *Synthesis*, 238 (1976).

4.433 Corey, E. J., Enders, D., and Bock, M. G., *Tetrahedron Lett.*, 7 (1976).

4.434 Meyers, A. I., and Knaus, G., *Tetrahedron Lett.*, 1333 (1974).

4.435   Meyers, A. I., Durandetta, J. L., and Munavu, R., *J. Org. Chem.*, <u>40</u>, 2025 (1975).

4.436   Schaefer, A. P., and Bloomfield, J. J., *Org. React.*, <u>15</u>, 1 (1967).

4.437   Krapcho, A. P., Jahgen, E. G. E., Jr., and Kashdan, D. S., *Tetrahedron Lett.*, 2721 (1974).

4.438   Krapcho, A. P., Kashdan, D. S., and Jahngen, E. G. E., Jr., *J. Org. Chem.*, <u>42</u>, 1189 (1977).

4.439   Couffignal, R., and Moreau, J. L., *J. Organometal. Chem.*, <u>127</u>, C65 (1977).

4.440   Seebach, D., and Ehrig, V., *Angew. Chem. Int. Ed. Engl.*, <u>11</u>, 127 (1972).

4.441   Bergmann, E. D., Ginsburg, D., and Pappo, R., *Org. React.*, <u>10</u>, 179 (1959).

4.442   Uda, H., et al., *J. C. S. Perkin I*, 214 (1978).

4.443   Rapson, W. S., and Robinson, R., *J. Chem. Soc.*, 1285 (1935).

4.444   Ramachandran, S., and Newman, M. S., *Org. Synth. Coll. Vol.*, <u>5</u>, 486 (1973).

4.445   Brewster, J. H., and Eliel, E. L., *Org. React.*, <u>7</u>, 99 (1953).

4.446   Stork, G., Brizzolara, A., Landesman, H., Szmuszkovicz, J., and Terrell, R., *J. Am. Chem. Soc.*, <u>85</u>, 207 (1963).

4.447   Stork, G., and Ganem, B., *J. Am. Chem. Soc.*, <u>95</u>, 6152 (1973); (b) Stork, G., and Singh, J., *J. Am. Chem. Soc.*, <u>96</u>, 6181 (1974). Also see Boeckman, R. K., Jr., *J. Am. Chem. Soc.*, <u>95</u>, 6867 (1973); <u>96</u>, 6179 (1974).

4.448   Lee, R. A., *Tetrahedron Lett.*, 3333 (1973). Also see Hagiwara, H., Nakayama, K., and Uda, H., *Bull. Chem. Soc. Japan*, <u>48</u>, 3769 (1975).

4.449   Danishefsky, S., Hatch, W. E., Sax, M., Abola, E., and Pletcher, J., *J. Am. Chem. Soc.*, <u>95</u>, 2410 (1973).

4.450   Stork, G., and Maldonado, L., *J. Am. Chem. Soc.*, <u>96</u>, 5272 (1974).

4.451   Stork, G., Leong, A. Y. W., and Touzin, A. M., *J. Org. Chem.*, <u>41</u>, 3491 (1976).

4.452   Grieco, P. A., and Yokoyama, Y., *J. Am. Chem. Soc.*, <u>99</u>, 5210 (1977).

4.453   Cregge, R. J., Herrmann, J. L., and Schlessinger, R. H., *Tetrahedron Lett.*, 2603 (1973).

4.454   Rosan, A., and Rosenblum, M., *J. Org. Chem.*, 40, 3621 (1975).

4.455   Fuchs, P. L., *J. Am. Chem. Soc.*, 96, 1607 (1974).

4.456   Danishefsky, S., Cain, P., and Nagel, A., *J. Am. Chem. Soc.*, 97, 380, 5282 (1975).

4.457   Herrmann, J. L., Kieczykowski, G. R., Romanet, R. F., Wepplo, P. J., and Schlessinger, R. H., *Tetrahedron Lett.*, 4711, 4715 (1973).

4.458   Romanet, R. F., and Schlessinger, R. H., *J. Am. Chem. Soc.*, 96, 3701 (1974).

4.459   (a) Creger, P. L., *J. Org. Chem.*, 37, 1907 (1972); (b) Fujita, T., Watanabe, S., and Suga, K., *Aust. J. Chem.*, 27, 2205 (1974).

4.460   Meyers, A. I., Mihelich, E. D., and Nolen, R. L., *J. Org. Chem.*, 39, 2783 (1974).

4.461   Larcheveque, M., Valette, G., Cuvigny, T., and Normant, H., *Synthesis*, 256 (1975).

4.462   McMurry, J. E., and Isser, S. J., *J. Am. Chem. Soc.*, 94, 7132 (1972).

4.463   (a) Stork, G., Cama, L. D., and Coulson, D. R., *J. Am. Chem. Soc.*, 96, 5268 (1974); (b) Stork, G., and Cohen, J. F., *J. Am. Chem. Soc.*, 96, 5270 (1974); (c) Stork, G., and Maldonado, L., *J. Am. Chem. Soc.*, 96, 5272 (1974).

4.464   (a) Meyers, A. I., *Heterocycles in Organic Synthesis*, Wiley-Interscience, 1974, 332 pp; (b) Meyers, A. I., and Milhelich, E. D., *Angew. Chem. Int. Ed. Engl.*, 15, 270 (1976); (c) Meyers, A. I., *Acc. Chem. Res.*, 11, 375 (1978).

4.465   (a) Collington, E. W., *Chem. Ind.*, 987 (1973); (b) Ap-Simon, J., and Holms, A., *Heterocycles*, 6, 731 (1977).

4.466   Meyers, A. I., and Mihelich, E. D., *J. Org. Chem.*, 40, 1186 (1976).

4.467   Meyers, A. I., and Witten, C. E., *Tetrahedron Lett.*, 1947 (1976).

4.468   Glaze, W. H., and Selman, C. M., *J. Organometal. Chem.*, 11, p3 (1968).

4.469   Smid, J., *J. Am. Chem. Soc.*, <u>87</u>, 655 (1965).

4.470   Huffman, J. W., and Charles, J. T., *J. Am. Chem. Soc.*, <u>90</u>, 6486 (1968).

4.471   (a) Martin, E. L., *Org. React.*, <u>1</u>, 155 (1942); (b) Vedejs, E., *Org. React.*, <u>22</u>, 401 (1975); (c) Also see Buchanan, J. G. St. C., and Woodgate, P. D., *Quart. Rev.*, <u>23</u>, 522 (1969).

4.472   Nakabayashi, T., *J. Am. Chem. Soc.*, <u>82</u>, 3900, 3906, 3909 (1960).

4.473   Cope, A. C., Barthel, J. W., and Smith, R. D., *Org. Synth. Coll. Vol.*, <u>4</u>, 218 (1963).

4.474   (a) EcElvain, S. M., *Org. React.*, <u>4</u>, 256 (1948); (b) Finley, K. T., *Chem. Rev.*, <u>64</u>, 573 (1964); (c) Bloomfield, J. J., Owsley, D. C., and Nelke, J. M., *Org. React.*, <u>23</u>, 259 (1976).

4.475   Schaefer, J. P., and Bloomfield, J. J., *Org. React.*, <u>15</u>, 1 (1967).

4.476   Bloomfield, J. J., *Tetrahedron Lett.*, 587, 591 (1968).

4.477   Corey, E. J., Danheiser, R. L., and Chandrasekaran, S., *J. Org. Chem.*, <u>41</u>, 260 (1976). Also see Mukaiyama, T., Sato, T., and Hanna, J., *Chem. Lett.*, 1041 (1973).

4.478   Whitesides, G. M., and Ehmann, W. J., *J. Org. Chem.*, <u>35</u>, 3563 (1970).

4.479   Bauld, N. L., *J. Am. Chem. Soc.*, <u>84</u>, 4345, 4347 (1962).

4.480   For a review, see Klimstra, P. D., *Intra-Sci. Chem. Rept.*, <u>3</u>, 61 (1969).

4.481   (a) Morris, S. G., Herb, S. F., Magidman, P., and Luddy, F. E., *J. Am. Oil Chem. Soc.*, 49 (1972); (b) Näf, F., Decorzant, R., Thommen, W., Willhalm, B., and Ohloff, G., *Helv. Chim. Acta*, <u>58</u>, 1016 (1975).

4.482   (a) Birch, A. J., *Quart. Rev.*, <u>4</u>, 69 (1950); (b) Birch, A. J., and Smith, H., *Quart. Rev.*, <u>12</u>, 17 (1958); (c) Harvey, R. G., *Synthesis*, 161 (1970); (d) Kaiser, E. M., *Synthesis*, 391 (1972); (e) Akhrew, A. A., Reshetova, I. G., and Titov, Y. A., *Birch Reduction of Aromatic Compounds*, Plenum, New York, 1972, 125 pp.

4.483   Clark, R. D., and Heathcock, C. H., *J. Org. Chem.*, <u>38</u>, 3658 (1973).

4.484   Hall, S. S., and McEnroe, F. J., *J. Org. Chem.*, <u>40</u>, 271 (1975).

4.485   Brown, H. C., Ikegami, S., and Kawakami, J. H., *J. Org. Chem.*, <u>35</u>, 3243 (1970).

4.486   Ireland, R. E., Muchmore, D. E., and Hengartner, V., *J. Am. Chem. Soc.*, <u>94</u>, 5098 (1972).

4.487   Rossi, R. A., and Bunnett, J. F., *J. Org. Chem.*, <u>38</u>, 2314 (1973).

4.488   (a) Marshall, J. A., and Lewellyn, M. E., *Synth. Comm.*, <u>5</u>, 293 (1975); (b) *J. Org. Chem.*, <u>42</u>, 1311 (1977).

4.489   Kondo, K., Negishi, A., Matsui, K., Tunemoto, D., and Masamune, S., *J. C. S. Chem. Comm.*, 1311 (1972).

4.490   Crossley, N. S., and Dowell, R., *J. Chem. Soc. (C)*, 2496 (1971).

4.491   Brasen, W. R., and Hauser, C. R., *Org. Synth. Coll. Vol.*, <u>4</u>, 508 (1963).

4.492   Axelrod, E. H., Milne, G. M., and van Tamelen, E. E., *J. Am. Chem. Soc.*, <u>92</u>, 2139 (1970).

4.493   Köbrich, G., and Trapp, H., *Chem. Ber.*, <u>99</u>, 670 (1966).

4.494   Köbrich, G., and Goyert, W., *Tetrahedron*, <u>24</u>, 4327 (1968).

4.495   Stotter, P. L., and Hill, K. A., *J. Org. Chem.*, <u>38</u>, 2576 (1973).

4.496   Greene, A. E., Muller, J. C., and Ourisson, G., *J. Org. Chem.*, <u>39</u>, 186 (1974).

4.497   Sosnovsky, G., and Brown, J. H., *Chem. Rev.*, <u>66</u>, 529 (1966).

4.498   Longone, D. T., and Wright, W. D., *Tetrahedron Lett.*, 2859 (1969).

4.499   Rauchschwalbe, G., and Schlosser, M., *Helv. Chim. Acta*, <u>58</u>, 1094 (1975).

4.500   Vedejs, E., *J. Am. Chem. Soc.*, <u>96</u>, 5944 (1974).

4.501   Hassner, A., Reuss, R. H., and Pinnick, H. W., *J. Org. Chem.*, <u>40</u>, 3427 (1975).

4.502   Ellison, R. A., Woessner, W. D., and Williams, C. E., *J. Org. Chem.*, <u>37</u>, 2757 (1972).

4.503   Trost, B. M., and Salzmann, T. N., *J. Am. Chem. Soc.*, <u>95</u>, 6840 (1973).

4.504 Seebach, D., and Teschner, M., *Chem. Ber.*, 109, 1601 (1976).

4.505 Lapkin, I. I., Abashev, G. G., and Saitkulova, F. G., *J. Org. Chem. USSR.* 12, 975 (1976).

4.506 Reich, H. J., Reich, I. L., and Renga, J. M., *J. Am. Chem. Soc.*, 95, 5813 (1973).

4.507 Brattesani, D. N., and Heathcock, C. H., *Tetrahedron Lett.*, 2279 (1974).

4.508 Grieco, P. A., and Miyashita, M., *J. Org. Chem.*, 39, 120 (1974).

4.509 Brocksom, T. J., Petragnani, N., and Rodrigues, R., *J. Org. Chem.*, 39, 2114 (1974). Also see Sharpless, K. B., Lauer, R. F., Teranishi, A. Y., *J. Am. Chem. Soc.*, 95, 6137 (1973).

4.510 Soretic, P. A., and Soja, P., *J. Org. Chem.*, 41, 3587 (1976).

4.511 Trost, B. M., and Leung, K. K., *Tetrahedron Lett.*, 4197 (1975).

4.512 Oppolzer, W., Mahalanabis, K. K., and Bättig, K., *Helv. Chim. Acta*, 60, 2388 (1977).

4.513 Stork, G., and Raucher, S., *J. Am. Chem. Soc.*, 98, 1583 (1976).

4.514 Trost, B. M., and Satzmann, T. N., *J. Org. Chem.*, 40, 148 (1975).

4.515 Schultz, A. G., Lee, Y. K., and Berger, M. H., *J. Am. Chem. Soc.*, 99, 8065 (1977). Also see Mukaiyama, T., Kobayashi, S., Kanno, K., and Takei, H., *Chem. Lett.*, 237 (1972).

4.516 Trost, B. M., and Tamaru, Y., *J. Am. Chem. Soc.*, 97, 3528 (1975); 99, 3101 (1977).

4.517 Baarschers, W. H., *Can. J. Chem.*, 54, 3056 (1976).

4.518 Quast, H., and Kees, F., *Synthesis*, 489 (1974).

4.519 Trost, B. M., *Acc. Chem. Res.*, 11, 453 (1978).

4.520 Barton, D. H. R., Bould, L., Clive, D. L. J., Magnns, P. D., and H se, T., *J. Chem. Soc. (C)*, 2204 (1971).

4.521 Yamada, S., Oguri, T., and Shioiri, T., *J. C. S. Chem. Comm.*, 623 (1972).

4.522   Reiff, H. F., and Rant, B. C., *J. Organometal. Chem.*, <u>17</u>, 165 (1969).

4.523   Gilman, H., and Stuckwisch, C. G., *J. Am. Chem. Soc.*, <u>63</u> 2844 (1941).

4.524   Baldwin, R. A., and Cheng, M. T., *J. Org. Chem.*, <u>32</u>, 1572 (1967).

4.525   Weiss, E., and Sauermann, G., *Chem. Ber.*, <u>103</u>, 265 (1970); *J. Organometal. Chem.*, <u>21</u>, 1 (1970).

4.526   Benkeser, R. A., *Synthesis*, 347 (1971).

4.527   Courtois, G., and Miginiac, L., *J. Organometal. Chem.*, <u>69</u>, 1 (1974).

4.528   Hill, E. A., *J. Organometal. Chem.*, <u>91</u>, 123 (1975).

4.529   Brown, C. A., and Yamashita, A., *J. Am. Chem. Soc.*, <u>97</u>, 891 (1975).

4.530   For a review on earlier results of acetylene isomerization, see Bushby, R. J., *Quart. Rev.*, <u>24</u>, 585 (1970).

4.531   (a) Brown, C. A., and Yamashita, A., *J. C. S. Chem. Comm.*, 959 (1976); (b) Lindhoudt, J. C., van Mourik, G. L., and Pabon, H. J. J., *Tetrahedron Lett.*, 2565 (1976); (c) For an application of this reaction to the synthesis of natural products, see Negishi, E., and Abramovitch, A., *Tetrahedron Lett.*, 411 (1977).

4.532   Brown, C. A., and Negishi, E., *J. C. S. Chem. Comm.*, 318 (1977).

4.533   Wittig, G., *Angew. Chem.*, <u>66</u>, 10 (1954).

4.534   For a review, see Grovenstein, E., Jr., *Adv. Organometal. Chem.*, <u>16</u>, 167 (1977).

4.535   Grovenstein, E., Jr., and Rhee, J. V., *J. Am. Chem. Soc.*, <u>97</u>, 769 (1975).

5

ORGANOBORONS AND ORGANOALUMINUMS (B, Al)

5.1 GENERAL CONSIDERATIONS

The Group IIIA elements consist of boron (B, 2.01), aluminum (Al, 1.47), gallium (Ga, 1.82), indium (In, 1.49), and thallium (Tl, 1.44). Their Allred-Rochow electronegativity values are indicated in parentheses. Whereas organometallics of boron and aluminum have found numerous applications in organic synthesis, there have been virtually no synthetically useful reactions of organogalliums and organoindiums. Gallium-containing compounds are far more expensive ($10 \sim 10^3$x) than the corresponding compounds containing other Group IIIA elements. Based on cost alone, it seems unlikely that organogallium compounds will find many synthetic applications in the near future. Although not inexpensive, indium-containing compounds are not prohibitively expensive either. Their costs are roughly comparable with those of the corresponding thallium compounds. It is therefore not inconceivable that organoindium compounds might soon find useful synthetic applications in organic synthesis. Indeed, it has recently been shown that the Reformatsky reaction can be carried out with activated indium metal; see Sect. 2.1 (*5.1*), although the results are inferior to those of the corresponding zinc reaction; see eq. 5.1 (*5.2*).

$$R^1COR^2 \ + \ BrCH_2COOEt \ \xrightarrow{\text{activated M}} \ R^1R^3\underset{\underset{OH}{|}}{C}CH_2COOEt \qquad (5.1)$$

34 to 55% (M = In)
95 to 98% (M = Zn)

The current limited scope of organoindium chemistry, however, does not permit further discussion of the subject in this book.

Within the last decade, organothalliums have emerged as uniquely useful organometallics. While thallium compounds share many features with the lighter members of the Group IIIA elements, their chemistry is dominated by certain features unique to some heavy elements, such as Hg, Pb, Pd, and Pt, and is therefore discussed with that of organomercuries and organoleads in Chap. 7.

5.2   ORGANOBORONS

Since 1972 at least four extensive monographs on organoboron chemistry pertinent to organic synthesis have been published (*5.3* to *5.6*). One by Brown (*5.5*) entitled *Organic Syntheses via Boranes* is unique in that it contains a number of detailed experimental procedures. There are a few earlier monographs (*5.7* to *5.9*) and a few more general treatises (*0.6*, *0.8*). Their usefulness, however, is severely limited due to the fact that most of the synthetically interesting reactions, with the notable exception of the hydroboration-oxidation reaction, were developed after their publication.

5.2.1   Fundamental Properties of the Boron Atom and Organoboron Compounds

Except for carboranes (*5.10*) and a very limited number of highly labile subvalent boron compounds, B-X (*5.11*), organoboron compounds generally exist as either tricoordinate or tetracoordinate species. The trisubstituted derivatives of boron are called *boranes* and exist either as essentially trigonal planar monomeric species or as aggregates in which the boron atoms occupy the central position in an essentially tetrahedral configuration. In fact, nearly all monoorganoboranes (RBH_2) and diorganoboranes (R_2BH), as well as the parent borane (BH_3), exist as dimers, although highly hindered members of diorganoboranes, such as <u>5.1</u> may exist as monomers (*5.12*). On the other hand, triorganoboranes (R_3B) are usually monomeric. For the sake of simplicity, however, organoboranes are often treated as monomers in cases where the degree of aggregation is not a significant factor.

5.1

5.2.1.1 Electronegativity and Atomic Size

The electronegativity of boron has been estimated at 2.0, which
is relatively close to the value of 2.5 for carbon, and is
greater than that of the alkali and alkaline earth metals and of
essentially all transition metals which range from 0.86 to 1.75
(Appendix II). According to an empirical equation by Pauling
(1.1), the boron-carbon bond may be estimated to be about 90%
covalent and only about 10% ionic. It is therefore not surprising that organoboranes generally do not react in the manner
of Grignard reagents or organolithium compounds with typical organic and inorganic electrophiles, such as (1) organic halides
and sulfonates, (2) various types of carbonyl compounds, (3)
epoxides, and (4) water and other "active" hydrogen compounds,
under the usual ionic reaction conditions. The relatively short
carbon-boron bond length, for example, 1.57 Å for Me_3B, must also
be responsible for the highly inaccessible nature of the bonding
electrons of the carbon-boron bond in these and other reactions
of organoboranes. It should be worth noting, however, that this
very property of organoboranes renders them highly tolerant of a
variety of electrophilic functional groups.

$$(5.2)$$

5.2.1.2 Empty p Orbital

The presence and/or ready availability of the empty p orbital makes organoboranes electrophilic or Lewis acidic. Thus organoboranes react readily with a variety of neutral or negatively charged bases to form the corresponding complexes. For example, organoboranes readily form 1:1 complexes with various amines and, in some cases, even with ethers, such as tetrahydrofuran (THF). Similarly, their reaction with a negatively charged species forms the corresponding *organoborates*, for which the monomeric ion pair with the essentially tetrahedral borate anion appears to be a reasonable representation (5.13).

$$R^1 - B\overset{\textstyle R^3}{\underset{\textstyle R^2}{\Huge\langle}} \quad + \quad M^+Y^- \quad \longrightarrow \quad \left[Y - \bar{B}\overset{\textstyle R^1}{\underset{\textstyle R^2}{\Huge\langle}} R^3 \right] M^+ \qquad (5.3)$$

As will become clear, this complexation reaction seems to be one of the crucial steps in essentially all of the ionic reactions of organoboranes.

The Lewis acidic property mentioned above is shared by many other organometallics including organoalanes, and is therefore not unique to organoboranes. Far more unique, however, is the ability of the boron p orbital to participate effectively in the p_π-p_π bonding with the adjacent p or sp^n (n = 1, 2, or 3) orbitals of other second-row elements, such as C, N, O, and F. This appears to be responsible, at least in part, for the highly unique 1,2-migration reactions and other reactions of organoboron compounds discussed later. Various types of p_π-p_π interactions pertinent to the present discussion are shown below.

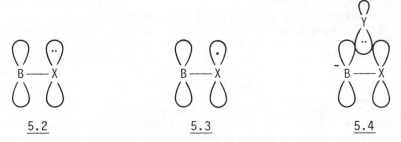

| 5.2 | 5.3 | 5.4 |

The low-lying empty p orbital can also accommodate readily either one electron or a free radical, thereby making organoboranes highly reactive with respect to free-radical reactions.

$$\left[R^1 - B \overset{\displaystyle \shortmid\shortmid\shortmid\shortmid R^3}{\underset{R^2}{}} \right]^{\bar{\cdot}} + X^+$$

$$R^1 - B \overset{\displaystyle \shortmid\shortmid\shortmid\shortmid R^3}{\underset{R^2}{}} + X\cdot \tag{5.4}$$

$$\left[C X \quad B \overset{\displaystyle R^1}{\underset{R^2}{\shortmid\shortmid\shortmid R^3}} \right]^{\cdot}$$

While quantitative interpretations of the chemistry of organo-boron compounds remain difficult, chemists are now in a position to be able to rationalize and, in many cases, even to predict the results on a qualitative basis in terms of a relatively few parameters such as those discussed above.

5.2.2   Preparation of Organoborons

5.2.2.1   Preparation of Organoboranes

Of the twelve general methods for the preparation of organometallics (Chap. 2), only hydrometallation (Method VI), transmetallation (Method XI), heterometallation (Method VII), and metal-hydrogen exchange (Method V) are readily applicable to the preparation of organoboranes with at least some degree of generality. When applicable, hydroboration (5.14) is usually by far the most convenient and is therefore the method of choice. On the other hand, various transmetallation reactions probably represent the most general approach, which has provided various routes to those organoboranes that are not readily accessible via hydroboration. In several significant respects, these two methods complement each other. Since the transmetallation route requires the intermediacy of another organometallic compound, the organoboron compound prepared by this method shares the same limitations and restrictions that the parent organometallic compound encounters. For example, if organolithiums are used as parent species, it is difficult to prepare organoboranes containing electrophilic functional groups. Organoboranes that have stereodefined carbon centers adjacent to the boron atom are also difficult to prepare by this method. On the other hand, a number of such organoboranes have been prepared via hydroboration, as discussed later. In any event, these two by far represent

the most significant methods for the preparation of organoboranes.

Before discussing the scope and limitations of these methods, however, a brief discussion of other direct and indirect
methods is presented below. For further details of this subject,
readers are referred to a few recent review articles (*5.15*,
5.16) and references therein.

Metal-Hydrogen Exchange.   Haloboranes can react with arenes to
form arylboranes; see eq. 5.5 (*2.21*, *5.17*).

$$ArH + BX_3 \longrightarrow ArBX_2 + HX \tag{5.5}$$

All of the trihaloboranes except BF_3 are useful in this reaction.
The reaction can be catalyzed by a number of reagents, such as
aluminum, or may be photochemically induced (*5.18*).

Heterometallation.   Haloboranes also react with alkenes and
alkynes; see, for example, eq. 5.6 (*5.19*, *5.20*).

$$RC{\equiv}CR + XBY_2 \longrightarrow R(X)C{=}CR(BY_2) \tag{5.6}$$

This reaction, however, has not yet been adequately developed
for use in organic synthesis. Highly electrophilic diboron compounds of the $X_2B\text{-}BX_2$ type add in a cis manner to unsaturated
carbon-carbon bonds or cyclopropanes to give 1,2- or 1,3-diboryl
compounds; see eq. 5.7 (*5.21*).

$$RC{\equiv}CR + B_2Cl_4 \longrightarrow \underset{Cl_2B}{\overset{R}{\diagdown}}C{=}C\underset{BCl_2}{\overset{R}{\diagup}} \tag{5.7}$$

There exist a number of indirect routes to organoboranes.
Inasmuch as most of the polar reactions of organoboranes involve
transformation of one organoborane into another via 1,2-migra-
tion as discussed later in more detail, they represent a number
of indirect routes to organoboranes. Then there are various
other organoborane interconversion reactions of which (1) disproportionation, (2) isomerization and displacement, and (3) substitution are worth mentioning here. Generally speaking, disproportionation is probably the most facile thermal process that
can take place even at 0°C or below. Isomerization and displacement, which presumably involve dehydroboration-rehydroboration,
usually become facile only at about 150°C. Substitution

generally requires even higher temperatures, and is significant only above 200°C.

Disproportionation. Although stereochemically quite rigid, organoboranes are generally far more labile than the corresponding carbon compounds and tend to undergo disproportionation, as represented by eq. 5.8.

$$\diagup_{\diagdown}\!B^1 \!-\! R^1 \ + \ R^2 \!-\! B^2\diagup_{\diagdown} \ \rightleftharpoons \ \diagup_{\diagdown}\!B^1 \!-\! R^2 \ + \ R^1 \!-\! B^2\diagup_{\diagdown} \qquad (5.8)$$

The reaction is markedly catalyzed by boron hydrides and alanes, and generally proceeds readily even at or below room temperature in the presence of a suitable catalyst. An important consequence is that hydroboration requiring the use of a boron hydride is usually accompanied by disproportionation. Although not yet established, the following four-center mechanism appears consistent with the results obtained with a boron hydride as a catalyst. Consistent with this mechanism is the observation that the disproportionation reaction proceeds with retention of configuration of the organic groups.

$$\diagup_{\diagdown}\!B^1\!-\!R^1 \ + \ H\!-\!B\diagup_{\diagdown} \ \rightleftharpoons \ \diagup_{\diagdown}\!B^{1\cdot\cdot\cdot R^1\cdots}B\diagup_{\diagdown} \ \rightleftharpoons \ \diagup_{\diagdown}\!B^1\!-\!H \ + \ R^1\!-\!B\diagup_{\diagdown}$$

$$\diagup_{\diagdown}\!B^1\!-\!H \ + \ R^2\!-\!B^2\diagup_{\diagdown} \ \rightleftharpoons \ \diagup_{\diagdown}\!B^{1\cdots R^2\cdots}B^2\diagup_{\diagdown} \ \rightleftharpoons \ \diagup_{\diagdown}\!B^1\!-\!R^2 \ + \ H\!-\!B^2\diagup_{\diagdown}$$
$$(5.9)$$
$$\diagup_{\diagdown}\!B^1\!-\!R^1 \ + \ H\!-\!B^2\diagup_{\diagdown} \ \rightleftharpoons \ \diagup_{\diagdown}\!B^{1\cdots R^1\cdots}B^2\diagup_{\diagdown} \ \rightleftharpoons \ \diagup_{\diagdown}\!B^1\!-\!H \ + \ R^1\!-\!B^2\diagup_{\diagdown}$$

In many cases, disproportionation represents an undesirable side reaction which is to be avoided. In many other cases, however, it can be used as a useful method of preparation, as exemplified in eq. 5.10 (5.22).

$$\bigcirc\!\!\!-\!B\!-\!(CH_2)_5\!-\!B\!-\!\!\!\bigcirc \ \xrightarrow{\ BH_3\ } \ 3 \ HB\!\!-\!\!\bigcirc \qquad (5.10)$$

Isomerization and Displacement. At elevated temperatures, generally at about 150°C, organoboranes undergo a facile isomerization that places the boron atom predominantly at the least hindered position of the alkyl groups. The boron atom, however,

will not migrate past a quaternary carbon center. The results
are consistent with the following dehydroboration-rehydroboration mechanism; see eq. 5.11 (*5.3*).

$$
\begin{array}{c}
\overset{\displaystyle H\ H}{\underset{\displaystyle B\ H}{R\text{-}C\text{-}C\text{-}H}} \rightleftharpoons
\overset{\displaystyle H\ H}{\underset{\displaystyle B\text{-}H}{R\text{-}C\text{=}C\text{-}H}} \rightleftharpoons
\overset{\displaystyle H\ H}{\underset{\displaystyle H\ B}{R\text{-}C\text{-}C\text{-}H}}
\end{array}
\qquad (5.11)
$$

Also consistent with this mechanism is the fact that the presence of another olefin of equal or greater reactivity causes
the original olefin to be displaced by the added olefin (*5.3*).

Substitution. The substitution reaction here refers to the ir-
reversible process represented by eq. 5.12.

$$
\overset{}{\underset{}{\diagdown}}B\text{-}H \ + \ H\text{-}C\overset{\diagup}{\underset{\diagdown}{}} \ \xrightarrow{\Delta} \ \overset{}{\underset{}{\diagdown}}B\text{-}C\overset{\diagup}{\underset{\diagdown}{}} \ + \ H_2 \qquad (5.12)
$$

Usually, the B-H species is generated in situ via dehydroboration at high temperatures. Although the intermolecular substitution reaction has been restricted largely to aromatic hydrocarbons, a number of intramolecular cyclization processes have
been observed with aliphatic derivatives (*5.23*).

5.2.2.2 Hydroboration

Hydroboration involves the addition of the B-H bond to the C=C
or C≡C bond (eqs. 5.13 and 5.14).

$$
\overset{}{\underset{}{\diagdown}}C\text{=}C\overset{\diagup}{\underset{\diagdown}{}} \ + \ H\text{—}B\overset{\diagup}{\underset{\diagdown}{}} \ \longrightarrow \ -\overset{|}{\underset{|}{C}}-\overset{|}{\underset{}{C}}- \qquad (5.13)
$$

$$
-C\equiv C- \ + \ H\text{—}B\overset{\diagup}{\underset{\diagdown}{}} \ \longrightarrow \ \overset{}{\underset{H}{\diagdown}}C\text{=}C\overset{\diagup}{\underset{B}{}} \qquad (5.14)
$$

As mentioned earlier (Sect. 2.6), various other metal hydrides
also undergo similar hydrometallation reactions. At the present time, however, hydroboration, discovered by H. C. Brown in
1956 (*5.24*), is by far the most general and versatile hydrometallation reaction for preparing organometallics as discrete products or intermediates.

Diborane (B_2H_6), which distils at -92°C, is most cleanly generated by adding $BF_3 \cdot OEt_2$ to $NaBH_4$ suspended in diglyme. The diborane gas generated is distilled through a couple of dry ice-acetone traps into THF to give the 1:1 complex ($BH_3 \cdot THF$) dissolved in THF (5.5).

$$3NaBH_4 \ + \ 4BF_3 \cdot OEt_2 \ \longrightarrow \ 4BH_3 \ + \ 3NaBF_4 \qquad (5.15)$$

Alternatively, this reaction may be carried out in the presence of olefins and acetylenes in THF or diglyme. Borane-THF is now commercially available. The scope and limitations of the reaction have been well delineated mainly by Brown and his associates, and several excellent reviews are available (5.3 to 5.5, 5.14, 5.25, 5.26).

<u>Scope and Limitations.</u> (1) Except for certain exceedingly hindered olefins virtually all olefins and acetylenes undergo hydroboration with reactive boron hydrides. In this sense, hydroboration seems roughly as general as bromination of these unsaturated compounds. (2) The general reactivity of $BH_3 \cdot THF$ and $BH_3 \cdot SMe_2$ towards typical organic functional groups may be summarized as follows. These organic functional groups are arranged in decreasing order of reactivity.

carboxylic acid	very fast →	alcohol
olefin and acetylene	fast →	organoborane
aldehyde	fast →	alcohol
ketones	moderate →	alcohol
nitrile	moderate →	amine
epoxide	slow →	alcohol
ester	slow →	alcohol
acyl chloride	very slow →	alcohol (?)
nitro	→	No reaction

carboxylic acid salt $\longrightarrow$ No reaction

Since hydroboration proceeds more rapidly than most of the other reactions listed, it is generally feasible to prepare organo-boranes containing various functional groups such as epoxide, ester, acyl chloride, and nitro groups. Although the cyano group can be accommodated, yields are generally modest. Carbo-xylic acid, aldehyde, and ketone must generally be protected. (3) No skeletal rearrangement of the organic moiety has been ob-served. As already mentioned, however, isomerization involving migration of the B-C bond occurs at moderately high temperatures. (4) One of the major limitations is that only certain types of "mixed" organoboranes are readily available. This stems largely from the fact that many types of monoalkylboranes (RBH_2) and di-alkylboranes (R_2BH and R^1R^2BH) are not readily available as pure substances via hydroboration. The reaction of unhindered ole-fins (such as ethylene), most of the monosubstituted olefins, and certain disubstituted olefins, with $BH_3 \cdot THF$ tends to produce fully alkylated boranes (R_3B) as major products regardless of the reactant ratio used.

$$RCH=CH_2 \xrightarrow[\text{fast}]{BH_3} (RCH_2CH_2)_3B \qquad (5.16)$$

$$\langle \text{cyclopentene} \rangle \xrightarrow[\text{fast}]{BH_3} \langle \text{cyclopentyl} \rangle_3 B \qquad (5.17)$$

Moderately hindered olefins, such as cyclohexene and 2-methyl-2-butene, tend to produce dialkylboranes, some of which can be prepared cleanly. In many cases the corresponding trialkylbor-anes can be prepared at or above room temperature.

$$\langle \text{cyclohexene} \rangle \xrightarrow[\text{fast}]{BH_3, 0°C} \langle \text{cyclohexyl} \rangle_2 BH \xrightarrow[\text{slow}]{25°C} \langle \text{cyclohexyl} \rangle_3 B \qquad (5.18)$$

$$\underset{H_3CC=CHCH_3}{\overset{H_3C}{|}} \xrightarrow[\text{fast}]{BH_3, 0°C} \underset{(Sia_2BH)}{(H_3C\overset{H_3C\ CH_3}{\underset{|\ \ \ |}{CCHCH}}-)_2BH} \xrightarrow[\text{slow}]{25°C} Sia_3B \qquad (5.19)$$

More highly hindered trisubstituted olefins, however, do not form the corresponding R_3B.

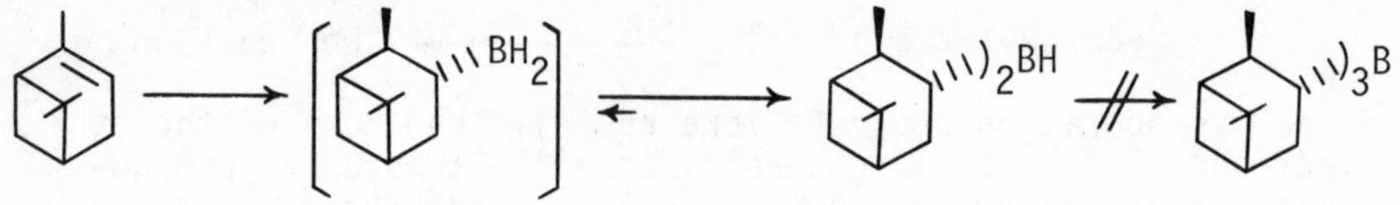

(dipinanylborane)   (5.20)

Tetrasubstituted olefins do not generally cleanly form even dialkylboranes, the usual products being monoalkylboranes.

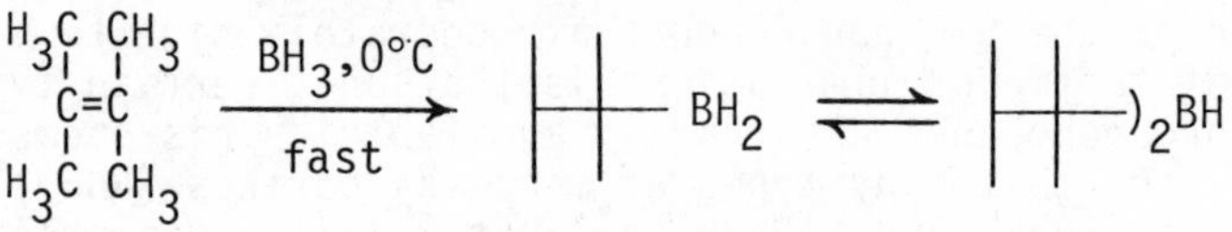

(thexylborane)

The chemistry of thexylborane has been reviewed comprehensively
(5.27). Some of these partially substituted boranes have proven
to be valuable reagents, as discussed later.

   Mono- and dialkylboranes react with unhindered and moderately hindered olefins to form mixed trialkylboranes of the type
$R^1R_2^2B$. Recent development of heterosubstituted hydroborating
agents, such as catecholborane (5.28), mono- and dihaloboranes
(5.29) have made this type of mixed organoboranes widely available. It is still difficult, however, to prepare totally mixed
organoboranes ($R^1R^2R^3B$) except for certain special ones, such as
thexyldialkylboranes (⊢⊢ BR^1R^2).

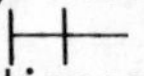

   Finally, various dienes and trienes have been converted to
cyclic organoboranes (5.26). It should be noted that in general, hydroboration has to be followed by disproportionation and/
or isomerization for the production of a single cyclic species
in high yield. Some of these, most notably 9-BBN, have proved
to be valuable reagents.

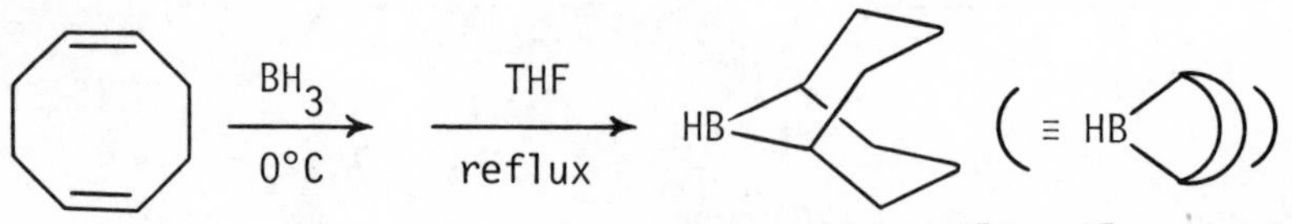

9-borabicyclo[3.3.1]nonane(≡9-BBN)

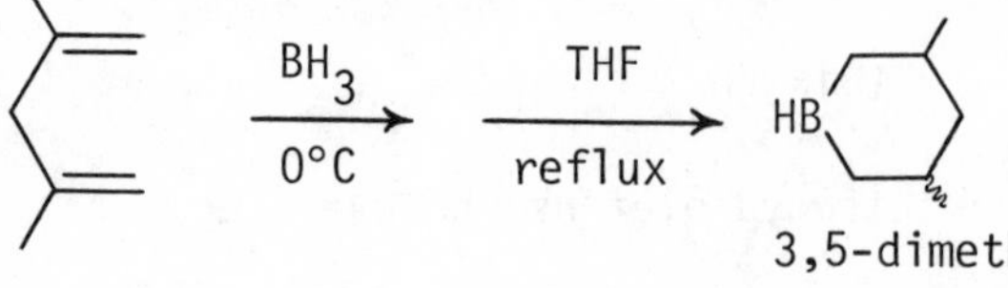

3,5-dimethylborinane

Regiochemistry (Directive Effect). Just as electrophilic aromatic substitution reactions are, hydroboration is influenced by (1) inductive (+I and -I), (2) resonance (+R and -R) and (3) steric (S) effects (IRS effects).

Presumably the B-H bond is polarized with the hydrogen atom having some hydride character. In addition, the boron-containing moiety clearly is sterically more demanding than the hydrogen atom. Some representative results showing the general trend in regiochemistry are summarized below. As a rule, the anti-Markovnikov addition of the B-H bond is observed with simple alkenes (R^1, R^2, and R^3 are alkyl groups).

Borane	$H_2C{=}CHR^1$		$H_2C{=}CR^1R^2$		$R^1HC{=}CR^2R^3$		Ref.
$BH_3{\cdot}THF$	94	6	99	1	98	2	5.14
$ClBH_2{\cdot}OEt_2$	>99.5	<0.5	>99.9	<0.1	99.7	0.3	5.29
⊢⊣—BH_2	94 to 95	5 to 6					5.30
Sia_2BH	99	1	>99	<1	>99	<1	5.14

These results are readily explained by a combination of inductive (+I) and steric effects exerted by the alkyl substituents.

$$H_2C^{\delta-}{=\!=}C^{\delta+}H{\leftarrow}R$$
$$\backslash B^{\delta+}{-\!-}H^{\delta-}/$$

Another point of significance is that the use of either dialkylboranes, such as Sia_2BH, or haloboranes, such as $ClBH_2{\cdot}OEt_2$, is advantageous in controlling the regiochemistry of hydroboration.

Hetero substituents on or near the carbon-carbon double or triple bond can exert strong directive influences. For example, the hydroboration of vinyl halides places the boron atom predominantly on the α-carbon atom, whereas the alkoxy group, which is directly bonded to the olefinic carbon, strongly directs the boron to the β-position (5.31). Some representative results obtained with borane-THF are shown below.

$$CH_3CH_2CH\!\!=\!\!CHCl$$
$$1585$$

$$CH_3CH_2CH\!\!=\!\!CHOEt$$
$$\sim100\sim0$$

$$(CH_3)_2C\!\!=\!\!CHCl$$
$$\sim0\sim100$$

$$(CH_3)_2C\!\!=\!\!CHOEt$$
$$\sim100\sim0$$

On a qualitative basis, these results are readily interpreted on a reasonable assumption that the -I effect is more significant than the +R effect in the cases of vinyl halides, whereas the hydroboration of enol ethers is totally dominated by the +R effect.

$$-\overset{|}{\underset{|}{C}}{}^{\delta+}\!\!=\!\!\overset{|}{\underset{|}{C}}{}^{\delta-}\!\!\rightarrow X \qquad -\overset{|}{C}\!\!=\!\!\overset{|}{C}\!\!-\!\!\overset{..}{O}\!\!-R \leftrightarrow -\overset{|}{C}{}^{-}\!\!-\!\!C\!\!=\!\!\overset{+}{O}\!\!-R \leftrightarrow -\overset{|}{C}{}^{-}\!\!-\!\!\overset{|}{C}{}^{+}\!\!-\!\!\overset{..}{O}\!\!-R$$

There is little doubt that these IRS effects are also operating in the hydroboration of dienes and polyenes. It has also become evident, however, that some additional effects operate in such cases. Anomalous results are often attributable to the cyclic nature of hydroboration. These results recently have been discussed in detail (5.26).

Stereochemistry. Hydroboration involves an exclusive cis addition of the B-H bond. The ready availability of stereodefined organoboranes shown in eqs. 5.20 and 5.24 makes organoboranes a class of uniquely valuable organometallics.

$$\tag{5.24}$$

Equally valuable are various stereodefined alkenylboranes readily obtainable from acetylenes (eq. 5.25).

$$RC\!\equiv\!CH(X)\tag{5.25}$$

X = halogen, alkyl

Although various disubstituted boranes, such as dicyclohexylborane, Sia_2BH, and catecholborane, as well as certain monosubstituted boranes, such as $ClBH_2 \cdot OEt_2$, are satisfactory for this transformation, borane-THF and many partially substituted boranes tend to react further with the alkenylboranes to produce gem-dibora derivatives.

   The exclusively cis nature of hydroboration provides strong support for the concerted mechanism discussed earlier (Sect. 2.6). When olefins possess two nonequivalent faces, hydroboration takes place predominantly, often nearly exclusively, on the less hindered side.

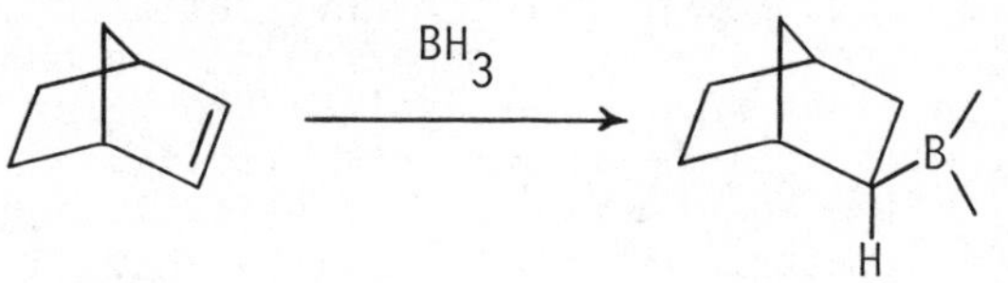

(99.5% exo)(5.26)

Asymmetric Hydroboration.   The reaction of α-pinene with borane takes place nearly exclusively on the less hindered side to give dipinanylborane. If α-pinene is optically pure, an optically pure sample of dipinanylborane with the opposite rotation sign can be obtained (5.32). This reagent permits a remarkably high degree of asymmetric induction (often >90% ee) in its reaction with certain olefins, such as cis-2-butene (eq. 5.27).

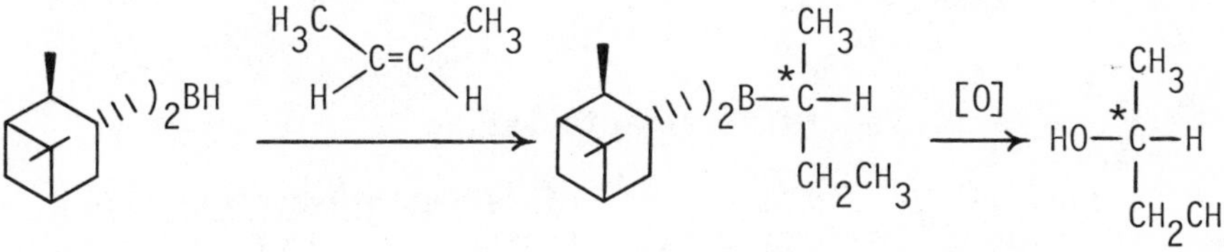

(5.27)

The following asymmetric synthesis of a prostaglandin intermediate in 92% optical purity demonstrates the synthetic utility of asymmetric hydroboration (5.33).

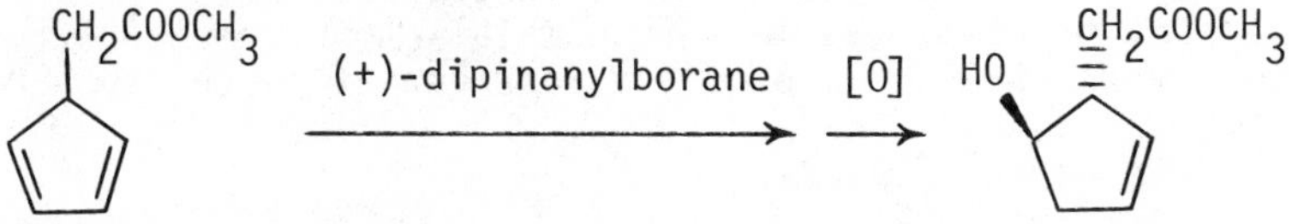

(5.28)

$[\alpha]_D$ -136°

(92% optically pure)

Although highly promising, there exists a serious limitation that must be overcome before the asymmetric hydroboration can become a method of wide applicability. Dipinanylborane, which, as the monomer, is a molecule of C_2 symmetry, is a highly hindered borane and has a tendency to form monopinanylborane (C_1 symmetry) via dehydroboration. The available information indicates that, whereas dipinanylborane is a satisfactory asymmetric hydroboration agent, monopinanylborane is not. Apparently, the C_2 symmetry, which renders both faces of dipinanylborane equivalent, is an essential requirement for a highly satisfactory asymmetric hydroborating agent. Because dehydroboration becomes a serious problem with hindered olefins, satisfactory results have been obtained only with unhindered internal olefins. An ideal asymmetric hydroborating agent would be an optically pure dialkylborane with C_2 symmetry and high stability that can react with a wide range of internal olefins, for example 5.5 or 5.6. The only question is how to make them conveniently.

5.5   5.6

5.2.2.3   Organoboranes via Transmetallation

Despite many unique advantages associated with hydroboration, there are various types of organoboranes that cannot be prepared readily by this method. Such compounds include those organoboranes in which the boron atom is bonded to aryl, *cis*-alkenyl, alkynyl, methyl, neopentyl, and many tertiary alkyl, for example, *t*-butyl, groups. Secondary alkyl groups, which can be derived from internal alkenes bearing two different groups of comparable steric and electronic requirements, present a difficult regiochemical problem. Various transmetallation reactions represented by eq. 5.29 have provided a number of routes to such derivatives.

$$BX_3 \xrightarrow{\ MR\ } MBRX_3 \xrightarrow[-MX]{} BRX_2 \xrightarrow{\ MR\ } MBR_2X_2$$

$$\downarrow {\scriptstyle -MX}$$

$$BR_3 \xleftarrow[-MX]{} MBR_3X \xleftarrow{\ MR\ } MR_2X \qquad (5.29)$$

The scope of the method is too broad to be discussed in detail here (see *0.6*, *0.8*, *5.6*, *5.15*). Since the electronegativity of boron is higher than that of most metals, the forward reactions are favorable with various types of organometallics including those of K, Na, Li, Mg, Al, Zn, Cd, and Sn. These metals are arranged more or less in decreasing order of reactivity. Organoalkali metals and Grignard reagents (*5.34*) are the most versatile organometallics in this reaction. Organotins, along with other types of organometallics, however, have provided certain unique routes to organoboranes (e.g., *5.35*).

$$\langle \overline{} \rangle SnR_2 \longrightarrow \langle \overline{} \rangle BPh \longrightarrow \langle \overline{} \rangle B^- \!\!-\! Ph \qquad (5.30)$$

Various groups can be used as the leaving group X (*5.6*, *5.15*). Their relative order of reactivity appears to be: halogen > OR > SR > NR_2. Boron hydrides can also be used (*5.6*, *5.15*). The transmetallation reaction provides a generally satisfactory route to triorganoboranes of the R_3B and $R^1R_2^2B$ types. As a method for the preparation of RBX_2 and R_2BX, however, it is only marginally satisfactory. R^1R^2BX and $R^1R^2R^3B$ are not generally obtained in high yields due to extensive disproportionation.

5.2.2.4   Preparation of Organoborates

As shown in the preceding section, organoborates are formed as intermediates in the transmetallation reaction. In this section, we are specifically concerned with thermally stable tetraorganoborates and triorganoborohydrides, two classes of organoboron compounds which have recently exhibited useful synthetic capabilities (*5.13*, *5.36*, *5.37*).

Although there are some other routes, the reaction of triorganoboranes with polar organometallics and metal hydrides represents the most general and convenient method by far for the preparation of tetraorganoborates and/or triorganoborohydrides (eqs. 5.31 and 5.32).

$$MR' + BR_3 \overset{\longrightarrow MBR'R_3}{\underset{\longrightarrow MBR_3H}{}} \qquad (5.31)$$

$$MH + BR_3 \longrightarrow MBR_3H \qquad (5.32)$$

These are simple acid-base reactions in which BR_3 acts as a
Lewis acid and MR' or MH as a Lewis base. Thus the formation of
stable "ate" complexes is favored by the use of strong bases
such as those containing K, Na, Li, and Mg. When these organometallics are used, the reactions appear to be quite general.
An extensive survey has been made by Negishi (5.38) using organolithiums. The results indicate the following: (1) Thermally stable tetraorganoborates can be obtained from a variety of
primary alkyllithiums, aryllithiums, alkenyllithiums, benzyl-
lithiums, and alkynyllithiums. Allyllithium forms relatively
weak organoborates in which the two terminal allylic carbons are
equivalent by NMR. On the other hand, lithium enolates and cy-
clopentadienyllithium do not form organoborates under the conditions used. Since sodium cyanide, which is a much weaker base
(pK_a 9 to 10) than those mentioned above, forms stable organoborates, the extent of charge delocalization is evidently far
more significant as a factor in controlling the stability of organoborates than in controlling the pK_a value. (2) Another significant observation is that, if MR' is sterically demanding and
contains one or more β-hydrogens, it tends to form the corresponding triorganoborohydride. This side reaction appears analogous to the reduction reaction frequently observed in the
reaction of organolithiums or Grignard reagents with ketones
(Sect. 4.3.2.2). As a rule, secondary or tertiary alkyllithiums
with β-hydrogens undergo either a predominant or an exclusive
borohydride formation even with moderately hindered triorganoboranes (5.38). This side reaction, however, has provided one
of the most general routes to triorganoborohydrides (5.39), as
exemplified by eq. 5.33.

$$\text{---}\overset{|}{\underset{|}{\text{}}}\text{Li} + B(\text{Bu-}sec)_3 \longrightarrow \text{LiB}(\text{Bu-}sec)_3\text{H} \qquad (5.33)$$

The reaction of triorganoboranes with metal hydrides, such as
LiH and KH, represents another general route to triorganoborohydrides; see eq. 5.32 (5.40). The reaction of LiH, however, is
sluggish when organoboranes are hindered. The use of KH, which
is far more reactive than LiH, is advantageous in such cases
(5.41).

5.2.3.   Reactions of Organoborons

5.2.3.1   General Reaction Patterns

Only in a relatively few cases have the mechanisms of organoboron reactions been fully established. Nevertheless, it is use-

ful to attempt to summarize and classify various organoboron reactions based on the available data including the reactant-product relations.

<u>Ionic Reactions.</u> The weakly nucleophilic but highly electrophilic nature of organoboranes has already been mentioned. The organoborates formed by the reaction of organoboranes with basic or nucleophilic species are often thermally stable and do not undergo any further spontaneous reactions in such cases. When the nucleophile is appropriately substituted, however, the 1,2-migration reactions represented by the following general equations take place.

$$R-B\diagdown \quad + \quad ^-Y-Z \quad \longrightarrow \quad \left[\begin{matrix} R \\ | \\ -B-Y-Z \\ | \end{matrix} \right] \quad \longrightarrow \quad \begin{matrix} R \\ | \\ -B-Y \\ | \end{matrix} \quad + \quad Z^- \qquad (5.34)$$

$$R-B\diagdown \quad + \quad ^-Y=Z \quad \longrightarrow \quad \left[\begin{matrix} R \\ | \\ -B-Y=Z \\ | \end{matrix} \right] \quad \longrightarrow \quad \begin{matrix} R \\ | \\ -B-Y-Z^- \\ | \end{matrix} \qquad (5.35)$$

$$R-B\diagdown \quad + \quad ^-Y\equiv Z \quad \longrightarrow \quad \left[\begin{matrix} R \\ | \\ -B-Y\equiv Z \\ | \end{matrix} \right] \quad \longrightarrow \quad \begin{matrix} R \\ | \\ -B-Y=Z^- \\ | \end{matrix} \qquad (5.36)$$

The key features common to all of these reactions are the formation of the borate anion as a transient species and the intramolecular 1,2-migration. Various minor variations of these mechanistic pathways, such as stepwise analogs and vinylogs, are also conceivable. It is important to note that the great majority of ionic reactions of organoboranes discovered and developed by H. C. Brown and others (5.3 to 5.6) can be rationalized in terms of these mechanistic pathways. Although there is little doubt about the intermediacy of the appropriately substituted organoborates, it has seldom been established. At least in one case, however, Matteson and Mah (5.42) have clearly established the intermediacy of an α-bromoorganoborate (<u>5.7</u>) as summarized in eq. 5.37.

$$\begin{matrix} Ph-B-CHR \\ | \quad | \\ BuO \quad Br \end{matrix} \underset{PhMgBr}{\overset{BuO^-}{\underset{H^+}{\rightleftharpoons}}} \begin{matrix} Ph \\ | \\ BuO-B-CHR \\ | \quad | \\ BuO \quad Br \end{matrix} \overset{25°C}{\longrightarrow} \begin{matrix} Ph \\ | \\ (BuO)_2B-CHR \end{matrix} \quad (5.37)$$

$$\begin{matrix} BuO-B-CHR \\ | \quad | \\ BuO \quad Br \end{matrix}$$

$$\underline{5.7}$$

The products of 1,2-migration reactions are new organoboranes which must be further converted into the desired organic compounds. This is accomplished in most cases by one of three methods: (1) protonolysis (eq. 5.38), (2) oxidation (eq. 5.39), and (3) elimination (eq. 5.40).

$$\text{\textbackslash}B\text{-}R \xrightarrow{HX} RH \ + \ \text{\textbackslash}B\text{-}X \qquad (5.38)$$

$$\text{\textbackslash}B\text{-}R \xrightarrow[H_2O]{[O]} ROH \ + \ \text{\textbackslash}B\text{-}OH \qquad (5.39)$$

$$-\overset{|}{\underset{B}{C}}-\overset{|}{\underset{X}{C}}- \longrightarrow \ \text{\textbackslash}C\text{=}C\text{/} \ + \ \text{\textbackslash}B\text{-}X \qquad (5.40)$$

These reactions are discussed later in more detail.

Ionic Reactions of Stable Organoborates. We have so far dealt with the ionic reactions of organoboranes which presumably go through unstable organoborates (eqs. 5.34 to 5.36). It has recently been recognized that thermally stable organoborate anions, which are negatively charged and coordinatively saturated, act as unique nucleophiles (5.13). They react with a variety of organic and inorganic electrophiles via either intermolecular or intramolecular transfer, as represented by eqs. 5.41 to 5.44.

Intermolecular transfer:

$$^-\overset{|}{\underset{|}{B}}\text{-}R \ + \ E^+ \longrightarrow \ \overset{|}{B} \ + \ R\text{-}E \qquad (5.41)$$

Intramolecular transfer:

$$^-\overset{R}{\underset{|}{B}}\text{-}Y\text{-}Z \ + \ E^+ \longrightarrow \ \overset{R}{\underset{|}{B}}\text{-}Y \ + \ Z\text{-}E \qquad (5.42)$$

$$^-\overset{R}{\underset{|}{B}}\text{-}Y\text{=}Z \ + \ E^+ \longrightarrow \ \overset{R}{\underset{|}{B}}\text{-}Y\text{-}Z\text{-}E \qquad (5.43)$$

$$^-\overset{R}{\underset{|}{B}}\text{-}Y\text{≡}Z \ + \ E^+ \longrightarrow \ \overset{R}{\underset{|}{B}}\text{-}Y\text{=}Z\text{-}E \qquad (5.44)$$

In the intermolecular transfer reaction the organoborate acts as the boron version of a Grignard-like reagent with very mild reactivity. On the other hand, the intramolecular transfer reactions are closely related to those of organoboranes. The only difference is that the organoborate reactions require activation by electrophiles, whereas the organoborane reactions generally occur spontaneously. It should also be emphasized that these intramolecular transfer reactions are not only highly unique as organometallic reactions, but tend to take place much more readily than the intermolecular reactions in cases where both can occur. These reactions may be best interpreted within the framework of the well-known anionotropic 1,2-migration reactions, such as the Wagner-Meerwein rearrangement. Significantly, the intramolecular transfer reactions of both organoboranes and organoborates have proceeded with retention of configuration of the migrating group. Mechanistic considerations also predict inversion of configuration at the migration terminus, which has been borne out in some cases; see eq. 5.45, for example (*5.43* to *5.45*).

$$(5.45)$$

Concerted Reactions. Considering the relatively covalent nature of the B-C bond and the size of the boron atom and its valence shell orbitals, which are comparable to those of the carbon atom, organoborons might be expected to participate readily in some concerted pericyclic reactions. The information available strongly indicates that the protonolysis of organoboranes with carboxylic acids is best viewed as a concerted $[2_s + 4_s]$ reaction in which the boranes act as 2-electron components. Although relatively few other reactions of this type have been observed, there exists a strong possibility that additional reactions, which may be represented by eq. 5.46 or its modifications, may be found in the future.

$$-\overset{\mid}{B}-R \quad W\,\rangle \underset{X-Y}{\overset{Z}{\rangle}} \longrightarrow R-Z \ + \ \rangle B-W \ + \ X=Y \qquad (5.46)$$

In fact, a recent study on the conjugate addition reaction of organoboranes (5.46) suggests that, at least in some cases, the reaction may proceed by a nonfree-radical mechanism. The following concerted mechanism appears plausible.

$$-\overset{\mid}{B}-R \quad O\,\rangle\underset{C-C}{\overset{C}{\rangle}} \longrightarrow -\overset{\mid}{B}\cdot \quad O\underset{C=C}{\overset{R}{\diagup}} \qquad (5.47)$$

Within the last decade or two, a considerable number of organoborane reactions, which appear to be best represented by the concerted $[4_s + 2_s]$ reaction in which the boranes act as four-electron components, have been observed.

$$\overset{X-C}{\underset{Y-Z}{\rangle B \langle}} \longrightarrow \left[\overset{X=C}{\underset{Y--Z}{\rangle B \langle}}\right] \longrightarrow \overset{X=C}{\underset{Y \quad Z}{\rangle B \langle}} \qquad (5.48)$$

Although the examples observed to date have been essentially restricted to some reactions of (1) allylic boranes and (2) alkenyloxyboranes (eqs. 5.49 to 5.51), it is likely that the general scheme will represent many other unknown reactions.

$$\overset{C-C}{\underset{O-H}{\rangle B \langle}} \longrightarrow \overset{C=C}{\underset{O \quad H}{\rangle B \langle}} \qquad (5.49)$$

$$\overset{O-C}{\underset{O=C}{\rangle B \langle}} \longrightarrow \overset{O=C}{\underset{O-C}{\rangle B \langle}} \qquad (5.50)$$

$$\begin{array}{c} \diagdown \diagup \diagup \\ C-C \\ \diagup \diagdown \diagup \\ B \qquad C \\ \diagup \diagdown \diagdown \\ -C \equiv C- \end{array} \longrightarrow \begin{array}{c} \diagdown \diagup \diagup \\ C = C \\ \diagdown \qquad \diagup \\ B \qquad C \\ \diagup \qquad \diagdown \\ C = C \\ \diagup \qquad \diagdown \end{array} \qquad (5.51)$$

Free-Radical Reactions. Recently it has been established that organoboranes readily participate in at least two distinctly different free-radical reactions. Although there has been essentially no firmly established example of the bimolecular homolytic substitution (s_H2) taking place at a simple saturated carbon center, organoboranes readily participate in this reaction, which may be represented by eq. 5.52 (5.47).

$$X \cdot \ + \ \begin{array}{c} \diagdown \\ B-R \\ \diagup \end{array} \longrightarrow \left[X---\overset{|}{\underset{|}{B}}---R \right]^{\cdot} \longrightarrow X-B\diagup + R \cdot \qquad (5.52)$$

Once an organic free radical is generated, it can participate in various free-radical reactions, as discussed later in detail. It should be pointed out that the s_H2 reaction is not unique to organoboranes and that a number of other metals, such as Li, Mg, Zn, Cd, Al, and Tl, have also participated in the s_H2 reaction (5.47).

Far more unique to organoborons is the α-abstraction reaction exemplified by the bromination reaction of organoboranes; see eq. 5.53 (5.48).

$$-\overset{|}{B}-\overset{|}{\underset{|}{C}}- \quad \xrightarrow[\text{h}\nu]{\text{Br}_2} \quad -\overset{|}{B}-\overset{|}{\underset{|}{C}}- \ + \ HBr \qquad (5.53)$$

The available data support the following free-radical chain mechanism (5.49).

$$-\overset{|}{B}-\overset{|}{\underset{|}{C}}- \ + \ Br \cdot \quad \longrightarrow \quad -\overset{|}{B}-\overset{|}{\underset{\cdot}{C}}- \ + \ HBr$$

$$\qquad \qquad \qquad \qquad \qquad \qquad \qquad \qquad \qquad (5.54)$$

$$-\overset{|}{B}-\overset{|}{\underset{\cdot}{C}}- \ + \ Br_2 \quad \longrightarrow \quad -\overset{|}{B}-\overset{|}{\underset{|}{C}}- \ + \ Br \cdot$$

The p_π-p_π interaction discussed earlier (5.3) must be responsible for promoting this reaction. A more recent study suggests that the α-C-H of B-alkyl-9-BBN is 5.5 times as reactive as that of cumene and some 600 times as reactive as the tertiary C-H of isobutane (5.49).

What are the factors controlling the point of attack by a free radical? Although this question has not been fully solved, nitrogen and oxygen free radicals have participated mainly in the S_H2 reaction, whereas the bromine free radical undergoes the α-abstraction. Based on these observations, we may tentatively conclude that those free radicals that can form thermodynamically highly stable bonds to boron, such as the B-O and B-N bonds, prefer the S_H2 reaction. With others the α-abstraction can be a dominant course of the reaction.

There can be a variety of other free-radical reactions for organoboron compounds. For example, the following photochemical reaction of trinaphthylborane has been interpreted in terms of the boron di-π-methane free-radical reaction (5.11). At present, however, the S_H2 reaction and the α-abstraction reaction represent the great majority of the free-radical reactions of organoboranes.

Np = α-naphthyl   (5.55)

5.2.3.2 Formation of Carbon-Hetero Atom Bonds via Organoborons

Organoboron compounds can be converted into various organic compounds containing C-H, C-O, C-X (X = halogen), C-N, and C-S bonds. Virtually nothing is known about the formation of the C-P bond via cleavage of the C-B bond. Organoboron compounds can serve as precursors of other organometallic compounds.

Protonolysis. Although the B-N, B-O, and B-S bonds are thermodynamically highly stable, they are solvolytically quite labile.
Thus, for example, hydrolysis of borane derivatives containing
these bonds readily produces the corresponding amines, alcohols,
and thioalcohols, respectively (eq. 5.56).

$$\diagdown\!\!\!>\!\!\text{B-X}\ +\ H_2O\ \longrightarrow\ \diagdown\!\!\!>\!\!\text{B-OH}\ +\ HX \qquad (5.56)$$

$$X = NR^1R^2,\ OR,\ or\ SR,\ \text{where } R = \text{an organic group}$$

On the other hand, the B-C bond of organoboranes is usually
quite resistant to hydrolysis. In most cases, however, it can
be cleaved by treating an organoborane with a carboxylic acid
(5.3 to 5.6). Other stronger acids, such as hydrochloric acid,
are less effective. The unique effectiveness of carboxylic
acids has been interpreted in terms of the following mechanism
(eq. 5.57).

$$\diagdown\!\!\!>\!\!\text{B-R}^*\ +\ RCOOH\ \longrightarrow\ \left[\begin{array}{c} H \\ R^* \quad O \\ | \qquad | \\ B \!=\! C \\ O \qquad R \end{array}\right]\ \longrightarrow\ R^*\text{-H}\ +\ \diagdown\!\!\!>\!\!\text{BOOCR} \quad (5.57)$$

The reaction may be viewed as a concerted $[2_s + 4_s]$ reaction.
It is significant to note that the reaction proceeds with retention of configuration of the R group in accordance with the proposed concerted mechanism. The order of the ease of protonolysis is: alkynyl > alkenyl or aryl > alkyl.
     Unlike organoboranes, certain tetraorganoborates undergo a
very facile cleavage of one of the four B-C bonds when treated
with various proton donors (eq. 5.58).

$$R^*\text{-}B^-R_3\ \xrightarrow{\ HX\ }\ R^*\text{-H}\ +\ BR_3 \qquad (5.58)$$

$$R^* = \text{alkyl, aryl, and so on}$$

Under acidic conditions, alkenyl- and alkynylborates undergo an
entirely different reaction which involves the formation of carbon-carbon bonds, as discussed later. These organoborates, however, can be cleaved stereospecifically with aqueous bases; see
eq. 5.59 (5.50). Under the same conditions, alkenylboranes are
unaffected. This basic hydrolysis nicely complements the conventional acidic protonolysis and should prove useful in cases
where organoborons contain acid-sensitive groups. As might be

$$\underset{\substack{\\ \text{H} \diagdown \\ \text{R}^1 \diagup}}{\text{C=C}}\underset{\substack{\diagup \text{H}(\text{R}^2) \\ \diagdown \text{B}^-\text{R}_3}}{} \xrightarrow{\text{eq NaOH}} \underset{\substack{\\ \text{H} \diagdown \\ \text{R}^1 \diagup}}{\text{C=C}}\underset{\substack{\diagup \text{H}(\text{R}^2) \\ \diagdown \text{H}}}{} \qquad (5.59)$$

98% stereospecific

expected, the basic hydrolysis of the alkynyl-boron bond of either organoboranes or organoborates is very facile. The allyl-boron bond can be cleaved readily even with water, presumably via the unique six-center mechanism shown in eq. 5.49. The reactions of organoborons with various proton donors are summarized in Table 5.1.

Table 5.1   Protonolysis of Organoborons

| Reagent | Organoborons | |
	Organoboranes	Organoborates
Acids	Protonolysis with carboxylic acids[a]	Facile protonoly-sis or 1,2-migra-tion
Bases (aqueous)	Stable[a]	Cleavage of un-saturated groups

[a]Ease of protonolysis varies. Alkynyl and allyl groups are cleaved under either condition.

Hydrogenolysis. The B-C bond is quite stable to hydrogenolysis. Since the hydrogenolysis reaction has been carried out at elevated temperatures, it is not clear if the reaction involves direct cleavage of the B-C bond with hydrogen or dehydroboration-hydrogenation (5.51).

Oxidation. Organoboranes are readily oxidized by a variety of reagents. The presence of an empty p orbital and the high thermodynamic stability of the B-O bond (115 to 135 kcal/mole) must be responsible for the ease of oxidation.

(a) Oxidation with alkaline hydrogen peroxide and tertiary amine N-oxides. The oxidation of organoboranes with 30% hydrogen peroxide and a suitable base, such as NaOH, NaOAc, or the NaH_2PO_4-K_2HPO_4 buffer solution, is by far the most dependable and convenient laboratory method (5.3 to 5.6). This reaction as well as the reaction with amine N-oxides is believed to proceed as shown in eq. 5.60 (5.52).

$$\ce{>B-R^* + ^-O-Z -> -\overset{R^*}{\underset{|}{B}}-O-Z -> -\overset{R^*}{\underset{|}{B}}-O + ^-Z}$$

$$Z = \ce{-OH}, \ce{-N^+R_3}, \text{ and so on} \qquad \downarrow H_2O \qquad (5.60)$$

$$\ce{>B-OH + R^*-OH}$$

Essentially all types of organoboranes can be converted into the corresponding alcohols with retention of configuration. Alkenylboranes are converted into ketones and aldehydes (eq. 5.61), and arylboranes into phenols; see eq. 5.62 (5.53). Under aqueous conditions, however, the alkynyl-boron bond undergoes protonolysis preferentially.

$$\ce{\underset{H}{\overset{R^1}{>}}C=C\underset{B}{\overset{R^2(H)}{<}}} \quad \xrightarrow[\text{NaOH}]{H_2O_2} \quad R^1CH_2COR^2(H) \qquad (5.61)$$

$$\ce{Ar-B<} \quad \xrightarrow[\text{NaOH}]{H_2O_2} \quad ArOH \qquad (5.62)$$

As might be expected based on the mechanism shown in eq. 5.60, the coordinatively saturated organoborates are not oxidized until they are converted into organoboranes via protonolysis or via some other reactions. Thus, for example, only three of the four alkyl groups (R) of $LiBR_4$ can be converted into alcohols.

(b) Other oxidation reactions. Organoboranes can be directly converted into ketones by oxidation with chromic acids (5.54). The reaction of organoboranes with oxygen involves an S_H2 reaction (5.47). The initial product is a peroxide; see eq. 5.63 (5.55), which reacts further with a second mole of oxygen to form a diperoxide (eq. 5.64).

$$R_3B \;+\; O_2 \longrightarrow R_2BOOR \qquad\qquad (5.63)$$

$$R_2BOOR \;+\; O_2 \longrightarrow RB(OOR)_2 \qquad\qquad (5.64)$$

Addition of 30% H_2O_2 liberates the alkyl hydroperoxide; see eq. 5.65 (*5.56*).

$$RB(OOR)_2 \;+\; H_2O_2 \;+\; 2H_2O \longrightarrow 2ROOH \;+\; ROH \;+\; B(OH)_3$$
$$(5.65)$$

A free-radical chain mechanism involving an s_H2 reaction is consistent with the results (eq. 5.66) (*5.57*).

$$R_3B \;+\; O_2 \longrightarrow R\cdot \;+\; R_2BOO\cdot$$

$$R\cdot \;+\; O_2 \longrightarrow ROO\cdot \qquad\qquad (5.66)$$

$$ROO\cdot \;+\; R_3B \longrightarrow R_2BOOR \;+\; R\cdot$$

When a trialkylborane is reacted with 1.5 molar equivalents of oxygen and then with aq NaOH, all three alkyl groups are converted into the corresponding alcohol in >90% yield. The course of the reaction may be depicted as shown in eq. 5.67.

$$2R_3B \;+\; 3O_2 \longrightarrow R_2BOOR \;+\; RB(OOR)_2$$

$$R_2BOOR \;+\; RB(OOR)_2 \;\xrightarrow[\;H_2O\;]{\;NaOH\;}\; R_2BX \;+\; RBX_2 \;+\; 3NaOOR$$

$$(5.67)$$

$$RBX_2 \;+\; NaOOR \longrightarrow \;\;\overset{R}{\underset{|}{-B}}\!-\!O\!-\!OR \;\xrightarrow{\;H_2O\;}\; 2HOR$$

$$R_2BX \;+\; 2NaOOR \;\xrightarrow{\;H_2O\;}\; 4HOR$$

One drawback of the free-radical oxidation is that it is not stereospecific, for example, eq. 5.68.

<u>Halogenolysis</u>. Organoboranes themselves are relatively inert to iodine. Under basic conditions, however, a facile iodinolysis reaction takes place (*5.58, 5.59*). There is little doubt that the actual reactive species are organoborates. Indeed, lithium

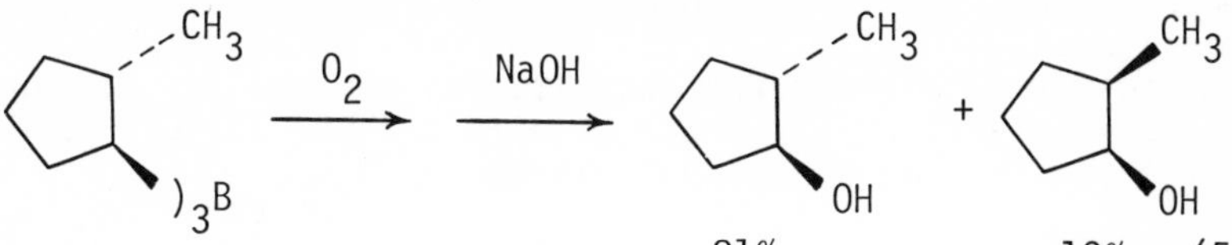
CH3
)3B
O2
NaOH
CH3
OH
+
CH3
OH

81% 19% (5.68)

tetraalkylborates react with I_2 to give alkyl iodides (1 mole) and trialkylboranes (*5.38*). A recent study indicates that the reaction proceeds with inversion of configuration of the alkyl group; see eqs. 5.69, 5.70 (*5.60*).

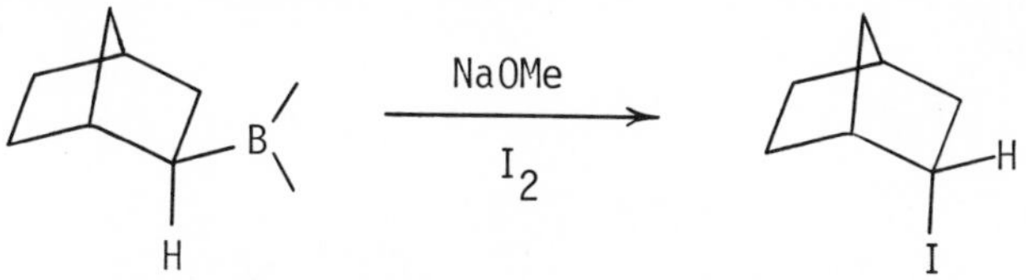
B
H
NaOMe
I2
H
I

(5.69)

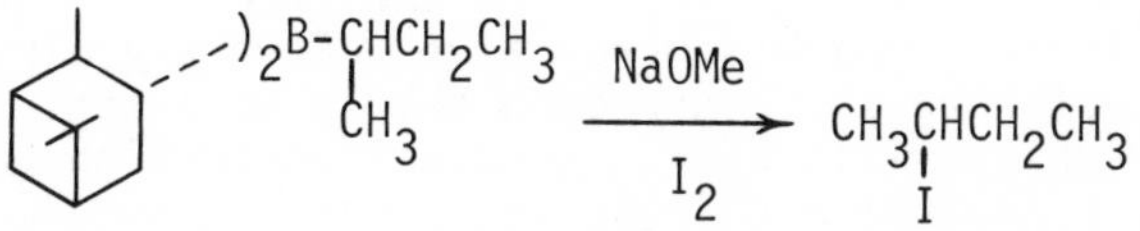
)2B-CHCH2CH3
CH3
NaOMe
I2
CH3CHCH2CH3
I

(5.70)

49% (84% optical purity)

The following S_H2 mechanism appears plausible.

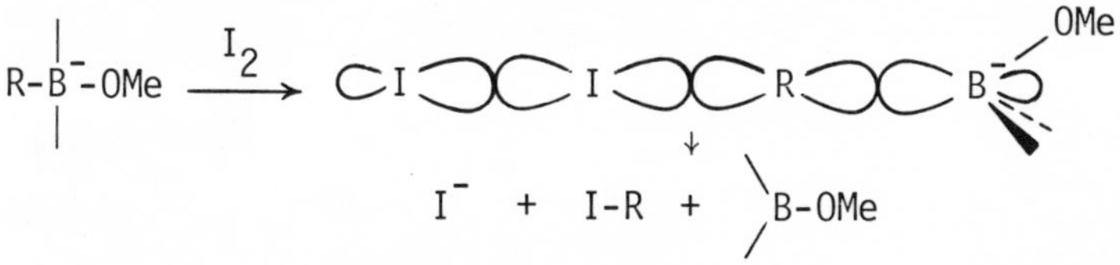
R-B-OMe
I2
I
I
R
B
OMe
I- + I-R + B-OMe

(5.71)

A very similar reaction takes place between organoboranes and Br_2; see eq. 5.72 (*5.61*).

$$R_3B + 3Br_2 + 4NaOMe \xrightarrow[THF]{0°C} 3RBr + 3NaBr + NaB(OMe)_4 \qquad (5.72)$$

Primary alkyl bromides are usually obtained in excellent yields. Secondary alkyl bromides can also be prepared by this reaction, although their yields are modest. This reaction also proceeds with inversion (5.62).

Under acidic or neutral conditions, an entirely different bromination reaction takes place, although the organic products are the same bromides, as shown in eq. 5.73.

$$R_3B + Br_2 \longrightarrow RBr + BrBR_2 \qquad (5.73)$$

A detailed study by Lane and Brown (5.63) has established that the reaction proceeds via the free-radical chain mechanism shown in eq. 5.54. The α-bromoalkylborane intermediate undergoes a slow protonolysis reaction with HBr formed under the reaction conditions (eq. 5.74). This protonolysis presumably is much faster than usual due to the presence of the α-bromine atom.

$$R_2B-\overset{\displaystyle |}{\underset{\displaystyle |}{C}}- + HBr \longrightarrow R_2BBr + H-\overset{\displaystyle |}{\underset{\displaystyle |}{C}}- \qquad (5.74)$$
$$\quad Br \qquad\qquad\qquad\qquad\qquad Br$$

Secondary alkyl groups are considerably more reactive than primary alkyl groups in sharp contrast with the order of reactivity observed in the ionic bromination reaction. Tertiary alkyl groups which do not have any α-hydrogen should be inert. The use of B-alkyl-9-BBN permits a near complete conversion (80 to 90%) of olefins into alkyl bromides; see eq. 5.75 (5.64). The 9-BBN moiety does not competitively react, because its rigid conformation is such that an effective p_π-p_π delocalization between the α-free radical and the empty boron p orbital is prohibited.

$$(5.75)$$

The α-bromoalkylboranes formed as intermediates undergo a fascinating carbon-carbon bond-forming reaction in the presence of a nucleophile, as discussed later.

A few reactions that convert organoboranes into alkyl chlorides are known (eq. 5.76). Their synthetic scope, however, is not very clear.

Alkenyl-, aryl-, and alkynylboranes can undergo entirely different reactions with halogens. Iodination and bromination

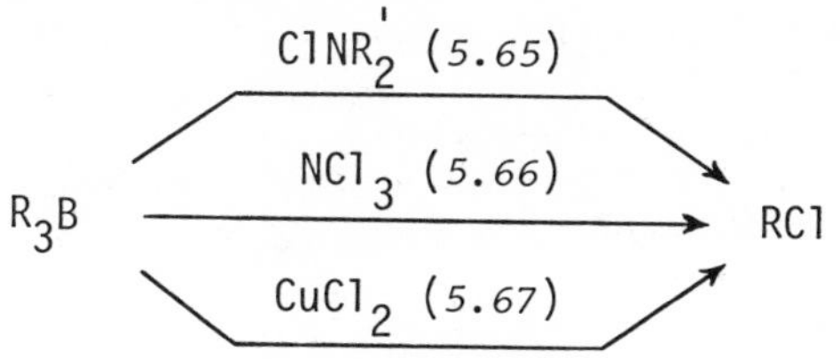

$$R_3B \xrightarrow{\begin{array}{c}ClNR'_2 \;(5.65)\\ NCl_3 \;(5.66)\\ CuCl_2 \;(5.67)\end{array}} RCl \qquad (5.76)$$

of alkenyl- and alkynylboranes can induce a highly unique carbon-carbon bond formation, as discussed later. When no carbon-carbon bond formation is possible, as in the reaction of alkeneboronic acids with I_2 or Br_2, the corresponding organic halides can be prepared in high yields. The following syntheses of alkenyl iodides and bromides (5.68) are attractive and unique in that the required alkeneboronic acids can be obtained readily via hydroboration (eq. 5.77).

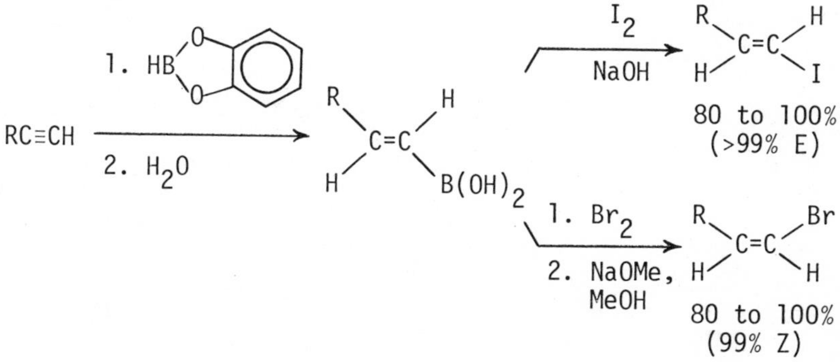

$RC\equiv CH \xrightarrow{\begin{array}{c}1.\; HB\\2.\; H_2O\end{array}}$ ... (5.77)

80 to 100% (>99% E)

80 to 100% (99% Z)

The diametrically opposed stereochemical results are intriguing, and have been interpreted as follows.

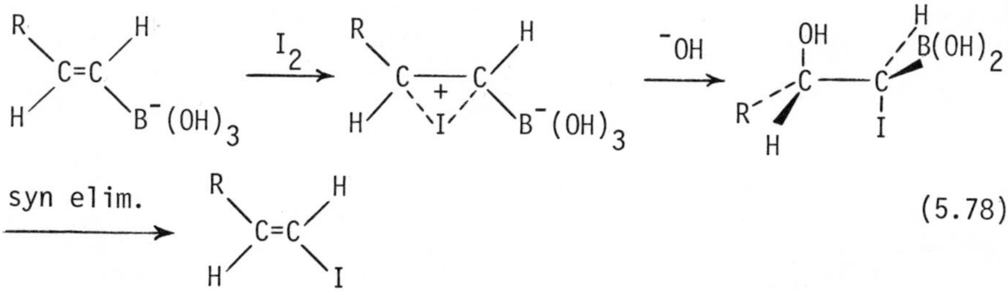

syn elim. (5.78)

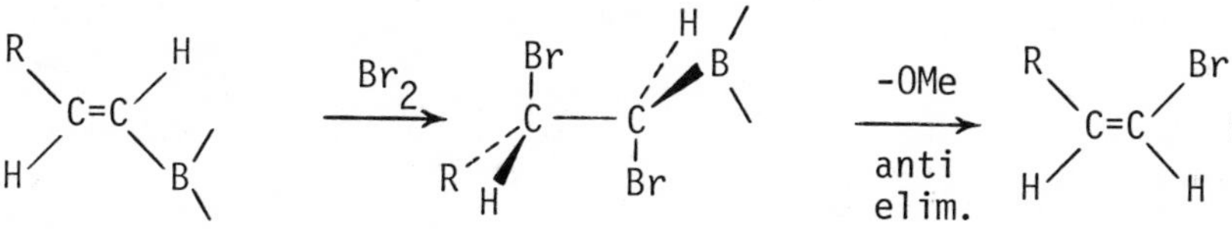

anti elim. (5.79)

In the absence of a good nucleophile, the bromination of alkenyl-boranes can produce (E)-alkenyl bromides in moderate yields (5.69). Unfortunately, the reaction is generally not highly stereospecific. There does not appear any obvious advantage in preparing aryl and alkynyl halides via organoboranes.

Amination. Amines containing good leaving groups such as NH_2Cl (5.70), NH_2OSO_3H (5.71), $2,4,6\text{-}Me_3H_2C_6SO_2ONH_2$ (5.72), and NMe_2Cl (5.65) can react with organoboranes (eqs. 5.80, 5.81).

$$R_3B \xrightarrow{NH_2X} R_2\overset{R}{\underset{X}{\overset{|}{B}}}{}^{-}\!\!-\!\overset{+}{N}H_2 \longrightarrow R_2\overset{R}{\underset{X}{\overset{|}{B}}}{}^{-}\!\!-\!\overset{+}{N}H_2 \xrightarrow{H_2O} RNH_2 \quad (5.80)$$

$$X = Cl,\ OSO_3H,\ 2,4,6\text{-}Me_3H_2C_6SO_3$$

$$R_3B \xrightarrow{ClNMe_2} RNMe_2 \quad\quad\quad\quad (5.81)$$

Olefins can be converted into stereo- and regiodefined amines via hydroboration-amination, for example, eq. 5.82.

$$58\% \quad (5.82)$$

Alkyl and aryl azides react analogously with organoboranes; see eq. 5.83 (5.73).

$$X_2BR \xrightarrow{R'N_3} X_2\overset{R}{\underset{\underset{N_2}{|}{+}}{\overset{|}{B}}}{}^{-}\!\!-\!NR' \longrightarrow X_2BNRR' \xrightarrow{H_2O} HNRR' \quad (5.83)$$

The reaction of 2-iodoalkyl azides with organoboranes followed by treatment of the intermediates with KOH produces N-alkyl- and N-arylaziridines; see eq. 5.84 (5.74).

$$\text{(5.84)}$$

More recently, the following double alkyl migration reaction was introduced; see eq. 5.85 (*5.75*).

$$\text{(5.85)}$$

DNP = 2,4-dinitrophenyl   50%

It is important to note that all of these amination reactions can be rationalized by the same common 1,2-migration mechanism.

Formation of the Carbon-Sulfur Bond via Organoboranes. The reaction of trialkylboranes with alkyl and aryl disulfides produces dialkyl and alkyl aryl sulfides (*5.76*).

$$R_3B \; + \; R'SSR' \longrightarrow RSR' \; + \; R_2BSR' \qquad (5.86)$$

The reaction has been shown to proceed by a free-radical chain mechanism (q. 5.86).

$$R_3B \xrightarrow{\text{initiator}} R\cdot$$

$$\qquad\qquad\qquad\qquad\qquad\qquad\qquad (5.87)$$

$$R\cdot \; + \; R'SSR' \longrightarrow RSR' \; + \; \cdot SR'$$

Transmetallation Reactions of Organoborons. Since boron is relatively electronegative, organoboron compounds do not readily serve as precursors of other organometallics. A significant exception is their reaction with mercuric salts to form organomercurials (Sect. 7.2.1). There have also been some indications

that other organometallics are formed as unstable intermediates.
Thus treatment of organoboranes with Ag(I) salts produces the
dimers of the alkyl groups, presumably via organosilver intermediates (5.77). More recently, it has been reported that the
products of the reaction of alkenylboron derivatives with either
Cu(I) salts or MeCu undergo reactions characteristic of organocoppers, as discussed in Sect. 9.1.

5.2.3.3   Formation of Carbon-Carbon Bonds via Reactions of Or-
ganoborons with Electrophilic Reagents

In Sect. 5.2.3.1 we briefly discussed the fact that organoboranes
do not readily react with electrophilic reagents, whereas the
corresponding organoborates can act as either Grignard-like reagents or undergo unique intramolecular transfer reactions in
their interaction with electrophiles. Under appropriate conditions, however, organoboranes can react with so-called electrophilic reagents via free-radical or concerted mechanisms. Some
typical carbon-carbon bond-forming reactions of organoboranes
and organoborates with electrophilic reagents are discussed in
this section.

<u>Reaction with Proton Donors</u>. Whereas alkyl-, aryl-, allyl-, and
benzylboron derivatives undergo cleavage of the B-C bond with
certain Brønsted acids, the corresponding reaction of alkenyl-
and alkynylborates generally leads to the formation of a carbon-
carbon bond between two boron-bound organic groups via 1,2-mi-
gration (eqs. 5.88, 5.89).

$$-\overset{|}{C}=\overset{R}{\underset{|}{C}}-\overset{|}{\underset{|}{B}}{}^{-}\ \overset{H^+}{\longrightarrow}\ -\overset{|}{C}-\overset{R}{\underset{H}{\overset{|}{C}}}{}^{+}-\overset{|}{\underset{|}{B}}{}^{-}\ \longrightarrow\ -\overset{|}{C}-\overset{R}{\underset{H}{\overset{|}{C}}}-\overset{|}{\underset{|}{B}}- \qquad (5.88)$$

$$-C\equiv C-\overset{R}{\underset{|}{\overset{|}{B}}}{}^{-}\ \overset{H^+}{\longrightarrow}\ -C=\overset{R}{\underset{H}{\overset{|}{C}}}{}^{+}-\overset{|}{\underset{|}{B}}{}^{-}\ \longrightarrow\ -C=\overset{R}{\underset{H}{\overset{|}{C}}}-\overset{|}{B}- \qquad (5.89)$$

The reaction provides a route to certain Markovnikov organoboranes which cannot be readily prepared by hydroboration (5.78).
The organoborane intermediates have been converted into alcohols,
aldehydes, or ketones via oxidation (5.78, 5.79) or olefins via
protonolysis (5.80) or dehydroboration (5.81).

Reaction with Iodine and Bromine. The reaction of alkenyl- and
alkynylborates can provide an efficient and selective method of
coupling two unlike boron-bound groups. Di- and trisubstituted
olefins, shown in eqs. 5.90 to 5.93 (5.82 to 5.85), acetylenes,
shown in eq. 5.94 (5.86 to 5.88), conjugated enynes, shown in
eq. 5.95 (5.89), conjugated diynes, shown in eq. 5.96 (5.90),
have been prepared in high yields.

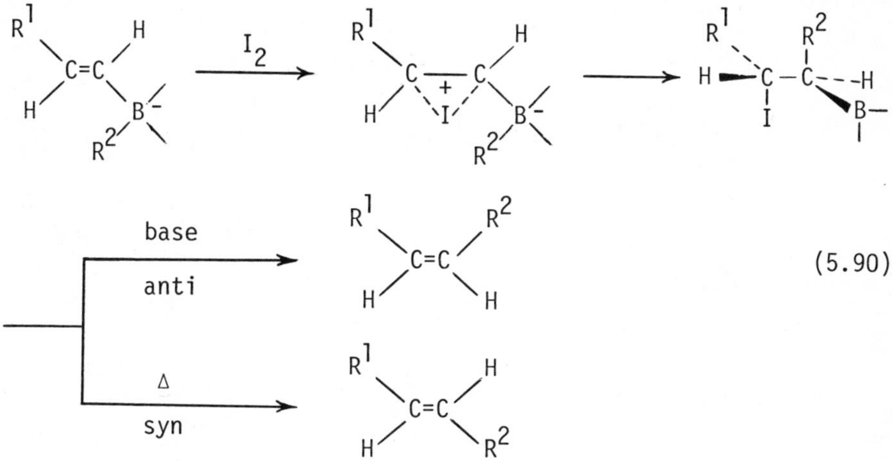

(5.90)

The use of thexylborane (5.83), 9-BBN (5.84), or alkaneboronic
acids (5.85) permits, in some favorable cases, a high-yield conversion of the alkyl or alkenyl group on the boron atom into the
desired product.

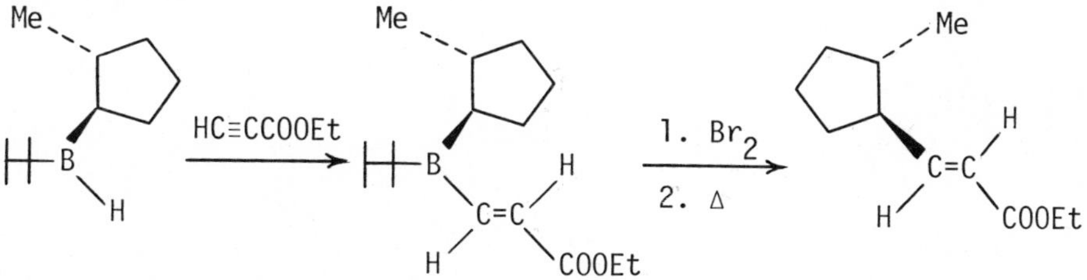

52% (95% E)

(5.91)

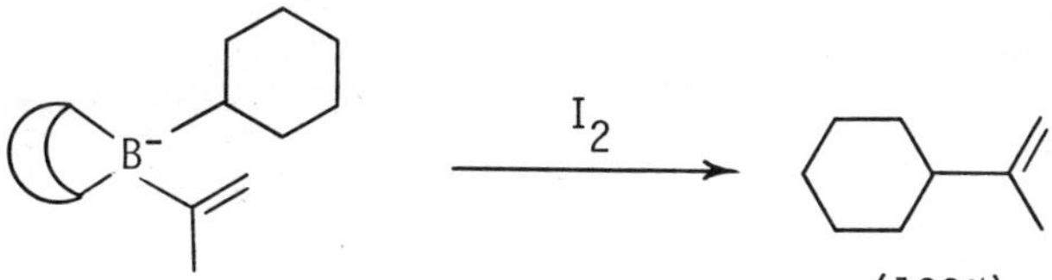

(100%)

(5.92)

$$(5.93)$$

58% (>99% E)

$$R^1C{\equiv}CB^-R^2_3Li^+ \xrightarrow{\text{I}_2} \quad \overset{R^1}{\underset{I}{\diagdown}}C{=}C\overset{R^2}{\underset{BR^2_2}{\diagup}} \longrightarrow R^1C{\equiv}CR^2 \qquad (5.94)$$

R^1 = H or an organic group, $R^2 = 1^0, 2^0$, alkyl, or aryl

$$60\% \qquad (5.95)$$

$$Li^+\left[Sia_2B^-\overset{C{\equiv}CR^1}{\underset{C{\equiv}CR^2}{\diagup}}\right] \xrightarrow[\text{NaOH}]{\text{I}_2} \quad R^1C{\equiv}C{-}C{\equiv}CR^2 \qquad (5.96)$$

It should be noted that these intramolecular coupling reactions of organoborons provide viable alternatives to more conventional cross-coupling reactions involving Li, Mg (Sect. 4.3.2.1), Cu (Sect. 9.1), and Ni or Pd (Sect. 9.2).

A few related reactions of tetraarylborates are known; see eq. 5.97 (5.91). Currently available procedures, however, do not appear suitable for selectively coupling two unlike aryl groups.

$$KB(-\bigcirc)_4 \xrightarrow{Br_2} \bigcirc-\bigcirc + \bigcirc-Br$$

$$(5.97)$$

Reactions with Alkyl Halides, Acyl Halides, and Epoxides. As mentioned earlier, organoboranes are generally inert to alkyl halides and acyl halides. While organoboranes may act as catalysts for the polymerization of epoxides, no carbon-carbon bond between these two reagents is readily formed. The nucleophilicity of the B-C bond can be markedly enhanced by converting organoboranes into organoborates. The reactions of organoborates with carbon electrophiles have recently been reviewed (5.13, 5.36). Therefore only a very brief discussion is presented here.

(a) Intermolecular transfer reactions. The Grignard-like reaction, that is, intermolecular transfer reaction, of organoborates is relatively rare. Acyl halides represent about the only class of carbon electrophiles that can readily participate in the intermolecular transfer reaction; see eq. 5.98 (5.92). This ketone synthesis is highly chemoselective.

$$Li^+[R^1-B^-R_3] + R^2COX \xrightarrow{25°} R^1COR^2 + BR_3 \qquad (5.98)$$

$$R^1 = alkyl, benzyl, aryl, MeSOCH_2$$

Even acyl halides preferentially undergo the intramolecular transfer reaction with cyanoborates (5.93), alkenyl- and alkynylborates (5.94), and thioalkoxymethylborates (5.95).
Alkyl halides and epoxides are generally quite reluctant to participate in the intermolecular transfer reaction. Only highly reactive alkyl halides, such as MeI and allyl bromides, react with lithium arylborates and sulfinyl- and sulfonyl-substituted lithium borates to form cross-coupled products, for example, eq. 5.99 (5.13).

$$(5.99)$$

More recently, however, it has been found that the inter-molecular transferrability of organoborates can be enhanced by forming copper(I) borates, as indicated in Scheme 5.1 (*5.96*). Despite these promising new developments, it may be said that the intermolecular transferrability of organoborates is generally low and that organoboranes are quite nonnucleophilic.

Scheme 5.1

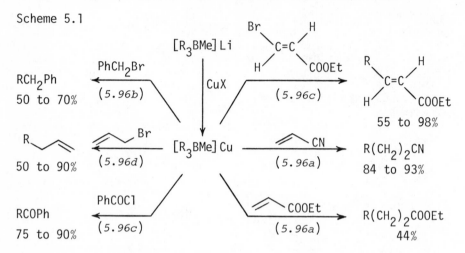

(b) Intramolecular Transfer Reactions. α,β-Unsaturated or-ganoborates, in particular alkynyl-, alkenyl-, and cyanoborates, have a strong tendency to undergo the intramolecular transfer reaction (eqs. 5.42 to 5.44). The organoborane products thus formed can be either oxidized or subjected to protonolysis to ob-tain organic products. Some typical results are summarized in Schemes 5.2 to 5.4.

Scheme 5.2

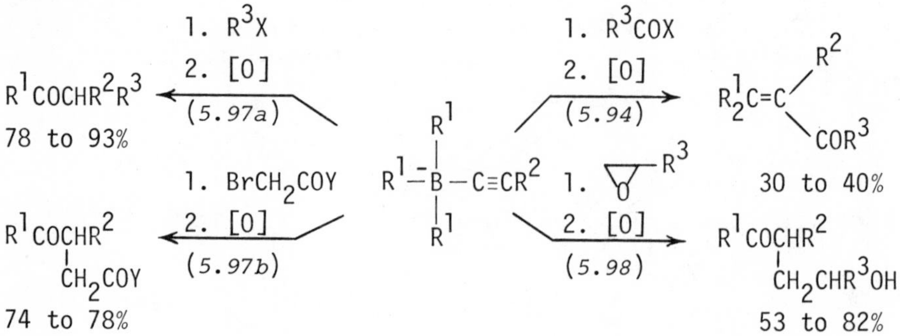

Scheme 5.3

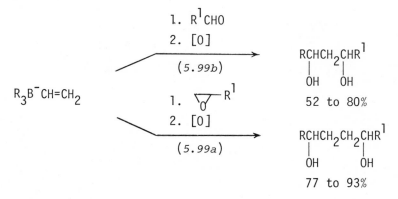

Unlike the reaction with simple alkyl halides, such as MeI, the reactions shown in Scheme 5.3 are reasonably stereoselective (>90%). The stereochemistry of these reactions, however, is somewhat unpredictable and difficult to rationalize.

Scheme 5.4

$$R_3B^-CH=CH_2$$

1. R^1CHO
2. [O]

(5.99b)

RCHCH$_2$CHR^1
 | |
 OH   OH

52 to 80%

1. (epoxide)–R^1
2. [O]

(5.99a)

RCHCH$_2$CH$_2$CHR^1
 | |
 OH   OH

77 to 93%

It is useful to note that alkynylborates can act as enolate equivalents in some of these reactions, for example, eq. 5.100. The alkylation of arylborates can proceed either via intermolecular transfer, for example, eq. 5.99, or via intramolecular transfer (5.100). For example, the reaction of LiBBu$_3$-(Naph) with CH$_3$OSO$_2$F produces the double alkylation product $\underline{5.8}^3$ in 70% yield after oxidation and the dihydroaromatic derivative

$$R^1COCHR^2R^3 \qquad (5.100)$$

<u>5.9</u> in 56% yield along with a 22% yield of <u>5.8</u> after hydrolysis. The results can be interpreted in terms of intramolecular 1,2-migration (eq. 5.101).

(5.101)

Reactions with Aldehydes, Ketones, and α,β-Unsaturated Carbonyl Compounds. Ordinary ketones are generally quite inert to organoborons, although allylboranes, shown in eq. 5.102 (*5.101*), enolboranes, shown in eq. 5.50 (*5.102*), and enolborates, shown in eq. 5.103 (*5.103*), react with ketones.

$$(5.102)$$

$$(5.103)$$

The reactions of allylboranes and enolboranes appear best viewed as $[2_s + 4_s]$ pericyclic processes. Aldehydes have participated in other reactions with organoborons as well. Their reaction with alkenylborates is shown in Scheme 5.4. Monomeric formaldehyde reacts with organoboranes to give one-carbon homologated alcohols in the presence of air; see eq. 5.104 (*5.104*).

$$R_3B \ + \ CH_2O \quad \begin{array}{c} \xrightarrow{O_2} \ R_2BOCH_2R \ \xrightarrow{H_2O} \ RCH_2OH \\ \\ \xrightarrow{no \ O_2} \ R_2BOCH_3 \ + \ olefin \end{array}$$

$$(5.104)$$

More recently, it has been shown that alkenyl-9-BBNs undergo a Grignard-like addition reaction with aldehydes at 65°C. The yields of the products, however, are generally modest (*5.105*).

$$(5.105)$$

Unlike simple aldehydes and ketones, their α,β-unsaturated derivatives are quite reactive toward organoboranes. Most of the earlier results have been thoroughly reviewed (*5.3* to *5.6*, *5.106*). Many, but not all, of these reactions have been shown to proceed by a free-radical chain mechanism (eq. 5.106). The reactions of β-unsubstituted enones are very facile and uniquely useful in that most of the other conjugate addition reactions, such as those involving copper (Sect. 9.4), tend to give poor results with these unhindered enones. Methyl vinyl ketone, acro-

$$R_3B \xrightarrow{\text{initiator}} R\cdot$$

$$R\cdot \ + \ \underset{\substack{| \ | \\ }}{C=C-C=O} \xrightarrow{\hspace{3cm}} R-\underset{|}{C}-C=C-O\cdot \qquad (5.106)$$

$$R-\underset{|}{C}-C=C-O\cdot \ + \ R_3B \xrightarrow{\hspace{2cm}} R-\underset{|}{C}-C=C-OBR_2 \ + \ R\cdot$$

lein, α-methyl-, and α-bromoacroleins have reacted satisfactorily
(*5.107*). On the other hand, the conjugate addition reaction of
β-monosubstituted enones generally requires some external free-
radical source, such as O_2, peroxides, and light. Under these
conditions, the following enones have reacted with organoboranes
to give 1,4-adducts (*5.108*).

One of the serious limitations of the 1,4-addition reactions of
organoboranes is that only one of the three organic groups is
utilized. This problem has been partially overcome by β-alkyl-
boracyclanes; see eq. 5.107 (*5.109*).

$$R-B\bigcirc \ + \ \underset{\substack{| \ | \ |}}{C=C-C-O} \xrightarrow{H_2O} R-\underset{|}{C}-CHC=O \qquad (5.107)$$

, and so on

It has recently been reported that alkenyl- and alkynyl-9-
BBNs react with enones to produce 1,4-adducts (*5.110*). Even some
β,β-disubstituted enones have reacted, albeit slowly. Enones
capable of assuming a cisoid conformation react, whereas transoid
enones have not. These results are not consistent with the free-
radical mechanism presented above. The following six-centered
pericyclic mechanism has been suggested as a plausible path (eq.
5.110).

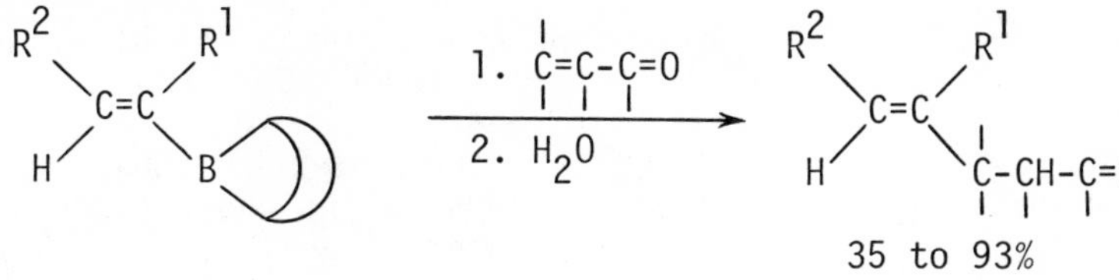

$$R^2 \diagdown \underset{H}{\overset{R^1}{C=C}} \diagdown B \quad \xrightarrow[\text{2. } H_2O]{\text{1. } C=C-C=O} \quad R^2 \diagdown \underset{H}{\overset{R^1}{C=C}} \diagdown \underset{}{\overset{}{C-CH-C=O}} \qquad (5.108)$$

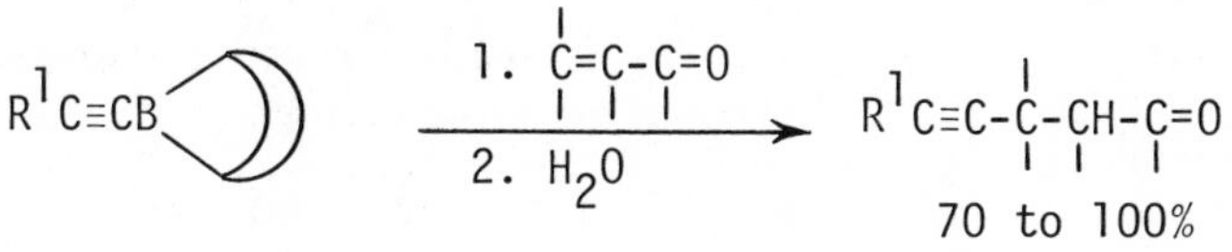

35 to 93%

$$R^1 C \equiv CB \diagdown \quad \xrightarrow[\text{2. } H_2O]{\text{1. } C=C-C=O} \quad R^1 C \equiv C - C - CH - C = O \qquad (5.109)$$

70 to 100%

$$R-B \diagdown \xrightarrow{C=C-C=O} \quad \text{[cyclic enol borate]} \longrightarrow R-C-C=C-O-B \diagdown$$

$$\xrightarrow{H_2O} R-C-CH-C=O \qquad (5.110)$$

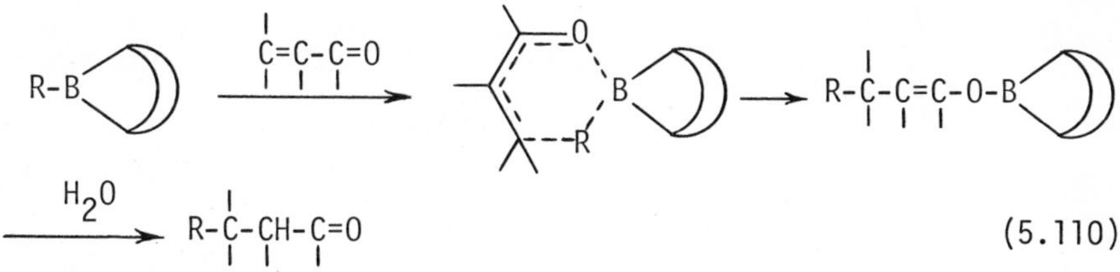

Organoborates should be capable of undergoing conjugate addition reactions. Indeed, a very facile conjugate addition takes place intramolecularly; see eq. 5.111 (*5.111*).

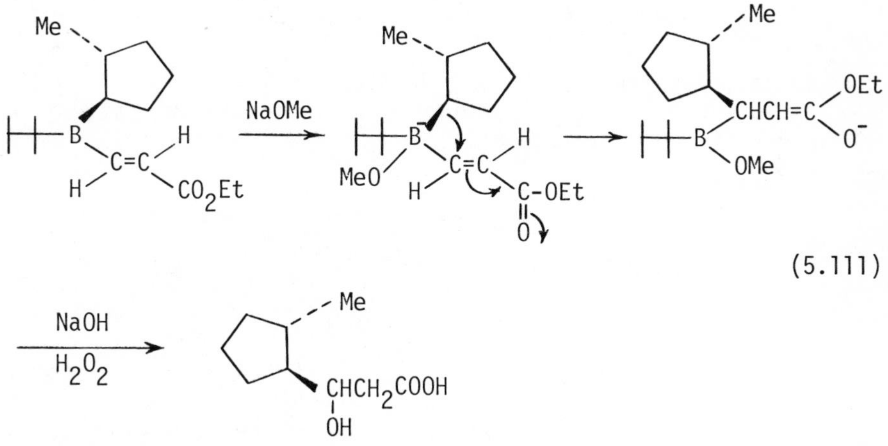

$$\xrightarrow{NaOMe} \longrightarrow \qquad (5.111)$$

$$\xrightarrow[H_2O_2]{NaOH} \quad \text{[cyclopentane derivative: CHCH}_2\text{COOH, OH]}$$

More recently, a few intermolecular conjugate addition reactions of organoborates involving $CuBR_4$ (*5.96a*) and alkynylborates (*5.112*) have been reported. The full scope of such reactions, however, remains to be further explored.

5.2.3.4 Formation of Carbon-Carbon Bonds via Reactions of Organoboranes with Nucleophilic Reagents

In Sect. 5.2.3.1 we discuss the fact that organoboranes react
with appropriately substituted nucleophiles to undergo the intramolecular transfer reactions represented by eqs. 5.34 to 5.36.
The scope of the carbon-carbon bond formation via the intramolecular transfer reactions of organoboranes is so broad that it is
not possible to discuss all of these reactions in detail in this
book. The reader is referred to monographs cited earlier (*5.3*
to *5.6*) and a general review (*5.113*). Specific reviews are cited
later.

Reactions with α-Hetero-Substituted α-Carbonyl Carbon Nucleophiles. The reaction of organoboranes with α-halo enolate anions
provides a useful method for α-alkylation and α-arylation of carbonyl compounds, for example, eq. 5.112 (*5.114*).

$$R_3B \xrightarrow{K^+C^-HBrCOOEt} R_2B\text{-}\overset{\overset{\displaystyle R}{|}}{\underset{\underset{\displaystyle Br}{|}}{C}}HCOOEt \longrightarrow R_2B\text{-}\overset{\overset{\displaystyle R}{|}}{C}HCOOEt$$

$$\text{(5.112)}$$

$$\xrightarrow{\text{tautomerization}} RCH=\overset{\overset{\displaystyle OEt}{|}}{C}\text{-}OBR_2 \xrightarrow{R'OH} RCH_2COOEt + R_2BOR'$$

The unusually facile protonolysis involves the reaction of enolboranes formed via tautomerization (*5.115*). A wide variety of
α-halo enolate anions, such as those derived from the following
compounds, undergo this alkylation reaction.

$BrCH_2COOEt$	$Br_2CHCOOEt$	$RCHBrCOOEt$
$BrCH_2COCH_3$	$BrCH_2COPh$	
$ClCH_2CN$	Cl_2CHCN	$RCHClCN$
$ClCH(CN)_2$	$BrCH(CN)COOEt$	$BrCH_2CH=CHCOOEt$

The following observations are worth noting: (1) Introduction
of hindered and/or stereo-defined alkyl groups and aryl groups in
the α-position of carbonyl compounds can now be achieved. (2)

α,α-Dihalo enolate anions can be converted into either α-halo-α-alkyl or α,α-dialkyl derivatives. (3) The use of B-alkyl-9-BBNs permits a high-yield conversion of the alkyl group into two-carbon homologated carbonyl compounds (5.116). (4) In some cases, the use of highly hindered phenolate bases, for example, potassium 2,6-di-t-butylphenolate, is either necessary or advantageous (5.117).

Related to these alkylation reactions are the reactions of organoboranes with α-diazo carbonyl compounds, for example, eq. 5.113 (5.118), and the corresponding sulfur ylides, for example, eq. 5.114 (5.119).

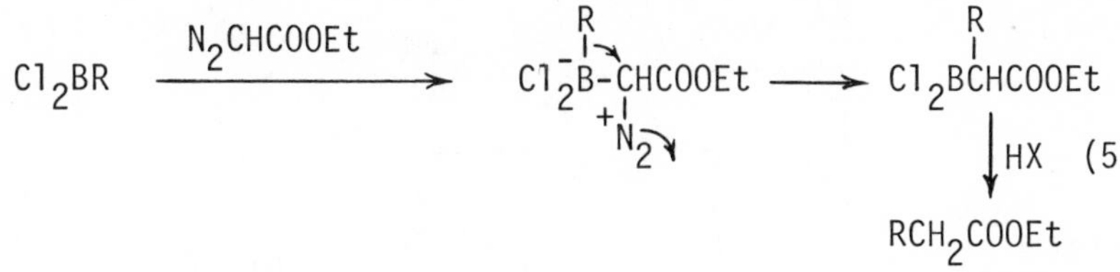

$$Cl_2BR \xrightarrow{\ N_2CHCOOEt\ } Cl_2\bar{B}\text{-}\overset{R}{\overset{|}{C}}HCOOEt \longrightarrow Cl_2BCHCOOEt$$

(5.113)

$$\downarrow HX$$

$$RCH_2COOEt$$

$$R_3B \xrightarrow{\ Me_2\overset{+}{S}\overset{-}{C}HCOOEt\ } R_2\bar{B}\text{-}\overset{R}{\overset{|}{C}}HCOOEt \xrightarrow{\ HX\ } RCH_2COOEt$$

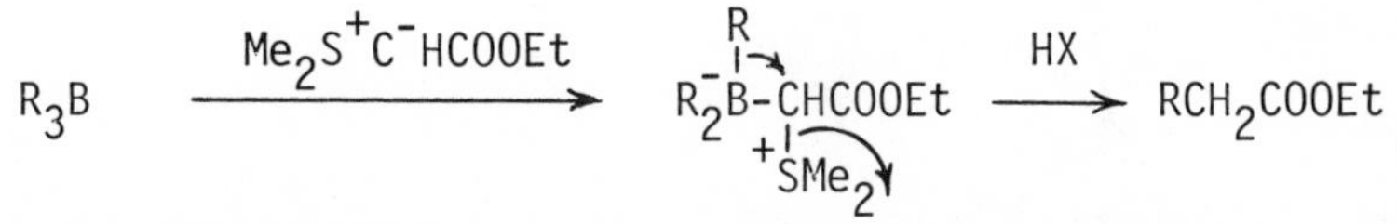

(5.114)

If the mechanism proposed for these reactions were correct, the presence of an α-carbonyl group is not a necessary requirement. Various other α-hetero-substituted carbon nucleophiles should also undergo similar intramolecular transfer reactions. Indeed, a number of such carbon nucleophiles, such as those shown below, have been shown to undergo the expected intramolecular transfer reactions.

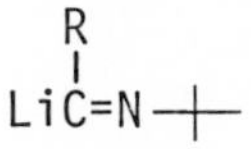

$$Li\overset{R}{\overset{|}{C}}{=}N\!\!-\!\!\!+ \qquad (5.120) \qquad Li\underset{OMe}{\overset{|}{C}}{=}CH_2 \qquad (5.121)$$

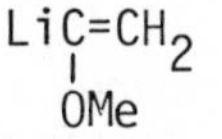

$$Li\text{—}\underset{N}{\diagup\!\!\diagdown}\text{—}Br \qquad (5.122) \qquad Li\text{—}\underset{O}{\diagup\!\!\diagdown} \qquad (5.123)$$

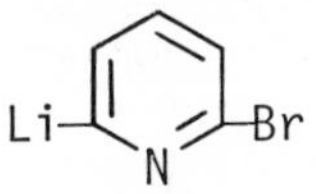

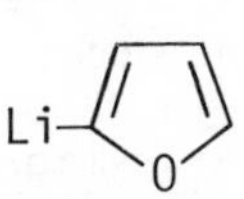

$$\bar{C}H_2\overset{+}{N}Me_3 \qquad (5.124) \qquad \bar{C}H_2\overset{+}{P}Ph_3 \qquad (5.125)$$

$$\bar{C}H_2\overset{+}{S}Me_2 \qquad (5.126)$$

<u>Reactions of α-Hetero-Substituted Organoboranes with Nucleo-</u>
<u>philes</u>. Inspection of the structure of the presumed key inter-
mediate (<u>5.10</u>) in the intramolecular transfer reaction of organoboranes discussed above suggests that, in addition to that route,
there are at least two other conceptually and practically different routes to <u>5.10</u> as shown in Scheme 5.5.

Scheme 5.5

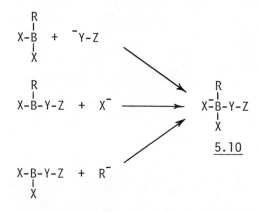

5.10

If the hypothetical species <u>5.10</u> were indeed intermediates
of these reactions, the same products should be obtained regardless of the method of their generation. There can be, however,
major differences in product yield, ease of generation of <u>5.10</u>,
and other significant aspects of the reaction.

The required appropriately α-hetero-substituted organoboranes can be prepared by various methods, of which the free-radical bromination of organoboranes (Sect. 5.2.3.2) and the hydroboration of haloalkenes and haloalkynes (Sect. 5.2.2.2) have been
most widely used. A brief discussion of the reactions of such
organoboranes is presented below.

(a) Reaction of α-bromoboranes obtained via bromination of
organoboranes. The reactions of triethylborane and tri-*sec*-butyl-
borane with Br_2 in the presence of H_2O have been shown to proceed
by the following mechanisms, in which the free-radical bromination and the intramolecular 1,2-migration alternate; see eqs.
5.115, 5.116 (*5.127*).

These reaction sequences provide a unique method for
coupling two boron-bound alkyl groups, as exemplified by eq.
5.117 and 5.118 (*5.128*, *5.129*).

$$Et_3B \xrightarrow{Br_2} \underset{\underset{Br}{|}}{Et\text{-}B\text{-}CHMe}^{Et} \xrightarrow{H_2O} \underset{\underset{HO}{|}}{Et\text{-}B\text{-}CHMe}^{Et} \qquad (5.115)$$

$$\xrightarrow{Br_2} \underset{\underset{HO\ Br}{||}}{\overset{EtEt}{B\text{-}CHMe}} \xrightarrow{H_2O} (HO)_2B\underset{\underset{Et}{|}}{\overset{\overset{Et}{|}}{\text{-}C\text{-}Me}} \xrightarrow{[O]} HO\underset{\underset{Et}{|}}{\overset{\overset{Et}{|}}{\text{-}C\text{-}Me}} \quad (85\%)$$

$$s\text{-}Bu_3B \xrightarrow{Br_2} \underset{\underset{Br}{|}}{s\text{-}BuB\overset{\overset{s\text{-}BuEt}{||}}{\text{-}C\text{-}Me}} \xrightarrow{H_2O} \underset{\underset{H\ OHMe}{|||}}{Me\text{-}\overset{\overset{Et\ Bu\text{-}s}{||}}{C\text{-}B\text{-}C}\text{-}Et} \xrightarrow{Br_2} \qquad (5.116)$$

$$\underset{\underset{BrOHMe}{|||}}{Me\text{-}\overset{\overset{Et\ Bu\text{-}s}{||}}{C\text{-}B\text{-}C}\text{-}Et} \xrightarrow{H_2O} \underset{\underset{MeMeMe}{|||}}{H\text{-}\overset{\overset{EtEtEt}{|||}}{C\text{-}C\text{-}C}\text{-}B(OH)_2} \xrightarrow{[O]} \underset{\underset{MeMeMe}{|||}}{H\text{-}\overset{\overset{EtEtEt}{|||}}{C\text{-}C\text{-}C}\text{-}OH} \quad (46\%)$$

$$(65\%) \qquad (5.117)$$

$$(5.118)$$

90%

(b) Reaction of α-haloboranes obtained via hydroboration. As discussed in Sect. 5.2.2.2, hydroboration of alkenyl and alkynyl halides produces α-haloboranes. Especially noteworthy is the

nearly exclusive formation of 1-halo-1-alkenylboranes (5.11) in a
regio- and stereoselective manner. Even more spectacular is the
reaction of 5.11 with nucleophiles, such as NaOMe, which induces
the intramolecular 1,2-migration proceeding with complete inversion of configuration at the migration terminus; see eq. 5.119
(*5.130*).

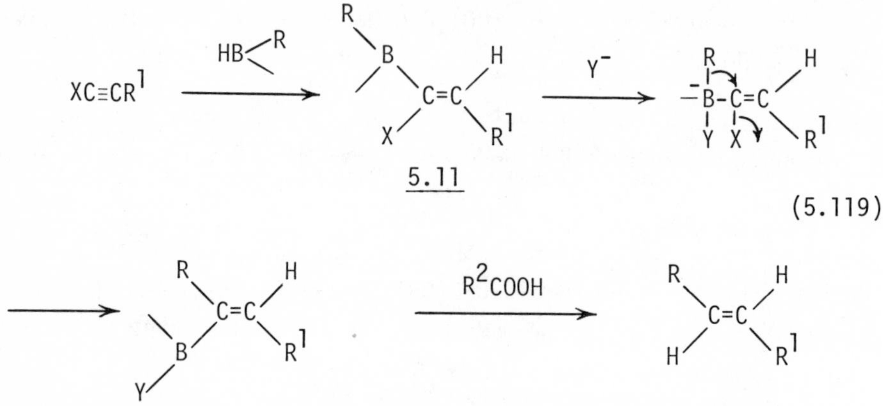

$$XC{\equiv}CR^1 \quad \xrightarrow{HB<^R_R} \quad \text{5.11} \quad \xrightarrow{Y^-} \quad$$

5.11

(5.119)

$$\longrightarrow \quad \xrightarrow{R^2COOH} \quad$$

Such a displacement reaction proceeding with clean inversion at
an sp^2 hybridized carbon center has rarely been observed in the
past. The configuration of the migrating group is completely re-
tained. Protonolysis of the alkenylboranes thus formed also proceeds with complete retention. Various highly stereoselective
procedures for the syntheses of monoolefins, shown in eq. 5.120
(*5.131*), conjugated (E,E)-dienes, shown in eq. 5.121 (*5.132*), and
even (E)-1,2,3-butatrienes, shown in eq. 5.122 (*5.133*), have been
developed based on the sequence discussed above.

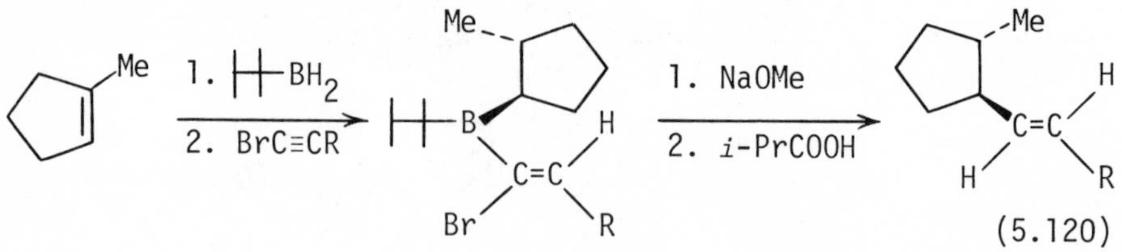

$$\xrightarrow[\text{2. BrC}{\equiv}\text{CR}]{\text{1. } \vdash\!\!-BH_2} \qquad \xrightarrow[\text{2. } i\text{-PrCOOH}]{\text{1. NaOMe}}$$

(5.120)

(5.121)

$$ClC{\equiv}CR \quad \xrightarrow[\text{2. HC}{\equiv}\text{CR'}]{\text{1. } \vdash\!\!-BH_2} \qquad \xrightarrow[\text{2. } i\text{-PrCOOH}]{\text{1. NaOMe}}$$

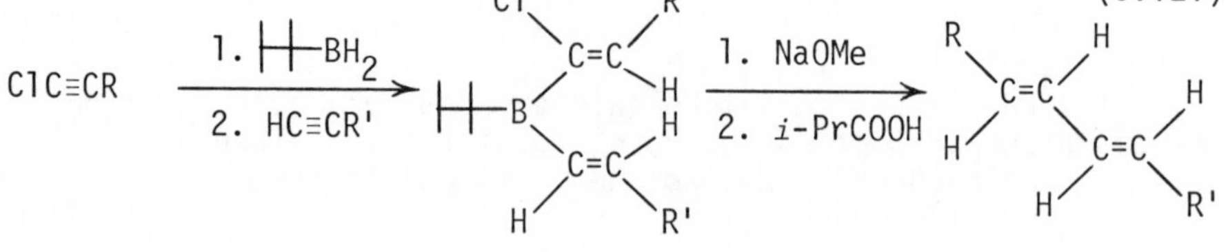

$$2IC \equiv CR \xrightarrow{\text{H-BH}_2} \text{(2,5-dibora structure)} \xrightarrow{\text{NaOMe}} \text{(cumulene structure)} \qquad (5.122)$$

Carbonylation and Related Reactions of Organoboranes. (a) Carbonylation. All of the 1,2-migration reactions of organoboranes discussed so far are represented by either eq. 5.34 or eq. 5.35. Let us now discuss some reactions that can be represented by eq. 5.36.

The reaction of organoboranes with CO is one of the most general and versatile reactions which organoboranes undergo. A variety of primary, secondary, and tertiary alcohols, aldehydes, and ketones have been synthesized by this reaction. Since this reaction has been extensively reviewed (*5.3* to *5.6*, *5.134*, *5.135*), only a brief summary is presented here.

In 1962 Hillmann (*5.136*) reported that trialkylboranes reacted with CO at about 10,000 psi at 25 to 75°C to form 2,5-dibora-1,4-dioxanes (5.12). At $\geq$140°C, 5.12 were smoothly converted into the corresponding boroxines 5.13. In the absence of protic solvents, mixtures of 5.12 and 5.13 are formed. No interconversion between 5.12 and 5.13 takes place even at 200°C.

$$6R_3B + 6CO \xrightarrow[50°C]{H_2O} 3 \underset{5.12}{\text{(5.12 structure)}} \xrightarrow[150°C]{H_2O} 2 \underset{5.13}{\text{(5.13 structure)}} \qquad (5.123)$$

Oxidation of 5.12 produces a 1:1 mixture of a ketone ($R_2C=O$) and an alcohol (ROH), whereas its hydrolysis gives a dialkylcarbinol (R_2CHOH). On the other hand, oxidation of 5.13 gives the corresponding trialkylcarbinol (R_3COH).

$$\underset{5.12}{\text{(5.12 structure)}} \begin{array}{l} \xrightarrow[\text{NaOH}]{H_2O_2} R_2C=O + ROH \\ \\ \xrightarrow[\text{NaOH}]{H_2O} R_2CHOH \end{array} \qquad (5.124)$$

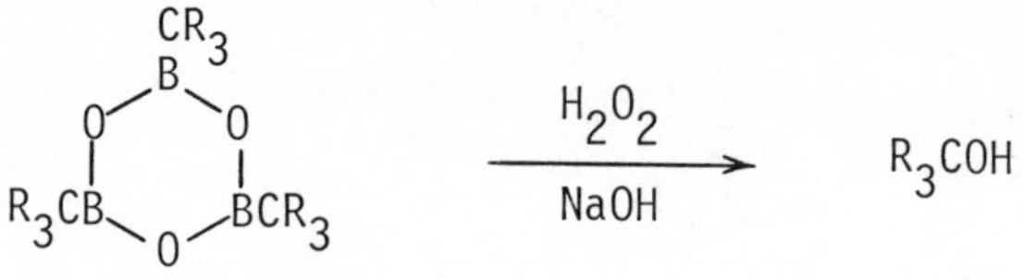

$$\text{5.13} \quad \xrightarrow[\text{NaOH}]{\text{H}_2\text{O}_2} \quad R_3\text{COH} \qquad (5.125)$$

Subsequent studies by Brown (5.137) have established that, in many cases, the reaction proceeds even at atmospheric pressure. Since the reaction of two different organoboranes does not give any crossover product, the transfer of the alkyl groups from boron to carbon must be an intramolecular process. The following mechanism is consistent with the available data (Scheme 5.6).

Scheme 5.6

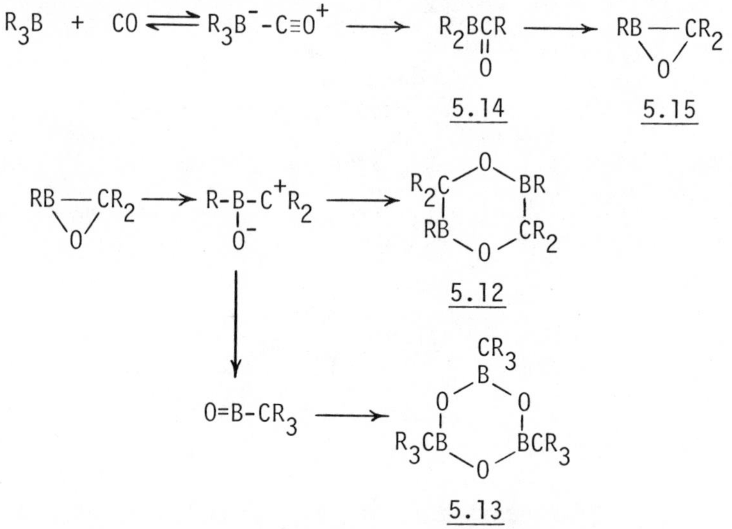

$$R_3B \ + \ CO \rightleftharpoons R_3B^--C\equiv O^+ \longrightarrow \underset{\underset{O}{\|}}{R_2BCR} \longrightarrow RB\overset{}{\underset{O}{\diagup}}CR_2$$

5.14   5.15

$$RB\overset{}{\underset{O}{\diagup}}CR_2 \longrightarrow R\text{-}B\text{-}\overset{+}{C}R_2 \longrightarrow \begin{matrix} R_2C\diagup O\diagdown BR \\ RB\diagdown O\diagup CR_2 \end{matrix}$$
$$\underset{O^-}{|}$$

5.12

$$O=B\text{-}CR_3 \longrightarrow \text{5.13}$$

Under the usual carbonylation conditions, it has not been possible to obtain the product of single alkyl migration 5.14. It has, however, been possible to stop the reaction after migration of one alkyl group by carrying out the reaction in the presence of LiAlH(OMe)$_3$. On oxidation of the intermediates thus obtained, aldehydes were formed in high yields (5.138). It has also been possible to obtain the corresponding methylols (5.139). The use of 9-BBN permits a high-yield conversion of olefins into aldehydes, while LiAlH(OBu-t)$_3$ makes this aldehyde synthesis

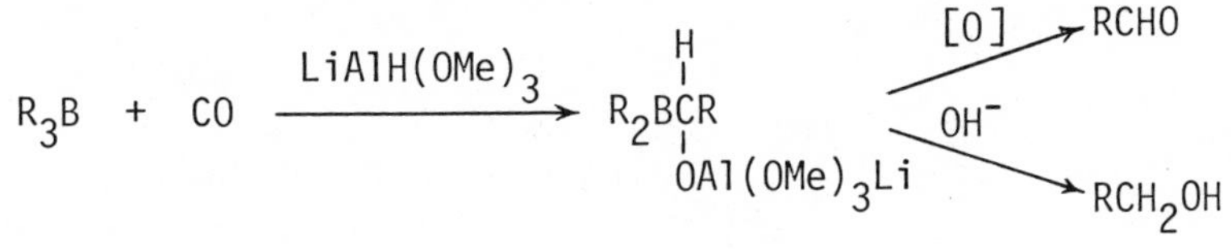

$$R_3B \; + \; CO \quad \xrightarrow{\text{LiAlH(OMe)}_3} \quad \underset{\underset{OAl(OMe)_3Li}{|}}{\overset{\overset{H}{|}}{R_2BCR}} \quad \overset{[O] \nearrow RCHO}{\underset{OH^- \searrow RCH_2OH}{}} \quad (5.126)$$

highly chemoselective, for example, eq. 5.127 (5.140).

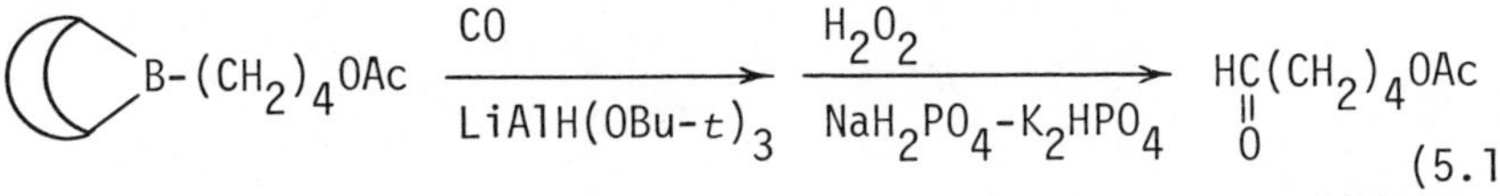

$$\xrightarrow[\text{LiAlH(OBu-}t)_3]{CO} \quad \xrightarrow[\text{NaH}_2\text{PO}_4\text{-K}_2\text{HPO}_4]{H_2O_2} \quad \underset{\underset{0}{\overset{\parallel}{}}}{HC(CH_2)_4OAc}$$

(5.127)

92%

The thexyl group exhibits a very low migratory aptitude for migration in the carbonylation of thexyldialkylboranes. This makes it possible to synthesize mixed and cyclic ketones in high yields; see eqs. 5.128, 5.129 (5.141).

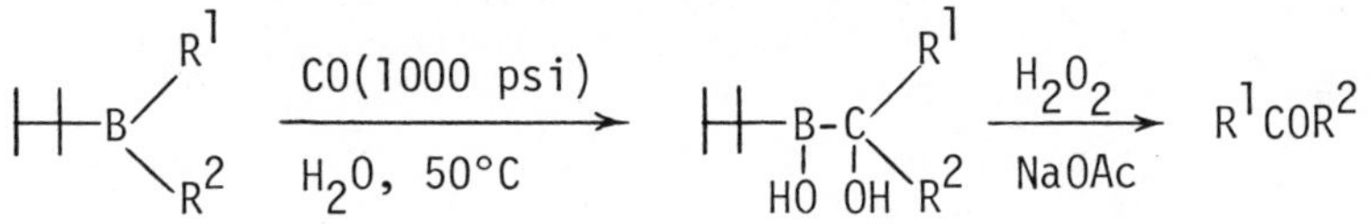

$$\underset{R^2}{\overset{R^1}{\diagdown}}{}_B \quad \xrightarrow[H_2O, \; 50°C]{CO(1000 \; psi)} \quad \underset{HO \;\; OH \;\; R^2}{\overset{R^1}{H\text{-}B\text{-}C\diagdown}} \quad \xrightarrow[\text{NaOAc}]{H_2O_2} \quad R^1COR^2 \quad (5.128)$$

1. CO,H$_2$O

2. H$_2$O$_2$,NaOAc

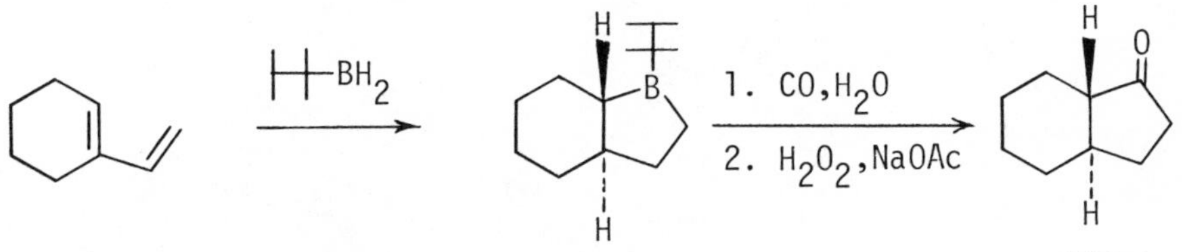

66%
(100% trans)

(5.129)

This ketone synthesis has been applied to the syntheses of juvabione, shown in eq. 5.130 (5.142), and a steroidal compound, shown in eq. 5.131 (5.143).

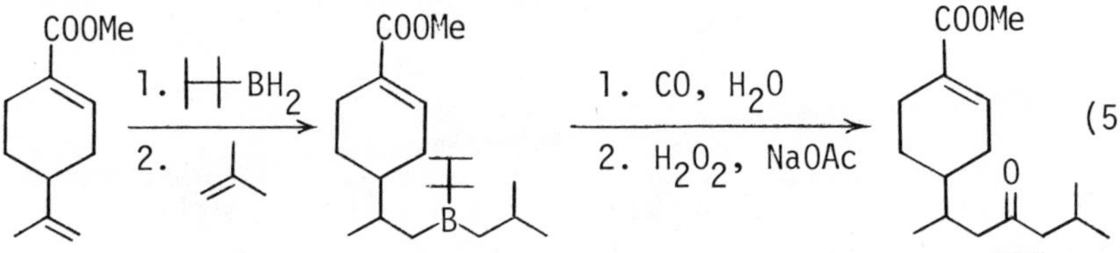

1. CO, H$_2$O

2. H$_2$O$_2$, NaOAc

(5.130)

78%

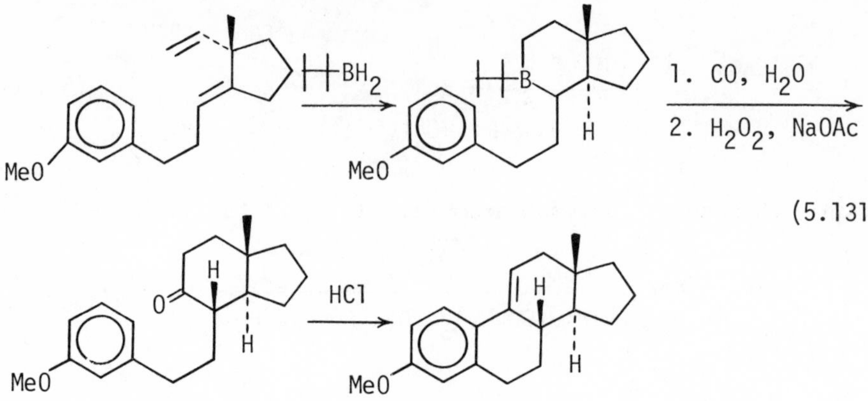

(5.131)

53%

The carbonylation of organoboranes at 150°C in the presence of a protic solvent followed by oxidation provides a means for converting organoboranes into the corresponding carbon structures with complete retention of the overall and stereochemical integrity of the organoboranes, as shown in eq. 5.132 (*5.144*).

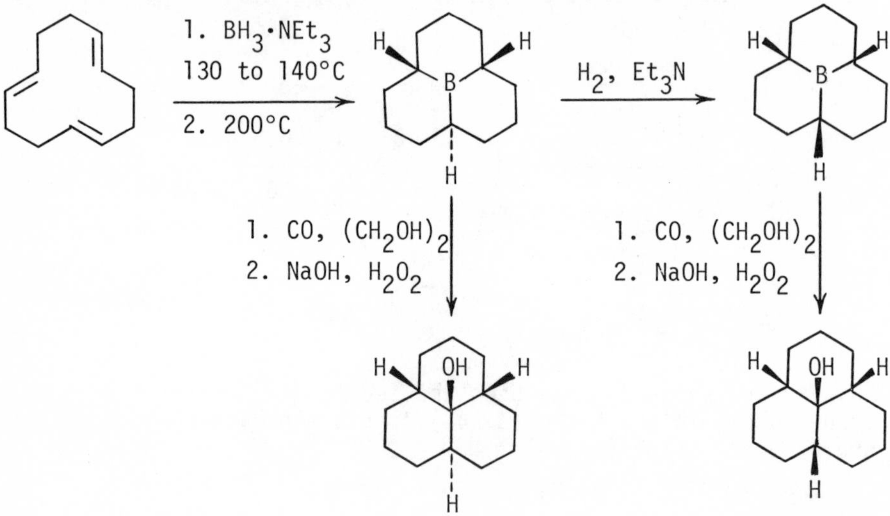

(5.132)

Although CO does not require a more complicated apparatus than that which catalytic hydrogenation requires, and is therefore not inconvenient to use, there exists an unfortunate but prevailing tendency among practicing chemists to avoid CO in laboratories. More recently, a few organoboron reactions, which allow transformations similar to those discussed above, have been developed.

(b) The reaction of cyanoborates with electrophiles. Treatment of trialkylboranes with KCN or NaCN produces the corresponding trialkylcyanoborates, which are isoelectronic with the trialkylborane-CO adducts in Scheme 5.6. Trialkylcyanoborates, however, do not undergo spontaneous 1,2-migration, presumably due to the lack of an electrophilic migration terminus. The earlier discovery by Köster (*5.145*) that the 1,2-migration of alkynylborates can be induced by their treatment with electrophiles led Pelter (*5.93*, *5.146*) to develop a closely related cyanoborate-electrophile reaction. Trifluoroacetic anhydride (TFAA) has been used most extensively as the required electrophile. The reaction is applicable to the synthesis of ketones and dialkyl- and trialkylcarbinols. Aldehydes have not yet been prepared by this reaction. The results obtained to date may be represented by the following scheme.

Scheme 5.7

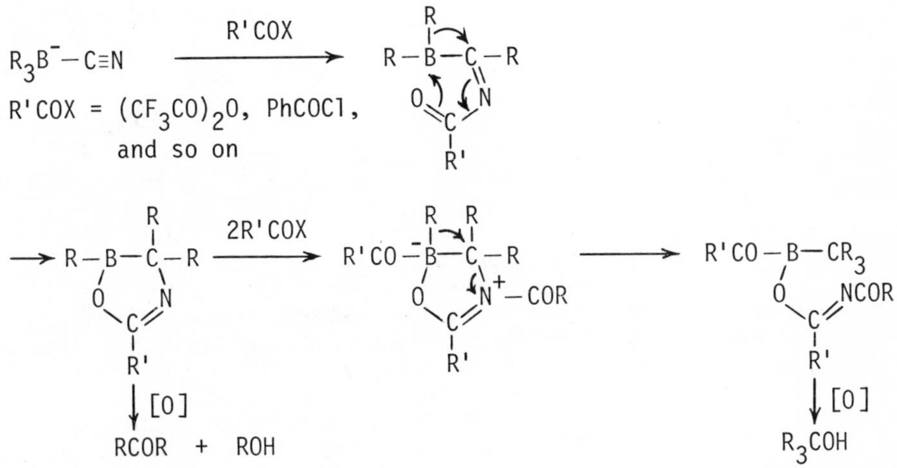

$$R_3B^- - C \equiv N \xrightarrow{R'COX}$$

$R'COX = (CF_3CO)_2O$, PhCOCl, and so on

$$\xrightarrow{} R-B-C-R \xrightarrow{2R'COX} R'CO-B-C-R \xrightarrow{} R'CO-B-CR_3$$

$$\downarrow [O] \qquad\qquad\qquad\qquad\qquad\qquad \downarrow [O]$$

$$RCOR + ROH \qquad\qquad\qquad\qquad\qquad\qquad R_3COH$$

It appears that virtually all of the compounds that have been prepared by this reaction can also be prepared by the carbonylation reaction. On the other hand, it is not clear to the author if the reverse is also true. It should be noted that the cyanoborate-electrophile reaction requires a strong Lewis acid, such as TFAA. Thus those functional groups that are destroyed by such reagents may not be tolerated.

(c) The reaction of organoboranes with polyhalocarbanions. One serious limitation associated with both the carbonylation and the cyanoborate-electrophile reaction is that only triorganoboranes can be used successfully. It is often advantageous to be able to use dialkylborane derivatives in the synthesis of ketones and dialkylcarbinols. This problem has been partially overcome by the development of the reaction of organoboranes with lithio-dichloromethyl methyl ether as shown in Scheme 5.8 (5.147).

Scheme 5.8

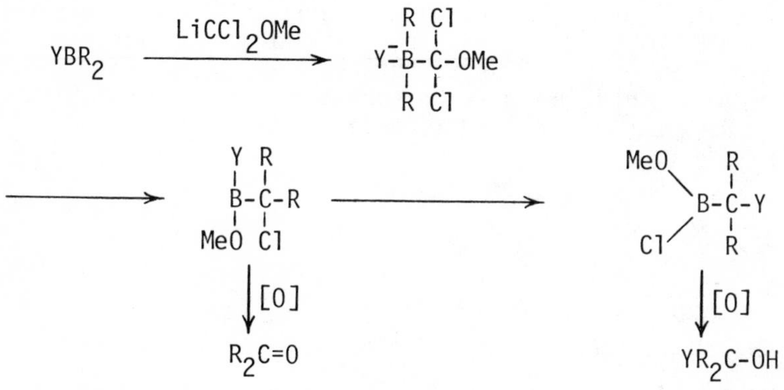

$$
YBR_2 \xrightarrow{\text{LiCCl}_2\text{OMe}} Y\overset{\displaystyle R\;\; Cl}{\underset{\displaystyle R\;\; Cl}{-B-C-OMe}}
$$

$$
\longrightarrow \underset{MeO\;\; Cl}{\overset{Y\;\; R}{B-C-R}} \longrightarrow \underset{Cl}{\overset{MeO}{}}B\underset{R}{\overset{R}{-C-Y}}
$$

$$
\downarrow [O] \qquad\qquad\qquad\qquad \downarrow [O]
$$

$$
R_2C{=}O \qquad\qquad\qquad\qquad YR_2C{-}OH
$$

Y = alkyl, OH, OR, and so on

Although this reaction is nothing but a triple-migration version of some of the reactions discussed earlier in this section, the exact order of the ligand migration is not clearly established. In the reaction of dialkylborinic acid esters, the intermediates have been identified as either α-chloroboronic acid esters or α-methoxyboronic acid esters (5.147c, 5.147e).

The reaction appears quite general as a route to trialkyl-carbinols, as indicated by eq. 5.133 (5.147d). The following synthesis of bicyclo[3.3.1]nonan-9-one demonstrates a unique advantage of the reaction over the carbonylation or the cyanoborate-electrophile reaction; see eq. 5.134 (5.148). At present, the synthesis of ketones by this reaction requires dialkylborane derivatives. Thus the scope of the ketone synthesis is limited by the lack of a general route to mixed dialkylborane derivatives.

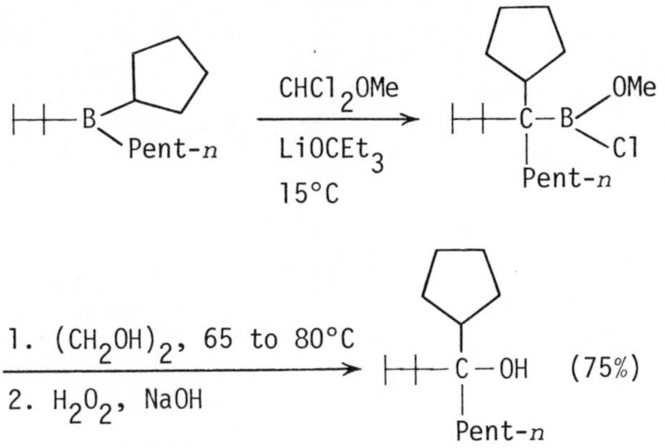

(5.133)

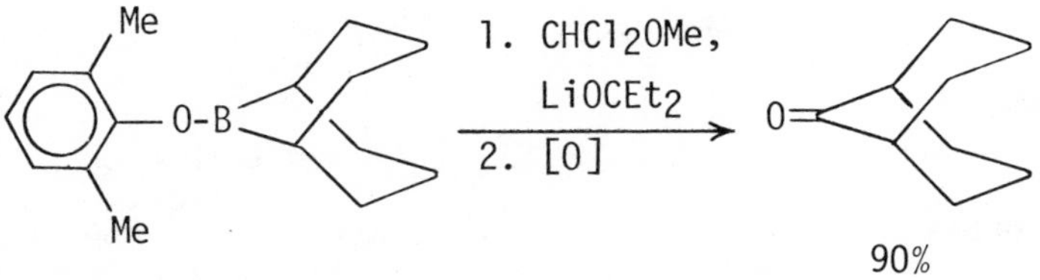

(5.134)

90%

5.2.3.5   Other Carbon-Carbon Bond-Forming and Bond-Cleaving Reactions of Organoborons

Miscellaneous Elimination Reactions.   It has been amply demonstrated that organoborons carrying a good leaving group in the α-position undergo the intramolecular transfer via 1,2-migration (eq. 5.34).   We have also learned that those carrying a good leaving group in the β-position undergo β-elimination (eq. 5.90). Usually, the base-catalyzed elimination is an anti process, whereas the thermal elimination is a syn process.   What about organoborons carrying a good leaving group in a more remote position?

(a) Cyclopropane synthesis via γ-elimination.   Organoborons carrying a good leaving group in the γ-position undergo a facile γ-elimination to form cyclopropane derivatives; see eq. 5.135 (5.149).

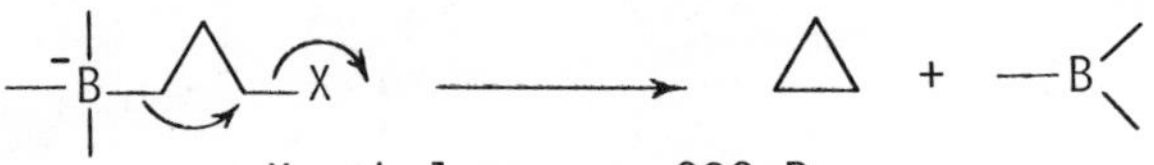

(5.135)

X = halogen or OSO_2R

(b) Fragmentation reactions of δ-hetero-substituted organoborons. δ-Hetero-substituted organoborons can undergo either
formation of cyclobutanes (*5.149d*) or a Grob-type fragmentation
reaction (*5.150*). The latter appears to require rigid and favorable conformational arrangements, as shown in eq. 5.136.

$$(5.136)$$

Little is known about the reaction of organoborons carrying more
remote leaving groups.

Generation and Reactions of Boron-Stabilized Carbanions. Just as
the carbonyl group stabilizes an adjacent carbanionic center, the
empty *p* orbital of the boron atom of organoboranes should also
exhibit a similar effect. The intrinsic difficulty is how to
generate such species without forming undesirable organoborate
complexes.

$$(5.137)$$

One successful approach involves the use of so-called highly
basic but nonnucleophilic bases (*5.151*). The boron-stabilized
carbanions thus generated can react with carbonyl compounds to
undergo a Wittig-like reaction (eq. 5.138).

$$(5.138)$$

An alternate approach involves generation of *gem*-dibora and *gem*-tribora compounds followed by treatment with a base, for example, eqs. 5.139 and 5.140 (*5.152, 5.153*).

$$R^1C\equiv CH \xrightarrow[\text{2. MeLi}]{\text{1. 2 eq HB}\langle} R^1CH_2CHB\langle \xrightarrow[\substack{\text{2. } H_2O_2, \\ \text{NaOH}}]{\text{1. } R^2X} R^1CH_2CHR^2OH \qquad (5.139)$$

with the intermediate shown as $R^1CH_2CHB\langle$ bearing Li.

$$\underset{\substack{O \\ O}}{\boxed{}}B-)_3CH \xrightarrow{\text{MeLi}} \underset{\substack{O \\ O}}{\boxed{}}B-)_2CHLi \xrightarrow{R^1COR^2} \underset{\substack{O \\ O}}{\boxed{}}B-CH=CR^1R^2$$

$$\downarrow [O] \qquad (5.140)$$

$$OHCCHR^1R^2$$

$$(63 \text{ to } 97\%)$$

Miscellaneous One-Electron Transfer Reactions. It has been amply demonstrated that organoborons are good sources of carbon free-radicals. It now appears that, under electrolytic conditions, even carbocations can be generated from organoboranes via free radicals. The anodic oxidation of organoboranes using graphite as the anode produces alkyl methyl ethers, when carried out in the presence of $NaClO_4$, NaOMe, and MeOH, and alkyl acetates, when carried out in the presence of NaOAc and HOAc (*5.154*). The formation of alkyl methyl ethers, for example, has been rationalized as follows.

$$R_3B \xrightarrow{^-OMe} R_3B^-OMe \xrightarrow{-e} R_3\overset{.}{B}OMe \longrightarrow R\cdot + R_2BOMe$$

$$R\cdot \xrightarrow{-e} R^+ \xrightarrow{MeOH} ROMe \qquad (5.141)$$

Electrolysis of organoboranes in CH_3CN (*5.155a*) and CH_3NO_2 (*5.155b*) has produced homologated nitriles and nitro compounds, respectively.

Carboboration Reactions. Unlike organoalanes, ordinary organoboranes do not readily undergo carbometallation with olefins and acetylenes. Allylboranes, however, are exceptional and react with acetylenes and olefins to form the corresponding addition products (5.101), as shown in Scheme 5.9. The results have been interpreted in terms of a $[2_s + 4_s]$ process which is analogous to the "ene" reaction.

Scheme 5.9

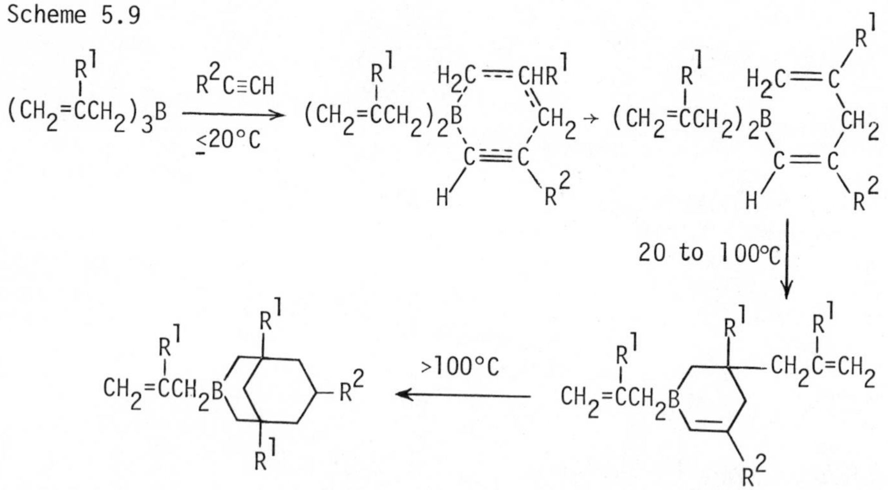

The formation of the monocyclic intermediate involves an allyl-boration reaction of the terminal olefin group, whereas that of the bicyclic product must be a rare example of the addition of an alkenyl-boron bond to an olefin.

Miscellaneous Reactions. There can be a variety of organoborane reactions in which the boron atom acts as a carbonyl-like elec-tron-withdrawing group. For example, there have been indications (5.156) that alkenyl- and alkynylboranes act as efficient dieno-philes in the Diels-Alder reaction. Although highly promising, the scope and synthetic utility of such reactions are yet to be explored.

5.2.3.6   Organoboranes and Organoborates as Reagents in Organic Synthesis

The single most important application of organoboranes and or-ganoborates to organic synthesis is the reduction of organic com-pounds with these reagents via hydride transfer. Some represen-tative organoboranes and organoborates that have been used as selective reducing agents include partially alkylated organobor-

anes, such as 9-BBN (*5.157*), disiamylborane (*5.158*), and thexylborane (*5.27*, *5.159*), trialkylboranes, such as B-alkyl-9-BBN's
(*5.160*), partially organo-substituted borates, such as NaBH$_3$CN
(*5.161*), LiBR$_3$H and KBR$_3$H (*5.162*), and tetraorganoborates, such
as lithium dialkyl-9-borabicyclo[3.3.1]nonanates (*5.163*). Since
there are many excellent reviews of these reagents, such as those
cited above, a very brief discussion is presented here.

Organoboranes and organoborates may be viewed as modified
borane (BH$_3$) and borohydrides, for example, NaBH$_4$, respectively.
The effect of an alkyl group in organoboranes appears twofold.
Introduction of an alkyl group not only decreases their electrophilicity via inductive effect but also exerts steric hindrance.
As a result, partially alkylated organoboranes are generally
milder reducing agents than BH$_3$ itself (*5.164*), as indicated in
Table 5.2. On the other hand, the presence of three or even two
electron-donating organic groups in organoborates make them some
of the most powerful nucleophiles known to organic chemists,
while NaBH$_3$CN, which contains the electron-withdrawing CN group,
is a milder reducing agent than NaBH$_4$. In light of the very low
nucleophilicity of tetraalkylborates in the intermolecular sense
(Sect. 5.2.3.3), the exceedingly high nucleophilicity of trialkylborohydrides might be puzzling. It may be noted that, whereas attack of the B-C bond of organoborates, front or back, by an
electrophile appears severely hindered by steric crowding (5.16),
the s orbital of the hydrogen atom of the B-H bond of borohydrides does not appear to suffer from similar steric hindrance
(5.17). It is even possible that the alkyl groups of organo-substituted borohydrides may exert steric acceleration in addition
to their inductive effect mentioned above.

Front attack

Back attack

5.16

5.17

As a rough guide, the reducing properties of some representative
boranes and borates are summarized in Table 5.2. Some noteworthy
reducing properties of organoboranes and orbanoborates are briefly discussed below.

Thexylborane (*5.27*, *5.159*). Under carefully controlled conditions, thexylborane can reduce free aliphatic and aromatic carboxylic acids to aldehydes.

Table 5.2. Selective Reduction with Boranes and Borates

Functional Group	$BH_3 \cdot THF$	Sia_2BH	9-BBN	$NaBH_4$	MBR_3H (M = Li or K)	$NaBH_3CN$ Neutral or Basic	$NaBH_3CN$ Acidic
Aldehyde	+	+	+	+	+	-	+
Ketone	+	+	+	+	+	-	+
Acid halide	-	-	+	+	+	-	+
Lactone	+	+	+	∓ (very slow)	∓ (very slow)	-	-
Ester	∓	-	+ (slow)	∓ (very slow)	∓	-	-
Epoxide	+ (slow)	∓ (very slow)	+ (slow)	∓ (very slow)	+ (very slow)	-	+
Carboxylic acid	+	react but no reduction	∓ (very slow)	react but no reduction		-	-
Carboxylic acid salt	-	-	-	-	-	-	-
t-Amide	+	+	+	-	-	-	-
Nitrile	+	∓ (very slow)	+	-	-	-	-
Nitro	-	∓ (very slow)	-	-	-	-	-
Olefin	+	+	+	-	∓+	-	-

Disiamylborane (*5.158*). Lactones can be reduced to hydroxyalde-
hydes with one mole of disiamylborane. Similarly, tertiary
amides can be reduced to aldehydes. In many cases, however, the
same transformations can also be achieved with diisobutylaluminum hydride (DIBAH); see Sect. 5.3.3.3. Disiamylborane can also
be used to protect free carboxylic acids.

9-BBN (*5.157*). Quite unexpectedly, 9-BBN reduces acyl chlorides
relatively rapidly. Another somewhat unexpected finding is the
reduction of conjugated enones to allylic alcohols with one mole
of 9-BBN (*5.165*). Although many other reducing agents can also
be used for this transformation, 9-BBN ranks among the best a-
long with DIBAH (Sect. 5.3.3.3).

NaBH$_3$CN (*5.161*). One of the most unique features of NaBH$_3$CN is
that it is reasonably stable in aq acidic media. Under neutral
or weakly basic conditions, practically all electrophiles, except
alkyl halides and sulfonates (*5.166*), are inert. As a reagent
for the reduction of alkyl halides, however, NaBH$_3$CN is generally
inferior to trialkylborohydrides. Under acidic conditions, a
wide variety of functional groups can be reduced. The reductive
amination of aldehydes and ketones with NaBH$_3$CN represents one of
the best methods for converting them into amines. Oximes and enamines can be reduced to *N*-alkylhydroxylamines and tertiary amines, respectively (*5.167*). Treatment of aliphatic ketone and
aldehyde tosylhydrazones with NaBH$_3$CN in acidic DMF-sulfolane
provides a selective and high-yield method for deoxygenation of
aldehydes and ketones; see eq. 5.142 (*5.168*).

$$CH_3CO(CH_2)_3COO(CH_2)_6CN \xrightarrow[\substack{2.\ NaBH_3CN,\ 100°C \\ DMF\text{-}sulfolane}]{1.\ TsNHNH_2} CH_3(CH_2)_4COO(CH_2)_6CN$$

$$75\% \quad (5.142)$$

This reaction represents a useful alternative to the Wolff-Kishner reduction. The reduction of α,β-unsaturated aldehydes and
ketones by this procedure is accompanied by complete allylic rearrangement, for example, eq. 5.143 (*5.169*).

$$\text{C}_6\text{H}_5\text{-CH=CHCHO} \xrightarrow[\substack{2.\ NaBH_3CN \\ 3.\ H^+}]{1.\ TsNHNH_2} \text{C}_6\text{H}_5\text{-CH}_2\text{CH=CH}_2 \quad (5.143)$$

$$85\%$$

Some of the unique reducing capabilities of $NaBH_3CN$ are summarized below.

Scheme 5.10

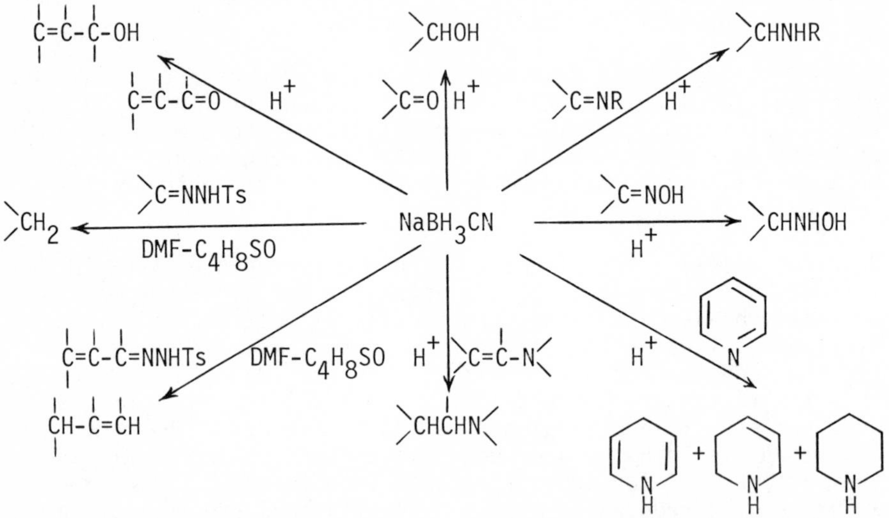

C=C-C-OH   >CHOH   >CHNHR

C=C-C=O H^+ >C=O H^+ >C=NR H^+

>C=NNHTs   >C=NOH
>CH$_2$ ←———— $NaBH_3CN$ ————→ >CHNHOH
   $DMF-C_4H_8SO$ H^+

C=C-C=NNHTs $DMF-C_4H_8SO$ H^+ >C=C-N< H^+

CH-C=CH   >CHCHN<

Trialkylborohydrides (5.162). Trialkylborohydrides have exhibited exceptional synthetic capabilities in the reduction of four types of organic compounds, that is, (1) alkyl halides and sulfonates (5.170), (2) epoxides and other strained cyclic ethers (5.171), (3) aldehydes and ketones (5.172), and (4) conjugated enones (5.173).

The reduction of primary and secondary alkyl halides can most efficiently be achieved with trialkylborohydrides (5.170). Some noteworthy examples follow.

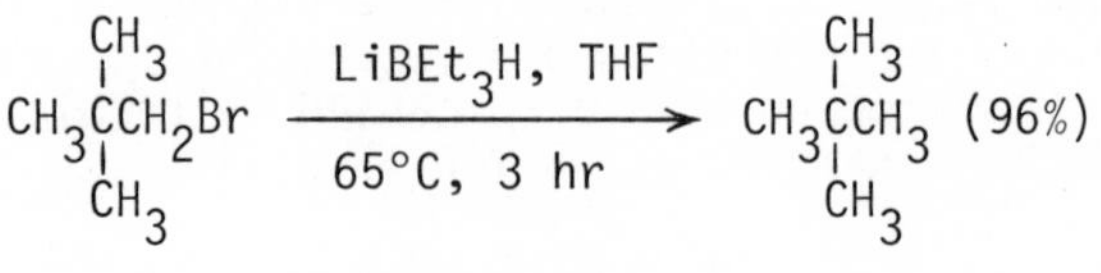

$$CH_3CCH_2Br \xrightarrow[65°C, 3\ hr]{LiBEt_3H,\ THF} CH_3CCH_3 \ (96\%) \quad (5.144)$$

$$\xrightarrow[65°C]{LiBEt_3D,\ THF} \quad (5.145)$$

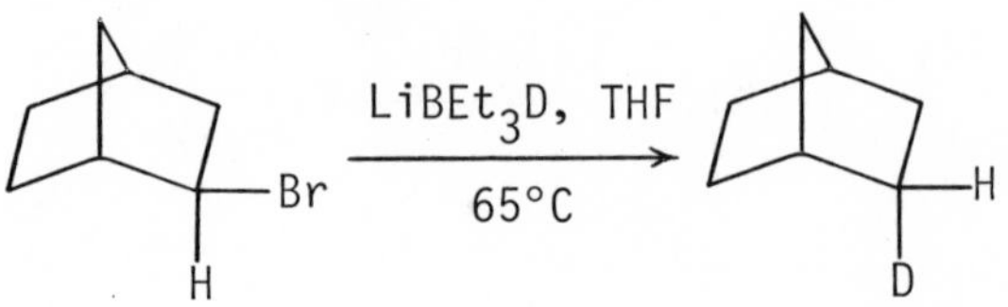

The available data indicate that the reaction is of the S_N2 type. The exceptional nucleophilicity of trialkylborohydrides may be indicated by the following second-order rate constants (1 mole^{-1} min^{-1}) for the reaction of various nucleophiles with n-octyl chloride in THF at 25°C (5.170): LiBEt$_3$H(2 x 10^{-1}), NaSPh(1 x 10^{-2}), LiAlH$_4$(5 x 10^{-3}), and LiBH$_4$(2 x 10^{-5}). Although trialkylborohydrides are inert to aryl halides, they can be activated toward aryl and alkenyl halides by the addition of CuI (5.174). The exceptional nucleophilicity of trialkylborohydrides is also evident in their reactions with alkyl tosylates (5.170b) and with epoxides (5.171).

$$\text{(5.146)}$$

Li$^+$ Me-B-H	88%	12%	-
LiAlH$_4$	54%	25%	19%

LiBEt$_3$H	93%	<0.1%
LiAlH$_4$	15%	85%

$$\text{(5.147)}$$

One of the most remarkable properties of trialkylborohydrides is their unprecedented ability to reduce ketones selectively and consistently from the less hindered side of the carbonyl group (5.172). Although some trialkylborohydrides of complex structure were used in earlier studies (5.172a, 5.172b), steric hindrance appears to be the single most important factor. Lithium trisiamylborohydride (5.172e) is probably the most stereoselective nonenzymatic reagent for the reduction of ketones.

(99.7% cis)   (5.148)

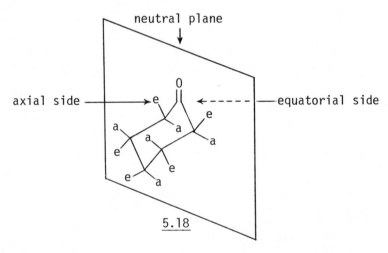

$$\text{(99.6\% trans)} \qquad (5.149)$$

$$\text{(99.0\% cis)} \qquad (5.150)$$

It should be clear that, except in rare situations where some bulky substituents occupy the 2- and/or 6-axial positions, the so-called equatorial side of cyclohexanones, which has only the 2- and 6-axial groups, is the less hindered side. The 2- and 6-equatorial groups as well as the carbonyl group are in the neutral plane, and everything else is on the axial side, as shown in 5.18.

neutral plane

axial side ——————→e ←-----—equatorial side

5.18

Trialkylborohydrides tend to undergo conjugate reduction of enones to form the corresponding regio-specified enolate which can be alkylated regiospecifically, for example, eq. 5.151 (*5.173b*).

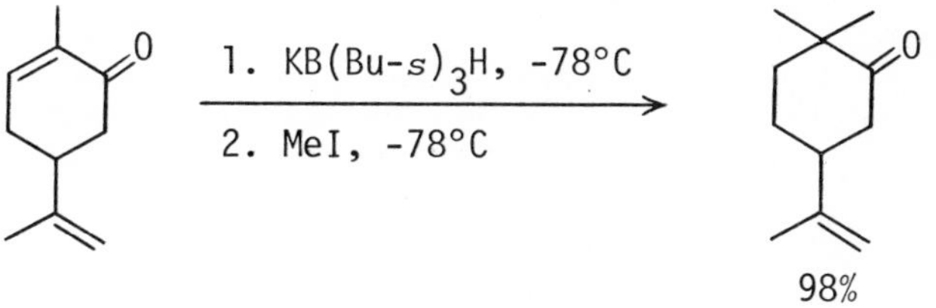

$$\text{1. KB(Bu-}s)_3\text{H, }-78°C$$
$$\text{2. MeI, }-78°C$$

(5.151)

98%

In some cases where the β-carbon atom of enones is relatively hindered, the carbonyl group has been reduced, for example, eq. 5.152 (*5.173a*).

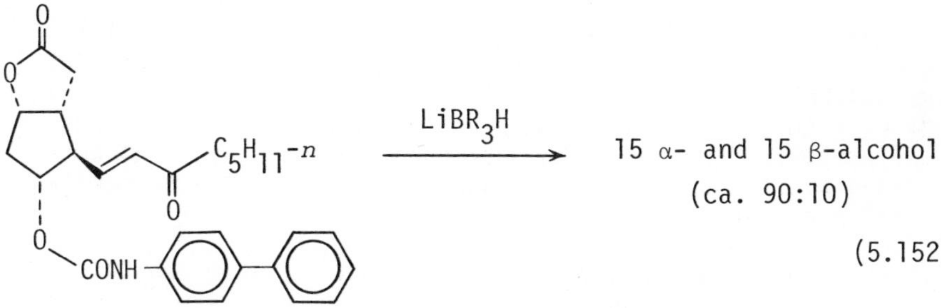

$$\xrightarrow{\text{LiBR}_3\text{H}}$$

15 α- and 15 β-alcohol
(ca. 90:10)

(5.152)

Lithium Dialkyl-9-borabicyclo[3.3.1]nonanates (*5.163*). Lithium di-*n*-butyl-9-borabicyclo[3.3.1]nananate reduces tertiary alkyl, benzylic, and allylic halides without reducing primary and secondary alkyl halides (*5.175*). It is one of the bridge-head hydrogens that acts as a hydride (eq. 5.153). Aldehydes, ketones, and epoxides can also be reduced (*5.176*).

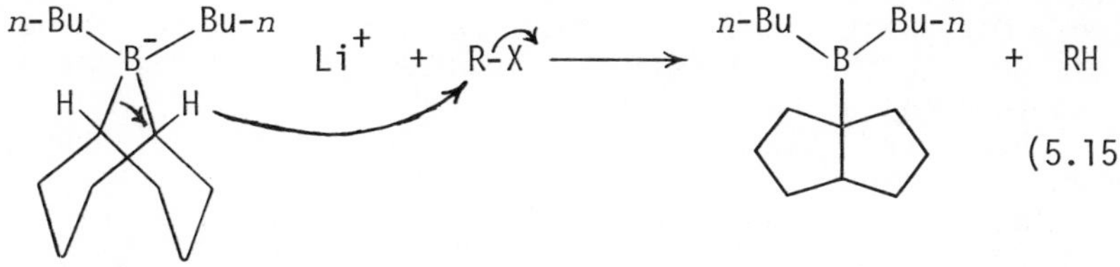

$$\text{Li}^+ + \text{R-X} \longrightarrow$$

+ RH

(5.153)

B-Alkyl-9-BBNs (*5.160*). B-Alkyl-9-BBNs having tertiary C-H bonds in the β-position, for example, B-(2-methylcyclopentyl)-9-BBN reduces aldehydes at 65°C (*5.160*). Benzyl-α-d alcohol of 90% enantiomeric excess has been obtained using B-α-pinanyl-9-BBN of 92%

enantiomeric excess; see eq. 5.154 (*5.160b*).

$$\text{(structure with } B \text{, } H) + Ph\overset{O}{\underset{}{\overset{\|}{C}}}D \longrightarrow Ph\overset{OH}{\underset{H}{\overset{|}{\underset{|}{C}}}}D \qquad (5.154)$$

92% ee   90% ee

5.3 ORGANOALUMINUMS

In addition to a few general treatises (*0.6, 0.8*) that discuss
the chemistry of organoaluminums established through the mid-
1960s, there is an extensive monograph on organoaluminum chemistry published in 1972 (*5.177*). An extensive review in German is
also presented in Houben-Weyl (*5.178*). The synthetic aspects of
organoaluminums have been reviewed by Bruno (*5.179*), Reinheckel
(*5.180*), and Negishi (*5.36*).

5.3.1 Fundamental Properties of the Aluminum Atom and Organoal-
uminum Compounds

Aluminum lies immediately below boron in the Periodic Table and
has the electronic configuration: $3s^2 3p$. As might be expected,
these two elements display similar properties. One of the most
significant properties common to both of these elements is their
ability to form Lewis-acidic sextet species as stable compounds
that have an empty p orbital available for coordination. Such
sextet species, which may exist as dimeric or polymeric aggregates, are called alanes, regardless of their degree of aggregation. Many of the reaction characteristics of organoboranes associated with the availability of the empty p orbital are also
found in organoalanes.

There are, however, a number of significant differences
which must be emphasized here.

The Allred-Rochow (*1.8*) and Pauling (*1.1*) electronegativity
values for aluminum are 1.47 and 1.61, respectively, which are
roughly 0.5 lower than the corresponding values for boron. Thus
the Al-C bond is considerably more ionic than the corresponding
B-C bond. Moreover, the former is considerably longer and more
polarizable than the latter. For example, the Al-C bond lengths
of Me_3Al, which exists as a bridged dimer 5.19, are 1.97 and 2.14
Å (*5.181*). The B-C bond length of Me_3B is 1.57 Å.

Me ⟵ 2.14 Å

Me

1.97 Å ⟶ Al — 75° — Al

Me

Me

Me

5.19

These properties combine to make organoaluminums better carbanion sources than the corresponding organoboranes. Indeed, it
has long been known that organoalanes undergo Grignard-like reactions, that is, intermolecular transfer reactions, with a variety
of organic and inorganic electrophiles. As will become clear
from the discussions presented later in this chapter, the intermolecular transfer is the predominant mode of reaction of organoaluminums.
There are, however, some significant differences between the
intermolecular transfer reactions of organoaluminums and those of
Grignard reagents or organolithiums. Just recall the case of
cross coupling (Sect. 4.3.2.1). The order of reactivity of alkyl
halides toward the highly basic organometallics is: primary >
secondary > tertiary. Moreover, hindered alkyl halides undergo
mainly β-elimination reactions. On the other hand, the order of
reactivity of alkyl halides toward organoalanes is: tertiary >
secondary > primary. In fact, typical primary alkyl halides,
such as ethyl iodide, are quite inert, indicating that the intrinsic nucleophilicity of the Al-C bond of organoalanes is rather low. These results indicate that the reaction of Grignard
reagents and organolithiums is essentially associative or S_N2-
like, whereas that of organoalanes is dissociative or S_N1-like.
Complexation between organoalanes with alkyl halides, in which
the organoalanes act as electrophiles, must be a crucial step of
the cross-coupling reaction of organoalanes, and appears far more
significant than in the corresponding reaction of Grignard reagents and organolithiums. The effect of complexation seems twofold or synergistic. It should increase not only the electrophilicity of the alkyl group of alkyl halides through polarization or ionization, but also the nucleophilicity of the Al-C bond
by weakening it electronically and sterically and making the
bonding electrons more readily available for interaction with electrophiles. This point is further elaborated later.
Alanes react with various bases to form the corresponding
octet species called aluminate or alanate anions as shown in eq.
5.155.

$$\text{(structure)} + \; ^-Y \longrightarrow \text{(structure)} Al^- - Y \qquad (5.155)$$

Recent studies have established that organoaluminates are reasonably good nucleophiles whose reaction characteristics resemble those of organolithiums and Grignard reagents (5.36). In particular, the ready availability of stereodefined alkenylaluminates makes such species uniquely useful, as the synthesis of the corresponding alkenylmetals containing lithium and magnesium is often cumbersome.

Quite interestingly, organoalanes and organoaluminates have rarely participated in the intramolecular transfer via 1,2-migration which dominates the reactions of the corresponding organoboranes and organoborates. The precise reason for this highly contrasting behavior is not yet clear. The ready availability of the intermolecular transfer paths might simply be blocking the intramolecular transfer paths. Alternatively, the $3p$ orbital of aluminum might not participate effectively in the p_π-p_π overlap with the $2p$ orbitals of C, N, and O, which are considerably smaller than the aluminum $3p$ orbital. Whatever the reason may be, there are only a few known 1,2-migration reactions of organoalanes, as exemplified by eq. 5.156 (5.182).

$$\underset{R_2^1 Al}{\overset{H}{\diagdown}} C=C \underset{R^2}{\overset{H}{\diagup}} \xrightarrow{CH_2N_2} R_2^1 Al^- \overset{H}{\diagdown} C=C \overset{H}{\diagup} R^2 \longrightarrow R_2^1 AlCH_2 \overset{H}{\diagdown} C=C \overset{H}{\diagup} R^2 \qquad (5.156)$$

The following tentative generalization may be useful:

1. Organoalanes can act as Lewis acids or electrophiles, as do the corresponding organoboranes.
2. Organoalanes can also act simultaneously as both electrophiles and nucleophiles ("amphophiles") in their reactions with alkyl halides and other types of organic and inorganic compounds.
3. Organoaluminates can act as unique nucleophiles in their reactions with various electrophilic compounds.
4. Organoaluminums appear quite reluctant to participate in the intramolecular transfer.

5.3.2 Preparation of Organoaluminums

5.3.2.1 Preparation of Organoalanes

Oxidative metallation (Method I), hydrometallation (Method VI), and transmetallation (Method XI) represent three general methods for preparing organoaluminums. Before discussing these reactions, however, some of the general organoalane reactions which convert one organoalane into another are discussed briefly.

Disproportionation. Organoalanes are considerably more reactive with respect to disproportionation than the corresponding organoboranes.

$$\rangle Al^1\text{-}R + X\text{-}Al^2\langle \rightleftharpoons \rangle Al^1 \underset{X}{\overset{R}{\langle}} Al^2 \langle \rightleftharpoons \rangle Al^1\text{-}X + R\text{-}Al^2 \langle \quad (5.157)$$

$$X = C, N, O, \text{ or halogen group}$$

Typical examples follow.

$$2R_3Al + AlCl_3 \rightleftharpoons 3R_2AlCl \quad\quad (5.158)$$

$$R_3Al + 2AlCl_3 \rightleftharpoons 3RAlCl_2 \quad\quad (5.159)$$

$$R_3Al + R_3Al_2Cl_3 \rightleftharpoons 3R_2AlCl \quad\quad (5.160)$$

The disproportionation reaction often takes place even under very mild reaction conditions, thereby providing a convenient indirect route to organoalanes. Although it can also represent a problem, the overall structural integrity of organoalanes has seldom been critical in organic synthesis. Consequently, disproportionation is usually not a deleterious factor.

Dissociation and Displacement. Organoalanes are considerably more labile than organoboranes with respect to dissociation into olefins and aluminum hydrides. In fact, it is difficult to prepare pure trialkylalanes uncontaminated with dialkylalanes from di- and polysubstituted olefins. Dissociation must involve the precise reversal of hydroalumination.

$$-\underset{\underset{H}{\overset{|}{C}}}{\overset{|}{C}}-\underset{\underset{Al\langle}{}}{\overset{|}{C}}- \rightleftharpoons \rangle C=C\langle + HAl\langle \quad\quad (5.161)$$

If an unhindered olefin is present in the reaction mixture undergoing dissociation, it can react with the aluminum hydride generated via dissociation. The overall transformation involves displacement of one olefin with another that can form the more stable organoalane of the two.

The displacement reaction provides a useful method for the synthesis of certain organoalanes; see eq. 5.162 (*5.183*).

$$(Me_2CHCH_2)_3Al \ + \ 3H_2C=CH_2 \longrightarrow Et_3Al \ + \ 3Me_2C=CH_2 \quad (5.162)$$

Carboalumination. At elevated temperatures, organoalanes can also undergo addition reactions with olefins, acetylenes, dienes, and polyenes (*5.184*). The reaction often takes place in competition with the displacement reaction.

$$\begin{array}{c}C=C\end{array} \quad \xrightarrow{\text{R-Al}} \quad R-\overset{|}{\underset{|}{C}}-\overset{|}{\underset{|}{C}}-Al \qquad (5.163)$$

$$-C{\equiv}C- \quad \xrightarrow{\text{R-Al}} \quad R-\overset{|}{C}{=}\overset{|}{C}-Al \qquad (5.164)$$

In cases where the organoalane products possess reactivities comparable to those of the starting organoalanes, polymerization involving multiple carbometallation takes place. Thus, for example, ethylene reacts with Et_3Al and n-Pr_3Al, but not with Me_3Al, at 90 to 120°C and 100 atm to form trialkylalanes containing long linear alkyl groups (*5.185*).

$$Al(C_2H_5)_3 \quad \xrightarrow{CH_2=CH_2} \quad Al\begin{array}{l} {\diagup}(C_2H_4)_k C_2H_5 \\ {-\!\!-}(C_2H_4)_l C_2H_5 \\ {\diagdown}(C_2H_4)_m C_2H_5 \end{array} \quad (5.165)$$

The chain length of the alkyl groups is a function of various factors. At 100 atm, the alkyl chain grows to about C_{200} and is detached as a long chain olefin via displacement.

The polymerization reaction can produce trialkylalanes containing C_4 to C_{30} linear alkyl groups by controlling the ratio of ethylene to Et_3Al. Controlled oxidation of the products produces, after hydrolysis, primary alcohols.

The reaction, on the other hand, can be markedly facilitated by the so-called Ziegler-Natta catalysts so as to product polyolefins of high molecular weight (*5.186*). Since this reaction involves the use of transition metal compounds, it is discussed later in more detail (Sect. 10.2).

Carboalumination of olefins is probably of little significance as a method for preparing monomeric organic compounds, since it is generally difficult to suppress undesirable side reactions, such as polymerization and displacement. Moreover, there is little incentive to prepare monomeric organic compounds via carboalumination of olefins. On the other hand, carboalumination of acetylenes offers an attractive possibility of synthesizing tri- or tetrasubstituted olefins in a selective manner. Until recently, however, the synthetic scope of carboalumination of acetylenes had been very limited. Some of the more useful examples are shown below.

Acetylene itself reacts with trialkylalanes, such as Et_3Al and $(i\text{-Bu})_3Al$, to form alkenylalanes (5.187).

$$HC{\equiv}CH \;+\; R_3Al \;\longrightarrow\; RCH{=}CHA1R_2 \qquad (5.166)$$

The synthetic utility of the reaction does not appear to have been well delineated.

The reaction of organoalanes with monosubstituted acetylenes is usually complicated by (1) metallation of acetylenes and (2) formation of the two possible regioisomers (5.188), as exemplified by the reaction shown in eq. 5.167.

$$Et_3Al + HC{\equiv}CPh \xrightarrow{\;70 \text{ to } 90°C\;} \qquad\qquad (5.167)$$

$$Et_2AlC{\equiv}CPh \;+$$
50%

The synthetic utility of the carboalumination of disubstituted acetylenes is severely limited by (1) the lack of high regioselectivity and (2) the competitive dimerization and polymerization. A typical example follows (5.188).

$$Et_3Al + PhC{\equiv}CPh \xrightarrow{\;85°C\;}$$

major   minor   (5.168)

The minor product presumably arises via dissociation, hydroalumination, and carboalumination (eq. 5.169).

$$Et_3Al \longrightarrow Et_2AlH \xrightarrow{PhC\equiv CPh} \begin{array}{c} Ph \\ \diagdown \\ C=C \\ H \diagup \quad \diagdown AlEt_2 \end{array} \quad (5.169)$$

$$\xrightarrow{PhC\equiv CPh} \begin{array}{c} Ph \qquad Ph \\ \diagdown \diagup \\ C=C \qquad AlEt_2 \\ H \diagup \quad \diagdown C=C \diagup \\ \qquad Ph \quad Ph \end{array}$$

Only in some favorable cases is the reaction both regio- and stereoselective (*5.189*).

$$Ph_3Al + MeC\equiv CPh \xrightarrow{90°C} \begin{array}{c} Me \qquad Ph \\ \diagdown \diagup \\ C=C \\ Ph \diagup \quad \diagdown AlPh_2 \end{array} \quad (5.170)$$

Recently, it has been discovered that mono- and disubstituted acetylenes react under very mild conditions with reagents obtained by mixing organoalanes with Cl_2ZrCp_2 to produce the corresponding alkenylmetal derivatives in a highly stereoselective manner; see eq. 5.171 (*5.190*).

$$R^1C\equiv CH(R^2) + R_3^3Al \cdot Cl_2ZrCp_2 \xrightarrow[(CH_2Cl)_2]{25 \text{ to } 50°C} \begin{array}{c} R^1 \qquad H(R^2) \\ \diagdown \diagup \\ C=C \\ R^3 \diagup \quad \diagdown AlR_2^3 \cdot Cl_2ZrCp_2 \end{array} \quad (5.171)$$

This reaction holds considerable promise as a selective route to tri- and tetrasubstituted olefins, as discussed later in more detail (Sect. 10.2).

It should be kept in mind that the organoalane interconversion reactions discussed above can accompany and complicate the preparation of organoalanes with the reactions discussed below.

5.3.2.2 Oxidative Metallation of Organic Halides with Aluminum Metal

Certain organic chlorides, bromides, and iodides, but not fluorides, react with aluminum to produce organoaluminum sesquihalides (*5.191*).

$$6RX \xrightarrow[\substack{\text{neat or} \\ \text{hydrocarbon}}]{4Al} 2 \underset{R'}{\overset{R}{\searrow}}\underset{X}{\overset{}{Al}}\underset{X}{\overset{X}{\searrow}}\underset{X}{\overset{R}{Al}} \underset{}{\rightleftharpoons} \underset{R'}{\overset{R}{\searrow}}\underset{X}{\overset{X}{Al}}\underset{R}{\overset{R}{Al}} + \underset{X'}{\overset{R}{\searrow}}\underset{X}{\overset{X}{Al}}\underset{R}{\overset{X}{Al}}$$

(5.172)

The reaction is analogous to the formation of Grignard reagents from organic halides but is of much more limited scope.

Organic halides which have been successfully converted to organoaluminum sesquihalides include: methyl, ethyl, n-propyl, allylic, propargylic, and benzylic halides. The reaction represents the most convenient route to organoalanes containing some of these groups. It is difficult to use higher alkyl halides. Phenyl iodide reacts with aluminum at 100°C. Aryl bromides and chlorides are unreactive toward ordinary aluminum metal, although they will probably react with the aluminum vapor and powder discussed earlier (Sect. 2.1).

These organoaluminum sesquihalides are satisfactory reagents in most cases. They can also be converted to mono-, di-, and triorganoalanes, however, via disproportionation.

5.3.2.3 Hydroalumination

<u>Hydroalumination with Hydroalanes.</u> Certain aluminum hydrides react with olefins and acetylenes to form the addition products (*5.187*, *5.192*).

$$\overset{\textstyle\diagdown}{\diagup}C=C\overset{\textstyle\diagup}{\diagdown} \; + \; HAl\overset{\textstyle\diagup}{\diagdown} \longrightarrow \; H-\overset{|}{\underset{|}{C}}-\overset{|}{\underset{|}{C}}-Al\overset{\textstyle\diagup}{\diagdown} \qquad (5.173)$$

$$-C{\equiv}C- \; + \; HAl\overset{\textstyle\diagup}{\diagdown} \longrightarrow \; \underset{H}{\overset{\textstyle\diagdown}{\diagup}}C=C\underset{Al\overset{\diagup}{\diagdown}}{\overset{\textstyle\diagup}{}} \qquad (5.174)$$

In most cases, the aluminum hydride reagents used are diorganoaluminum hydrides. Although they can be prepared by the reduction of various organoalanes of the R_2AlX type with saline metal hydrides, such as LiH and NaH, it is the so-called "direct" synthesis that has made certain dialkylaluminum hydrides industrially important chemicals.

The direct synthesis of organoalanes consists of the following two reactions (*5.193*).

$$4R_3Al \; + \; 2Al \; + \; 3H_2 \xrightarrow[\text{100 to 120°C}]{\text{50 to 300 atm}} 6R_2AlH \qquad (5.175)$$

$$6R_2AlH \quad 6 \,\rangle C=C\langle \quad \xrightarrow{60 \text{ to } 80°C} \quad 6R_3Al \qquad (5.176)$$

The net change is shown in eq. 5.177.

$$2Al \; + \; 3H_2 \; + \; 6 \,\rangle C=C\langle \quad \xrightarrow{\qquad} \quad 2R_3Al \qquad (5.177)$$

It is important to note that aluminum metal and hydrogen do not react even under these drastic conditions, unless a trialkylalane is present. The direct synthesis, which is also called the Ziegler synthesis, can be carried out either in one step or two steps. The one-step procedure is, of course, the more desirable of the two. There are, however, at least two side reactions, namely, hydrogenolysis of trialkylalanes and carboalumination of olefins, which must be taken into consideration. Both of these problems can be avoided in the two-step procedure. Triisobutylalane, which does not participate readily in either of the side reactions, is usually prepared by the one-step method, whereas triethylalane requires the two-step procedure.

It should also be noticed that typical alkylalanes are remarkably stable to hydrogen. The hydrogenolysis of R_3Al to form R_2AlH can be achieved only at ≥120°C and about 300 atm, unless catalyzed by some transition metal compounds.

Scope of Hydroalumination. The scope of hydroalumination of olefins is considerably more limited than that of hydroboration, mainly due to (1) competitive dissociation and displacement and (2) competitive carboalumination. The two hydrometallation reactions, however, share some significant common features. Both reactions involve a cis addition which predominantly proceeds in the anti-Markovnikov sense. It is therefore likely that the mechanisms of the two reactions are also very similar (Sect. 2.6.1).

(a) Olefins. The approximate order of reactivity of various types of olefins is: $H_2C=CH_2$ > $H_2C=CHR$ > $H_2C=CRR'$ > $RCH=CHR'$. Many exceptions exist, however. For example, norbornene is more reactive than 1-hexene. Triisobutylaluminum contains a small amount (~10%) of diisobutylaluminum hydride (DIBAH). It is therefore difficult to prepare by hydroalumination pure trialkylalanes, which are sterically more demanding than i-Bu_3Al.

(b) Dienes. The course of hydroalumination of dienes (5.194) is often considerably different from that of the corresponding hydroboration (5.16). The reaction is often followed by the intramolecular carboalumination, which provides an attractive method of forming carbocyclic derivatives, as shown in the following examples, see eqs. 5.178, 5.179 (5.195).

$$\text{(5.178)}$$

$$\text{(5.179)}$$

   (c) Acetylenes. Hydroalumination of acetylenes has proven to be of considerable synthetic interest in recent years, although its scope is much more limited than that of hydroboration. Under carefully controlled conditions, usually at about 50°C, dialkylaluminum hydrides react with acetylenes to give 1:1 cis-addition products (5.187, 5.196).

$$\text{(5.180)}$$

$$\text{(5.181)}$$

The reaction can, however, be complicated by (1) substitution of the methine hydrogen or other heterosubstituents, such as bromine and SnR_3, and (2) competitive carbometallation. Moreover, the regiochemistry and stereochemistry of hydroalumination of internal acetylenes are often nonselective. The representative results are summarized in Table 5.3.

   One of the serious limitations of hydroalumination is its incompatibility with various functional groups. As discussed later, various functional groups are readily reduced by aluminum hydrides. Even most of the etherial functionalities cannot be tolerated, although they are not usually destroyed. This represents a major disadvantage of hydroalumination in relation to other hydrometallation reactions.

Table 5.3.   Regio- and Stereochemistry of Hydroalumination of Acetylenes with DIBAH

Acetylene ($R^1C_\alpha \equiv C_\beta R^2$)	Conditions	Product Distribution (%)			Substitution	Ref.
		cis-α	*cis-β*	*trans-β*		
$n\text{-}C_4H_9C\equiv CH$	50°C	2	88		6	*5.197*
$PhC\equiv CH$	20°C		71		29	*5.198*
$n\text{-}C_4H_9C\equiv CCH_3$	70°C	33	67			*5.200*
$\text{—}{\mid}\text{—}C\equiv CCH_3$	70°C	15	85			*5.200*
$PhC\equiv CCH_3$	50°C	82	18			*5.200*
$PhC\equiv C\text{—}{\mid}\text{—}$	50°C	100	0			*5.199*
$PhC\equiv CBr$	100°C				100	*5.198*
$PhC\equiv CSiMe_3$	60°C,		96	4		*5.198*
	20°C		4	96		*5.198*
$PhC\equiv CSnMe_3$	20°C				100	*5.198*

$$cis\text{-}\alpha = \underset{R_2Al}{\overset{R^1}{\diagup}} C=C \underset{H}{\overset{R^2}{\diagdown}} \qquad cis\text{-}\beta = \underset{H}{\overset{R^1}{\diagup}} C=C \underset{AlR_2}{\overset{R^2}{\diagdown}} \qquad trans\text{-}\beta = \underset{H}{\overset{R^1}{\diagup}} C=C \underset{R^2}{\overset{AlR_2}{\diagdown}}$$

Hydroalumination with Hydroaluminates.   Unlike dialkylalanes (R_2AlH), hydroaluminates, such as $LiAlH_4$ and $LiAl(Bu\text{-}i)_2MeH$, react with internal acetylenes to form the (Z)-isomers via trans addition (*5.201*).

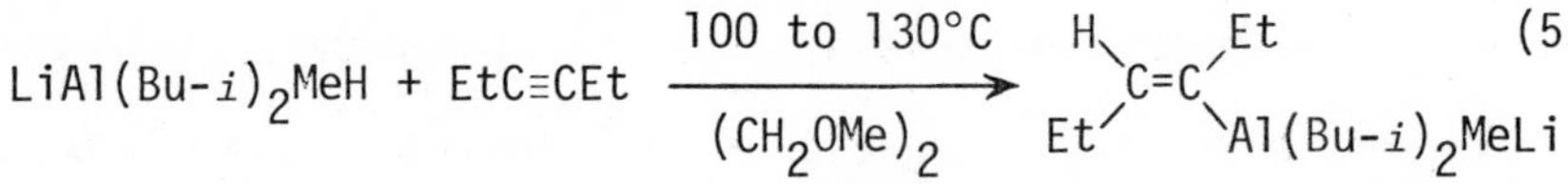

$$\text{LiAl(Bu-}i)_2\text{MeH + EtC≡CEt} \xrightarrow[\text{(CH}_2\text{OMe)}_2]{\text{100 to 130°C}} \begin{array}{c} \text{H} \quad \text{Et} \\ \text{C=C} \\ \text{Et} \quad \text{Al(Bu-}i)_2\text{MeLi} \end{array} \qquad (5.182)$$

The well-known reduction of propargylic alcohols with LiAlH$_4$ to form the trans isomers of allylic alcohols presumably involves the trans hydroalumination (*5.202*).

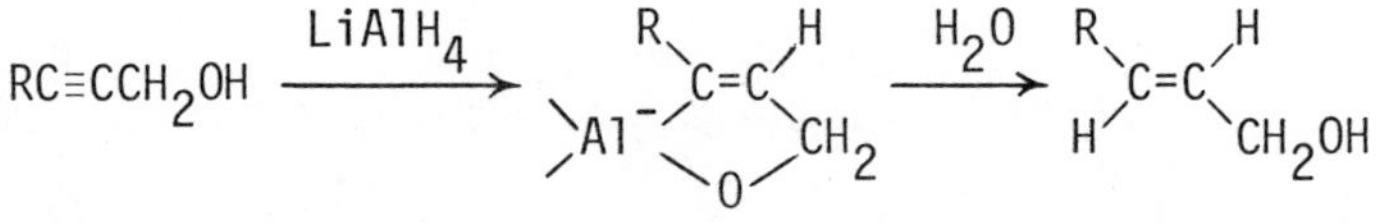

$$\text{RC≡CCH}_2\text{OH} \xrightarrow{\text{LiAlH}_4} \begin{array}{c} \text{R} \quad \text{H} \\ \text{C=C} \\ \text{Al}^{-} \quad \text{CH}_2 \\ \text{O} \end{array} \xrightarrow{\text{H}_2\text{O}} \begin{array}{c} \text{R} \quad \text{H} \\ \text{C=C} \\ \text{H} \quad \text{CH}_2\text{OH} \end{array} \qquad (5.183)$$

<u>5.20</u>

If the reaction involves the intermediacy of alkenylaluminates, such as <u>5.20</u>, it should also be applicable to the synthesis of trisubstituted allylic alcohols as well. Indeed, Corey (*5.203*) has developed such a procedure, which is regio- and stereoselective, and applied it to the synthesis of various natural products, such as farnesol (<u>5.21</u>), juvenile hormone (<u>5.22</u>), and a synthetic precursor (<u>5.23</u>) of α-santalol.

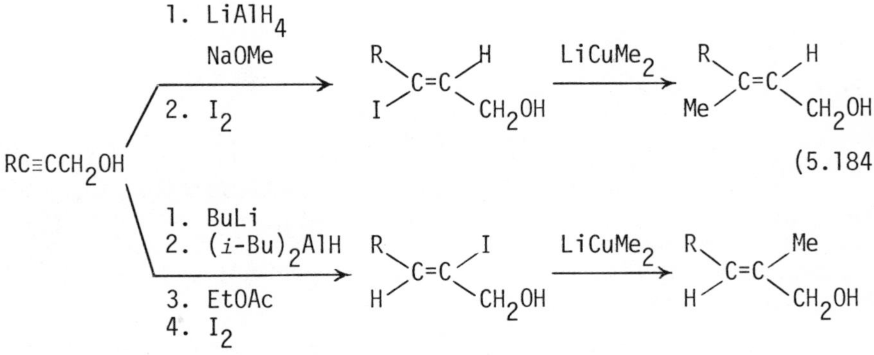

(5.184)

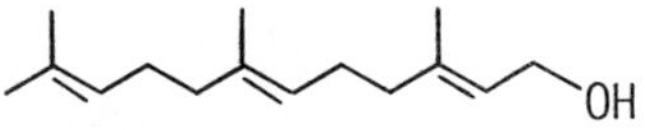

<u>5.21</u>

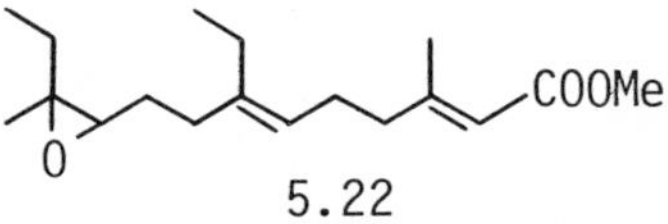

<u>5.22</u>

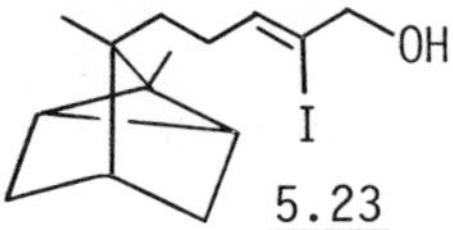

<u>5.23</u>

   One of the major limitations of the trans hydroalumination of acetylenes is that the terminal acetylenes cannot be used in this reaction, since their reaction with hydroaluminates produces alkynylaluminates. A similar reaction also occurs with hydroalanes complexed with tertiary amines (5.179).

$$NaAlR_3H \quad + \quad HC{\equiv}CR^1 \quad \xrightarrow{\text{50 to 60°C}} \quad NaAlR_3(C{\equiv}CR^1) \qquad (5.185)$$

$$R_2AlH{\cdot}NR_3^1 \quad + \quad HC{\equiv}CR^2 \quad \xrightarrow{\text{20 to 40°C}} \quad R_2AlC{\equiv}CR^2{\cdot}NR_3^1 \qquad (5.186)$$

5.3.2.4 Organoalanes via Transmetallation and Organoaluminates via Complexation

Organoalanes that cannot be prepared via oxidative metallation or hydroalumination can most conveniently be prepared by treating organometallics containing alkali metals and magnesium with aluminum halides (eq. 5.187).

$$RM \quad + \quad XAlY_2 \quad \longrightarrow \quad RAlY_2 \quad + \quad MX \qquad (5.187)$$

$$M = Li, Na, K, Mg, \text{ and so on}$$

Typical examples are shown below.

$$\text{⟨C}_6\text{H}_5\text{⟩–Li} \quad + \quad ClAl(Bu\text{-}i)_2 \quad \longrightarrow \quad \text{⟨C}_6\text{H}_5\text{⟩–Al}(Bu\text{-}i)_2 \qquad (5.188)$$

$$3CH_2{=}CHMgBr \quad + \quad AlCl_3 \quad \xrightarrow{OEt_2} \quad (CH_2{=}CH)_3Al{\cdot}OEt_2 \qquad (5.189)$$

$$3RC{\equiv}CNa \quad + \quad AlCl_3 \quad \xrightarrow{OEt_2} \quad (RC{\equiv}C)_3Al{\cdot}OEt_2 \qquad (5.190)$$

$$3RC{\equiv}CLi \quad + \quad AlCl_3 \quad \xrightarrow{\text{Hexane}} \quad (RC{\equiv}C)_3Al \qquad (5.191)$$

The immediate products of the reaction of alanes with polar organometallics are organoaluminate complexes ($MAlRXY_2$). If they decompose, organoalanes ($RAlY_2$) are formed. If they do not, the organoaluminates are obtained as the final products. As in the case of the synthesis of organoborates, this complexation method represents by far the most general route to organoaluminates. It is also possible to obtain organoaluminates by treating organo-alanes with either organometallics or inorganic salts; see eq. 5.192 (*5.204*).

$$MR + XAlY_2$$
$$MX + RAlY_2$$
$$\longrightarrow M[R-AlXY_2] \qquad (5.192)$$

A few representative examples are shown below.

$$LiH + AlEt_3 \longrightarrow LiAlEt_3H \qquad (5.193)$$

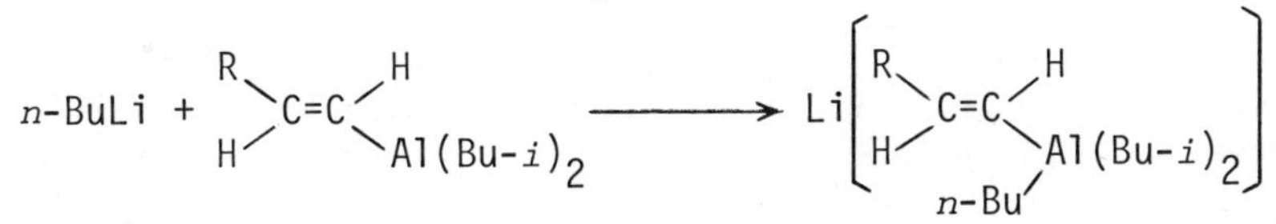

$$n\text{-BuLi} + \underset{H}{\overset{R}{\diagdown}} C=C \underset{Al(Bu-i)_2}{\overset{H}{\diagup}} \longrightarrow Li\left[\underset{H}{\overset{R}{\diagdown}} C=C \underset{\underset{n\text{-Bu}}{Al(Bu-i)_2}}{\overset{H}{\diagup}} \right] \qquad (5.194)$$

Finally, certain ether-free organoalanes can be prepared by treating organomercuries with aluminum metal (Method XII, Sect. 2.12), as in the following synthesis of trivinylalane (*5.205*).

$$3(CH_2=CH)_2Hg + 2Al \xrightarrow[-20°C]{Pentane} 2(CH_2=CH)_3Al + 3Hg \quad (5.195)$$

5.3.3   Reactions of Organoaluminums

The general patterns of the reactions of organoaluminum compounds are briefly discussed in Sect. 5.3.1. Various specific reactions of organoaluminums of synthetic interest are presented in this section.

5.3.3.1   Formation of Carbon-Hetero Atom Bonds via Organoaluminums

Reactions with Nonmetallic Reagents.   The Al-C bond can be cleaved readily with a variety of inorganic reagents to form the corresponding compounds containing the carbon-hetero atom bonds. These reactions represent a convenient and general means of converting organoaluminums into organic compounds. Some of the more useful transformations are summarized in Scheme 5.11 (*5.179*).

Scheme 5.11

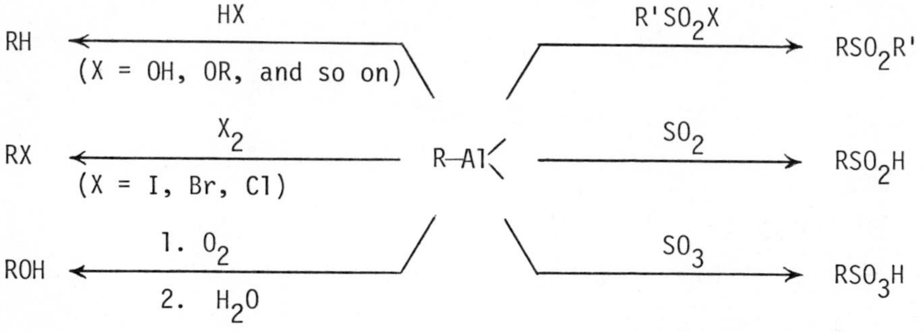

$$RH \xleftarrow{\text{HX}} \quad (X = OH, OR, \text{ and so on})$$

$$RX \xleftarrow{X_2} \quad (X = I, Br, Cl) \quad R\text{--}Al\!\!<$$

$$ROH \xleftarrow[\text{2. } H_2O]{\text{1. } O_2}$$

$$\xrightarrow{R'SO_2X} RSO_2R'$$

$$\xrightarrow{SO_2} RSO_2H$$

$$\xrightarrow{SO_3} RSO_3H$$

Under appropriate conditions these reactions provide the indicated products in high yields.

Reactions with Metallic Reagents. The aluminum-containing moiety of organoalanes can also be replaced with various metal-containing groups via transmetallation; see eq. 5.196 (*5.206*).

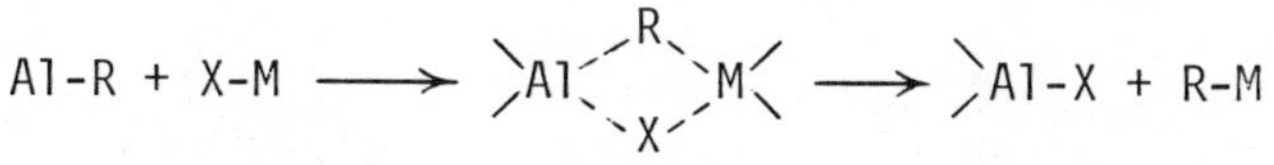

$$Al\text{-}R + X\text{-}M \longrightarrow \; >\!Al\!\!<^{\!R}_{\!X}\!\!>\!M\!\!< \; \longrightarrow \; >\!Al\text{-}X + R\text{-}M \qquad (5.196)$$

$$\underline{5.24}$$

The most noteworthy example of these types of reactions is the formation of an active catalyst in the Ziegler-Natta polymerization reaction, which is believed to be either organotitanium species or Al-Ti complexes represented by 5.24 (M = Ti); see Sect. 10.2.

Organometallics that can be prepared by this method include those that contain Be, Zn, Hg, B, Ga, Si, Sn, Pb, Ti, Zr, Ni, and Ru (*5.206*). The alkylating ability of organoalanes are, however, generally considerably lower than that of the corresponding alkali metal derivatives. Thus if one is only interested in the preparation of organometallics containing other metals as discrete products, there is little advantage in using organoalanes as starting material over organoalkali metals and Grignard reagents.

The above argument does not apply to the cases where only equilibrium amounts of organometallics are to be generated as intermediates. Recent studies in the author's laboratory indicate that alkenylalanes, readily obtainable via hydroalumination of acetylenes, react with nickel and palladium complexes to form the corresponding alkenyl derivatives of nickel and palladium that undergo reductive elimination to form olefins; see eq. 5.197 (5.207).

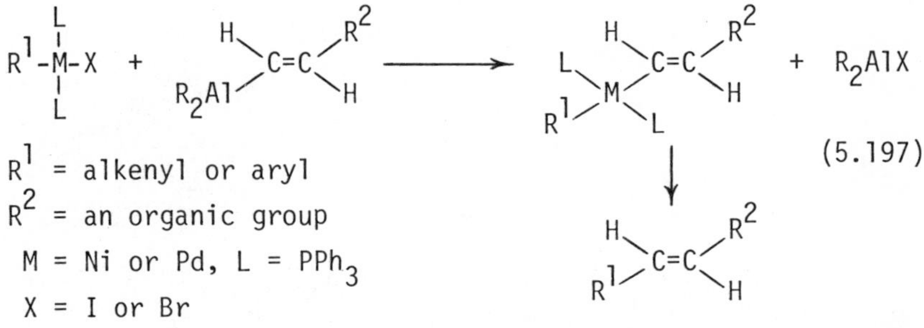

R^1 = alkenyl or aryl

R^2 = an organic group

M = Ni or Pd, L = PPh$_3$

X = I or Br

(5.197)

It is important to notice that the preparation of the corresponding alkenyl derivatives of lithium and magnesium is more cumbersome. Moreover, for reasons as yet unclear, the use of organolithiums and Grignard reagents in the nickel- or palladium-catalyzed cross-coupling reaction frequently fails to provide the desired products in high yields. The transmetallations of organoalanes with transition metal compounds is discussed later in greater detail (Sect. 9.2 and 9.3).

5.3.3.2 Carbon-Carbon Bond Formation via Organoaluminums

As discussed earlier, both organoalanes and organoaluminates react with a variety of carbon electrophiles. Organoalanes can act as "amphophilic" species (species that act as both a nucleophile and an electrophile in a given reaction), and often nicely complement the synthetic capabilities of the nucleophilic organoaluminates.

A general scheme for some representative reactions of organoaluminums with carbon electrophiles is shown below. Some of the products often react further. Clearly, organoaluminums act as "Grignard-like" reagents in these reactions, although the mechanistic details and outcomes of these reactions are often unique. Some of the unique features of these reactions are discussed in this section.

Scheme 5.12

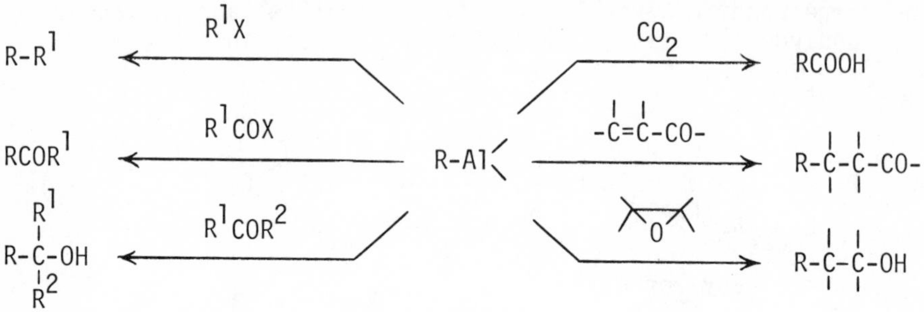

Reactions of Alkyl Halides and Sulfonates. (a) With organoalanes. Typical primary and secondary alkyl halides do not react
readily with organoalanes. On the other hand, tertiary alkyl
halides and secondary alkyl sulfonates react with certain organoalanes to form cross-coupled products (5.208).
   A few synthetically useful examples are shown below (5.209).

$$R-\underset{\underset{Me}{|}}{\overset{\overset{Me}{|}}{C}}-Cl \;+\; Me_3Al \;\xrightarrow[-78°C]{MeCl}\; R-\underset{\underset{Me}{|}}{\overset{\overset{Me}{|}}{C}}-Me \;+\; Me_2AlCl$$

(5.198)

(R = Me, Et, i-Pr, and t-Bu)

Particularly noteworthy is the reaction of alkynylalanes with
tertiary alkyl halides and secondary alkyl sulfonates, which permits a clean t-alkyl-alkynyl coupling (5.210).

$$RX \;+\; (R'C{\equiv}C)_3Al \;\xrightarrow{(CH_2Cl)_2,\;0°C}\; RC{\equiv}CR'$$

(5.199)

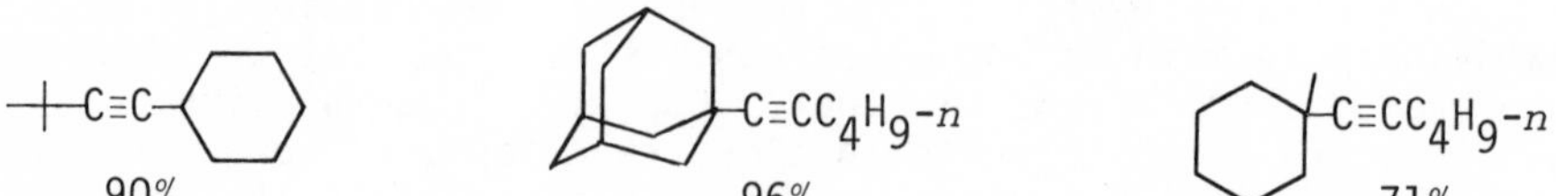

90%   96%   71%

The corresponding reaction of alkenylalanes has been practical
only with secondary alkyl sulfonates. The reaction, however, is
complicated by (1) stereochemical scrambling and (2) rearrangement of the secondary alkyl group to form a tertiary alkyl group,
when possible. The reaction most probably involves ionization of

the alkyl sulfonate to form an ion pair as a transient species. These results suggest the following addition-elimination path for the related alkyl-alkynyl coupling reaction.

$$\alpha\text{-attack} \qquad R'-C^+=C-Al\overset{\diagup}{\underset{\diagdown}{-}}X$$
$$\qquad\qquad\qquad\qquad\qquad\overset{|}{R}$$

$$R^+ + R'C\equiv CAl\overset{\diagup}{\underset{\diagdown}{-}}X \qquad\longrightarrow R'C\equiv CR \quad (5.200)$$

$$\beta\text{-attack} \qquad R'-C=C^+-Al\overset{\diagup}{\underset{\diagdown}{-}}X$$
$$\qquad\qquad\qquad\qquad\overset{|}{R}$$

In any event, the ability of organoalanes to undergo coupling reactions with hindered alkyl halides appears to exceed that of any other organometallics. This area clearly deserves further exploration.

Organoalanes also react with certain alcohols and their derivatives other than sulfonates. The reaction of tertiary alcohols requires high temperatures (80 to 200°C), and its scope with respect to the structure of organoalanes appears limited (5.211).

$$Ph_2MeCOH \xrightarrow[\text{benzene, reflux}]{Me_3Al} Ph_2CMe_2 \qquad (5.201)$$
$$\qquad\qquad\qquad\qquad\qquad\qquad\qquad 62\%$$

Various derivatives of allylic alcohols react readily with organoalanes, providing a promising procedure for allyl-alkyl coupling; see eq. 5.202 (5.212).

$$(5.202)$$

5.25   5.26

X	R	Total Yield, %	5.25	:	5.26
P(O)(OEt)$_2$	Me	>90	90	:	10
Ac	Me	74	92	:	8

Ac	Bu-*i*	66	97	:	3
CO$_2$Et	Bu-*i*	73	97	:	3
OTHP	Me	80	96	:	4
OTHP	Bu-*i*	3	-		

(b) With organoaluminates. The scope of the reaction of organoaluminates with organic halides appears quite limited. The reaction of alkenylaluminates with highly reactive primary alkyl halides, however, provides a facile stereoselective route to trans-disubstituted olefins; see eq. 5.203 (*5.213*).

$$R^1C{\equiv}CH(R^2) \xrightarrow[\quad 2. \quad RLi \quad]{1. \ i\text{-}Bu_2AlH} \quad \underset{H}{\overset{R^1}{\diagdown}}C{=}C\underset{Al(Bu\text{-}i)_2RLi}{\overset{H(R^2)}{\diagup}}$$

R^3X = MeI (60 to 65%),

     n-C$_8$H$_{17}$I (49%),

     PhCH$_2$Br (46%),

     H$_2$C=CHCH$_2$Br (50 to 73%)

$$\Big\downarrow R^3X \quad \underset{H}{\overset{R^1}{\diagdown}}C{=}C\underset{R^3}{\overset{H(R^2)}{\diagup}} \quad (5.203)$$

Reactions of Carbonyl Compounds. Just as organolithiums and Grignard reagents do, organoaluminums can participate in (1) alkylation, (2) reduction, and (3) enolization of various carbonyl compounds, such as aldehydes, ketones, carboxylic acids and their derivatives, and nitriles (*5.214*). In addition, organoaluminums undergo conjugate addition reactions with α,β-unsaturated carbonyl compounds. In this section, we focus our attention on the alkylation and conjugate addition reactions. In general, trialkylalanes possess a greater alkylating ability than alkylaluminum halides. Alkylaluminum alkoxides are even less reactive. Branched alkylalanes, such as triisobutylalane, are poor alkylating agents, and mostly undergo reduction.

(a) Reaction of acyl halides. The reaction of acyl halides with organoalanes represents one of the most convenient methods for the synthesis of ketones via acyl halides, along with the related reactions of organometallics containing Cd (Sect. 4.3.2.2), B (Sect. 5.2.3.3), Cu (Sect. 9.5), and Fe (Sect. 11.2).

$$R^1COX + R^2_nAlX_{3-n} \longrightarrow [R^1CO]^+[R^2_nAlX_{4-n}]^- \longrightarrow R^1COR^2$$

$$R^1 = \text{alkyl, aryl, alkenyl,} \quad R^2 = \text{alkyl or aryl} \qquad (5.204)$$

The reaction presumably proceeds via an acylium ion, as shown in eq. 5.204 (5.215). Thus organoalanes presumably act as "amphophilic" reagents. Acyl chlorides and monoorganoaluminum dichlorides ($RAlCl_2$) are most commonly used. Tri- and diorganoalanes also react, but afford only partial utilization of the organic groups. Moreover, these reagents have a greater tendency to react with the ketone products. The solvent of choice is methylene chloride. Aromatic hydrocarbons have also been used, but they tend to undergo Friedel-Crafts acylation. The reaction of organoalanes with acyl halides is chemoselective and general with respect to acyl halides. The scope of the reaction with respect to organoalanes is not clear, as most of those which have been used are alkylalanes.

(b) Reactions of aldehydes and ketones. Although organoaluminums can react with these carbonyl compounds to give addition products, these reactions are often complicated by reduction of carbonyl compounds and other side reactions and are generally inferior to the corresponding reactions of organolithiums and other polar organometallics. In some cases, however, organoaluminums offer certain unique advantages, as shown in the following examples.

The reaction of *trans*-alkenylaluminums with aldehydes produces stereodefined allylic alcohols. Organoaluminates (5.216) appear generally superior to the corresponding organoalanes (5.217). These reactions proceed with nearly complete retention of configuration.

30 to 50%

68 to 73%

(5.205)

Similarly, the reactions of alkenylaluminates with CO_2 (*5.216*), $(CN)_2$ (*5.218*), and ClCOOR (*5.219*) proceed with retention (Scheme 5.13). The corresponding reactions of alkenylalanes give much poorer results.

Scheme 5.13

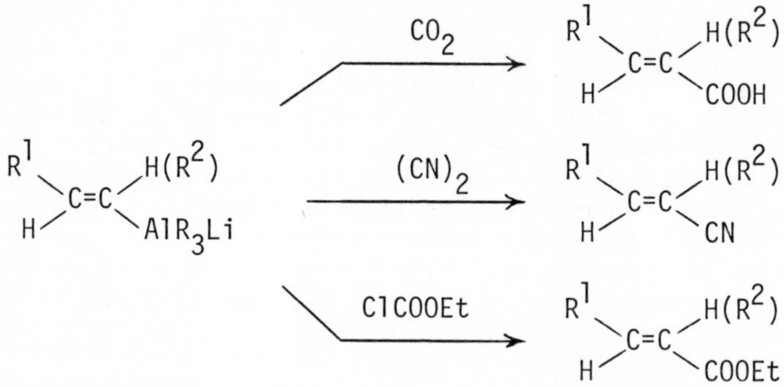

The stereochemistry of the reaction of cyclic ketones with organoalanes varies widely depending on the R_3Al/ketone ratio, thereby permitting the formation of certain stereoisomers which are obtainable only as minor products in the corresponding reactions with organolithiums and Grignard reagents (*5.220*).

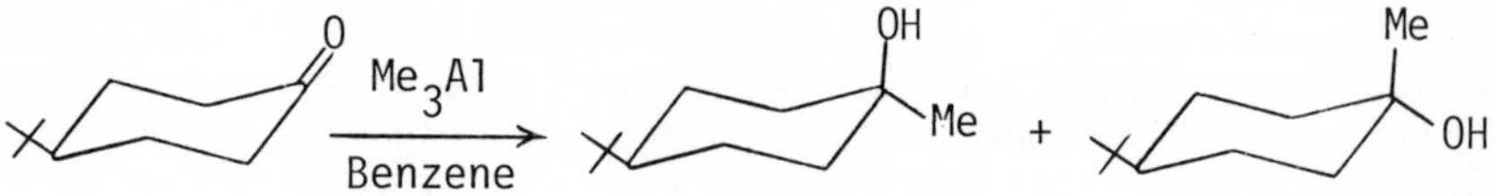

(5.206)

Me_3Al/Ketone	Axial	Equatorial
1.0	76	24
2.0	17	83
3.0	12	88

Other carbonyl and related compounds, such as esters and nitriles, cannot readily be alkylated but are readily reduced by various organoalanes, such as triisobutylalane.

(c) Exhaustive alkylation. A limited number of oragnoalanes, such as Me_3Al, can undergo a unique exhaustive alkylation of ketones, aldehydes, carboxylic acids, and carbinols at elevated temperatures of 100 to 200°C (*5.221*, *5.222*).

Scheme 5.14

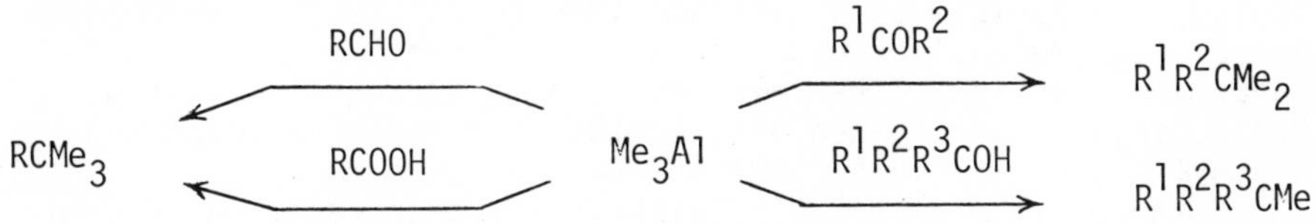
RCHO
RCOOH
$RCMe_3$
Me_3Al
R^1COR^2
$R^1R^2R^3COH$
$R^1R^2CMe_2$
$R^1R^2R^3CMe$

Reactions of α,β-Unsaturated Carbonyl Compounds. (a) With organoalanes. Organoalanes appear to react with conjugated enones by various mechanisms. Some of the reactions appear to proceed by a free-radical mechanism, shown in eq. 5.208 (*5.223*), as in the corresponding reaction of organoboranes while others occur only when enones can assume the cisoid conformation, indicating the cyclic transition state represented by 5.27; see eq. 5.209 (*5.224*).

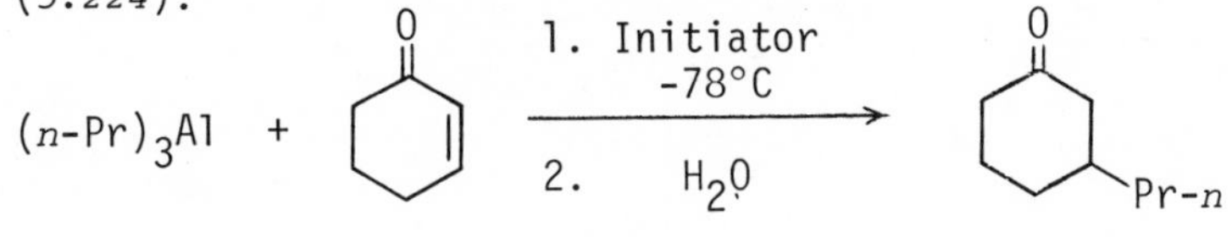
$(n\text{-}Pr)_3Al$ +
1. Initiator -78°C
2. H_2O
Pr-*n*
(75%)

(5.207)

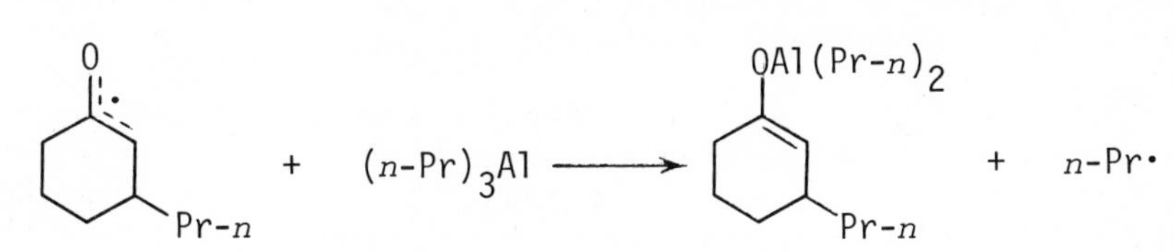
$(n\text{-}Pr)_3Al$ + Int· → *n*-Pr· + $(n\text{-}Pr)_2AlInt$
n-Pr· +
Pr-*n*
+ $(n\text{-}Pr)_3Al$
$OAl(Pr\text{-}n)_2$
Pr-*n*
+ *n*-Pr·

(5.208)

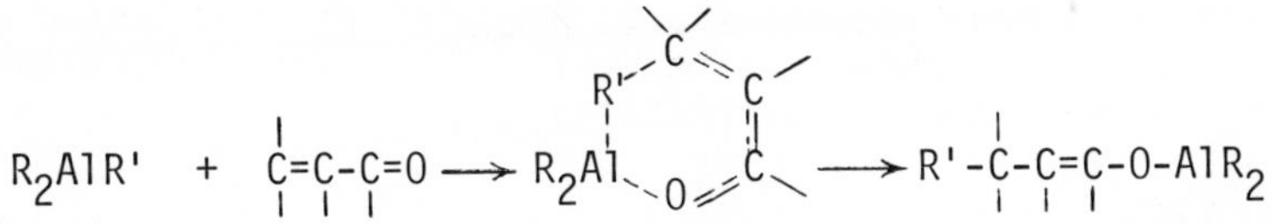
R_2AlR' + C=C-C=O
R_2Al
R'
$R'\text{-}C\text{-}C{=}C\text{-}O\text{-}AlR_2$

5.27

(5.209)

$$R = \text{Et or Bu-}i, \qquad R' = \text{alkenyl or alkynyl}$$

These results closely parallel those obtained more recently with alkenyl- and alkynyl-9-BBNs (Sect. 5.2.3.3). However, the scope of the free-radical conjugate addition reaction of organoboranes appears considerably broader than that of the corresponding organoalane reaction. The relative advantages and disadvantages of organoboranes and organoalanes in the transfer of alkenyl and alkynyl groups are less clear.

Although the alkenyl and alkynyl transfer reaction of organoalanes fails with simple transoid enones, such as cyclohexenone, producing only the 1,2-addition products, certain suitably functionalized transoid enones evidently react via 1,4-addition, as shown in eq. 5.210 (5.225). The stereochemistry of the products strongly suggests that the hydroxy group actively participates in this reaction.

(5.210)

The reaction of organoalanes, such as Me_3Al, with enones can also be markedly catalyzed by transition metal compounds, such as $Ni(acac)_2$, although the synthetic utility of the nickel-catalyzed reaction has not been well delineated (5.226).

(b) With organoaluminates. Certain transoid enones undergo a conjugate addition reaction with alkenylaluminates; see eq. 5.211 (5.227). No 1,2-addition product is formed. On the other hand, the corresponding reaction of alkenylalanes gives only the 1,2-products. These results should be carefully contrasted with those in eq. 5.210. Unfortunately, little is known about the mechanistic details of these reactions.

$$(5.211)$$

(c) The Nagata hydrocyanation reaction. The most extensively developed conjugate addition reaction of organoalanes by far
is that of dialkylcyanoalanes, such as Et_2AlCN, discovered and
developed by Nagata; see eq. 5.212 (*5.228*).

$$(5.212)$$

Although the reaction is restricted to the introduction of the
cyano group, it is as general as the conjugate addition reaction
of organocuprates (Sect. 9.4) with respect to the structure of
enones. The required reagent, for example, Et_2AlCN, may be either preformed or generated in situ by the reaction of $RAlX_2$ with
HCN. The reaction is carried out in either hydrocarbons, such as
toluene, or ethers, such as THF and dioxane, the reaction rate
being considerably faster in toluene than in THF. Although the
use of HCN is somewhat inconvenient, the observed results are
generally far superior to those observed in other conventional
hydrocyanation reactions that use more ionic metal cyanides, such
as KCN, which are often complicated by (1) base-catalyzed condensation, (2) hydrolysis, and (3) solvent participation.
    The precise mechanism of the Nagata hydrocyanation is un-
clear. At least in some cases, however, the kinetically favored
products are the 1,2-adducts which subsequently isomerize to give
the 1,4-adducts. It is therefore likely that the successful
formation of the 1,4-adducts largely, if not entirely, rests on
the following. Firstly, the 1,4-adducts are thermodynamically
more stable than the corresponding 1,2-adducts. Secondly, the
reversal of the 1,2-adduct formation is facile under the reaction
conditions.
    The stereochemistry of the reaction is somewhat difficult to
predict, although, in most cases, the predominant isomer appears
to be the one that is the more stable of the two. The following
examples (*5.228*) may be of help in predicting the stereochemistry
of a given reaction.

1. Polycyclic octalenones with a terminal ring enone ⟶ trans + cis (trans > cis)

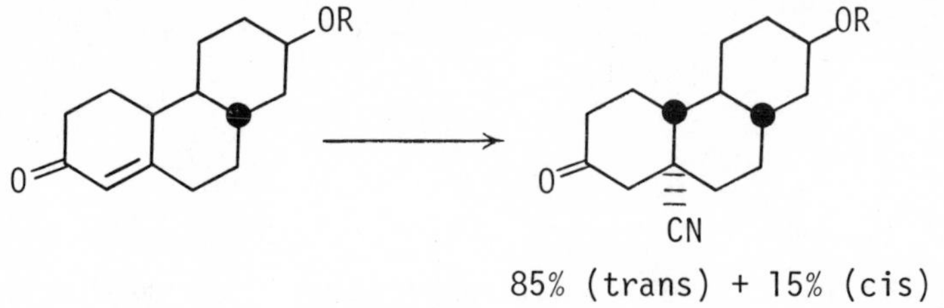

(5.213)

85% (trans) + 15% (cis)

2. Polycyclic methyloctalenones with a terminal ring enone ⟶ trans + cis (trans ≥ cis)

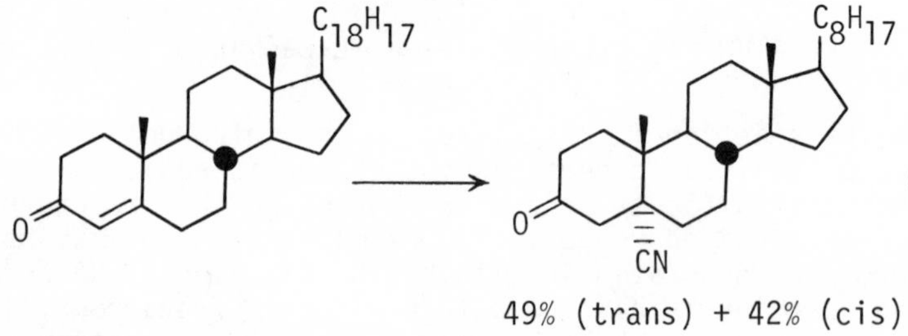

(5.214)

49% (trans) + 42% (cis)

3. Polycyclic compounds with a trans-fused internal enone ⟶ trans

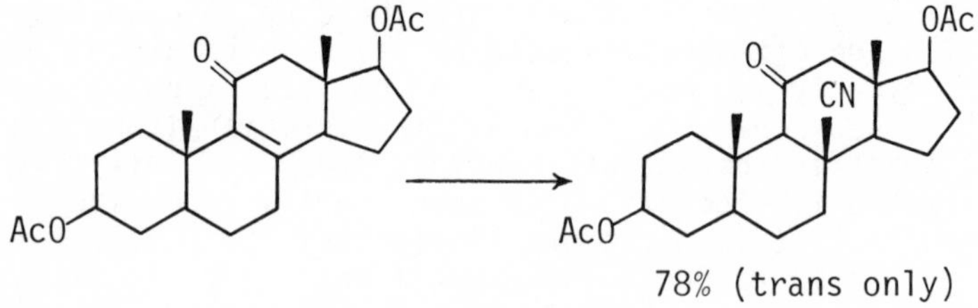

(5.215)

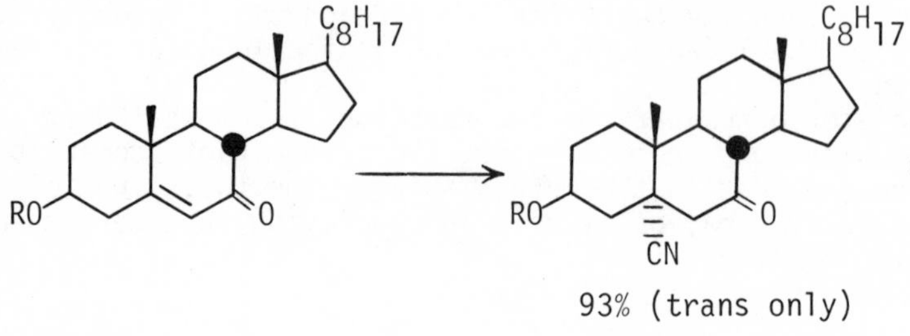

78% (trans only)

(5.216)

93% (trans only)

4. Polycyclic acetylhydrindenes
    ⟶ trans + cis (trans >> cis)

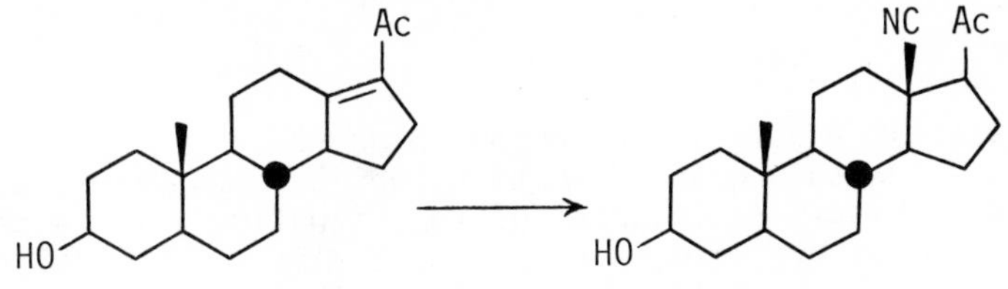

(5.217)

80% (trans only)

5. Polycyclic hydrindeneones
    ⟶ trans + cis (cis >> trans)

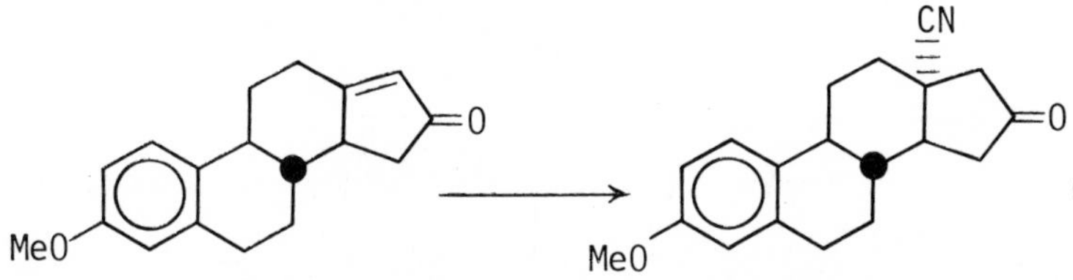

(5.218)

65% (cis) + 16% (trans)

Reactions of Epoxides. While organolithiums (Sect. 4.3.2.3) and
organocoppers (Sect. 9.5) remain the two most useful classes of
organometallic reagents in the reaction with epoxides, organoaluminums have exhibited some unique synthetic capabilities.
     The reaction of sterically unhindered epoxides with organo-
alanes tends to be complicated by (1) polymerization of epoxides,
(2) formation of the two possible regioisomers, and (3) reductive
opening of epoxides, although it can lead to the formation of the
Markovnikov products in high yields as shown in the following examples (5.179).

$$H_2C\overset{}{\underset{O}{\smile}}CHCH_3 \quad \xrightarrow[\text{80°C, 48 hr}]{\text{excess } Et_3Al} \quad HOCH_2\overset{Et}{\underset{|}{C}HCH_3}$$ (5.219)

98%

On the other hand, their reaction with organoaluminates cleanly
gives the anti-Markovnikov products (5.299). The unique advantage of the use of readily obtainable trans-alkenylaluminates
should be noted (eq. 5.220). Unfortunately, however, hindered
epoxides, such as cyclohexene oxide, do not react readily with
organoaluminates.

$$\text{(5.220)}$$

The available data indicate that hindered epoxides react more favorably with organoalanes than the corresponding organo-aluminates. This is reminiscent of the situation with respect to the reaction of alkyl halides discussed earlier. Organoalanes must be acting as amphophilic reagents in their reaction with epoxides as well.

The reaction of alkynylalanes with epoxides developed mainly by Fried (5.230) is particularly useful and often highly regio- and stereoselective, as indicated by the following examples (eq. 5.221).

$$\text{(5.221)}$$

$n = 2$	60%	0%
$n = 3$	50%	0%
$n = 7$	10%	50%

5.3.3.3 Reduction with Organoaluminums

Both acidic and basic aluminum hydrides can act as excellent reducing agents (5.179, 5.231). These reagents may be classified as follows:

Acidic aluminum hydrides -- AlH_3, H_2AlX, and $HAlX_2$

Basic aluminum hydrides -- MAl_4, $MAlH_3X$, $MAlH_2X_2$, and $MAlHX_3$

$X = C, N, O, S$, and halogen, $M = Li, Na$, and so on

In addition to these reagents, organoalanes that do not contain any Al-H bond, such as $(i\text{-Bu})_3Al$, can frequently act as reducing agents, as briefly mentioned earlier (Sect. 5.3.3.2).

By far the most widely used organoaluminum reducing agent is (i-Bu)$_2$AlH (DIBAH). While the use of nonhydridic organoalanes, such as (i-Bu)$_3$Al, might be advantageous in some special cases, such examples seem rare. As yet, relatively little is known about the unique capabilities of the organo-substituted complex aluminum hydrides. Consequently, our discussion in this section is restricted to some of the unique reducing capabilities of DIBAH. Its reaction with olefins and acetylenes is discussed in Sect. 5.3.2.3. Some representative results of the reduction with DIBAH are summarized in Table 5.4.

Table 5.4.   Reduction of Organic Compounds with DIBAH (5.179)

Organic Compound	Amount of DIBAH (eq)	Product	Typical Yields (%)
RCHO	1	RCH$_2$OH	High
RR'CO	1	RR'CHOH	High
RCOOH	1.35	RCHO	40 to 70
	2.35	RCH$_2$OH	60 to 90
RCOOR'	1	RCHO	70 to 90
	2	RCH$_2$OH	70 to 100
RCOCl	2	RCH$_2$OH	60 to 70
RC(OR')$_3$	1	RCH(OR')$_2$	90 to 95
RCONR'R"	2 to 3	RCH$_2$NR'R"	75 to 95
RCONHR'	3 to 4	RCH$_2$NHR'	65 to 95
RCN	1	RCHO	80 to 95
	2	RCH$_2$NH$_2$	60 to 90
RCH=NR'	1	RCH$_2$NHR'	70 to 95
R–⟨triangle⟩O	1	R(CH$_2$)$_2$OH and RCH(OH)CH$_3$	High

The following transformations are usually best carried out with DIBAH. Carboxylic acids, esters, lactones, and nitriles are readily converted to aldehydes with DIBAH, as exemplified by the following; see eq. 5.222 (*5.232*).

$$(5.222)$$

The observed success presumably depends on the relatively high stability of the reduction product 5.28 which does not readily decompose to the corresponding aldehyde in the reduction step, but can readily be hydrolyzed to form the aldehyde (eq. 5.223).

$$(5.223)$$

Selective reduction of carbonyl compounds and nitriles in the presence of olefins, conjugated or isolated, can readily be achieved with DIBAH as shown in the following example (eq. 5.224) (*5.179*).

$$(5.224)$$

REFERENCES

5.1 Chao, L. C., and Rieke, R. D., *J. Org. Chem.*, <u>40</u>, 2253 (1975).

5.2 Uhm, S. J., and Rieke, R. D., *Synthesis*, 452 (1975).

5.3 Brown, H. C., *Boranes in Organic Chemistry*, Cornell University Press, Ithaca, N.Y. 1972, 462 pp.

5.4 Cragg, G. M. L., *Organoboranes in Organic Synthesis*, Dekker, New York, 1973, 422 pp.

5.5 Brown, H. C., *Organic Synthesis via Boranes*, Wiley-Interscience, New York, 1975, 283 pp.

5.6 Onak, T., *Organoborane Chemistry*, Academic Press, New York, 1975, 360 pp.

5.7 Gerrard, W., *The Organic Chemistry of Boron*, Academic Press, New York, 1961, 308 pp.

5.8 Muetterties, E. L., *The Chemistry of Boron and Its Compounds*, John Wiley & Sons, New York, 1967, 699 pp.

5.9 Steinbeg, H., and Brotherton, R. J., *Organoboron Chemistry*, 2 vols., John Wiley & Sons, New York, 1964 and 1966.

5.10 For an extensive review see, for example, Grimes, R. N., *Carboranes*, Academic Press, New York, 1970, 272 pp.

5.11 Ramsey, B. G., and Anjo, D. M., *J. Am. Chem. Soc.*, **99**, 3182 (1977); cf. Timms, P. L., *Acc. Chem. Res.*, **6**, 118 (1973).

5.12 Negishi, E., Katz, J. J., and Brown, H. C., *J. Am. Chem. Soc.*, **94**, 4025 (1972).

5.13 See, for example, Negishi, E., *J. Organometal. Chem.*, **108**, 281 (1976) and references therein.

5.14 Brown, H. C., *Hydroboration*, W. A. Benjamin, New York, 1962, 290 pp.

5.15 Smith, K., *Chem. Soc. Rev.*, **3**, 443 (1974).

5.16 Brown, H. C., and Negishi, E., *Tetrahedron*, **33**, 2331 (1977).

5.17 Muetterties, E. L., and Tebbe, F. N., *Inorg. Chem.*, **7**, 2263 (1968).

5.18 Bowie, R. A., and Musgrave, O. E., *J. Chem. Soc. C*, 485 (1970).

5.19 Joy, F., Lappert, M. F., and Prokai, B., *J. Organometal. Chem.*, **5**, 506 (1966).

5.20 Blackborrow, J. R., *J. Chem. Soc.*, *Perkin II*, 1989 (1973).

5.21 Coyle, T. D., and Ritter, J. J., *Adv. Organometal. Chem.*, **10**, 273 (1972).

5.22 Brown, H. C., and Negishi, E., *J. Organometal. Chem.*, **26**, C67 (1971).

5.23   (a) Köster, R., *Adv. Organometal. Chem.*, 2, 257 (1964); (b) Köster, R., *Prog. Boron Chem.*, 1, 289 (1964).

5.24   Brown, H. C., and Subba Rao, B. C., *J. Am. Chem. Soc.*, 78, 5694 (1956).

5.25   Zweifel, G., and Brown, H. C., *Org. React.*, 13, 1 (1963).

5.26   Brown, H. C., and Negishi, E., *Pure Appl. Chem.*, 29, 527 (1972).

5.27   Negishi, E., and Brown, H. C., *Synthesis*, 77 (1974).

5.28   Brown, H. C., and Gupta, S. K., *J. Am. Chem. Soc.*, 93, 1816 (1971); 94, 4370 (1972).

5.29   Brown, H. C., and Ravindran, N., *J. Am. Chem. Soc.*, 94, 2112 (1972); 95, 2396 (1973); 98, 1785 (1976).

5.30   Zweifel, G., and Brown, H. C., *J. Am. Chem. Soc.*, 85, 2066 (1963).

5.31   Brown, H. C., and Sharp, R. L., *J. Am. Chem. Soc.*, 90, 2915 (1968).

5.32   Brown, H. C., Ayyangar, N. R., and Zweifel, G., *J. Am. Chem. Soc.*, 86, 397, 1071 (1964).

5.33   Partridge, J. J., Chadha, N. K., and Uskokovic, M. R., *J. Am. Chem. Soc.*, 95, 7171 (1973).

5.34   Kramer, G. W., and Brown, H. C., *J. Organometal. Chem.*, 73, 1 (1974).

5.35   Ashe, A. J., and Shu, P., *J. Am. Chem. Soc.*, 93, 1804 (1971).

5.36   Negishi, E., *J. Organometal. Chem. Libr.*, 1, 93 (1976).

5.37   Krishnamurthy, S., *Aldrichim. Acta*, 7, 55 (1974).

5.38   Negishi, E., Idacavage, M. J., Chiu, K. W., Yoshida, T., Abramovitch, A., Goettel, M. R., Silveira, A., Jr., and Bretherick, H. D., *J. C. S. Perkin II*, 1225 (1978).

5.39   (a) Corey, E. J., Albonico, S. M., Koelliker, U., Schaaf, T. K., and Varma, R. K., *J. Am. Chem. Soc.*, 93, 1491 (1971); (b) Corey, E. J., Becker, K. B., and Varma, R. K., *J. Am. Chem. Soc.*, 94, 8616 (1972); (c) Krishnamurthy, S., and Brown, H. C., *J. Am. Chem. Soc.*, 98, 3383 (1976).

5.40   (a) Brown, H. C., and Krishnamurthy, S., *J. C. S. Chem. Comm.*, 868 (1972); (b) Brown, H. C., and Krishnamurthy, S., *J. Am. Chem. Soc.*, 94, 7159 (1972).

5.41 Brown, C. A., *J. Am. Chem. Soc.*, <u>95</u>, 4100 (1973).

5.42 Matteson, D. S., and Mah, R. W. H., *J. Am. Chem. Soc.*, <u>85</u>, 2599 (1963).

5.43 Zweifel G., *Intra-Sci. Chem. Rep.*, <u>7</u>, 131 (1973), and references therein.

5.44 Corey, E. J., and Ravindranathan, T., *J. Am. Chem. Soc.*, <u>94</u>, 4013 (1972).

5.45 Negishi, E., Katz, J. J., and Brown, H. C., *Synthesis*, 555 (1972).

5.46 (a) Jacob, P., III, and Brown, H. C., *J. Am. Chem. Soc.*, <u>98</u>, 7832 (1976); (b) Sinclair, J. A., Molander, G. A., and Brown, H. C., *J. Am. Chem. Soc.*, <u>97</u>, 954 (1977).

5.47 Davies, A. G., and Roberts, B. P., *Acc. Chem. Res.*, <u>5</u>, 387 (1972).

5.48 (a) Lane, C. F., and Brown, H. C., *J. Am. Chem. Soc.*, <u>92</u>, 7212 (1970); (b) For a review see Lane, C. F., *Intra-Sci. Chem. Rep.*, <u>7</u>, 133 (1973).

5.49 Brown, H. C., and De Lue, N. R., *J. Am. Chem. Soc.*, <u>96</u>, 311 (1974).

5.50 Negishi, E., and Chiu, K. W., *J. Org. Chem.*, <u>41</u>, 3484 (1976).

5.51 Köster, R., Bellut, H., and Hattori, S., *Ann. Chem.*, <u>720</u>, 1 (1969).

5.52 Kuivila, H. G., *J. Am. Chem. Soc.*, <u>76</u>, 870 (1954); <u>77</u>, 4014 (1955); Kuivila, H. G., and Wiles, P. S., *J. Am. Chem. Soc.*, <u>77</u>, 4830 (1955); Kuivila, H. G., and Armour, A. G., *J. Am. Chem. Soc.*, <u>79</u>, 5659 (1957).

5.53 Breuer, S. W., and Broster, F. A., *J. Organometal. Chem.*, <u>35</u>, C5 (1972).

5.54 Brown, H. C., and Garg, C. P., *J. Am. Chem. Soc.*, <u>83</u>, 2951 (1961).

5.55 Davies, A. G., and Roberts, B. P., *J. Chem. Soc. B*, 311 (1969).

5.56 (a) Brown, H. C., and Midland, M. M., *J. Am. Chem. Soc.*, <u>93</u>, 4078 (1971); (b) Midland, M. M., and Brown, H. C., *J. Am. Chem. Soc.*, <u>95</u>, 4069 (1973).

5.57 (a) Brown, H. C., Midland, M. M., and Kabalka, G. W., *J. Am. Chem. Soc.*, <u>93</u>, 1024 (1971); (b) For a review, see Brown, H. C., and Midland, M. M., *Angew. Chem. Int. Ed.*

Engl., <u>11</u>, 692 (1972).

5.58 Brown, H. C., Rathke, M. W., and Rogić, M. M., *J. Am. Chem. Soc.*, <u>90</u>, 5038 (1968).

5.59 De Lue, N. R., Brown, H. C., *Synthesis*, 114 (1976).

5.60 Brown, H. C., De Lue, N. R., Kabalka, G. W., and Hedgecock, H. C., Jr., *J. Am. Chem. Soc.*, <u>98</u>, 1290 (1976).

5.61 Brown, H. C., and Lane, C. F., *J. Am. Chem. Soc.*, <u>92</u>, 6660 (1970).

5.62 Brown, H. C., and Lane, C. F., *J. C. S. Chem. Comm.*, 521 (1971).

5.63 Lane, C. F., and Brown, H. C., *J. Am. Chem. Soc.*, <u>92</u>, 7212 (1970).

5.64 Lane, C. F., and Brown, H. C., *J. Organometal. Chem.*, <u>26</u>, C51 (1971).

5.65 Davies, A. G., Hook, S. C. W., and Roberts, B. P., *J. Organometal. Chem.*, <u>23</u>, C11 (1970).

5.66 Brown, H. C., and De Lue, N. R., *J. Organometal. Chem.*, <u>135</u>, C57 (1977).

5.67 Lane, C. F., *J. Organometal. Chem.*, <u>31</u>, 421 (1971).

5.68 Brown, H. C., Hamaoka, T., and Ravindran, N., *J. Am. Chem. Soc.*, <u>95</u>, 5786, 6456 (1973).

5.69 Brown. H. C., Bowman, D. H., Misumi, S., and Unni, M. K., *J. Am. Chem. Soc.*, <u>89</u>, 4531 (1967).

5.70 Brown, H. C., Heydkamp, W. R., Breuer, E., and Murphy, W. S., *J. Am. Chem. Soc.*, <u>86</u>, 3565 (1964).

5.71 Rathke, M. W., Inoue, N., Varma, K. R., and Brown, H. C., *J. Am. Chem. Soc.*, <u>88</u>, 2870 (1966).

5.72 Tamura, Y., Minamikawa, J., Fujii, S., and Ikeda, M., *Synthesis*, 196 (1974).

5.73 (a) Suzuki, A., Sono, S., Itoh, M., Brown, H. C., and Midland, M. M., *J. Am. Chem. Soc.*, <u>93</u>, 4329 (1971); (b) Brown, H. C., Midland, M. M., and Levy, A. B., *J. Am. Chem. Soc.*, <u>94</u>, 2114 (1972); (c) Brown, H. C., Midland, M. M., and Levy, A. B., *J. Am. Chem. Soc.*, <u>95</u>, 2394 (1973).

5.74 Levy, A. B., and Brown, H. C., *J. Am. Chem. Soc.*, <u>95</u>, 4067 (1973).

5.75 Mueller, R. H., *Tetrahedron Lett.*, 2925 (1976).

5.76   Brown, H. C., and Midland, M. M., *J. Am. Chem. Soc.*, 93, 3291 (1971).

5.77   (a) Brown, H. C., and Snyder, C. H., *J. Am. Chem. Soc.*, 83, 1001 (1961); (b) Brown, H. C., Hebert, N. C., and Synder, C. H., *J. Am. Chem. Soc.*, 83, 1001 (1961); (c) Brown, H. C., Verbrugge, C., and Snyder, C. H., *J. Am. Chem. Soc.*, 83, 1002 (1961); (d) For a review, see Snyder, C. H., *Intra-Sci. Chem. Rep.*, 7, 169 (1973).

5.78   Brown, H. C., Levy, A. B., and Midland, M. M., *J. Am. Chem. Soc.*, 97, 5017 (1975).

5.79   Zweifel, G., and Fisher, R. P., *Synthesis*, 339 (1974).

5.80   Miyaura, N., Yoshinari, T., Itoh, M., and Suzuki, A., *Tetrahedron Lett.*, 2961 (1974).

5.81   Chiu, K. W., Negishi, E., Plante, M. S., and Silveira, A., Jr., *J. Organometal. Chem.*, 112, C3 (1976).

5.82   (a) Zweifel, G., Arzoumanian, H., and Whitney, C. C., *J. Am. Chem. Soc.*, 89, 3652 (1967); (b) For a review, see Zweifel, G., *Intra-Sci. Chem. Rep.*, 7, 181 (1973).

5.83   Negishi, E., Lew, G., and Yoshida, T., *J. Org. Chem.*, 39, 2321 (1974).

5.84   Miyaura, N., Tagami, H., Itoh, M., and Suzuki, A., *Chem. Lett.*, 1411 (1974).

5.85   Evans, D. A., Crawford, T. C., Thomas, R. C., and Walker, J. A., *J. Org. Chem.*, 41, 3947 (1976).

5.86   Suzuki, A., Miyaura, N., Abiko, S., Itoh, M., Brown, H. C., Sinclair, J. A., and Midland, M. M., *J. Am. Chem. Soc.*, 95, 3080 (1973).

5.87   Midland, M. M., Sinclair, J. A., and Brown, H. C., *J. Org. Chem.*, 39, 731 (1974).

5.88   Yamada, K., Miyaura, N., Itoh, M., and Suzuki, A., *Synthesis*, 679 (1977).

5.89   (a) Negishi, E., Lew, G., and Yoshida, T., *J. C. S. Chem. Comm.*, 874 (1973); (b) Negishi, E., and Abramovitch, A., *Tetrahedron Lett.*, 411 (1977).

5.90   Sinclair, J. A., and Brown, H. C., *J. Org. Chem.*, 41, 1078 (1976).

5.91   Eisch, J. J., and Wilcsek, R. J., *J. Organometal. Chem.*, 71, C21 (1974), and references therein.

5.92 (a) Negishi, E., Chiu, K. W., and Yoshida, T., *J. Org. Chem.*, 40, 1676 (1975); (b) Negishi, E., Abramovitch, A., and Merrill, R. E., *J. C. S. Chem. Comm.*, 138 (1975).

5.93 (a) Pelter, A., Hutchings, M. G., and Smith, K., *J. C. S. Chem. Comm.*, 1529 (1970); 1048 (1971); 186 (1973); (b) Pelter, A., et al., *J. Chem. Soc.*, 129, 138, 142, 145 (1975).

5.94 Naruse, M., Tomita, T., Utimoto, K., and Nozaki, H., *Tetrahedron*, 30, 835 (1974).

5.95 Negishi, E., Yoshida, T., Silveira, A., Jr., and Chiou, B. L., *J. Org. Chem.*, 40, 814 (1975).

5.96 (a) Miyaura, N., Itoh, M., and Suzuki, A., *Tetrahedron Lett.*, 255 (1976); (b) *Synthesis*, 618 (1976); (c) Miyaura, N., Sasaki, N., Itoh, M., and Suzuki, A., *Tetrahedron Lett.*, 173, 3369 (1977); (d) Miyaura, N., Itoh, M., and Suzuki, A., *Bull. Chem. Soc. Japan*, 50, 2199 (1977).

5.97 (a) Pelter, A., Harrison, C. R., and Kirkpatrick, D., *J. C. S. Chem. Comm.*, 544 (1973); (b) *Tetrahedron Lett.*, 4491 (1973); (c) Pelter, A., and Harrison, C. R., *J. C. S. Chem. Comm.*, 828 (1974); (d) Pelter, A., Subramanyam, C., Laub, R. J., Gould, K. J., and Harrison, C. R., *Tetrahedron Lett.*, 1633 (1975); (e) Pelter, A., Gould, K. J., and Harrison, C. R., *Tetrahedron Lett.*, 3327 (1975); (f) Pelter, A., et al., *J. Chem. Soc.*, *Perkin I*, 2419, 2428, 2435 (1976).

5.98 Naruse, M., Utimoto, K., and Nozaki, H., *Tetrahedron Lett.*, 2741 (1975); *Tetrahedron*, 30, 3037 (1974).

5.99 (a) Utimoto, K., Uchida, K., and Nozaki, H., *Tetrahedron Lett.*, 4527 (1973); (b) *Chem. Lett.*, 1493 (1974).

5.100 Negishi, E., and Merrill, R. E., *J. C. S. Chem. Comm.*, 860 (1974).

5.101 Mikhailov, B. M., *Organometal. Chem. Rev. A*, 8, 1 (1972).

5.102 (a) Mukaiyama, T., Inomata, K., and Muraki, M., *J. Am. Chem. Soc.*, 95, 967 (1973); (b) Inomata, K., Muraki, M., and Mukaiyama, T., *Bull. Chem. Soc. Japan*, 46, 1804 (1973); 48, 3200 (1975); (c) Mukaiyama, T., and Inoue, T., *Chem. Lett.*, 559 (1976); (d) Inoue, T., Uchimaru, T., Mukaiyama, T., *Chem. Lett.*, 153 (1977).

5.103 Unpublished results by Negishi, E., and Idacavage, M. J.

5.104   Miyaura, N., Itoh, M., Suzuki, A., Brown, H. C., Midland, M. M., and Jacob, P., III, *J. Am. Chem. Soc.*, <u>94</u>, 6549 (1972).

5.105   Jacob, P., III, and Brown, H. C., *J. Org. Chem.*, <u>42</u>, 579 (1977).

5.106   Kabalka, G. W., *Intra-Sci. Chem. Rep.*, <u>7</u>, 57 (1973).

5.107   (a) Suzuki, A., Arase, A., Matsumoto, H., Itoh, M., Brown, H. C., Rogić, M. M., and Rathke, M. W., *J. Am. Chem. Soc.*, <u>89</u>, 5708 (1967); (b) Brown, H. C., Rogić, M. M., and Kabalka, G. W., *J. Am. Chem. Soc.*, <u>89</u>, 5707 (1967); (c) Brown, H. C., Labalka, G. W., Rathke, M. W., and Rogić, M. M., *J. Am. Chem. Soc.*, <u>90</u>, 4165 (1968).

5.108   (a) Kabalka, G. W., Brown, H. C., Suzuki, A., Honma, S., Arase, A., Itoh, M., *J. Am. Chem. Soc.*, <u>92</u>, 710 (1970); (b) Brown, H. C., and Kabalka, G. W., *J. Am. Chem. Soc.*, <u>92</u>, 712, 714 (1970).

5.109   Brown, H. C., and Negishi, E., *J. Am. Chem. Soc.*, <u>93</u>, 3777 (1971).

5.110   (a) Jacob, P., III, and Brown, H. C., *J. Am. Chem. Soc.*, <u>98</u>, 7832 (1976); (b) Sinclair, J. A., Molander, G. A., and Brown, H. C., *J. Am. Chem. Soc.*, <u>99</u>, 954 (1977); (c) Molander, G. A., and Brown, H. C., *J. Org. Chem.*, <u>42</u>, 3106 (1977).

5.111   Negishi, E., and Yoshida, T., *J. Am. Chem. Soc.*, <u>95</u>, 6837 (1973).

5.112   Pelter, A., and Hughes, L., *J. C. S. Chem. Comm.*, 913 (1977).

5.113   See, for example, Negishi, E., *J. Chem. Educ.*, <u>52</u>, 159 (1975).

5.114   For a review, see (a) Brown, H. C., and Rogić, M. M., *Organometal. Chem. Syn.*, <u>1</u>, 305 (1972); (b) Rogić, M. M., *Intra-Sci. Chem. Rep.*, <u>7</u>, 155 (1972).

5.115   Pasto, D. J., and Wojtkowski, P. W., *Tetrahedron. Lett.*, 215 (1970).

5.116   (a) Brown, H. C., and Rogić, M. M., *J. Am. Chem. Soc.*, <u>91</u>, 2146 (1969); (b) Brown, H. C., Rogić, M. M., Nambu, H., and Rathke, M. W., *J. Am. Chem. Soc.*, <u>91</u>, 2147 (1969).

5.117   Brown, H. C., Nambu, H., and Rogić, M. M., *J. Am. Chem. Soc.*, <u>91</u>, 6855 (1969).

5.118   (a) Hooz, J., and Linke, S., *J. Am. Chem. Soc.*, 90, 5963, 6819 (1968); (b) Hooz, J., and Gum, D. M., *J. Am. Chem. Soc.*, 91, 6195 (1969); (c) Hooz, J., et al., *J. Org. Chem.*, 38, 2574 (1973).

5.119   Tufariello, J. J., Lee, L. T. C., and Wojtkowski, P., *J. Am. Chem. Soc.*, 89, 6804 (1967).

5.120   (a) Yamamoto, Y., Kondo, K., and Moritani, I., *Tetrahedron Lett.*, 793 (1974); (b) 2689 (1975); (c) *Bull. Chem. Soc. Japan*, 48, 3681 (1975); (d) *J. Org. Chem.*, 40, 3644 (1975).

5.121   Levy, A. B., and Schwartz, S. J., *Tetrahedron Lett.*, 2201 (1976).

5.122   Utimoto, K., Sakai, N., and Nozaki, H., *J. Am. Chem. Soc.*, 96, 5601 (1974).

5.123   Suzuki, A., Miyaura, N., and Itoh, M., *Tetrahedron*, 27, 2775 (1971).

5.124   Musker, W. K., and Stevens, R. R., *Tetrahedron Lett.*, 995 (1967).

5.125   Köster, R., and Rickborn, B., *J. Am. Chem. Soc.*, 89, 2782 (1967).

5.126   (a) Tufariello, J. J., and Lee, L. T. C., *J. Am. Chem. Soc.*, 88, 4757 (1966); (b) Tufariello, J. J., Wojtkowski, P., and Lee, L. T. C., *J. C. S. Chem. Comm.*, 505 (1967).

5.127   Lane, C. F., and Brown, H. C., *J. Am. Chem. Soc.*, 93, 1025 (1971).

5.128   (a) Brown, H. C., and Lane, C. F., *Synthesis*, 303 (1972); (b) Yamamoto, Y., and Brown, H. C., *J. Org. Chem.*, 39, 861 (1974).

5.129   (a) Brown, H. C., Yamamoto, Y., and Lane, C. F., *Synthesis*, 304 (1972); (b) Yamamoto, Y., and Brown, H. C., *J. C. S. Chem. Comm.*, 801 (1973).

5.130   Zweifel, G., and Arzoumanian, H., *J. Am. Chem. Soc.*, 89, 5086 (1967).

5.131   (a) Corey, E. J., and Ravindranathan, T., *J. Am. Chem. Soc.*, 94, 4013 (1972); (b) Negishi, E., Katz, J. J., and Brown, H. C., *Synthesis*, 555 (1972).

5.132   Negishi, E., and Yoshida, T., *J. C. S. Chem. Comm.*, 606 (1973).

5.133   Yoshida, T., Williams, R. M., and Negishi, E., *J. Am. Chem. Soc.*, 96, 3688 (1974).

5.134   Brown, H. C., *Acc. Chem. Res.*, 2, 65 (1969).

5.135   Negishi, E., *Intra-Sci. Chem. Rep.*, 7, 81 (1973).

5.136   Hillman, M. E. D., *J. Am. Chem. Soc.*, 84, 4715 (1962).

5.137   Brown, H. C., and Rathke, M. W., *J. Am. Chem. Soc.*, 89, 2737, 2738, 4528 (1967).

5.138   Brown, H. C., Coleman, R. A., and Rathke, M. W., *J. Am. Chem. Soc.*, 90, 499 (1968).

5.139   Rathke, M. W., and Brown, H. C., *J. Am. Chem. Soc.*, 89, 2740 (1967).

5.140   (a) Brown. H. C., Knights, E. F., and Coleman, R. A., *J. Am. Chem. Soc.*, 91, 2144 (1969); (b) Brown, H. C., and Coleman, R. A., *J. Am. Chem. Soc.*, 91, 4606 (1969).

5.141   (a) Brown, H. C., and Negishi, E., *J. Am. Chem. Soc.*, 89, 5285 (1967); (b) Negishi, E., and Brown, H. C., *Synthesis*, 196 (1972); (c) Brown, H. C., and Negishi, E., *J. Am. Chem. Soc.*, 89, 5477 (1967); (d) Brown, H. C., and Negishi, E., *J. C. S. Chem. Comm.*, 594 (1968).

5.142   Negishi, E., Sabanski, M., Katz, J. J., and Brown, H. C., *Tetrahedron*, 32, 925 (1976).

5.143   Bryson, T. A., and Pye, W. E., *J. Org. Chem.*, 42, 3214 (1977).

5.144   (a) Brown, H. C., and Negishi, E., *J. Am. Chem. Soc.*, 89, 5478 (1967); (b) Brown, H. C., and Dickason, W. C., *J. Am. Chem. Soc.*, 91, 1226 (1969); (c) Brown, H. C., and Negishi, E., *J. Am. Chem. Soc.*, 91, 1224 (1969).

5.145   Binger, P., Benedikt, G., Rotermund, G. W., and Köster, R., *Ann. Chem.*, 717, 21 (1968).

5.146   For a review, see Pelter, A., *Intra-Sci. Chem. Rep.*, 7, 73 (1973).

5.147   (a) Brown, H. C., Carlson, B. A., and Prager, R. H., *J. Am. Chem. Soc.*, 93, 2070 (1971); (b) Brown, H. C., and Carlson, B. A., *J. Org. Chem.*, 38, 2422 (1973); (c) Carlson, B. A., and Brown, H. C., *J. Am. Chem. Soc.*, 95, 6876 (1973); (d) Brown, H. C., Katz, J. J., and Carlson, B. A., *J. Org. Chem.*, 38, 3968 (1973); (e) Carlson, B. A., Katz, J. J., and Brown, H. C., *J. Organometal. Chem.*, 67, C39 (1974).

5.148   Carlson, B. A., and Brown, H. C., *Synthesis*, 776 (1973).

5.149   (a) Hawthorne, M. F., and DuPont, J. A., *J. Am. Chem. Soc.*, 80, 5830 (1958); (b) Hawthorne, M. F., *J. Am. Chem. Soc.*, 82, 1886 (1960); (c) Binger, P., and Köster, R., *Tetrahedron Lett.*, 156 (1961); (d) Brown, H. C., and Rhodes, S. P., *J. Am. Chem. Soc.*, 91, 2149, 4306 (1969); (e) Merrill, R. E., Allen, J. L., Abramovitch, A., and Negishi, E., *Tetrahedron Lett.*, 1019 (1977).

5.150   For a review, see Marshall, J. A., *Synthesis*, 229 (1971).

5.151   Rathke, M. W., and Kow, R., *J. Am. Chem. Soc.*, 94, 6854 (1972).

5.152   (a) Cainelli, G., DaBello, G., and Zubiani, G., *Tetrahedron Lett.*, 3429 (1965); (b) Zweifel, G., and Arzoumanian, H., *J. Am. Chem. Soc.*, 89, 291 (1967); (c) Zweifel, G., Fisher, R. P., and Horng, A., *Synthesis*, 37 (1973).

5.153   (a) Matteson, D. C., Moody, R. J., and Jesthi, P. K., *J. Am. Chem. Soc.*, 97, 5608 (1975); (b) Matteson, D. S., and Hagelee, L. A., *J. Organometal. Chem.*, 93, 21 (1975); (c) For a review, see Matteson, D. S., *Synthesis*, 147 (1975).

5.154   Taguchi, T., Takahashi, Y., Itoh, M., and Suzuki, A., *Chem. Lett.*, 1021 (1974).

5.155   (a) Takahashi, Y., Tokuda, M., Itoh, M., and Suzuki, A., *Chem. Lett.*, 523 (1975); (b) *Synthesis*, 616 (1976).

5.156   Matteson, D. S., and Waldbillig, J. O., *J. Org. Chem.*, 28, 366 (1963).

5.157   (a) Brown, H. C., Krishnamurthy, A., and Yoon, N. M., *J. Org. Chem.*, 41, 1778 (1976); (b) For a review, see Brown, H. C., and Lane, C. F., *Heterocycles*, 7, 453 (1977).

5.158   Brown, H. C., Bigley, D. B., Arora, S. K., and Yoon, N. M., *J. Am. Chem. Soc.*, 92, 7161 (1970).

5.159   Brown, H. C., Heim, P., and Yoon, N. M., *J. Org. Chem.*, 37, 2942 (1972).

5.160   (a) Midland, M. M., Tramontano, A., and Zderic, S. A., *J. Organometal. Chem.*, 134, C17 (1977); (b) *J. Am. Chem. Soc.*, 99, 5211 (1977).

5.161   For a review, see Lane, C. F., *Aldrichimica Acta*, 8, 3 (1975).

5.162   For a review, see Krishnamurthy, S., *Aldrichimica Acta*, 7, 55 (1974).

5.163   For a review, see Kramer, G. W., and Brown, H. C., *Heterocycles*, 7, 487 (1977).

5.164   Lane, C. F., *Chem. Rev.*, 76, 773 (1976).

5.165   Krishnamurthy, S., and Brown, H. C., *J. Org. Chem.*, 40, 1864 (1975); 42, 1197 (1977).

5.166   Hutchins, R. O., Maryanoff, B. E., and Milewski, C. A., *J. C. S. Chem. Comm.*, 1097 (1971).

5.167   Borch, R. F., Bernstein, M. D., and Durst, H. D., *J. Am. Chem. Soc.*, 93, 2897 (1971).

5.168   Hutchins, R. D., Maryanoff, B. E., and Milewski, C. A., *J. Am. Chem. Soc.*, 93, 1793 (1971); 95, 3672 (1973).

5.169   Hutchins, R. O., Kacher, M., and Rua, L., *J. Org. Chem.*, 40, 923 (1975).

5.170   (a) Brown, H. C., and Krishnamurthy, S., *J. Am. Chem. Soc.*, 95, 1669 (1973); (b) Krishnamurthy, S., and Brown, H. C., *J. Org. Chem.*, 41, 3064 (1976).

5.171   (a) Brown, H. C., Krishnamurthy, S., and Coleman, R. A., *J. Am. Chem. Soc.*, 94, 1750 (1972); (b) Brown, H. C., and Krishnamurthy, S., *J. C. S. Chem. Comm.*, 868 (1972); (c) Krishnamurthy, S., Schubert, R. M., and Brown, H. C., *J. Am. Chem. Soc.*, 95, 8486 (1973).

5.172   (a) Brown, H. C., and Dickason, W. C., *J. Am. Chem. Soc.*, 92, 709 (1970); (b) Corey, E. J., and Varma, R. K., *J. Am. Chem. Soc.*, 93, 7319 (1971); (c) Brown, H. C., and Krishnamurthy, S., *J. Am. Chem. Soc.*, 94, 7159 (1972); (d) Brown, H. C., *J. Am. Chem. Soc.*, 95, 4100 (1973); (e) Krishnamurthy, S., and Brown, H. C., *J. Am. Chem. Soc.*, 98, 3383 (1976).

5.173   (a) Corey, E. J., Becker, K. B., and Varma, R. K., *J. Am. Chem. Soc.*, 94, 8616 (1972); (b) Ganem, B., *J. Org. Chem.*, 40, 146 (1975); (c) Ganem, B., and Fortunato, J. M., *J. Org. Chem.*, 40, 2846 (1975).

5.174   Yoshida, T., and Negishi, E., *J. C. S. Chem. Comm.*, 762 (1974).

5.175   Yamamoto, Y., Toi, H., Murahashi, S.-I., and Moritani, I., *J. Am. Chem. Soc.*, 97, 2558 (1975).

5.176   (a) Yamamoto, Y., Toi, H., Sonoda, A., and Murahashi, S.-I., *J. Am. Chem. Soc.*, 98, 1965 (1976); (b) *J. C. S. Chem. Comm.*, 672 (1976).

5.177   Mole, T., and Jeffery, E. A., *Organoaluminum Compounds*, Elsevier, Amsterdam, 1972, 465 pp.

5.178   Lehmkuhl, H., Ziegler, K., and Gellert, H. G., in *Houben-Weyl*, *Methoden der Organischen Chemine*, 4th ed., Vol. XIII, Part 4, G. Thieme, Stuttgart, 1970, pp. 1 to 314.

5.179   Bruno, G., *The Use of Aluminum Alkyls in Organic Synthesis*, Ethyl Corp., Baton Rouge, La., 1970 and 1972, 113 pp.

5.180   Reinheckel, H., Haage, K., and Jahnke, D., *Organometal. Chem. Rev. A*, $\underline{4}$, 47 (1969).

5.181   Vranska, R. G., and Amma, E. L., *J. Am. Chem. Soc.*, $\underline{89}$, 3121 (1967).

5.182   Hoberg, H., *Ann. Chem.*, $\underline{656}$, 1 (1962); $\underline{695}$, 1 (1966); *Angew. Chem.*, $\underline{77}$, 1084 (1965); $\underline{78}$, 492 (1966).

5.183   Zieger, K., Martin, H., Krupp, F., *Ann. Chem.*, $\underline{629}$, 14 (1960).

5.184   (a) Ziegler, K., Kroll, W. R., Larbig, W., and Steudel, O. W., *Ann. Chem.*, $\underline{629}$, 53 (1960); For a review, see Ref. *5.177*, Chap. 5.

5.185   (a) Ziegler, K., in *Organometallic Chemistry*, Zeiss, H., ed., Chap. 5, Reinhold, New York, N. Y., 1960; (b) Köster, R., and Binger, P., *Adv. Inorg. Chem. Radiochem.*, $\underline{7}$, 263 (1965).

5.186   Ziegler, K., Holzkamp, E., Breil, H., and Martin, H., *Angew. Chem.*, $\underline{67}$, 426, 541 (1955).

5.187   Wilke, G., and Müller, H., *Ann. Chem.*, $\underline{629}$, 222 (1960).

5.188   Ref. *5.177*, Chap. 11.

5.189   (a) Eisch, J. J., and Kaska, W. C., *J. Am. Chem. Soc.*, $\underline{88}$, 2976 (1966).

5.190   Van Horn, D. E., and Negishi, E., *J. Am. Chem. Soc.*, $\underline{100}$, 2252 (1978).

5.191   For a review, see Ref. *5.177*, Chap. 2.

5.192   (a) Ziegler, K., Gellert, H. G., Martin, H., Nagel, K., and Schneider, J., *Ann. Chem.*, $\underline{589}$, 91 (1954); (b) For a review, see Ref. *5.177*, Chaps. 3 and 11.

5.193   Ziegler, K., Gellert, H. G., Zosel, K., Lehmkuhl, W., and Pfohl, W., *Angew. Chem.*, $\underline{67}$, 424 (1955).

5.194   For a review, see Ref. *5.177*, Chap. 3.

5.195   (a) Hata, G., and Miyake, A., *J. Org. Chem.*, $\underline{28}$, 3237 (1963); (b) Marcus, E., MacPeek, D. L., and Tinsley, S. W., *J. Org. Chem.*, $\underline{36}$, 381 (1971).

5.196   For a review, see (a) Ref. *5.177*, Chap. 11; (b) Ref. *5.179*.

5.197   Zweifel, G., in *Aspects of Mechanism and Organometallic Chemistry*, Brewster, J. H., ed., Plenum Press, 1978, pp. 229 to 249.

5.198   Eisch, J. J., Amtmann, R., and Foxton, M. W., *J. Organometal. Chem.*, 16, P55 (1969).

5.199   Eisch, J. J., and Foxton, M. W., *J. Org. Chem.*, 36, 3520 (1971).

5.200   Miller, R. L., Ph.D. Thesis, University of California, Davis, 1971.

5.201   (a) Zweifel, G., and Steele, R. B., *J. Am. Chem. Soc.*, 89, 5085 (1967); (b) Magoon, E. F., and Slaugh, L. H., *Tetrahedron*, 23, 4509 (1967).

5.202   See, for example, Raphael, R. A., *Acetylenic Compounds in Organic Synthesis* , Butterworth, London, 1955, p. 29.

5.203   (a) Corey, E. J., Katzenellenbogen, J. A., and Posner, G. H., *J. Am. Chem. Soc.*, 89, 4245 (1967); (b) Corey, E. J., Katzenellenbogen, J. A., Gilman, N. W., Roman, S. A., and Erickson, B. W., *J. Am. Chem. Soc.*, 90, 5618 (1968); (c) Corey, E. J., Kirst, H. A., and Katzenellenbogen, J. A., *J. Am. Chem. Soc.*, 92, 6314 (1970).

5.204   For a review, see Ref. *5.177*, Chap. 7.

5.205   Bartocha, B., Bilbo, A. J., Bublitz, D. E., and Gray, M. Y., *Z. Naturforsch.*, 16b, 357 (1961).

5.206   Ref. *5.177*, Chaps. 14 and 15.

5.207   (a) Negishi, E., and Baba, S., *J. C. S. Chem. Comm.*, 596 (1976); (b) Baba, S., and Negishi, E., *J. Am. Chem. Soc.*, 98, 6729 (1976).

5.208   Ref. *5.177*, Chap. 13.

5.209   Kennedy, J. P., *J. Org. Chem.*, 35, 532 (1970).

5.210   Negishi, E., and Baba, S., *J. Am. Chem. Soc.*, 97, 7385 (1975).

5.211   (a) Meisters, A , and Mole, T., *J. C. S. Chem. Comm.*, 595 (1972); (b) Harney, D. W., Meisters, A., and Mole, T., *Aust. J. Chem.*, 27, 1639 (1974).

5.212   (a) Kitagawa, Y., Hashimoto, S., Iemura, S., Yamamoto, H., and Nozaki, H., *J. Am. Chem. Soc.*, 98, 5030 (1976); (b)

Hashimoto, S., Kitagawa, Y., Iemura, S., Yamamoto, H., and Nozaki, H., *Tetrahedron Lett.*, 2615 (1976).

5.213   (a) Baba, S., Van Horn, D. E., and Negishi, E., *Tetrahedron Lett.*, 1927 (1976); (b) Eisch, J. J., and Damasevitz, G. A., *J. Org. Chem.*, **41**, 2214 (1976); (c) Uchida, K., Utimoto, K., and Nozaki, H., *J. Org. Chem.*, **41**, 2215 (1976).

5.214   For a review, see Ref. *5.117*, Chap. 12.

5.215   For reviews, see (a) Ref. *5.177*, Chap. 12; (b) Reinheckel, H., Haage, K., and Jahnke, D., *Organometal. Chem. Rev. A*, **4**, 47 (1969).

5.216   Zweifel, G., and Steel, R. B., *J. Am. Chem. Soc.*, **89**, 2754, 5085 (1967).

5.217   (a) Newman, H., *Tetrahedron Lett.*, 4571 (1971); (b) Newman, H., *J. Am. Chem. Soc.*, **95**, 4098 (1973).

5.218   Zweifel, G., Snow, J. T., and Whitney, C. C., *J. Am. Chem. Soc.*, **90**, 7139 (1968).

5.219   Zweifel, G., and Lynd, R. A., *Synthesis*, 625 (1976).

5.220   For a review, see Ashby, E. C., *Chem. Rev.*, **75**, 521 (1975).

5.221   Meisters, A., and Mole, T., *J. C. S. Chem. Comm.*, 595 (1972).

5.222   Meisters, A., and Mole, T., *Aust. J. Chem.*, **27**, 1655, 1665 (1974).

5.223   Kabalka, G. W., and Daley, R. F., *J. Am. Chem. Soc.*, **95**, 4428 (1973).

5.224   (a) Hooz, J., and Layton, R. B., *J. Am. Chem. Soc.*, **93**, 7320 (1971); (b) Hooz, J., and Layton, R. B., *Can. J. Chem.*, **51**, 2093 (1973).

5.225   (a) Pappo, R., and Collins, P. W., *Tetrahedron Lett.*, 2627 (1972); (b) Collins, P. W., Dajani, E. Z., Bruhn, M. S., Brown, C. H., Palmer, J. R., and Pappo, R., *Tetrahedron Lett.*, 4217 (1975).

5.226   (a) Jeffery, E. A., Meisters, A., and Mole, T. J., *Organometal. Chem.*, **74**, 365 (1974); (b) Ashby, E., and Heinsohn, G., *J. Org. Chem.*, **39**, 3297 (1974); (c) Bagnell, L., Jeffery, E. A., Meisters, A., and Mole, T., *Aust. J. Chem.*, **28**, 801 (1975).

5.227   (a) Bernady, K. F., and Weiss, M. J., *Tetrahedron Lett.*, 4038 (1972); (b) Bernady, K. F., Polletto, J. F., and

Meiss, M. J., *Tetrahedron Lett.*, 765 (1975).

5.228   (a) For a review with essentially all pertinent papers by Nagata and his coworkers, see Nagata, W., and Yoshioka, M., *Org. React.*, 25, 255 (1977); (b) Nagata, W., Yoshioka, M., and Hirai, S., *J. Am. Chem. Soc.*, 94, 4635 (1972); (c) Nagata, W., Yoshioka, M., and Murakami, M., *J. Am. Chem. Soc.*, 94, 4644, 4654 (1972); (d) Nagata, W., Yoshioka, M., and Terasawa, T., *J. Am. Chem. Soc.*, 94, 4692 (1972).

5.229   (a) Boireau, G., Abenheim, D., Bernardon, C., Henry-Basch, E., and Sabourault, B., *Tetrahedron Lett.*, 2521 (1975); (b) Warwel, S., Schmitt, G., and Ahlfaengen, B., *Synthesis*, 623 (1975); (c) Negishi, E., Baba, S., and King, A. O., *J. C. S. Chem. Comm.*, 17 (1976).

5.230   (a) Fried, J., et al., *J. C. S. Chem. Comm.*, 634 (1968); (b) Fried, J., Lin, C.-H., and Ford, S. H., *Tetrahedron Lett.*, 1397 (1969); (c) Fried, J., Mehra, M. M., Kao, W. L., and Lin, C.-H., *Tetrahedron Lett.*, 2695 (1970); (d) Fried, J., Mehra, M. M., and Kao, W. L., *J. Am. Chem. Soc.*, 93, 5594 (1971); (e) Fried, J, Sih, J. C., Lin, C.-H., and Dalven, P., *J. Am. Chem. Soc.*, 94, 4343 (1972); (f) Crosby, G. A., and Stephenson, R. A., *J. C. S. Chem. Comm.*, 287 (1975); (g) Danishefsky, S., Kitahara, T., Tsai, M., and Dynak, J., *J. Org. Chem.*, 41, 1669 (1976).

5.231   For an extensive review on reduction with basic aluminum hydrides, see Gaylord, N. G., *Reduction with Complex Metal Hydrides*, Wiley-Interscience, New York, 1956.

5.232   Corey, E. J., Weinshenker, N. M., Schaaf, T. K., and Huber, W., *J. Am. Chem. Soc.*, 91, 5675 (1969).

6
ORGANOSILICONS AND ORGANOTINS
(Silicon, Tin)

There are numerous monographs and reviews on the chemistry of organometallics that contain the Group IVA elements. In addition to a textbook by Coates and Wade (0.6), a two-volume treatise edited by MacDiarmid (6.1) provides extensive general coverage of the chemistry of organometallics of the Group IVA elements. A few excellent monographs on organosilicons (6.2) and organotins (6.3 to 6.5) discussing all aspects of these compounds are available. Unfortunately, applications of these organometallics to organic synthesis have been developed mainly after the publication of these monographs. There have been a number of reviews, however, discussing recent results, which are cited throughout this chapter. A series of reviews on recent advances in organosilicon chemistry have appeared in *Pure Appl. Chem.*, *13* (1969); *19* (1969), *Intra-Sci. Chem. Rep.*, *7* (1973), and *J. Organometal. Chem. Libr.*, *2*, (1976). New chemistry and applications of organotins have been extensively reviewed by various workers (6.5a).

6.1 SOME FUNDAMENTAL PROPERTIES OF SILICON AND TIN AND THE BONDS TO THESE ELEMENTS

Silicon (Si), germanium (Ge), tin (Sn), and lead (Pb) lie directly below carbon in the Periodic Table (Group IVA elements). As might be expected, the structure and reactivity of organometallics containing these elements are similar in many respects to those of the corresponding carbon compounds. In this book, however, we are mainly interested in the differences between these elements and carbon and some of the unique physical and chemical properties of organometallics containing these elements.

The chemistry of organogermanium compounds is not discussed in this book, since these compounds have not yet found many applications in organic synthesis. This must be, at least in part, due to the fact that germanium-containing compounds are considerably (10 to 10^2X) more expensive than the corresponding silicon and tin compounds. While organolead compounds share many noteworthy properties with the corresponding silicon and tin compounds, they also exhibit chemical properties common to some other organometallics containing heavy elements, such as mercury and thallium. Therefore, their chemistry is discussed in the following chapter along with that of organomercuries and organothalliums.

6.1.1 Electronegativity and Bond Energy

As shown in Table 6.1, the Group IVA elements are characterized by their relatively high electronegativity values, although they are more electropositive than carbon.

Table 6.1   Electronegativity of the Group IVA Elements

Element	Electronegativity	
	Pauling (*1.1*)	Allred-Rochew (*1.8*)
C	2.5	2.60
Si	1.8	1.90
Ge	1.8	2.00
Sn	1.8	1.93
Pb	1.8	2.45

Thus their bonds to carbon and hydrogen are highly covalent. The covalent radii for silicon and tin are 1.17 and 1.40 Å, respectively, and are reasonably independent of the nature of the ligands bonded to these metals. Due to their highly covalent nature, the intrinsic nucleophilicity of the Si-C and Si-H bond is low. Direct cleavage of the Si-C bond seldom takes place in the reaction of organosilicons with electrophiles, although certain functionalized organosilicons, such as α,β-unsaturated derivatives, can react with electrophiles by multistep mechanisms, as discussed later. On the other hand, the Sn-C bond, although highly covalent, can readily participate in ionic reactions thorugh polarization. The carbon group bonded to tin acts as a nucleophile and the tin moiety acts as an electrophile.

The great majority of silicon and tin compounds exist as tetracoordinate species, in which silicon or tin occupies the central position in an essentially tetrahedral configuration. They

might therefore be considered as coordinatively saturated octet
species. Unlike their carbon analogs, however, the empty d orbitals of the higher members of the Group IVA elements can provide
extra coordination sites. Thus organosilicons can act as weak e-
lectrophiles. The high polarizability of the Sn-C bond, together
with the availability of d orbitals, makes organotins reasonably
good electrophiles.

It may be pointed out that the generally low nucleophilicity
and electrophilicity of organosilicons make certain silicon-containing moieties good protecting groups (Sect. 6.4.2.4).

Some representative single-bond energy values for bonds to
C, Si, and Sn are summarized in Table 6.2.

Table 6.2   Single-Bond Energy Values for Bonds to C, Si, and Sn

	C $(6.6)^{\underline{a}}$	Si $(6.6)^{\underline{a}}$	Sn $(6.3)^{\underline{b}}$
H	99	70 to 76	35
C	83	69 to 76	46 to 60
Si	69 to 76	42 to 53	–
Sn	50 to 60	–	50
N	70	80	65 to 70
O	84 to 86	88 to 108	95
F	105 to 111	129 to 135	–
Cl	79 to 81	86 to 91	85
Br	66 to 68	69 to 74	76
I	58 to 61	51 to 56	62

$\underline{a}$ Page 25 of Ref. 6.6
$\underline{b}$ These values are for Me_3SnX (p. 9 of Ref. 6.3).

It is noteworthy that the Si-X and Sn-X bonds, where X = H and
C, are weaker than the corresponding C-X bonds, whereas the bonds
to silicon and tin are stronger than the bonds to carbon when
X = N, O, or halogen (except I). Since N, O, and halogens can do-
nate a lone pair of p electrons, these bond energy data can most
readily be interpreted in terms of p_π-d_π interaction involving
the empty d orbitals of silicon and tin.

Another important point to be noted is that even the Si-H
and Si-C bonds are thermodynamically relatively strong bonds. In
fact, the homolytic cleavage of the Si-C bond is generally not a
facile process. This property must be at least in part responsible for the development of various thermally stable organosilicon
polymers which have found a number of significant practical applications. Although much weaker than the Si-C bond, typical Sn-C
bonds do not readily react with a free radical either. On the

other hand, the Si-H and Sn-H bonds are quite reactive toward
free-radical species. The difference between the Si-C and Si-H
bonds cannot be interpreted in terms of their bond energies. The
high reactivity of the bonds to hydrogen relative to the reactivity of the corresponding bonds to carbon is reminiscent of the
situation with respect to the relative reactivities of the B-H
and B-C bonds (Sect. 5.2.3.6). Although not clear, these differences may qualitatively be attributable to the spherical nature
of the s orbital of hydrogen and its sterically least demanding
nature, which presumably permit an approach of the attacking species with minimum steric congestion.

6.1.2   Silicon-Hetero Atom and Tin-Hetero Atom Bonds

As mentioned earlier, the Si-X and Sn-X bonds, where X = N, O, F,
Cl, and Br, are thermodynamically quite stable. These bonds, however, are chemically highly labile and readily participate in
various substitution reactions. For example, they can be cleaved
readily with water or alcohols. This fact is also responsible for
the effectiveness of trialkylsilyl groups as hydroxy- and amino-
protection groups. The large size of silicon (covalent radius
1.17 Å) compared to that of carbon (covalent radius 0.77 Å), the
greater polarizability of the Si-X bond relative to that of the
C-X bond, and the possible participation of the silicon $3d$ orbitals appear to be the major factors responsible for the greater
ease of substitution reactions at silicon than at carbon.

The mechanistic pathways for substitution reactions taking
place at silicon centers have been studied and discussed in detail by Sommer (6.6). In brief, these substitution reactions proceed either with inversion or retention of configuration at the
silicon center. The S_N1 mechanism, which would lead to racemization, rarely operates. The inversion reactions most probably proceed by one or more types of S_N2 mechanisms, which may or may not
involve an unstable intermediate. On the other hand, formation of
a full-fledged pentacovalent silicon intermediate in a rate-
determining step appears to be highly unlikely based on the stereochemical and kinetic results. One of the $3d$ orbitals may participate in the inversion reactions, as for example, in 6.1. The
results, however, can also be accommodated by the conventional
S_N2 mechanism (6.2) involving no $3d$ orbital participation.

The retention reactions most probably involve quasicyclic transition states which are generally four-centered as in 6.3 and 6.4.

6.3   6.4

6.1.3 Effect of the Silicon or Tin Atom on the Stability of a Cation, Anion, or Free-Radical Center

The stability of a cation, anion, or free-radical center that is close to the silicon or tin atom can be greatly influenced by the metal atom. The availability of the d orbitals and the inductive and conjugative effect of the M–C bond appear to be mainly responsible for the observed influences. The available information allows us to make the following generalizations summarized in Table 6.3.

Table 6.3. Effect of the Silicon or Tin Atom on the Stability of a Cation, Anion, or Free-Radical Center

Type	Metal	Position α	Position β
Cation	Si	Slightly destabilized (?)	Stabilized (σ-π conjugation)
Cation	Sn	Stabilized (+I effect)	Stabilized (α-π conjugation)
Anion	Si	Somewhat stabilized (p_π-d_π interaction)	?
Anion	Sn	Somewhat stabilized (p_π-d_π interaction)	?
Free radical	Si	Stabilized (p_π-d_π interaction)	?
Free radical	Sn	Stabilized (p_π-d_π interaction)	Stabilized (σ-π conjugation)

6.2 STRUCTURE

6.2.1 Tetracoordinate Species and Optical Property

As mentioned earlier, the majority of organometallics containing silicon or tin exist as unassociated tetracoordinate species in which the metal atoms occupy the central position of an essentially tetrahedral configuration. In fact, tetraalkylsilanes and tetraalkylstannanes resemble branched paraffins of similar molecular weights with respect to boiling point, solubility, and appearance. When the four groups bonded to the metal atom are different, the metal atom provides a chiral center. A number of such chiral organosilanes and organostannanes have been prepared and isolated as optically active species, indicating that such chiral centers are stereochemically rigid (*6.6*).

6.2.1.1 Multiple Bonding Involving Silicon and Tin

$\underline{p_\pi\text{-}d_\pi \text{ Bond}}$ There is little doubt that $p_\pi\text{-}d_\pi$ bonding (<u>6.5</u>) plays an important role in many species containing the Group IVA metals. The bond parameters, such as dissociation energy and bond length, clearly point to the significance of the $p_\pi\text{-}d_\pi$ bonding in the bond between silicon or tin and N, O, or halogens.

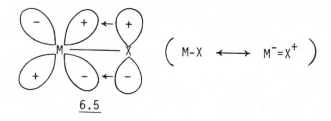

6.5

Stereochemical data also corroborate the above conclusion. Thus unlike trialkylamines, which are pyramidal, trisilylamines, such as $N(SiH_3)_3$, are planar. The $p_\pi\text{-}d_\pi$ bonding between nitrogen and silicon must be responsible for planarity. Despite the partial double-bond character, however, the rotation about the Si-X or Sn-X bond, where X = N, O, or halogen, is usually not frozen at ambient temperature.

pyramidal   planar

p_π-p_π Interaction Despite the great abundance of stable C=C and C≡C compounds, there has been no isolated example of organometallics containing the Si=X or Sn=X bond of the p_π-p_π type where X = C, O, N, and so on. The existence of the Si=X species, where X = C, N, O, S, or Si, as short-lived species, however, has been amply demonstrated in recent years (6.7). Thus, for example, the following transformations and trapping experiments are readily interpreted only if one assumes the intermediacy of $Me_2Si=CH_2$ (6.7).

$$Me_2Si \quad \xrightarrow{400°C} \quad Me_2Si=CH_2 \longrightarrow Me_2Si \quad SiMe_2 \qquad (6.1)$$

$$+ \quad CH_2=CH_2 \qquad + \quad (Me_2SiCH_2)_n$$

$$Me_2Si=CH_2 \quad \xrightarrow{H_2C=CHCH=CH_2} \quad Me_2Si \qquad (6.2)$$

$$(6.3)$$

One of the most facile thermal routes to Si=C and Si=Si compounds involves pyrolysis of the corresponding bicyclo[2.2.2]octadienes, for example, 6.6 (6.8) and 6.8 (6.9).

$$\text{6.6} \xrightarrow{400^{\circ}C} Me_2Si=CH_2 \longrightarrow \text{6.7} \qquad (6.4)$$

$$\text{6.8} \xrightarrow{260^{\circ}C} Me_2Si=SiMe_2 \xrightarrow{PhCH=CHCH=CHPh} \text{6.9} \qquad (6.5)$$

The exact reason for the instability of these species containing Si=X bonds is not clear. Pitzer has postulated that the low π-bond strength is due to the repulsion between bonding orbitals and filled inner shells (6.10). On the other hand, Mulliken (6.11) has proposed that the low π-bond stability might merely mean that the π-bonded products, that is, cyclodimers and polymers, are more stable than the Si=X compounds and that the kinetic barriers to such interconversions are low.

6.2.1.2 Penta- and Hexacoordinate Complexes

The availability of empty d orbitals for coordination suggests the existence of complexes in which the coordination number is greater than four. In addition to various presumed intermediates, which are believed to be pentacoordinate species, a number of penta- and hexacoordinate species containing Group IVA elements, such as 6.10 and 6.11, have been characterized (6.12, 6.13)

$$K(Ph_4SnNH_2) \qquad\qquad R_2SnCl_2 \cdot 2Py$$

$$\text{6.10} \qquad\qquad\qquad \text{6.11}$$

6.2.1.3 Tricoordinate Anions, Cations, and Free Radicals

As described in the following section, a number of bimetallic complexes of the M(M'R₃) type, where M = alkali metal and M' = Si, Ge, Sn, or Pb, have been prepared (6.14). The ⁻M'R₃ moiety may be viewed as a tricoordinate anionic species. Although thermodynamically stable, they are chemically highly active as nucleophiles, and have demonstrated some useful synthetic capabilities, as discussed later in this chapter.

The silicon and tin free radicals have been postulated as intermediates in various reactions of organosilicons and organotins and have been detected spectroscopically (*6.15*, *6.16*). Bond energy data suggest that silyl radicals are poor abstractors of hydrogen from organic molecules, but are good abstractors of halogens.

Highly elusive has been the tricoordinate silicon cation, called either siliconium ion or silelinium ion (*6.17*). For gaseous silicon atom ($3p^23p^2$) the first ionization energy to form silicon anion is 187.9 kcal/mole, whereas the corresponding value for carbon is 259.5 kcal/mole (*6.6*). In solution, however, solvation plays an important role, thereby making the formation of carbocations a relatively facile process. The formation of silicon cations in solution appears to be a highly unfavorable process. On the other hand, certain Sn-X bonds are easily polarizable so that organotins with strongly negative ligands, such as halogens, dissociate to a minor extent. For example, the specific conductance of Et_3SnCl is 1.8×10^{-9} ohm^{-1}cm^{-1} at 20°C in benzene. The following dissociation must be responsible for the conductivity.

$$Et_3SnCl \quad \rightleftharpoons \quad Et_3Sn^+ + Cl^- \qquad (6.6)$$

6.2.1.4   Discoordinate Species

Discoordinate species, MX_2, containing Si, Ge, Sn, and Pb are analogs of carbenes. Divalent silicon derivatives (silylenes) have been known for many decades and the thermodynamic stability of the divalent state of silicon has been long recognized. Especially noteworthy is the stability of SiF_2, the half-life of which is about 150 sec at low pressure and ambient temperature (*6.18*). Although less stable, diorganosilylenes, such as Me_2Si, have also been generated as reactive intermediates. The chemistry of silylenes has been reviewed extensively (*6.18*, *6.19*). Through the use of bulky ligands, for example, $(Me_3Si)_2CH^-$, diorganometals containing Ge, Sn, and Pb, which are monomeric in benzene or cyclohexane, have been obtained (*6.20*). These carbene analogs of the Group IVA metals exhibit chemical behavior similar to that of carbenes. Thus they undergo insertion reactions, as shown in eq. 6.7 (*6.19*), and formation of complexes with transition metal compounds (*6.20*, *6.21*).

$$
Me_2SiSiMe_2 \xrightarrow[225°C]{MeC\equiv CMe}
\left[
\begin{array}{c}
SiMe_2 \\
\diagup \diagdown \\
MeC = CMe
\end{array}
\right]
\qquad (6.7)
$$

The ring intermediate reacts with MeOH to give two products:

$$
\underset{Me \quad Me}{Me_2Si \diagup \diagdown SiMe_2}
$$

$$
\underset{H}{Me}C=C\underset{SiMe_2(OMe)}{Me}
$$

6.3 PREPARATION OF ORGANOSILICONS AND ORGANOTINS

Transmetallation (Method XI) and hydrometallation (Method VI) represent the two most common laboratory methods for the synthesis of organosilicons and organotins, while oxidative metallation (Method I) of organic halides ("direct synthesis") is the most significant industrial method for the synthesis of organosilicons, which is also applicable to the synthesis of certain organotin compounds. The reactions of silyl anions ($^-SiX_3$) and stannyl anions ($^-SnX_3$) with organic halides and α,β-unsaturated carbonyl compounds (Method II) appear highly promising as routes to organosilicons and organotins. The great majority of organosilicons and organotins have been prepared by these four general methods. Although much less general, oxidative coupling involving divalent silicon and tin compounds, discussed in the preceding section, provides a unique method of preparation.

6.3.1 Oxidative Metallation with Silicon and Tin

The so-called direct synthesis of organosilicons and organotins consists of the high-temperature reaction (100 to 550°C) of alkyl or aryl halides, usually chlorides or bromides with metallic silicon or tin usually in the presence of a catalyst, such as metallic Cu, Ag, or Zn. This method provides a basis for the preparation in industry of Me_2SiCl_2, Ph_2SiCl_2, and other organohalosilanes. Despite its industrial significance, however, the method is not frequently used in research laboratories presumably due to the rather drastic conditions required. Therefore, no detailed discussion of the method is presented here. Interested readers are referred to pertinent monographs (6.1 to 6.5, 6.22) and reviews (6.23).

6.3.2 Organosilicons and Organotins via Transmetallation

The transmetallation reactions shown in eqs. 6.8 and 6.9 provide the most general route by far to organosilicons and organotins, respectively.

$$R^1_m SiX_{4-m} + \frac{4-m}{n}MR^2_n \longrightarrow R^1_m SiR^2_{4-m} + \frac{4-m}{n}MX_n \quad (6.8)$$

$$R^1_m SnX_{4-m} + \frac{4-m}{n}MR^2_n \longrightarrow R^1_m SnR^2_{4-m} + \frac{4-m}{n}MX_n \quad (6.9)$$

m = 0,1,2, or 3 n = 1,2,$\cdots$ X = F, Cl, Br, alkoxy, and so on.

The scope of these reaction is quite broad, and organosilicons and organotins containing various types of organic groups have been prepared by these reactions. A variety of organometallic reagents (MR^2_n), such as those containing Group IA and Group IIA elements, zinc and aluminum, can be used as starting compounds. The ability to transfer organic ligands to the silicon or tin atom, however, is roughly proportional to the electropositivity of M. This factor and their ready availability make organolithiums and organomagnesiums the reagents of choice. Although Grignard reagents have been most widely used in the past, it appears that organolithiums often offer distinct advantages over Grignard reagents, as detailed below.

Thus if one uses organolithiums, even highly hindered tetra-organotins, such as 6.12, 6.13, and 6.14, can be readily prepared. These tin compounds, however, cannot be cleanly obtained in high yiels by the reactions of the corresponding Grignard reagents; see eqs. 6.10 to 6.12 (*6.1, 6.3 to 6.5*).

$$\underset{\text{6.12}}{\qquad} \quad \xleftarrow{\quad +\text{MgCl}\quad} \quad SnX_4 \quad \xrightarrow{\quad +\text{Li}\quad} \quad Sn(\text{—}{\big|})_4 \qquad (6.10)$$

6.12

$$(\text{C}_6\text{H}_{11})_3\text{SnSn}(\text{C}_6\text{H}_{11})_3 \quad \xleftarrow{\ \text{C}_6\text{H}_{11}\text{-MgBr}\ } \quad \text{SnX}_4 \quad \xrightarrow{\ \text{C}_6\text{H}_{11}\text{-Li}\ } \quad \text{Sn}(\text{C}_6\text{H}_{11})_4$$

$$+ \ \underline{6.13} \qquad\qquad\qquad\qquad\qquad\qquad\qquad \underline{6.13}$$

$$(6.11)$$

$$(\text{biphenylyl})_3\text{SnBr} \quad \xleftarrow{\ \text{biphenylyl-MgBr}\ } \quad \text{SnBr}_4 \quad \xrightarrow{\ \text{biphenylyl-Li}\ } \quad (\text{biphenylyl})_4\text{Sn}$$

$$\underline{6.14}$$

$$(6.12)$$

All attempts to prepare the corresponding organosilicon compounds 6.15, 6.16, and 6.17 have failed, the structures of the most highly substituted products being those shown in eqs. 6.13 to 6.15 (6.22).

$$\text{Si}(t\text{-Bu})_4 \qquad\qquad \text{Si}(\text{C}_6\text{H}_{11})_4 \qquad\qquad (\text{biphenylyl})_4\text{Si}$$

$$\underline{6.15} \qquad\qquad\qquad \underline{6.16} \qquad\qquad\qquad \underline{6.17}$$

$$t\text{-Bu-Li} + \text{SiCl}_4$$
$$\qquad\qquad\qquad\qquad\searrow$$
$$\qquad\qquad\qquad\qquad\qquad (t\text{-Bu})_2\text{SiCl}_2 \qquad\qquad (6.13)$$
$$\qquad\qquad\qquad\qquad\nearrow$$
$$t\text{-Bu-MgCl} + \text{SiCl}_4$$

$$\text{C}_6\text{H}_{11}\text{-Li (or } \text{C}_6\text{H}_{11}\text{-MgBr}) + \text{SiCl}_4 \longrightarrow (\text{C}_6\text{H}_{11})_3\text{SiCl} \qquad (6.14)$$

$$\text{biphenylyl-Li} + \text{SiCl}_4 \longrightarrow (\text{biphenylyl})_3\text{Si} \qquad (6.15)$$

The reaction of SnX_2 with polar organometallics, such as those containing Li, Mg, and Zn produces polymeric products, as shown in eq. 6.16, while their reaction with organometallics of heavy metals, such as Hg, Tl, and Pb, involves oxidative-reductive transmetallation (Method XII), as shown in eq. 6.17 (*6.1*, *6.3* to *6.5*).

$$n \ SnX_2 \ + \ 2n \ RLi \ \longrightarrow \ (R_2Sn)_n \ + \ 2n \ LiX \qquad (6.16)$$

$$SnX_2 \ + \ R_2Hg \ \longrightarrow \ R_2SnX_2 \ + \ Hg \qquad (6.17)$$

6.3.3 Hydrosilation and Hydrostannation

The Group IVA metal hydrides react with the C=C and C≡C bonds to form the corresponding addition products as shown in eq. 6.18 and 6.19.

$$\overset{|}{\underset{|}{C}}=\overset{|}{\underset{|}{C}} \ + \ HMX_3 \ \longrightarrow \ H-\overset{|}{\underset{|}{C}}-\overset{|}{\underset{|}{C}}-MX_3 \qquad (6.18)$$

$$-C\equiv C- \ + \ HMX_3 \ \longrightarrow \ H-\overset{|}{C}=\overset{|}{C}-MX_3 \qquad (6.19)$$

$$M = Si, \ Ge, \ Sn, \ or \ Pb$$

The reaction of hydrosilanes, discovered in 1947 (*6.24*), is usally called hydrosilylation, whereas both hydrostannation and hydrostannylation have been used for the H-Sn reaction (*6.25*). If we were to adopt hydrostannylation for the H-Sn reaction, the reaction of boron hydrides, for example, should be called hydroborylation. For the sake of consistency and simplicity, we tentatively adopt the simpler of the two conventions in this book. Thus the M-H addition reactions, where M = Si, Ge, Sn, and Pb, will be termed hydrosilation, hydrogermation (*6.26*), hydrostannation, and hydroplumbation (*6.27*), respectively. In addition to several monographs covering all aspects of organosilicon and organotin chemistry (*6.1* to *6.5*), there is a monograph devoted to the reactions of the Group IVA metal hydrides with the C=C and C≡C bonds (*6.28*).

6.3.3.1 Free-Radical Hydrosilation

Hydrosilanes add to unsaturated compounds at temperatures of about 300°C without a catalyst. The reaction can be promoted by γ-radiation, UV irradiation, and electric discharges and catalyzed by

various free-radical initiators, such as peroxides and azonitri-
les. Under these conditions, hydrosilation appears to proceed
generally by a free-radical mechanism shown in eq 6.20.

$$HMX_3 \longrightarrow \cdot MX_3$$

$$\overset{|\;\;|}{\underset{|\;\;|}{C=C}} + \cdot MX_3 \longrightarrow \cdot \overset{|\;\;|}{\underset{|\;\;|}{C-C}}-MX_3 \qquad\qquad (6.20)$$

$$\cdot \overset{|\;\;|}{\underset{|\;\;|}{C-C}}-MX_3 + HMX_3 \longrightarrow H-\overset{|\;\;|}{\underset{|\;\;|}{C-C}}-MX_3 + \cdot MX_3$$

$$M = Si \text{ or } Sn$$

In accordance with the mechanism presented above, the silicon-containing group adds to the C=C or C≡C group such that the
more stable of the two possible free radicals is formed either
exclusively or predominantly (6.28).

$$n\text{-}C_6H_{13}CH=CH_2 \xrightarrow[\text{(AcO)}_2,\ 45^\circ C]{HSiCl_3} n\text{-}C_6H_{13}CH_2CH_2SiCl_3 \qquad (6.21)$$

$$99\%$$

$$(CH_3)_2C=CH_2 \xrightarrow[250 \text{ to } 300^\circ C]{HSiCl_3} (CH_3)_2CHCH_2SiCl_3 \qquad (6.22)$$

$$80\%$$

Even highly hindered olefins can be readily hydrosilated with appropriate hydrosilanes (6.28).

$$(CH_3)_2C=C(CH_3)_2 \xrightarrow[\text{(AcO)}_2,\ 50^\circ C]{HSiCl_3} (CH_3)_2CH\overset{CH_3}{\underset{CH_3}{\overset{|}{\underset{|}{C}}}}SiCl_3 \qquad (6.23)$$

$$59\%$$

There are a few serious limitations. Firstly, the hydrosilation reaction carried out under the free-radical conditions is
nonstereoselective (eq. 6.24)

$$n\text{-C}_5\text{H}_{11}\text{C}{\equiv}\text{CH} \quad \xrightarrow[\text{(PhCOO)}_2,\ 60\ \text{to}\ 70^\circ\text{C}]{\text{HSiCl}_3} \quad \begin{array}{c} n\text{-C}_5\text{H}_{11} \\[2pt] \diagdown \\ \text{H} \end{array}\!\!\text{C}{=}\text{C}\!\!\begin{array}{c} \diagup \text{H} \\[2pt] \diagdown \text{SiCl}_3 \end{array} \qquad (6.24)$$

47% (cis:trans = 75:25)

Secondly, those olefins and acetylenes that readily undergo free-radical polymerization, such as styrene and acrylonitrile, generally produce polymers. Fortunately, both of these difficulties can be readily overcome by certain Group VIII metals and their complexes, as discussed in the following section.

6.3.3.2 Transition Metal-Catalyzed Hydrosilation

The hydrosilation reaction can also be catalyzed by certain Group VIII metals and their complexes. One of the most widely used catalysts is $H_2PtCl_6\cdot 6H_2O$, which induces an exclusive cis addition of the Si–H bond to the C=C (*6.29*) and C≡C (*6.30*) bonds.

The reaction of simple terminal acetylenes with $HSiCl_3$ in the presence of a catalytic amount of $H_2PtCl_6\cdot 6H_2O$ produces the *trans*-β-adducts (6.18) as major products and α-adducts (6.19) as minor products. No *cis*-β-adducts are formed; see Table 6.4 (*6.30*).

$$\text{RC}{\equiv}\text{CH} \quad \xrightarrow[\text{H}_2\text{PtCl}_6\ 6\text{H}_2\text{O}]{\text{HSiCl}_3} \quad \begin{array}{c} \text{R} \\ \diagdown \\ \text{H} \end{array}\!\!\text{C}{=}\text{C}\!\!\begin{array}{c} \diagup\text{H} \\ \diagdown\text{SiCl}_3 \end{array} \quad + \quad \begin{array}{c} \text{R} \\ \diagdown \\ \text{Cl}_3\text{Si} \end{array}\!\!\text{C}{=}\text{C}\!\!\begin{array}{c} \diagup\text{H} \\ \diagdown\text{H} \end{array} \qquad (6.25)$$

6.18 (major)   6.19 (minor)

Table 6.4 Hydrosilation of Terminal Acetylenes

Acetylene	Yield of β-Adducts (trans:cis), %	
	Peroxide	$H_2PtCl_6\cdot 6H_2O$
n-PrC≡CH	35(21:79)	84(100:0)
i-PrC≡CH	37(28:72)	79(100:0)
t-BuC≡CH	39(100: 0)	50(100:0)

Although highly stereoselective, the platinum-catalyzed hydrosilation is not regiospecific, and the reaction of olefins often results in the formation of silanes in which the silicon

atom is bonded to the least hindered carbon atom of the organic
groups regardless of the position of the C=C bond (6.29). Thus
both 1- and 2-pentenes give n-pentylsilanes, and 1-methylcyclo-
hexene is converted to the cyclohexylmethyl derivative (eq. 6.26).

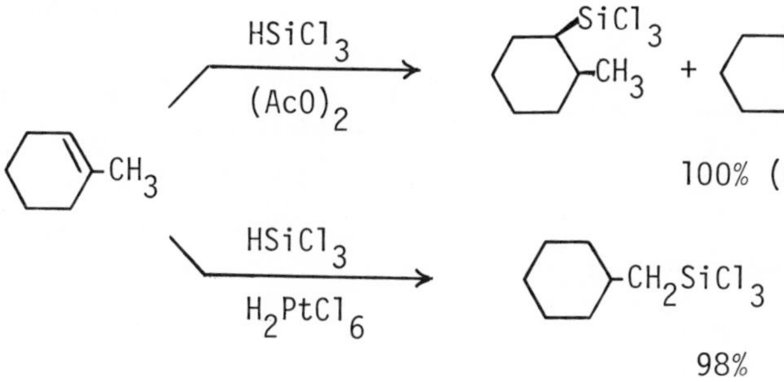

100% (85:15)   (6.26)

98%

Recently, the rhodium-catalyzed hydrosilation reaction has
been developed as a convenient method for reducing α,β-unsaturated
carbonyl compounds (6.31). The reaction takes places under mild
conditions, in a highly regioselective manner, and appears to be
superior to most of the other procedures involving the use of
Group VIII metals and their complexes. Interestingly, triorganosilanes undergo a 1,4-addition reaction, whereas diorganosilanes
undergo a 1,2-addition reaction (eq. 6.27).

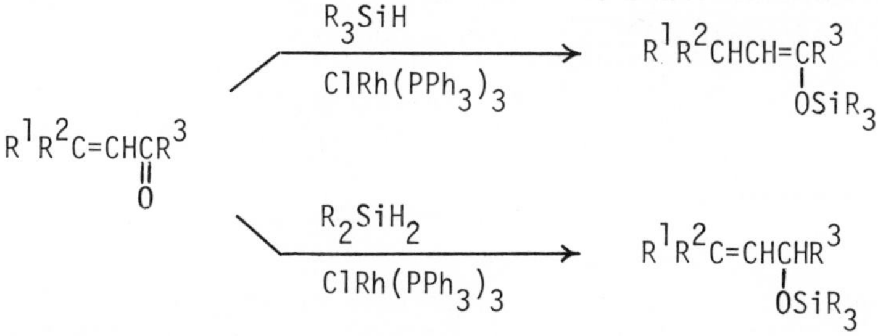

(6.27)

The products can be readily hydrolyzed to the corresponding saturated ketones and allylic alcohols, respectively. The scope of
the rhodium-catalyzed 1,4-addition reaction is indicated by the
following examples (6.31).

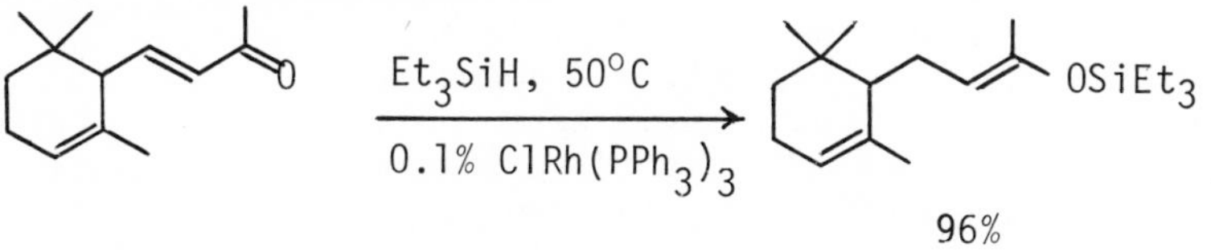

96%

$$\text{Et}_3\text{SiH, 80}^\circ\text{C}$$
$$0.1\% \text{ ClRh(PPh}_3)_3$$

(89%)

(6.29)

+ (6%)

As discussed later (Sect. 10.1), the transition metal-catalyzed hydrosilation most probably involves a sequence consisting
of (1) formation of a transition metal hydride species, (2) its
addition to the C=C or C≡C bond, and (3) transmetallation to produce the desired organosilanes.

6.3.3.3 Miscellaneous Hydrosilation Reactions

The hydrosilation reaction can also be catalyzed by various other
types of compounds, such as tertiary amines (6.28), triorganopho-
sphines, and various metals and their derivatives, such as Na,
Zn, ZnCl$_2$, TiCl$_4$, and ZrCl$_4$ (6.28). Some of these catalysts appear capable of inducing Si-H-addition reactions which proceed by
totally different mechanisms than those discussed above. For example, the addition of SiHCl$_3$ to phenylacetylene in the presence
of a tertiary amine evidently involves a nucleophilic addition of
$^-$SiCl$_3$ generated by the reaction of SiHCl$_3$ with the added tertiary
amine (6.32). At present it is not clear, however, if any of
these less-known procedures will provide any unique advantages
over the ones discussed earlier.

6.3.3.4 Hydrostannation

The hydrostannation reaction takes place considerably more readily
than the corresponding hydrosilation reaction, and is usually carried out at 80 to 100°C either in the absence of a catalyst or
in the presence of catalytic amounts of free-radical initiators.
The applicability of various catalysts, such as those used in the
hydrosilation reaction, does not appear to have been well delineated.
    Both olefins and acetylenes can readily be hydrostannated.
The reaction of acetylenes with trialkylstannanes producing the
corresponding alkenyltrialkylstannanes, however, is of particular
synthetic interest, since the alkenyl group of such products can
be selectively transferred from tin to various other atoms including C, H, halogen, and Li while retaining the structural inte-

grity of the alkenyl group.

The regio- and stereochemistry of the triorganostannane-acetylene reaction has been investigated extensively, and some representative results are summarized in Table 6.5 (6.33 to 6.36).

Table 6.5 The Reaction of Triorganostannanes with Acetylenes

R	R¹	Reaction Cond.		Yield (%)				Ref.
		Temp.	Time	Total	I	II	III	
Me	Bu	60°C,	7 hr	85	2	29	69	6.33
Ph	Bu	50°C,	3 hr	>80	0	85	15	6.33
Et	Ph	60°C,	7 hr	75	0	70	30	6.33
Ph	Ph	60°C,	1 hr	90	0	28	72	6.33
Bu	CH₂C(H)(CH₃)OTHP	95°C,	-	92	-	15	85	6.34
Bu	CH₂OTHP	80°C,	2 hr	89	0	0	100	6.35
Et	CH₂OH	100°C,	6 hr	>70	25	45	30	6.33
Pr	COOMe	55°C,	6 hr	72	31	36	33	6.33
Et	CN	20°C,	1 hr	100	100	0	0	6.33
Et	OEt	50°C,	1.5 hr	100	4	91	5	6.33
Bu	SnBu₃	-	-	88	0	0	100	6.36

Before attempting to interpret these results, a word of caution may be in order. It has been found that, under the hydrostannation conditions, the hydrostannated products can undergo isomerization. Therefore, the observed product distribution does not usually represent that of the initial kinetic products. Despite such a complication, the following tentative generalization can be made. Firstly, an electron-donating substituent (R^1) places the tin atom predominantly or exclusively on the β-carbon atom, while an electron-withdrawing substituent directs the tin atom more to the α-carbon atom. Secondly, hydrostannation is generally not highly stereoselective. It appears that, under kinetic conditions, trans addition predominates, whereas the trans isomers (i.e., cis addition products) tend to predominate under thermodynamic conditions.

The mechanism of hydrostannation does not appear to have been well-established. In some cases, the reaction is facilitated by the addition of free-radical initiators, such as azobisisobutylonitrile (AIBN), suggesting that the reaction proceeds at least partially by a free-radical mechanism (*6.28*). On the other hand, examination of the effects of substituents, solvents, and isotopes and of the kinetics suggests that the reaction involves a stepwise trans addition of a hydride and a triorganostannyl cation (R_3Sn^+) (*6.33*). It is clear that further investigation is needed to clarify the mechanistic details.

$$R^1C\equiv CR^2 \xrightarrow{R_3SnH} \underset{H}{\overset{R^1}{>}}C=C\overset{-}{\underset{R^2}{<}} \;+\; R_3Sn^+ \longrightarrow \underset{H}{\overset{R^1}{>}}C=C\overset{SnR_3}{\underset{R^2}{<}} \qquad (6.30)$$

6.3.4   Organosilicons and Organotins via Silyl and Stannyl Anions

Various methods for preparing organometallics with metal-alkali metal bonds have been reviewed in detail (*6.14*). There are several routes to bimetallic compounds of the $M(M'R_3)$ type, where M = alkali metal, and M' = Si, Ge, Sn, or Pb (eqs. 6.31 to 6.33).

$$XM'R_3 \;+\; 2M \longrightarrow M(M'R_3) \;+\; MX \qquad (6.31)$$

$$(M' = Si, Ge, Sn, or Pb)$$

$$HM'R_3 \;+\; 2M \longrightarrow M(M'R_3) \;+\; MH$$
or $\qquad\qquad\qquad\qquad\qquad\qquad\qquad\qquad (6.32)$
$$HM'R_3 \;+\; M \longrightarrow M(M'R_3) \;+\; \tfrac{1}{2}H_2$$

$$(M' = Si, Ge, or Sn)$$

$$R_3M'-M'R_3 \quad + \quad 2M \longrightarrow 2M(M'R_3) \tag{6.33}$$

$(M' = Si \text{ or } Ge)$

These three methods require an alkali metal as a reagent. More recently, the following convenient method involving the use of organolithiums has been developed (6.37, 6.38).

$$R_3M'-M'R_3 \quad + \quad R^1Li \longrightarrow LiM^1R_3 \quad + \quad M'R_3R' \tag{6.34}$$

$(M' = Si \text{ or } Sn)$

In many respects triorganosilylmetals ($MSiR_3$) and triorgano-stannylmetals ($MSnR_3$), where M = alkali metals, behave just like organoalkali metals (6.14). They are strong bases as well as excellent nucleophiles. Only those reactions that involve the Si-C or Sn-C bond formation are discussed here. The general patterns of their reactions with typical carbon electrophiles are summarized in eqs. 6.35 to 6.40.

$$M(M'R_3) \quad + \quad R'X \longrightarrow R'M'R_3 \quad + \quad MX \tag{6.35}$$

$$M(M'R_3) \quad + \quad R^1COR^2(H) \longrightarrow \begin{array}{c} R^1 \\ \diagdown \\ (H)R^2 \diagup \end{array}\!\!C\!\!\begin{array}{c} M'R_3 \\ \diagup \\ \diagdown OH \end{array} \tag{6.36}$$

$$M(M'R_3) \quad + \quad R'COCl \longrightarrow R'COM'R_3$$
$$M(M'R_3) \quad + \quad R'COM'R_3 \longrightarrow R'\underset{\underset{OH}{|}}{C}(M'R_3)_2 \tag{6.37}$$

$$M(M'R_3) \quad + \quad \underset{O}{\overset{\diagdown}{C}\!-\!\overset{\diagup}{C}} \longrightarrow R_3M'-\overset{|}{\underset{|}{C}}-\overset{|}{\underset{|}{C}}-OH \tag{6.38}$$

$$M(M'R_3) \quad + \quad \overset{|}{\underset{|}{C}}=C-\overset{|}{C}=O \longrightarrow R_3M'-\overset{|}{\underset{|}{C}}-CHC=O \tag{6.39}$$

$$M(M'R_3) \quad + \quad \overset{|}{\underset{|}{C}}=\overset{|}{\underset{|}{C}} \longrightarrow R_3M'-\overset{|}{\underset{|}{C}}-\overset{|}{\underset{|}{C}}-M \tag{6.40}$$

All of these reactions provide routes to organosilicons and organotins.

The reaction with organic halides is often complicated by the alkali metal-halogen exchange reaction (6.14).

$$M(M'R_3) \quad + \quad R^1X \quad \longrightarrow \quad XM^1R \quad + \quad MR^1 \qquad (6.41)$$

At present, the synthetic utility of the reaction of $M(M^1R_3)$ with organic halides is not clear, since the same products usually can be obtained in a more satisfactory manner by the reaction of $XM'R_3$ with MR'.

The reaction with aldehydes and ketones provides a convenient route to α-hydroxysilanes and α-hydroxystannanes, for example, eq. 6.42, although the reaction is often complicated by abnormal addition (eq. 6.43) and enolization of the carbonyl compounds.

$$LiSiPh_3 \quad + \quad Me_2CO \quad \longrightarrow \quad Ph_3SiC(OH)Me \qquad (6.42)$$
$$52\%$$

$$M(M^1R_3) \quad + \quad R^1COR^2(H) \quad \longrightarrow \quad \underset{(H)R^2}{\overset{R^1}{\diagdown}}C\underset{OM^1R_3}{\overset{M}{\diagup}} \qquad (6.43)$$

Due to various competing side reactions, the reaction with acyl halides is generally poor as a route to acylsilanes and acylstannanes (6.14).

Both acyclic and cyclic ethers react with $MSiR_3$ and $MSnR_3$. The reaction of epoxides with these organometallics is facile, producing β-hydroxysilanes and β-hydroxystannanes. The reaction generally involves a highly stereospecific trans attack of the least hindered carbon atom. Thus the reaction presumably occurs by an S_N2 mechanism. As detailed later, β-heterosubstituted organosilicons and organostannes readily undergo β-elimination. Thus epoxides can be readily deoxygenated by the following sequence, which involves an overall inversion of configuration. Since olefins can be converted to epoxides with retention of configuration, the epoxidation-silylation-elimination sequence (eq. 6.44) achieves inversion of configuration of olefins (6.39).

$$(6.44)$$

99% yield based on epoxide

>99% E

Table 6.6 The Reaction of $LiSiMe_3$ and $LiSnMe_3$ with α,β-Unsaturated Carbonyl Compounds

α,β-Unsaturated Carbonyl Compound	Product	Yield (%) M = Si	Yield (%) M = Sn
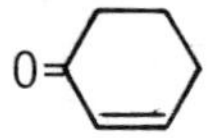	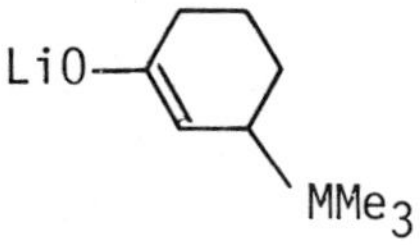	97	96
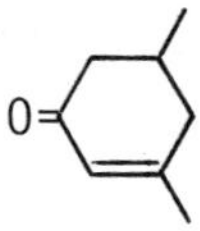	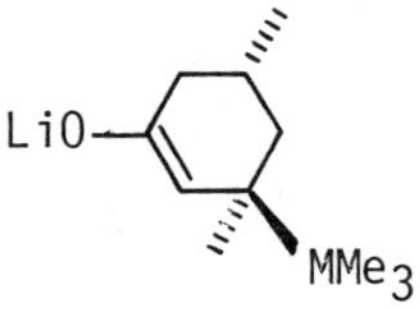	78	93
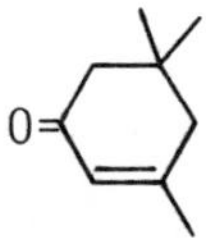	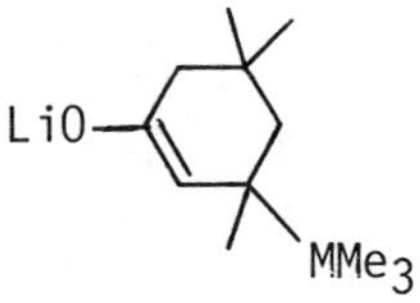	0	77
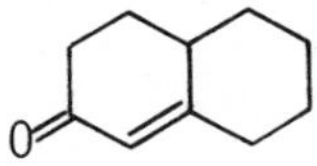	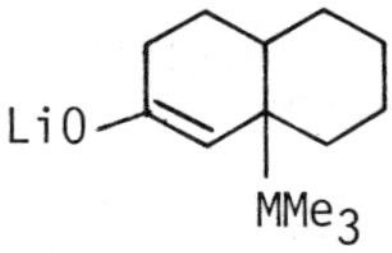	-	94
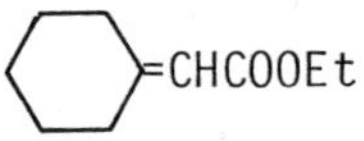	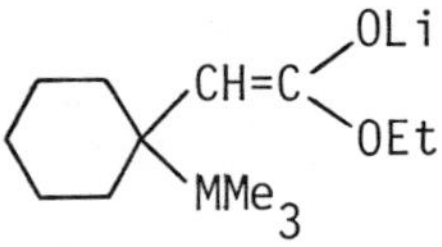	-	80
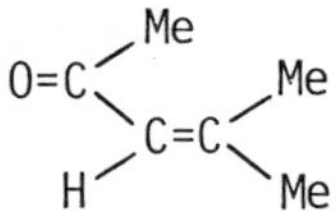	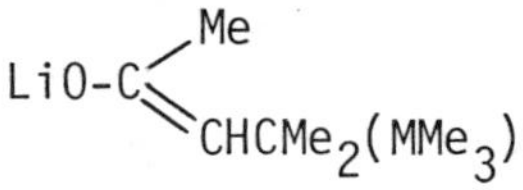	-	93

The bimetallic compounds of the $MSiR_3$ or $MSnR_3$ type react with various olefins and acetylenes to give the corresponding addition products. The results observed with unactivated olefins and acetylenes, such as ethylene, are not encouraging. The addition reaction, however, proceeds nicely with certain activated olefins and acetylenes, especially with α,β-unsaturated carbonyl compounds, to give the corresponding 1,4-addition products. The scope and limitation of the reaction of $LiSiMe_3$ (6.37) and (6.38) with α,β-unsaturated carbonyl compounds are indicated in Table 6.6. It should be noted that the scope of the conjugate addition reaction with $LiSnMe_3$ is one of the broadest in that high steric hindrance at the β-carbon can be tolerated. The axial attack appears to be the preferred mode of reaction. The products behave as typical lithium enolates in their protonation, silylation, and alkylation (6.14). Some of their unique synthetic applications are discussed in Sect. 6.4.3.

6.4   REACTIONS OF ORGANOSILICONS AND ORGANOTINS

6.4.1   General Reaction Patterns

In the preceding section, methods for the formation of the Si-C and Sn-C bonds are discussed. If the Si-C and Sn-C bonds were to directly participate in the formation of organic compounds that do not contain silicon or tin, these bonds would somehow have to be cleaved.

The discussion presented in Sect. 6.1. permits the following tentative generalization concerning the intrinsic reactivity of the Si-C and Sn-C bonds. Firstly, the typical unactivated Si-C bond, such as the Si-Et bond, is reluctant to participate directly in either nucleophile-electrophile or free-radical reactions. Secondly, the Sn-C bond is considerably more reactive toward both electrophiles and nucleophiles. Thirdly, homolytic cleavage of the Sn-C bond is not a facile process, although certain reactions of organotins involving Sn-C cleavage presumably proceed by free-radical mechanisms, as shown in eq. 6.45 (6.40, 6.41).

$$Br\cdot \ + \ R_4Sn \ \longrightarrow \ R_3SnBr \ + \ R\cdot$$

$$R\cdot \ + \ Br_2 \ \longrightarrow \ RBr \ + \ Br\cdot$$

$$(6.45)$$

6.4.1.1   Free-Radical Reactions Involving the Cleavage of the Si-C or Sn-C Bond

Based on the generalization presented above, it is not surprising that there are very few synthetically useful reactions involving homolytic cleavage of the Si-C or Sn-C bond.

6.4.1.2   Reaction with Electrophiles -- Significance of σ-π Conjugation

Certain functionalized organometallics of the Group IVA elements are highly reactive toward a variety of electrophiles. Those organometallics that readily participate in reactions with electrophiles include (1) α,β-unsaturated derivatives, such as alkenyl, aryl, and alkynyl derivatives, and (2) β,γ-unsaturated derivatives, such as allyl and benzyl derivatives. Mechanistic studies by Eaborn (6.42) and Traylor (6.43) have revealed that these reactions proceed stepwise without involving direct M-C bond cleavage, and that the key intermediate is the β-silyl or β-stannyl carbocationic species (6.20) stabilized though σ-π conjugation (hyperconjugation).

6.20

Quite significantly, the β-silyl or β-stannyl carbocation 6.20 undergoes a spontaneous elimination reaction to form the corresponding unsaturated organic compounds (eq. 6.46).

$$-\overset{|}{\underset{|}{C}}{}^{+}-\overset{\overset{\textstyle MX_3}{|}}{\underset{|}{C}}- \quad\longrightarrow\quad -\overset{|}{C}=\overset{|}{C}- \qquad (6.46)$$

6.20

Several representative reaction schemes involving the intermediacy of 6.20 are shown in eqs. 6.47 to 6.51.

$$-\overset{|}{\underset{|}{C}}=\overset{|}{\underset{|}{C}}-MR_3 \xrightarrow{\ E^+\ } -\overset{|}{\underset{|}{C}}{}^{\pm}\overset{\overset{\textstyle MR_3}{|}}{\underset{|}{C}}-E \longrightarrow -\overset{|}{C}=\overset{|}{C}-E \qquad (6.47)$$

$$-C\equiv C-MR_3 \xrightarrow{\ E^+\ } -C{}^{\pm}\overset{\overset{\textstyle MR_3}{|}}{C}-E \longrightarrow -C\equiv C-E \qquad (6.48)$$

$$\text{(6.49)}$$

$$\text{(6.51)}$$

There can be various other routes to 6.20, for example, eq. 6.52. It appears that in each case we can expect to obtain the corresponding elimination product.

$$\text{(6.52)}$$

The significance of σ-π conjugation and of 6.20 as a key intermediate in the reactions of organometallics containing the Group IVA elements must be stressed. The magnitude of the σ-π conjugation effect may be appreciated by comparing the electron-donating ability of silicon- and tin-containing groups with that of more familiar organic groups, such as Me and MeO groups (6.43c). As shown in Table 6.7, the electron-donating ability of the CH_2SnPh_3 and CH_2SnMe_3 groups, which has been shown to be largely due to σ-π conjugation rather than inductive effects, is either comparable to or greater than that of the MeO group. Other factors being comparable, the magnitude of the conjugation effect increases in the order: Si < Ge << Sn << Pb. Thus, for example, the relative rates of cleavage of $PhMEt_3$ by aqueous methanolic perchloric acids are: Si, 1; Ge, 36; Sn, 3.5×10^5; and Pb,

2×10^8 (*6.44*). The relative reactivity appears to be a function of (1) M-C bond energy, (2) polarizability, and (3) steric hindrance. The scope of the reactions proceeding via 6.20 is discussed further later in this chapter.

Talbe 6.7 Substituent Constants (σ^+)

Substituent	σ^+ Constant
-Me	-0.30
-OMe	-0.78
$-CH_2SiMe_3$	-0.62
$-CH_2SnMe_3$	-0.92
$-CH_2SnPh_3$	-0.75
$-CH_2PbPh_3$	-1.00
$-CH_2HgPh$	-1.20

As mentioned earlier, unactivated Si-C bonds are unreactive toward electrophiles. With certain highly reactive electrophiles, however, even organosilicons containing such unactivated Si-C bonds can participate in reactions in which σ-π conjugation presumably plays a key role (*6.45*).

$$R_2^1CHCR_2^2MR_3 \xrightarrow{Ph_3CBF_4} R_2^1\overset{MR_3}{\overset{|}{C^{\pm}CR_2^2}} \rightarrow R_2^1C=CR_2^2 \qquad (6.53)$$

The Sn-C bond is considerably more susceptible to attack by electrophiles. Water and aliphatic alcohols are generally inert, but phenols, mercaptans, and carboxylic acids readily cleave the Sn-C bond. The reaction with halogens, such as Br_2 and Cl_2, is also a facile process. Hydrogenolysis with hydrogen is generally difficult.

$$R^1-SnR_3 \quad \begin{array}{c} \xrightarrow{HX} \quad R^1H \;+\; XSnR_3 \\[2em] \xrightarrow{X_2} \quad R^1X \;+\; XSnR_3 \end{array} \qquad (6.54)$$

6.4.1.3 Reaction with Nucleophiles

The reaction of the Si-C or Sn-C bond with nucleophiles (Y^-) by direct displacement is another general scheme involving the cleavage of the Si-C or Sn-C bond (eq. 6.55).

$$R-MZ_3 + Y^- \rightleftharpoons R^- + Y-MZ_3 \qquad (6.55)$$

$$M = Si \text{ or } Sn$$

The reaction is considerably more facile with organotins than with the corresponding organosilicons. The greater reactivity of organotins relative to that of organosilicons must be due to the higher polarizability and the greater length of the Sn-C bond. The reaction can take place, if (1) the R group is less basic than the Y group, shown in eq. 6.56, (6.46), or (2) the M-Y bond is thermodynamically highly stable and the equilibrium can be shifted by some means, as shown in eq. 6.57, (6.47).

$$\diagdown\diagup\text{SnBu}_3 + \text{BuLi} \longrightarrow \diagdown\diagup\text{Li} + \text{SnBu}_4 \qquad (6.56)$$

$$RC{\equiv}CSiMe_3 + R_4^1N^+F^- \longrightarrow RC{\equiv}C^-N^+R_4^1 + FSiMe_3 \qquad (6.57)$$

These reactions presumably proceed via the penta-coordinate transition states 6.1 to 6.4 discussed in Sect. 6.1.

6.4.1.4 Intramolecular Interaction of the Si-C and Sn-C Bonds with Nucleophiles and Electrophiles -- Elimination and Rearrangements

The reaction of the Si-C and Sn-C bonds with polar reagents can also take place intramolecularly. In fact, the intramolecular version of the reaction is often far more facile than the intermolecular counterparts. Such intramolecular reactions usually involve either an elimination or a rearrangement.

By far the most significant reaction from the viewpoint of organic synthesis is the β-elimination reaction of β-heterosubstituted organosilanes and organostannanes (eq. 6.58), which is the E2 version of the β-elimination reaction discussed earlier in this section.

$$X_3M-\overset{|}{\underset{|}{C}}-\overset{|}{\underset{|}{C}}-Y \longrightarrow \overset{|}{\underset{|}{C}}=\overset{|}{\underset{|}{C}} + X_3MY \qquad (6.58)$$

The reaction involves either syn or anti elimination. As a rule of thumb, syn elimination occurs if (1) the reaction is thermal and spontaneous, that is, unassisted (eq. 6.59), or (2) a highly nucleophilic center is generated at the substituent Y by the action of an added nucleophile (eq 6.60).

$$\underset{\substack{\displaystyle R \quad\quad R^1}}{\underset{\displaystyle H^{\prime\prime\prime\prime}C - C^{\prime\prime\prime\prime}H}{Me_3Si \quad O^-}} \xrightarrow{\quad syn \quad} \underset{\substack{\displaystyle R \quad\quad R^1}}{\overset{\displaystyle H \quad\quad H}{C=C}} \qquad (6.59)$$

$$\underset{\substack{\displaystyle R \quad\quad H}}{\underset{\displaystyle H^{\prime\prime\prime\prime}C - C^{\prime\prime\prime\prime}R^1}{Me_3Si \quad OH}} \begin{array}{c} \xrightarrow{\quad KH \quad} \underset{\substack{\displaystyle R \quad\quad H}}{\overset{\displaystyle H \quad\quad R^1}{C=C}} \qquad (6.60a) \\[2em] \xrightarrow{\quad BF_3 \quad} \underset{\substack{\displaystyle R \quad\quad R^1}}{\overset{\displaystyle H \quad\quad H}{C=C}} \qquad (6.60b) \end{array}$$

On the other hand, anti elimination takes place, if (1) an added electrophile can activate the Y group (eq. 6.60b), or (2) an added nucleophile can interact with the silicon or tin more readily than with the Y group (eq. 6.61).

$$\underset{\substack{\displaystyle H \quad\quad\quad Br}}{\underset{\displaystyle Br^{\prime\prime\prime}C - C^{\prime\prime\prime\prime}R}{Me_3Si}} \xrightarrow{\quad NaOMe \quad} \underset{\substack{\displaystyle Br \quad\quad H}}{\overset{\displaystyle H \quad\quad R}{C=C}} \qquad (6.61)$$

The role of the β-eliminarion reaction as a key step in various synthetically useful reactions of organosilicons and organotins is discussed further later. It is important to recognize that β-elimination is but one of a series of intramolecular substitution reactions represented by the following general equation (6.62).

$$X_3M-\overset{|}{C}- \underset{\displaystyle C_n}{\overbrace{\quad}} -\overset{|}{C}-Y \longrightarrow -\overset{|}{C}\underset{\displaystyle C_n}{\overbrace{\quad\quad}}\overset{|}{C}- \; + \; X_3MY \qquad (6.62)$$

$$n = 0, 1, 2, \ldots$$

In fact, the γ-elimination reaction of γ-halosilanes to form cyclopropanes is also a facile process; see eq. 6.63 (*6.48*). When an organosilane contains a nucleophilic group bonded to an α-carbon atom, attack of the silicon atom by the nucleophilic moiety induces a 1,2-migration of the silicon-containing group that migrates as a cationic species (cationotropic rearrangement), see eq. 6.64 (6.49).

$$Cl_3Si \underset{CH_2}{\overset{\displaystyle CH_2 \quad CH_2-Cl}{\diagdown\quad\diagup}} \xrightarrow{\text{NaOH}} \triangledown \qquad (6.63)$$

$$\underset{R_2C-OH}{\overset{Ph_3Si}{|}} \xrightarrow{\text{Na/K}} \underset{R_2C-O^-}{\overset{Ph_3Si}{\searrow}} \longrightarrow \underset{R_2C-O}{\overset{SiPh_3}{|}} \qquad (6.64)$$

This cationotropic rearrangement is closely related to the Wittig rearrangement and the Stevens rearrangement. However, its synthetic usefulness is not clear at present.

6.4.1.5 Miscellaneous Reactions Involving Cleavage of the Si-C or Sn-C Bond

Although the great majority of Si-C or Sn-C cleavage reactions are represented by those discussed above, there are various other modes of cleaving the Si-C or Sn-C bond, of which the following are worth mentioning here: (1) the reaction of the allylic and β-carbonyl derivatives of organosilicons and organotins with organic electrophiles (Sect. 6.4.2.1), (2) oxidation of the Sn-C bond with $CrO_3 \cdot 2Py$ to form the corresponding alcohol or ketone (Sect. 6.4.3.1), and (3) the anionotropic 1,2-rearrangement of the silicon-containing group (Sect. 6.4.2.1).

6.4.2 Reactions of Organosilicons

The use of organosilicons in organic synthes is may be classified as follows: (1) the reactions involving cleavage of the Si-C bond, (2) the reactions of the Si-O derivatives, and (3) the reactions of the Si-H bond. As in previous chapters, those reactions that do not involve either cleavage of the Si-C bond or formation of the C-C bond are discussed only briefly.

6.4.2.1 Synthetic Reactions Involving Cleavage of the Si-C bond

As discussed earlier, the intrinsic reactivity of the Si-C bond toward either polar or free-radical reagents is generally low. Thus nonfunctionalized organosilanes rarely participate in the Si-C cleavage reactions. If a functional group is present in an appropriate position, however, usually α, β, and/or γ to the silicon atom, an added reagent may interact with the functional

group. The silicon atom may or may not exert significant influence in this step. In many cases, the product is a silicon-containing compound.

In many other cases, however, the product or intermediate
formed in such a reaction can undergo either a thermal or a rea-
gent-assisted reaction involvong cleavage of the Si-C bond. As
discussed in the preceding section, such intermediate species are
represented by either 6.21 or 6.22 in a large number of reactions.

$$
\begin{array}{c}
R_3Si\text{-}\overset{|}{\underset{|}{C}}\text{-}\overset{|}{\underset{|}{C}}{}^{+} \\[2pt]
\underline{6.21} \\[10pt]
R_3Si\text{-}\overset{|}{\underset{|}{C}}\text{-}\overset{|}{\underset{|}{C}}\text{-}Y \\[2pt]
\underline{6.22}
\end{array}
\longrightarrow
\;\;\diagup\!\!\!\!\diagup\;\;
\begin{array}{c}
\diagdown \quad \diagup \\
C=C \\
\diagup \quad \diagdown
\end{array}
\qquad (6.65)
$$

<u>Electrophilic Substitution Reactions</u> The intermediate 6.21 is
generally formed by adding an electrophilic species (E^+) to either α,β- or β,γ-unsaturated species. Such reactions are acce-
lerated by the σ-π conjugation effect and are usually highly regioselective (6.42). It is important to note that organostannanes also participate in the same reaction, and that the tin reaction is more facile than the silicon reaction. Some representative electrophiles (E-Y) and their reaction products are summarized in Table 6.8.

Table 6.8   Reaction of Unsaturated Organosilanes ($R^1SiR_3^2$) and
Organostannanes ($R^1SnR_3^2$) with Electrophiles

Electrophile	Product	Electrophile	Product
$RCOX+AlCl_3$	R^1COR	HNO_3+Ac_2O	R^1NO_2
$(RCO)_2O+AlCl_3$	R^1COR	$NOCl$	R^1NO
⟨⟩-$CH_2Br+AlCl_3$	R^1CH_2-⟨⟩	$RSO_2X+AlCl_3$	R^1SO_2R
$HX(HCl, HOAc)$	R^1H	SO_3	R^1SO_3H
Br_2	R^1Br	$Hg(OAc)_2$	H^1HgOAc
ICl	R^1I	BCl_3	R^1BCl_2

The reaction offers various advantages over the corresponding reaction of the parent nonmetallated species or of polar organometallics, such as those containing lithium or magnesium.
The following specific advantages may be cited for the reactions
of arylsilanes: (1) formation of a single regioisomer (ipso isomer), (2) formation of isomers that cannot be obtained by the
direct aromatic substitution reaction of the parent compounds,
and (3) successful utilization of deactivated aromatics. The following otherwise difficult-to-obtain aromatic iodide can now be
prepared by this reaction; see eq. 6.66 (*6.50*).

(6.66)

The reaction of alkenylsilanes with electrophiles are not
only regioselective but frequently highly stereospecific. In
cases where the elimination step does not require addition of a
reagent, the overall process proceeds with retention, whereas
either retention or inversion may result if the elimination step
is assisted by an added reagent (*6.51*).

(6.67)

$EY = HX, DX, ClCH_2OMe, ClCHOHCCl_3, RCOCl, Br_2, ICl$

It should be pointed out here that the cationic intermediate
6.21 can also be derived from various organosilicon compounds
other than the α,β- and β,γ-unsaturated derivatives. For example,
the epoxysilanes 6.23, readily obtainable by epoxidation of alkenylsilanes, can be converted readily to the corresponding ketones.
The latter conversion presumably proceeds via 6.21; see eq. 6.68
(*6.52*). The overall sequence povides a convenient method for the
conversion of alkenylsilanes into the corresponding carbonyl compounds.

β-Heterosubstituted Organosilanes as Intermediates   It has become
increasingly clear in recent years that a number of organosilicon
reactions involve the intermediacy of β-heterosubstituted organo-

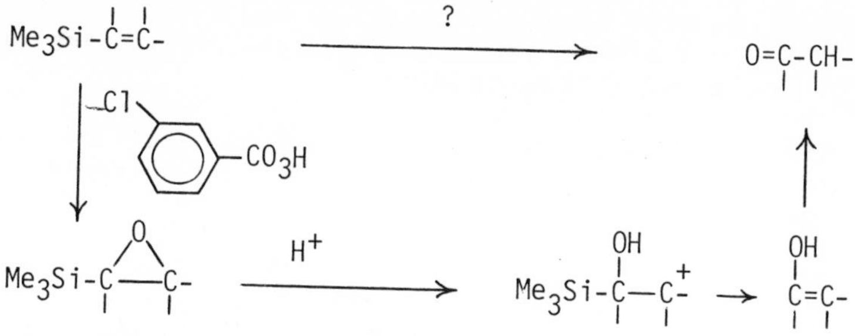

$$Me_3Si-\overset{|}{C}=\overset{|}{C}- \xrightarrow{\quad ? \quad} O=C-CH-$$

(6.68)

$$Me_3Si-\overset{O}{\overset{/\backslash}{C}-\overset{|}{C}-} \xrightarrow{\ H^+\ } Me_3Si-\overset{OH}{\overset{|}{C}}-\overset{+}{\overset{|}{C}}- \rightarrow \overset{OH}{\overset{|}{C}}=\overset{|}{C}-$$

6.23   6.21

silanes (6.22), which can be converted to the corresponding olefins as discussed earlier. Inspection of 6.22 suggests that it
can be generated by forming one or more of the five bonds (i) to
(v).

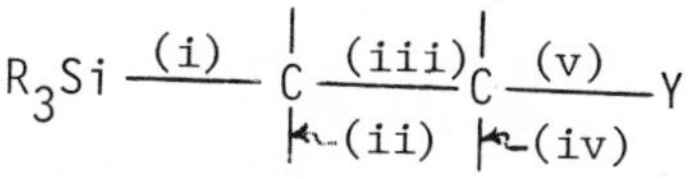

$$R_3Si \xrightarrow{(i)} \overset{|}{\underset{\downarrow -(ii)}{C}} \xrightarrow{(iii)} \overset{|}{\underset{\downarrow -(iv)}{C}} \xrightarrow{(v)} Y$$

6.22

(a) Via formation of bond (i) The reaction of triorganosilyl anions with epoxides produces β-silylalkoxy anions which
undergo a facile syn elimination. This sequence can be applied
to configurational inversion of olefins, as shown in eq. 6.44
(Sect. 6.3.4).

(b) Via formation of bond (ii) The reaction of epoxysilanes with appropriate nucleophiles produces β-hydroxylilanes;
see eq. 6.69 (6.53).

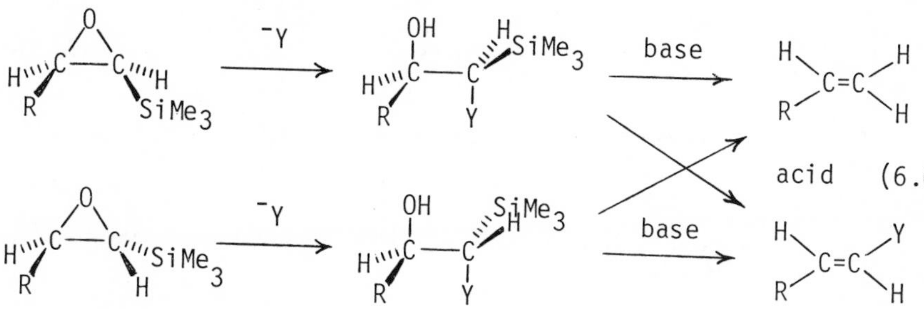

$$\text{(6.69)}$$

Y = R(LiCuR₂), Br, OAc, OMe, NHAc

Whereas $LiCuR_2$ are satisfactory alkylating agents in this reaction, LiR and $RMgX$ are not. The stereochemical results are in accord with the generalization made earlier.

(c) Via formation of bond (iii) A variety of silicon-stabilized carbanion of the $^-CH(Z)SiR_3$ type react with aldehydes and ketones, providing a general route to 6.22 (6.54, 6.51).

$$
\begin{array}{c}
R^1 \\
R^2
\end{array}\!\!\!\!\!\!> C=O \quad \xrightarrow{\;\overset{\displaystyle ^-CHSiR_3}{\underset{\displaystyle Z}{\vert}}\;} \quad R^1R^2\overset{O^-}{\underset{\vert}{C}}\!\!-\!\!\overset{SiR_3}{\underset{\vert}{CHZ}} \quad \xrightarrow[\substack{acid \\ or \\ base}]{H_2O} \quad R^1R^2\overset{OH}{\underset{\vert}{C}}\!\!-\!\!\overset{SiR_3}{\underset{\vert}{CHZ}}
$$

$$
\Big\downarrow
$$

$$
\longrightarrow R^1R^2C=CHZ \tag{6.70}
$$

This reaction provides an alternative to the Wittig olefination reaction. Unfortunately, mixtures of cis and trans isomers are usually obtained. A wide variety of functional groups have been used as the Z group. They include: COOR, COOH, CN, $SiMe_3$, PPh_2, $PO(OEt)_2$, SR, and SOR (6.51). It is interesting to note that the ability of a trialkylsilyl group to interact with a β-oxy anion appears greater than that of the $PO(OEt)_2$ group.

(d) Via formation of bond (iv) Unlike the reaction of trimethysilylethylene oxide with $LiCuR_2$, which involves an α attack by $LiCuR_2$, the corresponding reaction of $RMgBr$ takes an anomalous course, as shown in eq. 6.71.

$$
\begin{array}{ccc}
Me_3SiCH\!\!-\!\!CH_2 & \xrightarrow[\text{2.}\quad H_3O^+]{\text{1.}\quad RMgBr} & Me_3SiCH_2\overset{}{\underset{\vert}{CHR}} \\
\;\;\;\diagdown\!\!\diagup & & OH \\
\quad O & & \\
\Big\downarrow RMgBr & & \Big\uparrow H_3O^+ \\
& \xrightarrow{RMgBr} & \\
Me_3SiCH_2CHO & & Me_3SiCH_2\overset{}{\underset{\vert}{CHR}} \\
& & OMgBr
\end{array} \tag{6.71}
$$

Although such a reaction provides a route to 6.22, its scope and synthetic utility are not clear.

(e) Via formation of bond (v) Hydroboration-oxidation of

alkenylsilanes produces mixtures of α- and β-hydroxysilanes, as shown in eq. 6.72. As expected, the β-isomers are trans. On the other hand, oxymercuration-demercuration of alkenylsilanes produces only β-isomers, but the reaction is not stereoselective; see eq. 6.72 (6.55).

$$(6.72)$$

(f) Via formation of bonds (ii) and (v) Although we have tentatively interpreted the reactions of alkenylsilanes with electrophiles in terms of the intermediacy of 6.20, some of these reactions may form 6.22 as intermediates, as shown in the following example; see eq. 6.73 (6.56).

$$(6.73)$$

(g) Other β-heterosubstituted organosilanes All of the β-heterosubstituted organosilanes discussed above are represented by 6.22. It may be anticipated, however, that there can be other types of β-heterosubstituted organosilanes represented by 6.24 and 6.25 which would also readily undergo β-elimination.

$$R_3Si-\overset{|}{\underset{|}{C}}-X-Y \longrightarrow \overset{\diagdown}{\underset{\diagup}{}}C=X + R_3SiY \qquad (6.74)$$

6.24

$$R_3Si-X-\overset{|}{\underset{|}{C}}-Y \longrightarrow X=C\overset{\diagup}{\underset{\diagdown}{}} + R_3SiY \qquad (6.75)$$

6.25

Indeed, as discussed in the following sections, the great majority of the reactions of silyl enol ethers that produce organic products can be interpreted in terms of eq. 6.75. On the other hand, there have been relatively few reactions that can be represented by eq. 6.74.

The following oxidation reaction (eq. 6.76) which provides a convenient method for the conversion of terminal acetylenes to carboxylic acids has been reported recently (6.57) and interpreted in terms of an anionotropic 1,2-migration of a trialkylsilyl group (path a). Although the proposed mechanism is plausible, it appears equally plausible to interpret the results in terms of the β-elimination reaction of <u>6.24</u> (path b).

$$RC\equiv CH \longrightarrow \begin{array}{c} R \\ \diagdown \\ H \end{array} C=C \begin{array}{c} SiMe_3 \\ \diagup \\ BR_2^1 \end{array} \xrightarrow[NaOH]{H_2O_2}$$

$$\underset{\displaystyle RCH_2\overset{\displaystyle \overset{O}{\parallel}}{C}SiMe_3}{} \longrightarrow \underset{\displaystyle O-OH}{RCH_2\overset{\displaystyle \overset{O^-}{|}}{C}-SiMe_3} \xrightarrow{(a)} RCH_2\overset{\displaystyle \overset{O}{\parallel}}{C}_{OSiMe_3} \qquad (6.76)$$

$$\downarrow$$

$$RCH_2\overset{\displaystyle \overset{OH}{|}}{C}-SiMe_3 \xrightarrow{(b)} RCH_2COOH$$

with $\downarrow H_2O$ on the right path leading to RCH_2COOH.

Anionotropic Rearrangements The possibility that the organosilyl group might undergo anionotropic 1,2-migration in addition to cationotropic 1,2-rearrangement (Sect. 6.4.1) is mentioned above. Indeed, studies by Brook and others have revealed that the organosilyl group generally has a considerably greater aptitude than a typical carbon group, such as phenyl, to migrate as an anionic moiety (6.49c), as indicated by the following example (eq. 6.77).

$$Ph_3SiCOPh \xrightarrow{CH_2N_2} Ph_3Si\overset{\displaystyle \overset{O^-}{|}}{\underset{\displaystyle \underset{CH_2N_2}{+}}{C}}-Ph \longrightarrow Ph_3SiCH_2COPh$$

III

$$(6.77)$$

$$Ph_3Si\overset{\displaystyle \overset{-O}{|}}{\underset{\displaystyle \underset{CH_2N_2}{+}}{C}}-Ph \longrightarrow Ph_3SiO\overset{\displaystyle \overset{Ph}{|}}{C}=CH_2$$

6.26

No competitive phenyl migration takes place. It is of interest to note that the presumed intermediate 6.26 evidently undergoes both anionotropic and cationotropic rearrangements competitively. Although very interesting from the mechanistic viewpoint, the synthetic usefulness of such anionotropic rearrangements has not yet been well delineated. The key question here is how to generate the required intermediates in a more practical manner.

<u>Addition Reactions of Allylsilanes</u>   Although allylsilanes do not readily react with carbonyl compounds, such as aldehydes and ketones, these reactions can be markedly accelerated by the action of $TiCl_4$. The markedly rate-enhancing ability of $TiCl_4$ was originally investigated in the related reactions of silyl enol ethers, as discussed later (Sect. 6.4.2.2).

The scope of the titanium-promoted allylsilane reactions, being investigated mainly by Sakurai (6.58), has not yet been well delineated. Even so, the following useful procedures have been developed.

$$Me_3SiCH_2CH=CR^1R^2$$

$$\xrightarrow[\text{TiCl}_4]{R^3R^4C=O}\quad CH_2=CHC\underset{\underset{R^2}{|}}{\overset{\overset{R^1}{|}}{-}}C\underset{\underset{R^4}{|}}{\overset{\overset{R^3}{|}}{O}}H$$

(44 to 96%)

$$\xrightarrow[\text{TiCl}_4]{R^3R^4C(OR)_2}\quad CH_2=CHC\underset{\underset{R^2}{|}}{\overset{\overset{R^1}{|}}{-}}C\underset{\underset{R^4}{|}}{\overset{\overset{R^3}{|}}{O}}R \qquad (6.78)$$

(71 to 98%)

$$\xrightarrow[\text{TiCl}_4]{R^3R^4C=CHCOR^5}\quad CH_2=CHC\underset{\underset{R^2}{|}}{\overset{\overset{R^1}{|}}{-}}CH_2CH_2COR^5$$
$$\underset{R^4}{|}$$

(59 to 96%)

These reactions proceed with a complete allylic rearrangement. The conjugate addition reaction appears to be highly general with respect to the structural type of enones.

The mechanism of the reaction is not clear. It may be, however, that these reactions merely represent additional examples of the reaction of allylsilanes with electrophiles in which σ-π conjugation plays a dominant role. If so, the α,β-unsaturated organosilanes should also react in a similar manner, although this point does not appear to have been clearly established.

6.4.2.2 Reactions of Traialkylsilyl Enol Ethers

The chemistry of silyl enol ethers has been extensively reviewed recently (6.51, 6.59). When enolate anions are treated with trialkylhalosilanes, such as Me_3SiCl, the corresponding trialkylsilyl enol ethers are formed in a highly regiospecific manner and generally in excellent yields. Since the bond energies for the Si-C and Si-O bonds are approximately 70 and 90 kcal/mole, respectively, the oxygen-silylated products are formed exclusively (eq. 6.79).

$$[-\underset{|}{C}\dot{=}\underset{|}{C}\dot{=}O]^-M^+ \ + \ XSiR_3 \ \longrightarrow \ -\underset{|}{C}=\underset{|}{C}-OSiR_3 \quad (6.79)$$

It is feasible, however, to obtain carbon-silylated derivatives by treating α-carbonylorganomercuries with trialkylhalosilanes (6.60). The products are, however, thermally unstable and readily isomerize to form the oxygen-silylated derivatives.

$$2CH_2=C=O + HgX_2 + 2R^1OH \ \longrightarrow \ Hg(CH_2COOR^1)_2 + 2HX$$

$$(6.80)$$

$$Hg(CH_2COOR^1)_2 + R_3^2SiX \ \longrightarrow \ R_3^2SiCH_2COOR^1$$

Alternatively, silyl enol ethers can be prepared by treating aldehydes and ketones with either R_3SiCl-NEt_3 in DMF (6.61) or R_3SiCl-NEt_3-$ZnCl_2$ (6.62). The more stable regioisomers are formed predominantly by this method.

There also exist various other routes to silyl enol ethers, of which the hydrosilation of α,β-unsaturated carbonyl compounds is of reasonable generality (6.63).

$$-\underset{|}{C}=\underset{|}{C}-\underset{|}{C}=O \ \xrightarrow[\quad ClRh(PPh_3)_3 \quad]{R_{4-n}SiH_n} \ (-\underset{|}{C}H\underset{|}{C}=\underset{|}{C}-O)_nSiR_{4-n} \quad (6.81)$$

Sylyl enol ethers can be treated with a variety of strong bases, such as MeLi, to form the corresponding enolate anions in a highly regiospecific manner (eq. 6.82).

$$-\underset{|}{C}=\underset{|}{C}-OSiR_3 + MeLi \ \longrightarrow \ [-\underset{|}{C}\dot{=}\underset{|}{C}\dot{=}O]^-Li^+ + MeSiR_3 \quad (6.82)$$

Both Si-C and Si-O bonds of trialkylsilyl enol ethers share essentially the same chemical properties as those of other trialkylsilyl ethers discussed in the preceding section. Thus although the Si-O bond can be cleaved easily by hydrolysis and other methods, it is inert to many other reagents. It is soon noted that niether Si-O nor Si-C bond is directly involved in the initial stages of most of the reactions of trialkylsilyl enol ethers. The Si-O bond, however, exerts a strong influence on the chemical properties of the conjugated double and triple bonds mainly through the resonance effect.

$$-\overset{\curvearrowright}{\underset{|}{C}}=\overset{\curvearrowright}{\underset{|}{C}}-\overset{..}{O}-SiR_3 \longleftrightarrow -\underset{|}{\overset{-}{C}}-\underset{|}{C}=\overset{+}{O}-SiR_3 \longleftrightarrow -\underset{|}{\overset{-}{C}}-\underset{|}{\overset{+}{C}}-O-SiR_3$$

Due to the electron-donating +R effect of the trialkylsilyloxy group, the C=C bond is more nucleophilic than the more usual C=C bond, but it is less nucleophilic than the more ionic enolates, such as lithium enolates. In other words, the general order of reactivities toward electrophiles is:

$$-\underset{|}{C}=\underset{|}{C}-OLi \; > \; -\underset{|}{C}=\underset{|}{C}-OSiR_3 \; > \; -\underset{|}{C}=\underset{|}{C}R \quad (R = H \text{ or alkyl})$$

It may thus be said that trialkylsilyl enol ethers act as (1) trialkylsilyl-moderated enolate anions and (2) activated and functionalized olefins toward electrophiles.

Reactions with Various Halides Various electrophilic halides including halogens, such as Cl_2 and Br_2, have reacted with trialkysilyl enol ethers according to the following general equation consisting of an addition-elimination sequence (eq. 6.83).

$$-\underset{|}{\overset{\delta-}{C}}=\underset{|}{\overset{\delta+}{C}}-O-SiR_3 \xrightarrow{\;\;Y^{\delta+}-X^{\delta-}\;\;} -\underset{|}{\overset{Y}{C}}-\underset{|}{\overset{X}{C}}-O-SiR_3 \longrightarrow -\underset{|}{\overset{Y}{C}}-\underset{|}{C}=O \quad (6.83)$$

The halide reagents that have been successfully used and the products obtained are summarized in Table 6.9.

Halogenation of silyl enol ethers produces α-halocarbonyl compounds in good yields. It allows regiospecific halogenation of ketones and provides a highly satisfactory method for preparing α-haloaldehydes; see eq. 6.84 (6.64, 6.65d), which are otherwise difficult to obtain.

$$n\text{-}C_8H_{17}CH=CHOSiMe_3 \xrightarrow{\;\;Br_2\;\;} n\text{-}C_8H_{17}\underset{|}{\overset{}{C}}HCHO \;(94\%) \quad (6.84)$$
$$\qquad\qquad\qquad\qquad\qquad\qquad Br$$

Table 6.9 Reactions of Trialkylsilyl Enol Ethers with Halogans
and Halides

Halogen or Halide	Group Y of Product	Ref.
Cl_2	-Cl	6.64
Br_2	-Br	6.65
PhSCl	-SPh	6.66
$ArSO_2Cl$	$-SO_2Ar$	6.67
NOCl	=NOH	6.68

It should be noted that all of the reagents listed in Table 6.9
are those that also react readily with more usual C=C bonds. On
the other hand, acyl halides and alkyl halides, which react readily with highly ionic enolates but not with usual C=C bonds, do
not react readily with silyl enol ethers except for some poly-
halogenated derivatives, such as Cl_3CCOCl and $NCCCl_3$. The reaction with acyl halides can be markedly catalyzed by $HgCl_2$ (6.69).
The products, however, are oxygen-acylated enols.

Reactions with Brønsted Acids The reaction of silyl enol ethers
with strong mineral acids generally produces the parent carbonyl
compounds via addition-elimination. With some weaker Brønsted
acids, such as HCN and HN_3, the addition products can be obtained
as isolable compounds (6.59, 6.70).

Titanium(IV)-Promoted Reactions Recent extensive investigations
of the reactions of silyl enol ethers in the presence of $TiCl_4$,

conducted mainly by Mukaiyama and his associates, have led to various synthetically useful reactions which amount to the trialkylsilyl-moderated reactions of enolate anions. In general, these reactions provide the following advantages over the conventional enolate anion reactions:

1. The titanium-promoted reactions of silyl enol ethers are generally highly regiospecific.
2. Scrambling of two or more reactants does not take place extensively.
3. The titanium-promoted reactions of silyl enol ethers not only exhibit different chemoselectivity patterns but are generally more highly chemoselective than the corresponding enolate anion reactions.

(a) Cross-aldol reaction   Aldehydes and ketones undergo a remarkably selective cross-aldol reaction with silyl enol ethers in the presence of $TiCl_4$ (6.71). The reaction is regiospecific. The corresponding reaction of ketone alkyl (trialkylsilyl) acetals gives β-hydroxy esters (6.72). The following scheme has been proposed to account for the role of $TiCl_4$ (eq. 6.87).

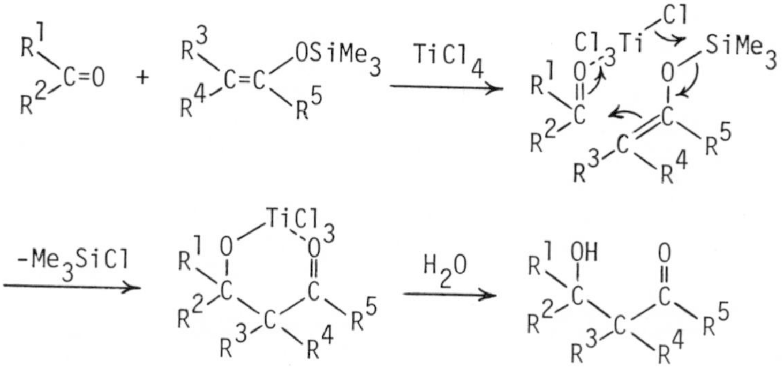

(6.86)

(b) Michael reaction   The titanium-promoted reaction of silyl enol ethers with α,β-unsaturated ketones and esters provides a novel procedure for the Michael reaction (6.73), which often appears superior to the conventional base-catalyzed procedures. Various other related titanium-promoted reactions of silyl enol ethers have also been reported by Mukaiyama (6.74).

Cycloaddition Reactions   As we have already seen, silyl enol ethers can act as silyloxy-substituted olefins. Therefore, the scope of their chemistry can be as broad as that of the chemistry of olefins, although the number of examples reported to date is

$$\underset{\underset{PhC=CH_2}{|}}{OSiMe_3} + (CH_3)_2C=CHCX \xrightarrow{TiCl_4} \underset{\underset{CH_3}{|}}{PhCCH_2CCH_2CX} \qquad (6.87)$$

$$X = CH_3 \ (76\%), \quad OCH_3 \ (68\%)$$

still limited. Even so, it is not possible to discuss all representative types in a book of this nature. Only a few selected examples are discussed here.

(a) [4+2] Cycloaddition (Diels-Alder reaction) *trans*-1-Methoxy-3-trimethylsilyloxy-1,3-butadiene undergoes facile Diels-Alder reactions, while simultaneously incorporating various functional groups into the products (*6.75*).

$$(6.88)$$

(b) [3+2] Cycloaddition The 1,3-dipolar cycloaddition of arenesulfonyl azides to silyl enol ethers permits a unique ring contraction reaction of cyclic ketones; see eq. 6.89 (*6.76*).

(c) [2+2] Cycloaddition Triplet-sensitized irradiation of a mixture of a silyl enol ether and a Michael acceptor gives the corresponding [2+2] cycloaddition product which readily hydrolyzes to give the Michael adduct; see eq. 6.90 (*6.77*).

The reaction of isocyanates with silyl enol ethers gives β-lactams, which readily solvolyze to form β-ketoamides; see eq. 6.91 (6.78).

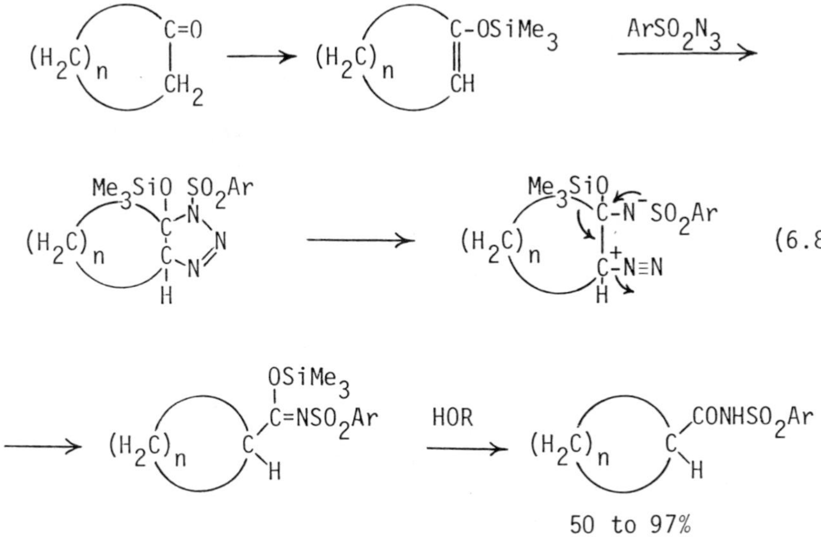

$$(H_2C)_n \underset{CH_2}{\overset{C=O}{\diagdown}} \longrightarrow (H_2C)_n \underset{CH}{\overset{C-OSiMe_3}{\diagdown}} \xrightarrow{ArSO_2N_3}$$

$$(H_2C)_n \overset{Me_3SiO}{\underset{H}{\diagdown}}\overset{SO_2Ar}{\underset{C-N_2}{\diagdown}} \longrightarrow (H_2C)_n \overset{Me_3SiO}{\underset{H}{\diagdown}}\overset{C-N^-SO_2Ar}{\underset{C-N\equiv N}{\diagdown}} \qquad (6.89)$$

$$\longrightarrow (H_2C)_n \overset{OSiMe_3}{\underset{H}{\overset{C=NSO_2Ar}{\diagdown}}} \xrightarrow{HOR} (H_2C)_n \underset{H}{\overset{CONHSO_2Ar}{\diagdown}}$$

50 to 97%

n = 4, 5, 6, and 10

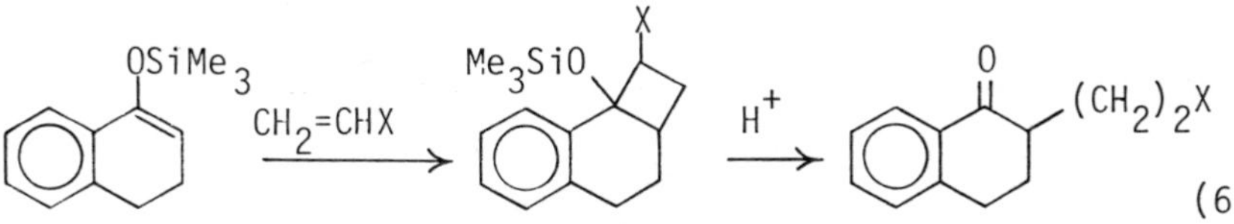

X = COOCH₃ (79%), CN (86%), COCH₃ (51%)

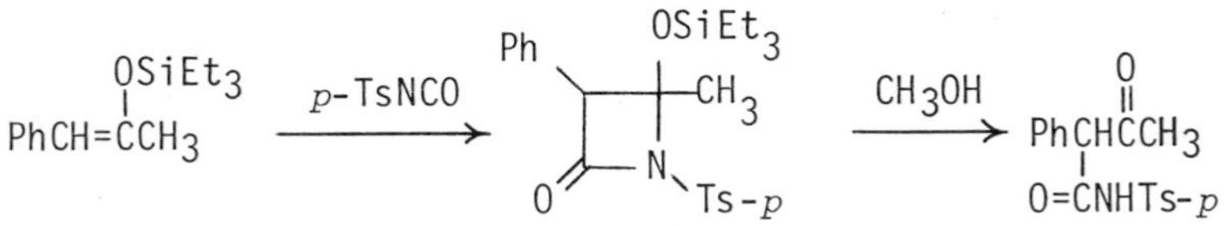

$$(6.90)$$

$$(6.91)$$

(d) [2+1] Cycloaddition (cyclopropanation)   Cyclopropanation of silyl enol ethers via addition of carbenoid reagents provides a facile route to cyclopropanol derivatives with can be converted to a variety of derivatives including α-methylcarbonyl compounds, cyclobutanones, cyclopentanones, and α,β-unsaturated carbonyl compounds (6.79).

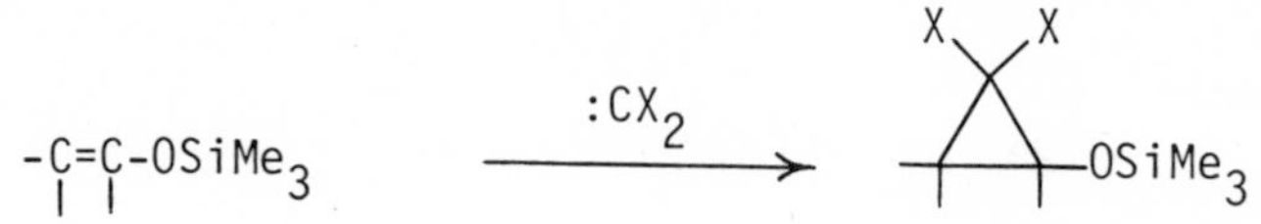

$$-\underset{|}{C}=\underset{|}{C}-OSiMe_3 \quad \xrightarrow{:CX_2} \quad \text{(structure)}-OSiMe_3 \qquad (6.92)$$

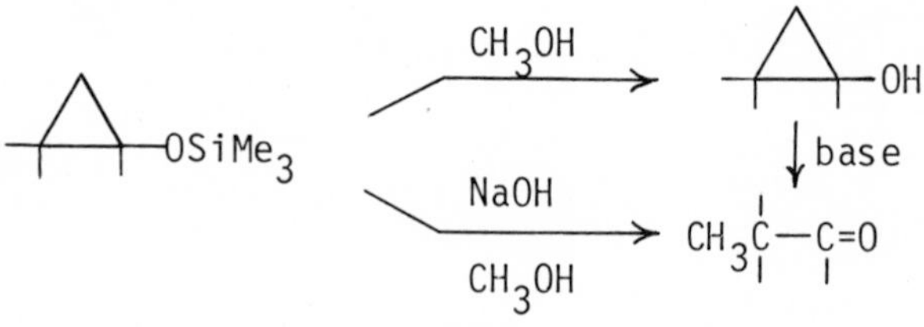

(6.93)

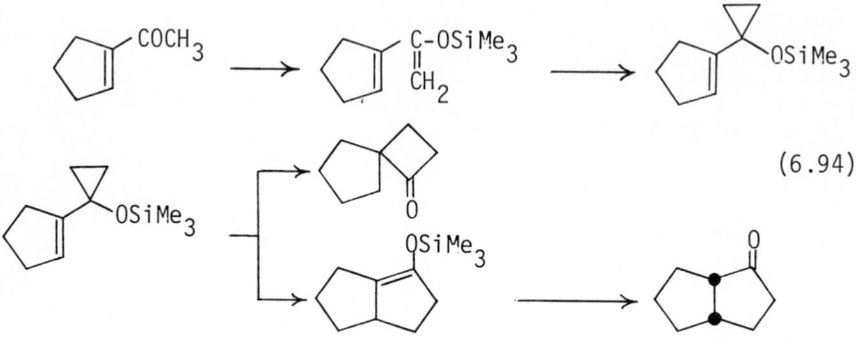

(6.94)

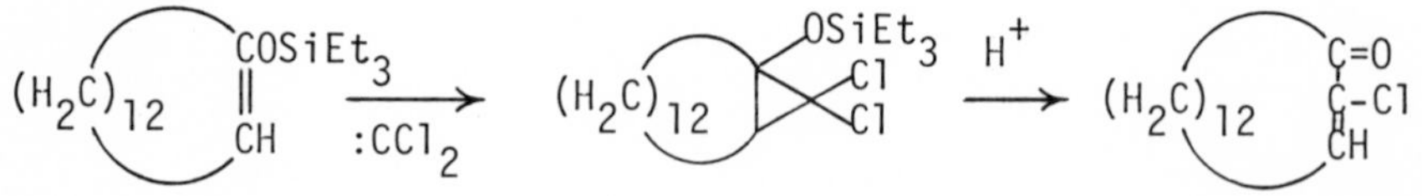

(6.95)

Oxidation Reaction of Silyl Enol Ethers   Various oxidation reactions of silyl enol ethers have been observed. The oxidizing agents that have been used include a combination of a boron hydride and alkaline hydrogen peroxide (6.80), singlet oxygen

(6.81), ozone (6.82). A few synthetically useful examples are shown below.

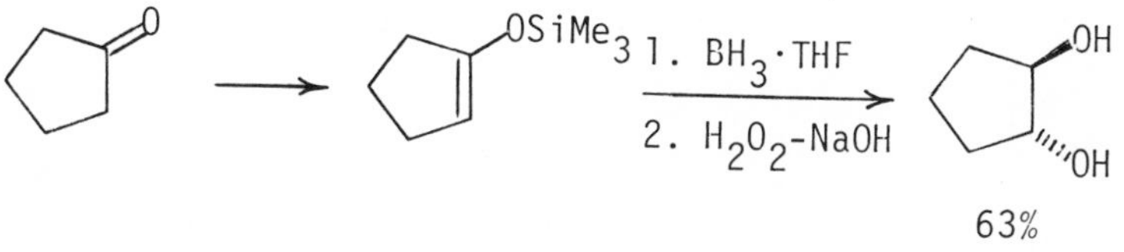

This reaction is regio- and stereoselective (6.80).

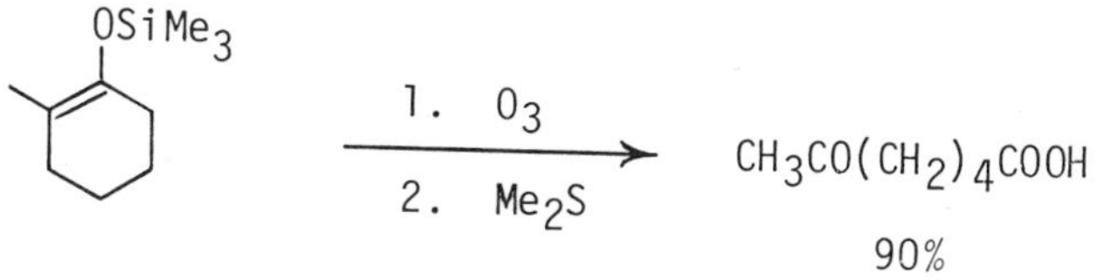

$$CH_3CO(CH_2)_4COOH \qquad (6.97)$$

90%

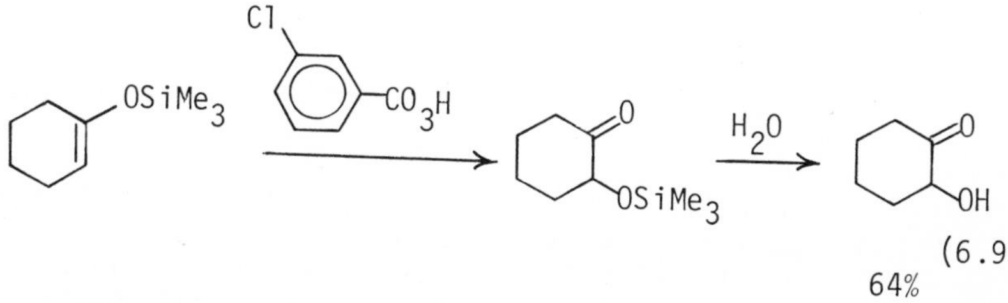

(6.98)

64%

All these transformations are either difficult or impossible to achieve with the corresponding enolate anions or the parent carbonyl compounds. The examples discussed above represent only a fraction of the known examples (6.51, 6.59). Many more useful reactions of silyl enol ethers undoubtedly will be developed. It may be said, however, that we are now in a position to understand reasonably well and in many instances even predict the courses of various silyl enol ether reactions.

6.4.2.3 α-Functional Silyl Ethers

Trialkylsilyl derivatives of the general structure R_3SiZ, where Z is CH_2CN or CH_2COOEt (6.83), Ar (6.84), CN (6.85), NR_2 (6.86), NHCOR or $N(COR)_2$ (6.87), N_3 (6.88) or SR (6.89), react with aldehydes and ketones, sometimes in the presence of acid or base catalysts, according to eq. 6.99.

$$\underset{/}{\overset{\backslash}{>}}C=O \quad \xrightarrow{\quad R_3SiZ \quad} \quad \overset{\backslash}{\underset{/}{>}}C\overset{OSiR_3}{\underset{Z}{\diagdown}} \qquad (6.99)$$

The trimethylsilyl cyanide adducts of ketones can be reduced to β-aminoalcohols, which may be used in the Tiffeneau-Demjanov reaction (6.90). The trimethylsilyl cyanide adducts of aldehydes have been deprotonated to give useful acyl anion equivalents (6.91).

$$ArCHO \longrightarrow Ar\overset{OSiMe_3}{\underset{|}{C}}HCN \quad \xrightarrow{\quad LiN(Pr\text{-}i)_2 \quad} \quad Ar^-\overset{OSiMe_3}{\underset{|}{C}}CN$$

$$\downarrow RX \qquad (6.100)$$

$$ArCOR \quad \xleftarrow{\quad R_3NHF \quad} \quad Ar\text{-}\overset{OSiMe_3}{\underset{|}{\underset{R}{C}}}CN$$

6.4.2.4 Triorganosilyl Groups as Protecting and Volatilizing Agents

One of the first applications of organosilicons to organic synthesis was the conversion of polar organic compounds, such as alcohols and amines, to their more volatile trialkylsilyl derivatives for mass spectral and gas chromatographic analyses and for the protection of active hydrogen-containing functionalities. This subject has been extensively reviewed (6.92), and is not discussed in detail in this book.

The synthetic usefulness of trialkylsilyl groups as protecting agents stems largely from (1) the ease with which the required protected derivatives are formed, (2) their stability in the presence of many carbon nucleophiles (RLi, RMgX, and so on) and oxidizing and reducing agents, and (3) the ease with which they are hydrolyzed with acids, bases, or fluorides.

Until recently, in most instances the trimethylsilyl group had been used. Its hydrolytic instability, however, can cause various difficulties. In order to circumvent such difficulties, various triorganosilyl groups with greater steric requirements have been synthesized and utilized. For example, the approximate relative rates of hydrolysis for a few representative trialkyl groups are as follows: $Me_3Si(1) \gg Et_3Si(10^{-2}) \gg (t\text{-Bu})Me_2Si$

(10^{-4}). Although highly hindered triorganosilyl groups are not readily removed by hydrolysis, they can still be readily removed by the action of R_4NF. The high thermodynamic stability of the Si-F bond (129 kcal/mole) undoubtedly must be responsible for the high efficiency of the fluoride ion.

6.4.3 Reactions of Organotins

As discussed earlier, the bonds to tin are longer and more polarizable than the corresponding bonds to silicon. Thus they can readily participate in a variety of polar reactions. They are also more reactive toward free radicals than the corresponding bonds to silicon. The available data indicate that organotins not only undergo many of those reactions that organosilicons undergo, but also participate in various other reactions. At present, their use in organic synthesis is still considerably more limited than that of organosilicons. It is not inconceivable, however, that in the future organotins might prove more versatile than organosilicons.

6.4.3.1 Reactions Involving Cleavage of the Sn-C Bond

Polar Substitution Reactions   The scope of direct polar substitution reactions of the Si-C bond is very limited. On the other hand, a variety of polar substitution reactions of the Sn-C bond are known. In cases where the Sn-C bond is unactivated, for example, Sn-Et, such reactions appear to involve direct substitution mechanisms proceeding via 6.1 to 6.4 and other related transition states and intermediates. Some of these reactions may also proceed partly or entirely by one-electron transfer processes.

(a) Protonolysis and halogenolysis   As mentioned in Sect. 6.4.1, protic acids, but not water or alcohols, and halogens readily cleave even "unactivated" Sn-C bonds (eq. 6.54). These reactions are discussed in detail in various monographs and reviews (6.1, 6.3, to 6.5). They represent some of the most straighforward methods for converting organotins into the corresponding organic compounds. Alghough protonolysis of the Sn-C bond can be achieved with both acids and bases, the tin-alkyl bonds are quite stable toward alkali. No clear-cut generalization can be made regarding the stereochemistry of protonolysis. Both retention and inversion have been observed.

We have already discussed the fact that certain halogenolysis reactions of organotins might involve at least partly free-radical processes (eq. 6.45). A recent study (6.93), however, reveals that the brominolysis of some optically active tetraalkyltins proceeds with either partial retention or partial inversion. Thus such a reaction appears to be at least partly polar. It has been tentatively suggested that retention of configuration is probably the main stereochemical course of the brominolysis in methanol and that, in special cases, large steric requirements of the ligands would induce a predominant inversion mechanism. The following transition states have been suggested for the retention reaction.

6.27

6.28

(b) Oxygenolysis of the Sn-C bond Relatively few reactions are known for the conversion of organotins into the corresponding alcohols and carbonyl compounds in part due to the fact that the Sn-C bond is quite resistant to attack by free radicals. Chromic oxide in acetic acid (6.94) and $CrO_3 \cdot 2Py$, Py = pyridine (6.38), however, achieve the desired transformation. A cyclic five-center mechanism involving 6.29 has been suggested.

6.29

Secondary and tertiary alkyltins are converted to ketones and tertiary alcohols, respectively, using 10 to 15 equivalents of $CrO_3 \cdot 2Py$ (eq. 6.102 to 6.104).

$$\ce{>C-Sn<} \xrightarrow{CrO_3 \cdot 2Py} \ce{>C-OH} \xrightarrow{CrO_3 \cdot 2Py} \ce{>C=O} \quad (6.101)$$

$$\text{(6.102)}$$

$$\text{(6.103)}$$

This reaction and the conjugate addition reaction of LiSnR$_3$ have been utilized in the development of the following dialkylative enone transposition sequence (*6.38*).

$$\text{(6.104)}$$

71% (overall)

(c) Cleavage of the Sn-C bond via transmetallation. As discussed in Sect. 6.4.1, strongly basic organolithiums readily displace less basic ligands, thereby providing a convenient method for preparing organolithiums, such as those containing allyl, benzyl, alkenyl, aryl, and cyclopropyl groups (*6.46*). A combination of hydrostannation and transmetallation provides a convenient stereoselective route to olefins and even a route to acetylenes, as shown in eqs. 6.105 and 6.106 (*6.35, 6.36*).

(d) Electrophilic substitution reactions of α,β- and β,γ-unsaturated organotins. These reactions are discussed in Sects. 6.4.1 and 6.4.2. Additional information may be found in a few reviews on these reactions (*6.42, 6.95*). It may be emphasized

here again that this addition-elimination reaction takes place
considerably more readily with organotins than with the corresponding organosilicons.

(6.105)

(6.106)

Addition Reactions of Allyltins. Allyltins undergo thermal addition reactions with aldehydes, perhaloketones, and electrophilic olefins and acetylenes (6.95). Although promising, the
current scope and synthetic utility of these reactions are rather
limited.

Reactions via β-Heterosubstituted Organotins. The high thermodynamic stability of the Sn-O (95 kcal/mole), Sn-Cl (85 kcal/mole),
and other Sn-hetero atom bonds suggests that, as in the cases of
organosilicons, certain β-heterosubstituted organotins can also
undergo facile β-elimination reactions to form olefins.

$$R_3Sn-\overset{|}{\underset{|}{C}}-\overset{|}{\underset{|}{C}}-X \longrightarrow >C=C< + XSnR_3 \qquad (6.107)$$

Indeed some olefination reactions, which are the tin-counterparts of some of the organosilicon reactions discussed earlier, have been reported, as shown in eq. 6.108 (6.96).

$$\text{Me}\overset{}{\underset{O}{\triangle}}\text{Me} \xrightarrow[\text{2. } H_2O]{\text{1. NaSnPh}_3} \overset{Ph_3SnOH}{\underset{MeCHCHMe}{}} \xrightarrow{\text{HOAc, } H_2O} \text{MeCH=CHMe}$$

$$\text{(6.108)}$$

$$+ \quad Ph_3SnOAc$$

(100% stereospecific)

Although the current scope of the olefination reaction via β-heterosubstituted organotins is limited, its potential scope appears to be broad.

6.4.3.2   Reactions of Organotins Containing the Sn-O and Other Sn-Hetero Atom Bonds

Various reactions of the Sn-O and other Sn-hetero atom bonds have been reviewed recently (6.95, 6.97). In general, the Sn-O derivatives can act as analogs of the alkoxides and enolates of alkali and alkaline earth metals with greatly reduced reactivities. They are considerably more reactive, however, than the corresponding Si-O derivatives.

Organotin alkoxides can be prepared by various methods including: (1) the reaction of alkyl carbonates with organotin oxides (eq. 6.109), (2) the reaction of metal alkoxides with organotin halides (eq. 6.110), and (3)transesterification (eq. 6.111).

$$R_3SnOSnR_3 + (R'O)_2CO \longrightarrow 2R_3SnOR' + CO_2 \qquad (6.109)$$

$$R_3SnX \quad + MOR' \quad \longrightarrow \quad R_3SnOR' + MX \qquad (6.110)$$

$$R_3SnOR^1 \quad + HOR^2 \quad \longrightarrow \quad R_3SnOR^2 + HOR^1 \qquad (6.111)$$

In addition to these methods, the conjugate addition reaction of triorganotin hydrides (Sect. 6.3.3) provides a convenient route to the enol ether derivatives (eq. 6.112).

$$\overset{|}{C}=\overset{|}{C}-\overset{|}{C}=O \xrightarrow{R_3SnH} H\overset{|}{C}-\overset{|}{C}=\overset{|}{C}-OSnR_3 \qquad (6.112)$$

In the following discussion, our attention is mainly focused on those reactions that involve the formation of C-C and C-O bonds.

Intramolecular Reactions. Typical unactivated organotin alkoxides are relatively inert to organic electrophiles, such as organic halides, carbonyl compounds, and epoxides. The intramolecular substitution reaction with organic halides, however, will take place at elevated temperatures; see eq. 6.113 (6.95, 6.97).

$$n = 1, 2, 3, \text{ or } 4$$

Intermolecular Reactions. (a) Reactions with α-haloketones. The alkoxy and amino derivatives of organotins react with α-chloroketones to produce the corresponding α-alkoxy- and α-aminoketones (6.95, 6.97). The corresponding reaction of α-bromoketones is complicated by side reactions.

Since alkali metal alkoxides and amides mainly undergo the Favorskii rearrangement with α-haloketones, the direct substitution reaction of the organotin derivatives appears to be of considerable synthetic value.

(b) Reactions of triorganostannyl enol ethers. Triorganostannyl enol ethers are more reactive than the corresponding silyl enol ethers. Their synthetic utility, however, has not been well delineated. At elevated temperatures, stannyl enol ethers react with more reactive alkyl halides, such as methyl iodide and allyl bromide (6.95, 6.97). As might be expected, polyalkylation does not occur, although regiochemical scrambling may take place to some extent. This reaction might provide some

advantages over the more classical alkylation reactions with enolate anions.

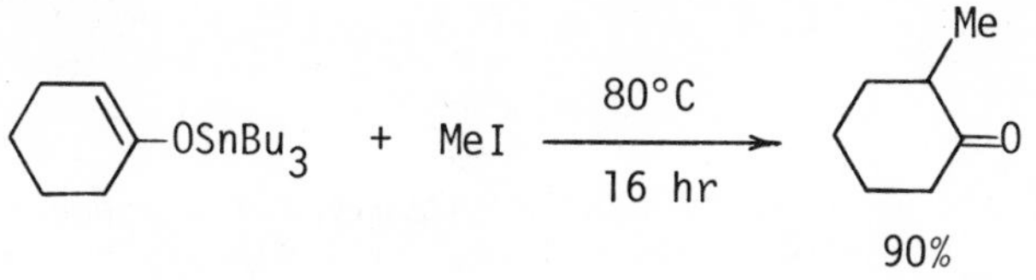

$$
\text{(cyclohexenyl)}-\text{OSnBu}_3 \quad + \quad \text{MeI} \quad \xrightarrow[\text{16 hr}]{80°C} \quad \text{(2-methylcyclohexanone)} \qquad (6.115)
$$

90%

$$
\text{Et}_2\text{C=CHOSnBu}_3 \quad + \quad \text{MeI} \quad \xrightarrow[\text{14 hr}]{90°C} \quad \text{Et}_2\text{MeCHO} \qquad (6.116)
$$

82%

Although not yet well-delineated, the reactions of stannyl enol
ethers with carbonyl compounds and electrophilic olefins and
acetylenes, under appropriate conditions, might be developed into synthetically useful reactions.

   (c) Oxidation of organotin alkoxides. It has been found
that organotin alkoxides serve as useful intermediates for the
oxidation of the corresponding alcohols (6.95, 6.97). A few representative examples follow.

$$
\text{PhMeCHOSnBu}_3 \quad \xrightarrow[\text{80°C, 14 hr}]{\text{BrCCl}_3,\ h\nu} \quad \text{PhCOMe} \qquad (6.117)
$$

83%

$$
\text{C}_7\text{H}_{15}\text{CH}_2\text{OSnEt}_3 \quad \xrightarrow{\text{Br}_2,\ \text{Et}_3\text{SnOMe}} \quad \text{C}_7\text{H}_{15}\text{CHO} \qquad (6.118)
$$

85%

The reaction shown in eq. 6.117 must involve a free-radical mechanism (eq. 6.119), while the reaction with bromine might proceed
by a polar cyclic mechanism; see eq. 6.120 (6.98). The use of
one equivalent of Et₃SnOMe is necessary to trap HBr.

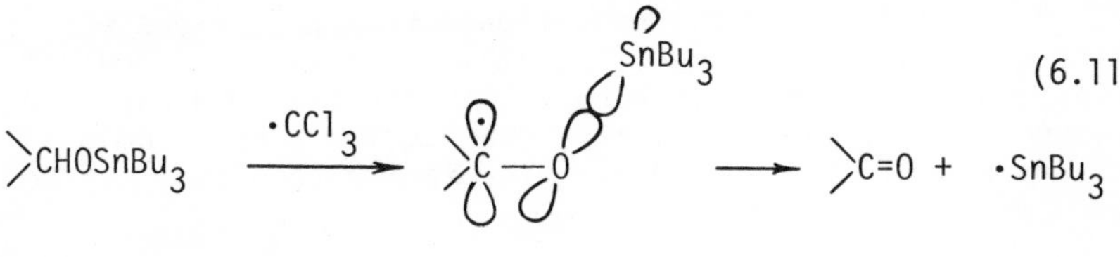

$$(6.119)$$

$$\begin{array}{c} \text{SnEt}_3 \\ \text{Br} \\ \text{Br} \end{array} \quad \longrightarrow \quad \text{>C=O} \; + \; \text{BrSnEt}_3 \; + \; \text{HBr} \qquad (6.120)$$

Regardless of the precise mechanisms involved, it appears certain that these reactions are assisted considerably by the σ-π conjugation effect. The oxidation reaction with bromine has provided a convenient method for the conversion of 1,2-glycols into acyloins; see eq. 6.121 (6.99).

$$(H_2C)_n \begin{array}{c}\text{CHOH} \\ \text{CHOH}\end{array} \xrightarrow{(Bu_2SnO)_n} (H_2C)_n \begin{array}{c}\text{CHO} \\ \text{CHO}\end{array}\!\!SnBu_2 \xrightarrow{Br_2} (H_2C)_n \begin{array}{c}\text{C=O} \\ \text{CHOH}\end{array}$$

$$(6.121)$$

6.4.3.3 Reactions of Organotin Hydrides

The chemistry of organotin hydrides is not discussed in detail here. It has been extensively reviewed by Kuivila (6.100).

Although a wide variety of functional groups can be reduced by organotin hydrides, the reduction of organic halides and that of olefins and acetylenes (hydrostannation) are the two most important transformations from the viewpoint of organic synthesis. The latter reaction is discussed in Sect. 6.3.3.

The reduction of organic halides has been shown to proceed by the following free-radical mechanism (eq. 6.122).

$$R_3SnH \xrightarrow{\cdot In} R_3Sn\cdot \; + \; HIn$$
$$R_3Sn\cdot \; + \; R^1X \longrightarrow R_3SnX \; + \; R^1\cdot \qquad (6.122)$$
$$R^1\cdot \quad R_3SnH \longrightarrow R_3Sn\cdot \; + \; R^1H$$

Since most of the reduction reactions of organic halides with Group IIIA metal hydrides proceed by polar mechanisms (Sect. 5.2.3.6), the free-radical reduction with tin hydrides and the polar reduction with Group IIIA metal hydrides are often complementary to each other. In accord with the free-radical mechanism shown above, the reactivity sequence for alkyl bromides is: tertiary > secondary > primary. The facile reduction of tertiary and secondary alkyl halides is probably the single most important feature of the Sn-H reduction. It is even possible to reduce in a stepwise manner, geminal dibromides first to the monohalides and then to the corresponding hydrocarbons, as shown in eq. 6.123 (6.100).

$$82\% \; (1:2.5) \tag{6.123}$$

Various other types of organic halides including alkynyl, allyl, benzyl, alkenyl, and aryl halides can also be reduced, although the Sn-H reduction may not represent the most satisfactory method in these cases. Various unsaturated functional groups other than olefins and acetylenes are also reduced. The reduction of acyl halides is fast, but mixtures of reduced products are usually obtained. Aldehydes react with organotin hydrides at moderate rates to give primary alcohols. On the other hand, ketones, esters, lactones, tertiary amides, and ethers react with trialkyltin hydrides only sluggishly, if at all. Thus these functional groups can be tolerated in the reduction of more reactive functional groups, as shown in eq. 6.124 (6.101).

$$\tag{6.124}$$

In this example, the actual reactive species is Bu₃SnH generated in situ by the reaction of NaBH₄ and a catalytic amount of Bu₃SnCl.

REFERENCES

6.1   MacDiarmid, A. G., Ed., *Organometallic Compounds of the Group IV Elements*, 2 vols., Marcel Dekker, New York, 1968 and 1972: Vol. 1, Part I, 603 pp.; Vol. 1, Part II, 261 pp.; Vol. 2, Part I, 374 pp.; and Vol. 2, Part II, 234 pp.

6.2   Eaborn, C., *Organosilicon Compounds*, Academic, New York, 1960, 530 pp.

6.3   Neumann, W. P., *The Organic Chemistry of Tin*, Wiley-Interscience, New York, 1970, 282 pp.

6.4   Poller, R. C., *The Chemistry of Organotin Compounds*, Logos Press, London, 1970, 315 pp.

6.5   Sawyer, A. K., Ed., *Organotin Compounds*, 3 vols., Marcel Dekker, New York, 1971 and 1972.

6.5a   Zuckerman, J. J., Ed., *Organotin Compounds: New Chemistry and Applications*, American Chemical Society, Washington, DC, 1976, 299 pp.

6.6   (a) Sommer, L. H., *Stereochemistry, Mechanism, and Silicon*, McGraw-Hill, New York, 1965, 189 pp; (b) Sommer, L. H., *Intra-Sci. Chem. Rep.*, **7**, No. 4, 1 (1973); (c) Gielen, M., Hoogzand, C., Simon, S., Tondeur, Y., Van den Eynde, I., and Vande Steen, M., *Adv. Chem. Ser.*, **157**, 249 (1976).

6.7   Guselnikov, L. E., Nametkin, N. S., and Vdovin, V. M., *Acc. Chem. Res.*, **8**, 18 (1975).

6.8   Barton, T. J., and Kline, E., *J. Organometal. Chem.*, **42**, C21 (1972).

6.9   Peddle, G. J. D., Roark, D. N., Good, A. M., and McGeachin, S. G., *J. Am. Chem. Soc.*, **91**, 2807 (1969).

6.10   Pitzer, K. S., *J. Am. Chem. Soc.*, **70**, 2140 (1948).

6.11   Mulliken, R. S., *J. Am. Chem. Soc.*, **72**, 4493 (1950); **77**, 884 (1955).

6.12   Gielen, M., and Sprecher, N., *Organometal. Chem. Rev.*, **1**, 455 (1966).

6.13   Okawara, R., and Wada, M., *Adv. Organometal. Chem.*, **5**, 137 (1967).

6.14   Vyazankin, N. S., Razuvaev, G. A., and Kruglaya, O. A., *Organometal. React.*, **5**, 101 (1975).

6.15   Davidson, I. M. T, *Quart. Rev.*, **25**, 111 (1971).

6.16   Sakurai, H., in *Free Radicals*, Kochi, J. K., Ed., Vol. 2, Chap. 25, John Wiley & Sons, New York, 1973.

6.17   (a) West, R., *Pure Appl. Chem.*, **13**, 1 (1966); (b) Corriu, R. J. P., and Henner, M., *J. Organometal. Chem.*, **74**, 1 (1974).

6.18 Margrave, J. L., and Wilson, P. W., *Acc. Chem. Res.*, **4**, 145 (1971).

6.19 (a) Atwell, W. H., and Weyenberg, D. R., *Angew. Chem. Int. Edit.*, **8**, 469 (1969); (b) Atwell, W. H., and Weyenberg,D.R. *Intra-Sci. Chem. Rep.*, **7**, No. 4, 139 (1973).

6.20 (a) Creemers, H. M. J. C., and Noltes, J. G., *Rec. Trav. Chim.*, **84**, 590, 1589 (1965); (b) Creemers, H. M. J. C., Verbeek, F., and Noltes, J. G., *J. Organometal. Chem.*, **8**, 469 (1967).

6.21 Blaauw, H. J. A., Nivard, R. J. F., and van der Kerk, G J. M., *J. Organometal. Chem.*, **2**, 326 (1964).

6.22 Petrov, A. D., Mironov, B. F., Ponomarenko, V. A., and Chernyshev, E. A., *Synthesis of Organosilicon Monomers*, Consultants Bureau, New York, 1964, 492 pp.

6.23 (a) Bazant, V., Joklik, J., and Rathousky, J., *Angew. Chem. Int. Ed. Engl.*, **7**, 112 (1968); (b) Bazant, V., *Pure Appl. Chem.*, **19**, 473 (1969).

6.24 Sommer, L. H., Pietrusza, E. W., and Whitmore, F. C., *J. Am. Chem. Soc.*, **69**, 188 (1947).

6.25 Van der kerk, G. J. M., Luijten, J G. A., and Noltes, J. G., *Chem. Ind.*, 352 (1956).

6.26 Fisher, A. K., West, R. C., and Rochow, E. G., *J. Am. Chem. Soc.*, **76**, 5878 (1954).

6.27 Becker, W. E., and Cook. S. E., *J. Am. Chem. Soc.*, **82**, 6264 (1960).

6.28 Lukevits, E. Ya., and Voronkov, M. G., *Organic Insertion Reactions of Group IV Elements*, Consultants Bureau, New York, 1966, 413 pp.

6.29 Speier, J. L., Webster, J. A., and Barnes, G. H., *J. Am. Chem. Soc.*, **79**, 974 (1957).

6.30 (a) Benkeser, R. A., Burrons, M. L., Nelson, L. E., and Swisher, J. V., *J. Am. Chem. Soc.*, **83**, 4385 (1961); (b) Benkeser, R. A., *Pure Appl. Chem.*, **13**, 133 (1966).

6.31 Ojima, I., in *Organotransition-Metal Chemistry*, Ishii, Y., and Tsutsui, M., Eds., Plenum, New York, 1975, p. 255.

6.32 Benkeser, R. A., *Acc. Chem. Res.*, **4**, 94 (1971).

6.33 (a) Leusink, A. J., Budding, H. A., and Marman, J. W., *J. Organometal. Chem.*, **9**, 285 (1967); (b) Leusink, A. J., Budding, H. A., and Drenth, W., *J. Organometal. Chem.*, **9**, 295 (1967).

6.34   Corey, E. J., Ulrich, P., and Fitzpatrick, J. M., *J. Am. Chem. Soc.*, <u>98</u>, 222 (1976).

6.35   Corey, E. J., and Wollenberg, R. H., *J. Org. Chem.*, <u>40</u>, 2265 (1975).

6.36   (a) Corey, E. J., and Wollenberg, R. H., *J. Am. Chem. Soc.*, <u>96</u>, 5582 (1974); (b) Nesmeyanov, A. N., and Borison, A. E., *Dokl. Akad. Nauk SSSR*, <u>174</u>, 96 (1967).

6.37   Still, W. C., *J. Org. Chem.*, <u>41</u>, 3063 (1976).

6.38   Still, W. C., *J. Am. Chem. Soc.*, <u>99</u>, 4836 (1977).

6.39   Dervan, P. B., and Shippey, M. A., *J. Am. Chem. Soc.*, <u>98</u>, 1265 (1976).

6.40   Davies, A. G., *Adv. Chem. Ser.*, <u>157</u>, 26 (1976).

6.41   Boué, S., Gielen, M., and Nasielski, J., *J. Organometal. Chem.*, <u>9</u>, 461 (1967).

6.42   Eaborn, C., *J. Organometal. Chem.*, <u>100</u>, 43 (1975).

6.43   (a) Ware, J. C., and Traylor, T. G., *J. Am. Chem. Soc.*, <u>89</u>, 2304 (1967); (b) Hosomi, A., and Traylor, T. G., *J. Am. Chem. Soc.*, <u>97</u>, 3682 (1975); (c) Hartman, G. D., and Traylor, T. G., *J. Am. Chem. Soc.*, <u>97</u>, 6147 (1975).

6.44   Eaborn, C., and Pande, K. C., *J. Chem. Soc.*, 1566 (1960).

6.45   Jerkunica, J. M., and Traylor, T. G., *J. Am. Chem. Soc.*, <u>93</u>, 6278 (1971).

6.46   Seyferth, D., and Weiner, M. A., *J. Org. Chem.*, <u>24</u>, 1395 (1959); <u>26</u>, 4797 (1961); (b) Seyferth, D., and Cohen, H. M., *Inorg. Chem.*, <u>2</u>, 625 (1963); (c) Seyferth, D., and Vaughan, L. G., *J. Am. Chem. Soc.*, <u>86</u>, 883 (1964); (d) Seyferth, D., Vaughan, L. G., and Suzuki, R., *J. Organometal. Chem.*, <u>1</u>, 437 (1964); (e) Seyferth, D., Suzuki, R., Murphy, C. J., and Sabet, C. R., *J. Organometal. Chem.*, <u>2</u>, 431 (1964).

6.47   (a) Kraihanzel, C. S., and Poist, J. E., *J. Organometal. Chem.*, <u>8</u>, 239 (1967); (b) Bøe, B., *J. Organometal. Chem.*, <u>107</u>, 139 (1976).

6.48   Sommer, L. H., Van Strien, R. E., and Whitmore, F. C., *J. Am. Chem. Soc.*, <u>71</u>, 3056 (1949).

6.49   (a) Brook, A. G., *J. Am. Chem. Soc.*, <u>80</u>, 1886 (1958); (b) West, R., *Pure Appl. Chem.*, <u>19</u>, 291 (1969); (c) Brook, A. G., *Acc. Chem. Res.*, <u>7</u>, 77 (1974).

6.50   Eaborn, C., Najam, A. A., and Walton, D. R. M., *Chem. Commun.*, 840 (1972), *J. Chem. Soc., Perkin I*, 2481 (1972).

6.51   Hudrlik, P. F., *J. Organometal. Chem. Libr.*, 1, 127 (1976).

6.52   (a) Stork, G., and Colvin, E., *J. Am. Chem. Soc.*, 93, 2080 (1971); (b) Stork, G., and Jung, M. E., *J. Am. Chem. Soc.*, 96, 3682 (1975).

6.53   (a) Hudrlik, P. F , Peterson, D., and Rona, R. J., *J. Org. Chem.*, 40, 2263 (1975); (b) Hudrlik, P. F., Hudrlik, A. M., Rona, R. J., Misra, R. N., and Withers, G. P., *J. Am. Chem. Soc.*, 99, 1993 (1977).

6.54   Peterson, D. J., *J. Org. Chem.*, 33, 780 (1968).

6.55   Musker, W. K., and Larson, G. L., *Tetrahedron Lett.*, 3481 (1968).

6.56   Miller, R. B., and Reichenbach, T., *Tetrahedron Lett.*, 543 (1974).

6.57   Zweifel, G., Backlund, S. J., *J. Am. Chem. Soc.*, 99, 3184 (1977).

6.58   (a) Hosomi, A., and Sakurai, H., *Tetrahedron Lett.*, 1295 (1976); (b) Hosomi, A., Endo, M., and Sakurai, H., *Chem. Lett.*, 941 (1976); (c) Hosomi, A., and Sakurai, H., *J. Am. Chem. Soc.*, 99, 1673 (1977).

6.59   Rasmussen, J. K., *Synthesis*, 91 (1977).

6.60   Nesmeyanov, A. N., *J. Organometal. Chem.*, 100, 161 (1975).

6.61   House, H. O., Czuba, L. J., Gall, M., and Olmstead, H. D., *J. Org. Chem.*, 34, 2324 (1969); 36, 2361 (1971).

6.62   Hall, H. K., Jr., and Ykman, P., *J. Am. Chem. Soc.*, 97, 800 (1975).

6.63   (a) Ojima, I., Kogure, T., Nihonyanagi, M., and Nagai, Y., *Bull. Chem. Soc. Japan*, 45, 3506 (1972); (b) Ojima, I., Kogure, T., and Nagai, Y., *Tetrahedron Lett.*, 5035 (1972).

6.64   Blanco, L., Amice, P., and Conia, J. M., *Synthesis*, 194 (1976).

6.65   (a) Strating, J., Reiffers, S., and Wynberg, H., *Synthesis*, 209, 211 (1971); (b) Rubottom, G. M., Vazquez, M. A., and Pelegrina, D. R., *Tetrahedron Lett.*, 4319 (1974); (c) Larson, G. L., Hernandez, D., and Hernandez, A., *J. Organometal. Chem.*, 76, 9 (1974); (d) Wilson, S. R., Walters, M. E., and Orgaugh, B., *J. Org. Chem.*, 41, 378 (1976).

6.66 Murai, S., Kuroki, Y., Hasegawa, K., and Tsutsumi, S., *Chem. Commun.*, 946 (1972).

6.67 Kuroki, Y., Murai, S., Sonoda, N., and Tsutsumi, S., *Organometal. Chem. Synth.*, 1, 465 (1972).

6.68 Rasmussen, J. K., and Hassner, A., *J. Org. Chem.*, 39, 2558 (1974).

6.69 Kramarova, E. N., Baukov, Yu. I., and Lutsenko, I. F., *Zh. Obshch. Khim.*, 43, 1857 (1973).

6.70 Parham, W. E., and Roosevelt, C. S., *Tetrahedron Lett.*, 923 (1971).

6.71 (a) Mukaiyama, T., Banno, K., and Narasaka, K., *J. Am. Chem. Soc.*, 96, 7503 (1974); (b) Banno, K., and Mukaiyama, T., *Chem. Lett.*, 741 (1975).

6.72 Saigo, K., Osaki, M., and Mukaiyama, T., *Chem. Lett.*, 989 (1975).

6.73 (a) Narasaka, K., Soai, K., Akiyama, Y., and Mukaiyama, T., *Bull. Chem. Soc. Japan*, 49, 779 (1976); (b) Saigo, K., Osaki, M., and Mukaiyama, T., *Chem. Lett.*, 15 (1976).

6.74 (a) Mukaiyama, T., and Hayashi, M., *Chem. Lett.*, 15 (1974); (b) Mukaiyama, T., and Ishida, A., *Chem. Lett.*, 319 (1975); (c) Ishida, A., and Mukaiyama, T., *Chem. Lett.*, 1167 (1976); (d) Mukaiyama, T., Ishihara, H., and Inomata, K., *Chem. Lett.*, 527, 531 (1975).

6.75 (a) Danishefsky, S., and Kitahara, T., *J. Am. Chem. Soc.*, 96, 7807 (1974); (b) Danishefsky, S., and Kitahara, T., *J. Org. Chem.*, 40, 538 (1975).

6.76 Wohl, R. A., *Helv. Chim. Acta*, 56, 185 (1973).

6.77 Mizuno, K., Okamoto, H., Pac, C., Sakurai, H., Murai, S., and Sonoda, N., *Chem. Lett.*, 237 (1975).

6.78 (a) Ojima, I., Inaba, S., and Nagai, Y., *Chem. Lett.*, 1069 (1974); (b) Ojima, I., Inaba, S., and Nagai, Y., *Tetrahedron Lett.*, 4271 (1973).

6.79 Conia, J. M., *Pure Appl. Chem.*, 43, 317 (1975), and references therein.

6.80 (a) Klein, J., Levene, R., and Dunkelblum, E., *Tetrahedron Lett.*, 2845 (1972); (b) Larson, G. L., Hernandez, D., and Hernandez, A., *J. Organometal. Chem.*, 76, 9 (1974).

6.81 Rubottom, G. M., and Nieves, M. I. L., *Tetrahedron Lett.*, 2423 (1972).

6.82   Clark, R. D., and Heathcock, C. H., *Tetrahedron Lett.*, 1713 (1974).

6.83   Birkofer, L., Ritter, A., and Wieden, H., *Chem. Ber.*, 95, 971 (1962).

6.84   Webb, A. F., Sethi, D. S., and Gilman, H., *J. Organometal. Chem.*, 21, p61 (1970).

6.85   (a) Evans, D. A., Truesdale, L. K., and Carroll, G. L., *Chem. Commun.*, 55 (1973); (b) Evans, D. A., Hoffman, J. M., and Truesdale, L. K., *J. Am. Chem. Soc.*, 95, 5822 (1973); (c) Evans, D. A., and Truesdale, L. K., *Tetrahedron Lett.*, 4929 (1973).

6.86   Itoh, K., Fukui, M., and Ishii, Y., *Tetrahedron Lett.*, 3867 (1968).

6.87   Birkofer, L., and Dickopp, H., *Chem. Ber.*, 101, 3579 (1968).

6.88   Birkofer, L., and Kaiser, W., *Ann. Chem.*, 266 (1975).

6.89   Evans, D. A., Grimm, K. G., and Truesdale, L. K., *J. Am. Chem. Soc.*, 97, 3229 (1975).

6.90   Evans, D. A., Carroll, G. L., and Truesdale, L. K., *J. Org. Chem.*, 39, 914 (1974).

6.91   (a) Deuchert, K., Hertenstein, U., and Hünig, S., *Synthesis*, 777 (1973); (b) Hünig, S., and Wehner, G., *Synthesis*, 180, 391 (1975).

6.92   (a) Pierce, A. E., *Silylation of Organic Compounds*, Pierce Chemical Co., Rockford, Ill. 1968; (b) Klebe, J. F., *Adv. Org. Chem.*, 8, 97 (1972).

6.93   Rahm, A., and Pereyre, M., *J. Am. Chem. Soc.*, 99, 1672 (1977).

6.94   Deblandre, C., Gielen, M., and Nasielski, J., *Bull. Soc. Chem. Belges*, 37, 214 (1964).

6.95   Pereyre, M., and Pommier, J. C., *J. Organometal. Chem. Libr.*, 1, 161 (1976).

6.96   (a) Davis, D. D., and Gray, C. E., *J. Organometal. Chem.*, 18, P1 (1969); (b) Davis, D. D., and Gray, C. E., *J. Org. Chem.*, 35, 1303 (1970).

6.97   Pommier, J. C., and Pereyre, M., *Adv. Chem. Ser.*, 157, 82 (1976).

6.98   Saigo, K., Morikawa, A., and Mukaiyama, T., *Chem. Lett.*, 145 (1975).

6.99   David, S., *C. R. Acad. Sci., Ser. C.*, 1051 (1974).

6.100   (a) Kuivila, H. G., *Adv. Organometal. Chem.*, 1, 47 (1964); (b) Kuivila, H. G., *Acc. Chem. Res.*, 1, 299 (1968); (c) Kuivila, H. G., *Synthesis*, 499 (1970).

6.101   Corey, E. J., and Suggs, J. W., *J. Org. Chem.*, 40, 2554 (1975).

7
ORGANOMETALLICS OF
HEAVY MAIN GROUP METALS
(Hg, T1, Pb)

The early chemistry of organomercuries and organothalliums is
discussed in textbooks by Coates and Wade (0.6) and by Nesmeyanov
and Kocheshkov (0.8). The former also contain a detailed discussion of organolead chemistry. Houben-Weyl's *Methoden der*
Organischen Chemie (0.9) contains extensive discussion on these
organometallics.

The preparations, structures, and properties of organothalliums and organoleads are also discussed in monographs by
Lee (7.1) and by Shapiro and Frey (7.2), respectively. Various
reactions of organomercury compounds have been reviewed by
Makarova (7.3) and Larock (7.4), and the recent advances in organothallium chemistry have been discussed by Taylor and McKillop
(7.5) and by Lee (7.6).

7.1 GENERAL CONSIDERATIONS

7.1.1 Some Fundamental Properties of Mercury, Thallium, and Lead and the Bonds to These Elements

Mercury, thallium, and lead are all relatively electronegative,
the electronegativity values for these elements being approximately 1.9 (1.1), 1.8 (1.1), and 1.8 (7.7), respectively (Appendix II). Thus the bonds between these elements and carbon are
highly covalent. As a result, even though the M-C bonds, where
M = Hg, T1, and Pb, are polarized such that the carbon atom

455

assumes some carbanion character, the intrinsic nucleophilicity
of these M-C bonds is very low. These heavy elements and the
bonds to these elements, however, are highly polarizable. The
Hg-X, Tl-X, and Pb-X bonds, where the X group is highly electronegative, for example, F, NO_3, ClO_4, can behave as highly ionic
bonds. Thus, for example, organomercurials of the RHgX type containing the highly electronegative X group tend to be more solu-
ble in water than in nonpolar solvents. These highly polarizable ("soft") bonds react readily with "soft" organic substrates, such as olefins, acetylenes, and aromatic compounds
(heterometallation). Thus compounds containing these bonds are
good electrophiles. This ability to interact readily with soft
nucleophiles is one of the most significant properties of compounds containing Hg, Tl, and Pb, and permits a unique and facile entry into organometallics containing these metals.
     Another significant feature associated with these elements
is the ease with which compounds containing these elements undergo redox reactions involving two oxidation states, as shown in
eqs. 7.1 to 7.4 (7.8).

$$Hg^{2+} + 2e \longrightarrow Hg^0 \qquad\qquad -0.85\ V \qquad (7.1)$$

$$2Hg^{2+} + 2e \longrightarrow Hg_2^{2+} \qquad\qquad -0.92\ V \qquad (7.2)$$

$$Tl^{3+} + 2e \longrightarrow Tl^+ \qquad\qquad -1.25\ V \qquad (7.3)$$

$$Pb^{4+} + 2e \longrightarrow Pb^{2+} \qquad\qquad -1.6\ V \qquad (7.4)$$

As the oxidation potentials for these reactions indicate, Pb(IV),
Tl(III), and Hg(II) are all good oxidizing agents, the relative
oxidizing ability being: Pb(IV) > Tl(III) > Hg(II). Although
the reduction of organomercuries usually requires suitable reduction agents, organothalliums of the $RTl(III)X_2$ type and organoleads of the $RPb(IV)X_3$ type tend to be quite unstable and
usually decompose spontaneously. The spontaneous or nucleo-
phile-assisted two-electron transfer process which $RTlX_2$ and
$RPbX_3$ undergo may be represented by eqs. 7.5 and 7.6.

$$R^1Tl(III)X_2 \xrightarrow{Y^-} R^2Y + Tl(I)X + X^- \qquad (7.5)$$

$$R^1Pb(IV)X_3 \xrightarrow{Y^-} R^2Y + Pb(II)X_2 + X^- \qquad (7.6)$$

R^1 may be the same as R^2;   Y = added nucleophile or X

In addition to the intrinsic instability of $R^1Tl(III)X_2$ and $R^1Pb(IV)X_3$, the thermodynamically stable nature of $Tl(I)X$ and $Pb(II)X_2$, where X = electronegative groups such as halogen, OR, OOCR, and so on, must also be providing a driving force for these reactions. A highly intriguing aspect of these decomposition reactions is that $R^1Tl(III)X_2$ and $R^1Pb(IV)X_3$ are now acting as sources of carbocations and/or carbocationic species. This property is unique only to a relatively small number of heavy metals, such as Tl, Pb, and Pd.

The Hg-C, Tl-C, and Pb-C bonds are thermodynamically rather weak, the bond energy values for these bonds, where the carbon group is a simple alkyl group, being approximately 50, 25 to 30, and 30 kcal/mole, respectively. Therefore, these organometallics should, in principle, be good sources of carbon-free radicals. Indeed, organomercuries and organoleads have acted as efficient sources of carbon-free radicals. On the other hand, organothalliums usually participate only in two-electron transfer processes. This is presumably due to the fact that thallium (II) is highly unstable with respect to disproportionation to thallium(III) and thallium(I).

Finally, although not yet fully exploited, it appears potentially significant to note that the weak and highly polarizable nature of the M-C bonds, where M = Hg and Pb, renders these bonds some of the most effective participants in $\sigma-\pi$ conjugation, as discussed in Sect. 6.4.1. It appears almost certain that many additional synthetic capabilities based on these unique properties of heavy metals await future exploration.

7.1.2 Structural Aspects

As already discussed, Hg, Tl, and Pb all can exist in two stable oxidation states in addition to the neutral oxidation state. There is one significant aspect to be pointed out. Whereas a number of inorganic derivatives of Hg, Tl, and Pb, in which the metal atoms exist in the lower oxidation state, that is, Hg(I), Tl(I), and Pb(II), are known as isolable species, there are relatively few compounds that contain the Hg(I)-C, Tl(I)-C, and Pb(II)-C bonds. Organothallium(I) compounds, in which the Tl(I)-C is largely covalent, do not appear to have been isolated. Cyclopentadienylthallium (7.1) and its derivatives, in which the Tl(I)-C is believed to be largely ionic, are the only group of organothallium(I) compounds known to date as isolable species (7.5b, 7.6).

Diorganolead(II) compounds are carbene analogs, and the probable structure of the singlet species may be represented by 7.2.

7.1

7.2

It is generally assumed that diorganoleads are formed as intermediates in the preparation of tetraorganoleads by the reaction
of PbX_2 with organometallic compounds (eq. 7.7).

$$PbX_2 + 2RMgX \longrightarrow PbR_2 + 2MgX_2 \qquad (7.7)$$

In general, diorganoleads are highly unstable (7.2). As mentioned in Sect. 6.2, however, certain diorganoleads containing bulky
ligands have been obtained and characterized (6.20). Although
they represent an interesting class of organometallics, their
utility in organic synthesis remains to be explored.

It follows from the discussion presented above that the
great majority of organomercuries, organothalliums, and organoleads contain Hg(II), Tl(III), and Pb(IV).

$$RHgX \qquad RTlX_2 \qquad RPbX_3$$
$$R_2Hg \qquad R_2TlX \qquad R_2PbX_2$$
$$R_3Tl \qquad R_3PbX$$
$$R_4Pb$$

Both RHgX and R_2Hg exist as linear or nearly linear species,
for example, 7.3, in which the mercury atom utilizes the sp hy-
dridized orbitals for bonding. This tendency to form linear

$$H_3C \longrightarrow Hg \longrightarrow Cl$$
$$2.06\text{Å} \quad 2.28\text{Å}$$

7.3

molecules is so strong that the formation of common rings containing one or more mercury atoms is quite unfavorable, as indicated by the following example; see eq. 7.8 (7.9).

Organothallium(III) compounds share many of the structural
features with organometallics of the lower members of the Group
IIIA elements (Sect. 5.2.1). Thus many of them exist as monomeric trigonal planar species, whereas others exist as bridged dimers and polymers. The main point of interest here is that $RTlX_2$,

R_2TlX, and R_3Tl differ vastly in their thermal and chemical stability. As discussed earlier, monoorganothallium(III) compounds
are thermally quite labile. On the other hand, diorganothallium-
(III) derivatives are both thermally and chemically quite stable.
They are not readily attacked by water and oxygen. Triorgano-
thalliums can readily be protonolyzed by various reagents, such
as HCl, water, alcohols, and thiols, in most cases producing di-
organothallium derivatives; see eq. 7.9 (7.6).

$$R_3Tl \ + \ HX \longrightarrow RH \ + \ R_2TlX \qquad (7.9)$$

Although a number of species represented by R_3Tl and R_2TlX are
now known (7.6), the very limited nature of their application in
organic synthesis does not warrant a detailed discussion of
these species here.

The most important application of organoleads by far is the
use of tetraorganoleads as antiknock agents, despite the current
effort to curtail this application. Tetraorganoleads are also
used in the preparation of organomercury fungicides, as shown in
eq. 7.10.

$$Et_4Pb \ + \ 2HgCl_2 \longrightarrow Et_2PbCl_2 \ + \ 2EtHgCl \qquad (7.10)$$

Other than these commercial applications, relatively little is
known about the usefulness of R_4Pb, R_3PbX, and R_2PbX_2 in organic
synthesis. On the other hand, monoorganolead(IV) derivatives
(RPbX3) have been postulated as intermediates in a number of oxidation reactions of organic compounds with $Pb(OAc)_4$, as discussed later. As to the structures of these organolead compounds,
suffice it to say that organolead compounds share many of the

structural features with the corresponding organotin compounds briefly discussed in Sect. 6.2. Interested readers are referred to the monographs and reviews cited earlier (0.6, 0.9, 7.2, 7.3).

7.2 PREPARATION OF ORGANOMERCURIES, ORGANOTHALLIUMS, AND ORGANO-LEADS

As in many other cases, transmetallation (Method XI) represents by far the most general method for the synthesis of organometallics containing Hg, Tl, and Pb. On the other hand, heterometallation (Method VII), especially oxymetallation, provides a unique and convenient route to these organometallics. Closely related to this method is the metal-hydrogen exchange reaction (Method V), which provides a direct route to the arylmetals containing mercury and thallium. While these three reactions represent the most convenient laboratory methods, the reaction of organic halides with metal alloys (Method I) is by far the most significant industrial method for preparing organoleads (7.2). This so-called "alloy method" is also applicable to the preparation of organomercuries (7.10), and organothalliums (7.1). In the presence of light, highly reactive organic iodides, such as those containing the methyl, methylene, benzyl, allyl, and propargyl groups, react with metallic mercury to produce the corresponding organomercuric iodides (7.10). Organic bromides and chlorides, however, are generally unreactive.

7.2.1 Transmetallation

7.2.1.1 Organomercuries

The reaction of organolithiums and Grignard reagents with mercuric halides produces either organomercuric halides or diorganomercuries (eqs. 7.11 and 7.12).

$$RM \ + \ HgX_2 \longrightarrow RHgX \ + \ MX \qquad (7.11)$$

$$RM \ + \ RHgX \longrightarrow R_2Hg \ + \ MX \qquad (7.12)$$

$$M = Li \ or \ MgX$$

Although generally less useful, various other polar organometallics, such as those containing Na, K, Zn, and Al, have also been used in this reaction. For example, various polyhalomethylmercury derivatives, which have proven to be valuable divalent carbon transfer agents (7.11), are readily obtained by the following reaction (eq. 7.13).

$$RHgX + HCX_3 + KOBu\text{-}t \xrightarrow{\text{THF}} RHgCX_3 + HOBu\text{-}t + KX \quad (7.13)$$

The reaction of organoboranes with mercuric salts has recently been developed into a valuable route to organomercuries (7.12). Primary trialkylboranes, readily available via hydroboration of terminal olefins (Sect. 5.2.2.2), undergo a rapid reaction with $Hg(OAc)_2$ at room temperature to produce alkylmercuric acetates (7.13), which are readily convertible to the corresponding alkylmercuric chlorides by treatment with aqueous sodium chloride.

$$(RCH_2CH_2)_3B \xrightarrow{3Hg(OAc)_2} 3RCH_2CH_2HgOAc \quad (7.14)$$

$$RCH_2CH_2HgOAc \xrightarrow{aq\ NaCl} RCH_2CH_2HgCl + NaOAc \quad (7.15)$$

It is important to note that various functionally substituted alkylmercury compounds are obtained by this method, as shown in eq. 7.16.

$$CH_2=CH(CH_2)_8COOMe \xrightarrow[\substack{2.\ Hg(OAc)_2 \\ 3.\ NaCl}]{1.\ BH_3} ClHg(CH_2)_{10}COOMe \quad (7.16)$$
$$97\%$$

Secondary alkyl-boron bonds do not react with mercuric salts under the same conditions. Thus they can serve as useful "dummy" groups allowing various selective transformations, as shown in eq. 7.17.

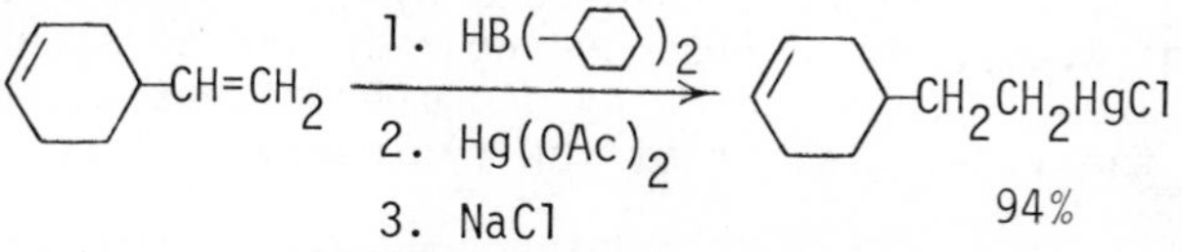

$$(7.17)$$
$$94\%$$

Secondary alkylmercury compounds can, however, be prepared by the reaction of secondary alkylboranes with mercury alkoxides (7.14).

$$(sec\text{-}R)_3B \xrightarrow[n = 1 \text{ or } 2]{Hg_n(OR')_2} sec\text{-}RHgOR' \qquad (7.18)$$

Alkynes can be converted to either alkenylmercuries or *gem*-dimercurialkanes (*7.15*).

$$RC\equiv CH \xrightarrow{HB(-\bigcirc)_2} \underset{H}{\overset{R}{\diagdown}}C=C\underset{B(-\bigcirc)_2}{\overset{H}{\diagup}} \qquad (7.19)$$

$$\xrightarrow[2.\ NaCl]{1.\ Hg(OAc)_2} \underset{H}{\overset{R}{\diagdown}}C=C\underset{HgCl}{\overset{H}{\diagup}}$$

$$RC\equiv CH(R') \xrightarrow{HB\overset{O}{\underset{O}{\diagup}}\bigcirc} \underset{H}{\overset{R}{\diagdown}}C=C\underset{B}{\overset{H(R')}{\diagup}}\overset{O}{\underset{O}{\diagdown}}\bigcirc \qquad (7.20)$$

$$\xrightarrow[2.\ NaCl]{1.\ Hg(OAc)_2} \underset{H}{\overset{R}{\diagdown}}C=C\underset{HgCl}{\overset{H(R')}{\diagup}}$$

$$RC\equiv CH \xrightarrow[2.\ MeOH]{1.\ 2BH_3} RCH_2CH\overset{B(OMe)_2}{\underset{B(OMe)_2}{\diagdown}} \qquad (7.21)$$

$$\xrightarrow[2.\ 3NaOH]{1.\ 2HgCl_2} RCH_2CH\overset{HgCl}{\underset{HgCl}{\diagdown}}$$

Whereas internal alkynes can be nicely converted to the corresponding alkenylmercuric chlorides by the catecholborane procedure (eq. 7.20), the corresponding reaction of alkenyldicyclohexylboranes is complicated by a competitive side reaction involving the 1,2-migration of the cyclohexyl group.

7.2.1.2 Organothalliums and Organoleads

Organothalliums of the R_3Tl and R_2RlX types (*7.1*, *7.6*) and organoleads of the R_4Pb type (*7.2*) can readily be prepared by the transmetallation reaction via organolithiums and Grignard reagents, while partially organosubstituted lead compounds of the R_3PbX and R_2PbX_2 types are most commonly prepared via R_4Pb (*7.2*). Although the limited usefulness of these compounds in organic synthesis does not permit detailed discussion of this subject, one peculiar aspect of organolead chemistry may be mentioned here. The inorganic salts of the divalent lead are generally more stable and easier to handle than those of the tetravalent lead, whereas diorganoleads are much less stable than tetraorganoleads. Furthermore, diorganoleads readily decompose to the corresponding tetraorganoleads and metallic lead. The ease with which inorganic divalent lead compounds (PbX_2) can be handled and the fact that R_2Pb can be converted readily into R_4Pb make the following method involving the use of PbX_2 often more convenient than the corresponding reaction of PbX_4 (*7.2*).

$$2PbX_2 \ + \ 4RM \longrightarrow 2R_2Pb \ + \ 4MX \qquad (7.22)$$

$$2R_2Pb \longrightarrow R_4Pb \ + \ Pb$$

7.2.2   Oxymetallation and Related Heterometallation Reactions

7.2.2.1   Oxymercuration

Mercuric salts, such as $Hg(OAc)_2$ and $Hg(OOCCF_3)_2$, react with olefins and acetylenes in the presence of active hydrogen compounds (HY) to produce a variety of β-heterosubstituted organomercuries (eq. 7.23).

$$\text{\textbackslash}C=C\text{\textbackslash} \ + \ HgX_2 \ + \ HY \longrightarrow Y\text{-}\overset{|}{C}\text{-}\overset{|}{C}\text{-}HgX \ + \ HX \qquad (7.23)$$

The reaction not only provides a highly convenient route to organomercuries, but also permits the Markovnikov functionalization of olefins (Sect. 7.3.2.1).

Scope, Regiochemistry, and Stereochemistry. The scope, regiochemistry, and stereochemistry of the oxymercuration reaction have been examined extensively (*2.27*). The available data indicate the following:
    (1) Although certain highly hindered olefins fail to give oxymercurated products, various mono-, di-, tri-, and tetrasubstituted olefins have participated in the reaction.

(2) The rates of oxymercuration generally decrease in the
order: $R_2C=CH_2$ > $RCH=CH_2$ > cis-RCH=CHR > trans-RCH=CHR >
$R_2CH=CHR$ > $R_2C=CR_2$ (2.27, 7.16). It is clear that the reactivity
order is governed by both steric and electronic effects.
(3) Oxymercuration proceeds such that, after replacement
of the mercury-containing group with a hydrogen atom, the Markovnikov addition products are obtained exclusively or predominantly (2.27, 7.16).
(4) The reaction with unstrained olefins generally involves
a trans addition, as shown in eq. 7.24 (7.17, 7.18). It should
be noted that comformationally rigid cyclohexenes, such as 4-
tert-butyl-1-cyclohexene, give almost exclusively the axially disubstituted products (7.18).

$$\text{(7.24)}$$

48%   41%

Certain strained olefins, such as norbornene and 7,7-dimethyl-
norbornene, give cis-addition products, as shown in eqs. 7.25 and
7.26 (7.17, 7.18). Particularly interesting is the nearly exclusive formation of the cis-exo products even in the case of
7,7-dimethylnorbornene.

95% (81:19)   (7.25)

80% (75:25)   (7.26)

(5) Oxymercuration of olefins containing suitably located
nucleophilic groups results in cyclization (7.21, 7.22).

$$CH_2=CH(CH_2)_2CH_2OH \xrightarrow{Hg(OAc)_2} AcOHgCH_2 \text{-} \underset{O}{\bigcirc} \qquad \text{(7.27)}$$

$$\text{(7.28)}$$

(6) Oxymercuration of acetylenes produces β-oxyalkenylmercury derivatives. The stereochemistry of the products seems to depend on both the nature of substituents and the reaction conditions (7.23, 7.24).

$$CH_3C\equiv CCH_3 \xrightarrow[\text{2. NaCl}]{\text{1. Hg(OAc)}_2} \quad \text{(7.29)}$$

$$PhC\equiv CPh \xrightarrow[\text{2. NaCl}]{\text{1. Hg(OAc)}_2} \quad \text{(7.30)}$$

(8) The most commonly employed nucleophile is water. Other oxy nucleophiles, however, can also be used. Using alcohols as solvents, β-alkoxyorganomercury compounds have been obtained; see eq. 7.31 (7.25).

$$\text{>C=C<} \xrightarrow[\text{ROH}]{\text{Hg(O}_2\text{CCF}_3)_2} RO-\overset{|}{\underset{|}{C}}-\overset{|}{\underset{|}{C}}-HgO_2CCF_3 \quad \text{(7.31)}$$

Mercuric trifluoroacetate, $Hg(O_2CCF_3)_2$, appears to be superior to $Hg(OAc)_2$ in this reaction (7.25). Peroxymercuration has also been achieved using hydrogen peroxide (7.26, 7.27) and alkylhydroperoxides (7.27, 7.28). Although $Hg(OAc)_2$ has been commonly used, $Hg(O_2CCF_3)_2$ also appears to be a superior reagent in this reaction (7.29).

$$\text{>C=C<} \xrightarrow[\text{ROOH}]{\text{HgX}_2} ROO-\overset{|}{\underset{|}{C}}-\overset{|}{\underset{|}{C}}-HgX \quad \text{(7.32)}$$

$$X = OAc \text{ or } O_2CCF_3, \qquad R = H \text{ or alkyl}$$

Formation of esters via oxymercuration has already been mentioned. By using $Hg(OAc)_2$, either in the absence of a protic solvent

or in the presence of acetic acid, olefins can be converted to
the corresponding acetates in high yields, as shown in eq. 7.33
(7.19b).

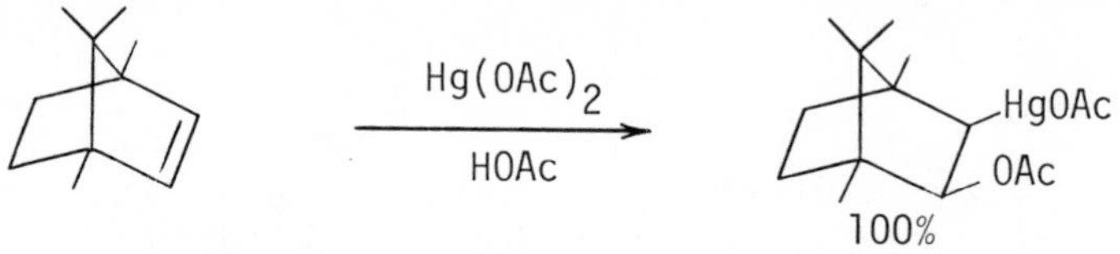

$$\text{(7.33)}$$

100%

Mechanism of Oxymercuration.   The mechanism of the oxymercuration reaction is still not very clear.   It is conceivable that
more than one mechanism might operate in this reaction.   The
trans oxymercuration reaction, which generally proceeds without
being accompanied by the Wagner-Meerwein rearrangement, has been
interpreted by many workers in terms of the intermediacy of mer-
curinium ions (7.4) as shown in eq. 7.34.

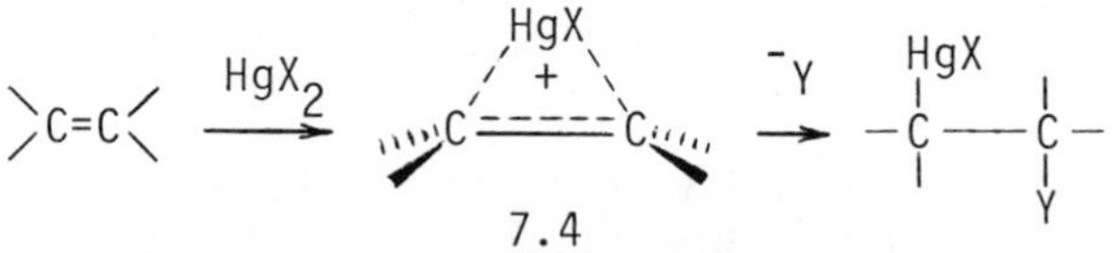

$$\text{(7.34)}$$

7.4

Spectroscopic data indicating the existence of the ethylenemer-
curinium ion (7.5) in $FSO_3H-SbF_5-SO_2$ have recently been presented (7.30).

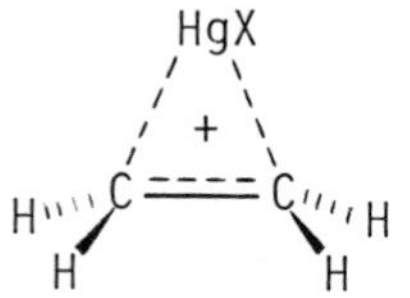

7.5

On the other hand, the cis oxymercuration reaction observed
with certain strained olefins, such as norbornene and 7,7-di-
methylnorbornene, is not consistent with the above mechanism.
Several mechanisms proposed by various workers are reviewed by
Traylor (2.28), who suggested that the cis oxymercuration reaction might proceed through a concerted molecular addition, as
shown in eq. 7.35.   It appears difficult, however, to accommodate the results observed with 7,7-dimethylnorbornene by this
mechanism, since certain addition reactions that are known to
proceed by concerted molecular addition mechanisms, such as hy-

droboration, have been shown to produce predominantly the endo isomers (7.31).

$$(7.35)$$

On the other hand, the stereochemical results observed with both norbornene and 7,7-dimethylnorbornene can be accommodated by the open carbonium ion mechanism proposed by Brown (7.32). This mechanism assumes that the attack of an olefin by an electrophile which can take place at one end of the C=C bond is not hindered by the presence of the *syn*-7-methyl group (eq. 7.36).

$$(7.36)$$

What this mechanism fails to explain convincingly is the fact that the reaction is not complicated by the Wagner-Meerwein rearrangement, although it is not necessarily inconsistent with the mechanism. At this point, we must conclude that the mechanism of oxymercuration still remains a subject of controversy. Additional information on this subject can be found in a few recent reviews of the mechanism of oxymercuration (0.10, 2.27, 2.28).

7.2.2.2   Other Heteromercuration Reactions

Various nucleophiles other than the oxy nucleophiles have been used in the heteromercuration reaction. Aminomercuration can be achieved using amines (7.33), acetonitrile (7.34), sodium azide (7.35), and sodium nitrite (7.36).

Halomercuration of olefins appears to be a thermodynamically unfavorable process. Certain acetylenes, however, have been converted to *trans*-ß-chloroalkenylmercury derivatives, as discussed in Sect. 2.7.

7.2.2.3 Oxythallation and Oxyplumbation

As mentioned in Sect. 2.7, several other heavy metals, such as
Tl, Pb, Pd, and Pt, also participate in oxymetallation (2.27).
Oxypalladation and oxyplatination are discussed in a later chapter (Sect. 10.3). Although oxythallation and oxyplumbation represent the most convenient routes to organothalliums of the
$RTlX_2$ type and organoleads of the $RPbX_3$ type, respectively, these
species are generally unstable and tend to decompose under the
oxymetallation conditions without permitting their isolation.
These reactions are therefore discussed in a later section of
this chapter dealing with various reactions of organometallics
containing Hg, Tl, and Pb.

7.2.3 Metal-Hydrogen Exchange -- Mercuration and Thallation
of Aromatic Compounds

The reaction of olefins and acetylenes with mercuric salts in
the presence of appropriate protic nucleophiles (HY) gives addition products, as discussed in the preceding section. On the
other hand, the corresponding reaction of arenes usually undergoes the following addition-elimination reaction (metal-hydrogen
exchange) to produce arylmercury compounds; see eq. 7.37 (7.10).

$$\tag{7.37}$$

Thallium salts, such as $Tl(OAc)_3$ and $Tl(O_2CCF_3)_3$, also undergo a
closely related reaction; see eq. 7.38 (7.5, 7.6).

$$\tag{7.38}$$

They are facilitated by highly ionic mercury and thallium salts,
such as $Hg(O_2CCF_3)_2$ and $Tl(O_2CCF_3)_3$.
These reactions exhibit the characteristics of typical electrophilic aromatic substitution reactions. Thus the difficulties
and limitations associated with the electrophilic aromatic substitution reactions have been encountered. These reactions often result in mixtures of o-, m-, and p-isomers, and deactivated
aromatic compounds fail to undergo these reactions. The regiochemistry of the thallation reaction is summarized in Table 7.1
(7.37).

Table 7.1. Regiochemistry of the Thallation of Aromatic Compounds at Room Temperature

Z	Isomer Distribution, (%)		
	o	m	p
Me	9	4	87
Pr-i	1	5	94
OMe	7	-	93
CH_2OMe	>99	-	-
CH_2CH_2OMe	85	3	12
$(CH_2)_3OMe$	27	6	67
CH_2OH	>99	-	-
CH_2CH_2OH	83	6	11
OAc	21	-	79
CH_2OAc	50	21	29
CH_2CH_2OAc	3	13	84
COOH(or COOMe)	95	5	0
CH_2COOH(or CH_2COOMe)	92	3	5
CH_2CH_2COOH	29	13	58
CH_2CH_2COOMe	53	7	40

The results summarized in Table 7.1 indicate the following: (1) Simple alkyl groups direct the thallium-containing group mainly to the para position. (2) When a substituent contains a nucleophilic moiety in such a position as to permit either a five- or a six-membered (but not four-membered) transition state, as shown in eqs. 7.39 and 7.40, the o-isomers tend to dominate the product mixtures.

$$\text{(structures for eq. 7.39)}$$

$$\tag{7.39}$$

$$\text{(structures for eq. 7.40)}$$

$$\tag{7.40}$$

Since deactivated aromatic compounds are very reluctant to participate in the thallation reaction, m-substituted aryl-thalliums are seldom obtained as major products under kinetic conditions. By taking advantage of the reversible nature of the thallation reaction, however, the m-isomers can be obtained as major products under equilibrating conditions, as shown in eq. 7.41.

$$\text{(structures for eq. 7.41)}$$

$$\tag{7.41}$$

major

7.3   REACTION OF ORGANOMERCURIES

At present, there are four major uses of organomercuries in organic synthesis in the laboratories: (1) hydration and related reactions of olefins via heteromercuration-demercuration, (2) carbenoid transfer reactions, (3) transmetallation reactions producing more reactive organometallics, and (4) reactions of α,β-unsaturated organomercuries with electrophiles. There are, of course, many other reactions of organomercuries. Most of them, however, either are not unique to mercury or are not very useful from the viewpoint of organic synthesis. Before discussing the four above-mentioned reactions, some general methods for the cleavage of the Hg-C bond are discussed.

7.3.1   Cleavage of the Hg-C Bond

7.3.1.1   Reactions with Carbon Electrophiles

The highly covalent nature of the Hg-C bonds makes them poor nucleophiles. Indeed, unactivated Hg-C bonds are generally unreactive toward typical organic electrophiles, such as organic halides, carbonyl compounds, and epoxides.

7.3.1.2   Protonolysis and Halogenolysis

Certain inorganic electrophiles, such as proton acids and halogens, readily react with various types of organomercuries to produce organic products via cleavage of the Hg-C bond (eqs. 7.42, 7.43). These reactions have been thoroughly reviewed by Jensen and Rickborn (7.38).

$$RHgY \xrightarrow{HX} RH + HgXY \qquad (7.42)$$

$$RHgY \xrightarrow{X_2} RX + HgXY \qquad (7.43)$$

The Hg-C bonds are generally stable to water and alcohols. Thus the protonolysis of these bonds requires stronger acids, such as HCl and H_2SO_4. Carboxylic acids are much less effective. In general the acid cleavage of dialkylmercuries occurs much more readily than the corresponding reaction of alkylmercuric salts. The Hg-aryl bonds are cleaved more readily than the Hg-alkyl bonds, the order of decreasing ease of cleavage by HCl being: anisyl > tolyl > phenyl > chlorophenyl > methyl > n-butyl > isoamyl > benzyl > cyclohexyl. The stereochemistry of the protonolysis of the Hg-C bond is not very clear. The available data, however, indicate that retention is the preferred course.

It may be pointed out here that the Hg-C bond is not cleaved readily by alkalis or other common bases.

Halogenolysis of the Hg-C bond is also a highly general reaction which takes place with all halogens (7.38). Although dialkylmercuries appear more reactive than the corresponding alkylmercuric aslts (RHgY), it is difficult to make a definitive generalization due to a rapid competing side reaction shown in eq. 7.44.

$$R_2Hg + HgX_2 \longrightarrow 2RHgX \qquad (7.44)$$

While the reaction of cis- or trans-4-methylcyclohexylmercuric bromide with bromine in pyridine proceeds with 100% retention, the corresponding reaction in CCl_4 or CS_2 involves essentially complete loss of configuration. The available information indicates that the halogenolysis of the Hg-C bond can proceed by both polar and free-radical mechanisms (7.38).

7.3.1.3 Oxidation and Reduction

The Hg-C bond is generally quite stable toward oxygen. On the other hand, organomercuries can readily be oxidized by ozone. Secondary alkylmercuries give good yields of ketones, while primary alkylmercuries are converted to mixtures of carboxylic acids (7.39). The following oxymercuration-ozonation sequence has been developed as a method for the conversion of olefins to α-oxy ketones (eq. 7.45).

$$\text{(reaction scheme)} \qquad (7.45)$$

with labels: $HgX_2\text{-}MeOH$, HgX, OMe, O_3, O, OMe

In general, the Hg-C bond is stable toward many other oxidizing agents, such as CrO_3.

   The oxidation of the alkenyl group bonded to mercury can readily be achieved by thermolysis, shown in eq. 7.46 (7.4, 7.40). This is a redox reaction in which the mercury is reduced to metallic mercury.

$$\underset{HgX}{>C=C<} \quad \xrightarrow{\Delta} \quad >C=C<_X \quad + \quad Hg \qquad (7.46)$$

$$X = OOCR, OAr, OTs, \overset{O}{\overset{\|}{OP}}(OR)_2, \overset{O}{\overset{\|}{SCOR}}, SR, SAr, SCN, \text{ and so on}$$

Although the reaction is nonstereospecific, it appears to be of some synthetic utility, and might be compared with a related transformation involving alkenylsilicon compounds (Sect. 6.4.2.1).

   In contrast to the oxidation of the Hg-C bond, its reduction is a facile process. A variety of reducing agents including $LiAlH_4$, $NaBH_4$, Na/Hg, and RNH_2 (7.4) have been used to convert organomercuries into the corresponding hydrocarbons.

$$RHgX \xrightarrow{\text{reduction}} RH \qquad (7.47)$$

Ready availability and high selectivity make $NaBH_4$ the reagent of choice. The reaction of organomercuries with $NaBH_4$ usually produces reduced products in high yields. The reduction of alkylmercuric halides with $NaBH_4$ appears to proceed via the corresponding alkylmercuric hydrides which decompose to alkanes and metallic mercury by a free-radical mechanism (7.41). One plausible mechanism is shown in eq. 7.48.

$$RHgX + NaBH_4 \longrightarrow RHgH$$

$$\text{(7.48)}$$

$$RHgH + R\cdot \longrightarrow [RHg\cdot + HR] \longrightarrow R\cdot + Hg + HR$$

In accord with the above free-radical mechanism, both exo- and endo-2-norbornylmercuric bromides react with $NaBD_4$, Et_2AlD, $CuD\cdot PBu_3$, or Bu_3SnD to produce in each case a 90:10 mixture of exo- and endo-2-deuterionorbornane. The intermediacy of the norbornyl radical was demonstrated by trapping it with the di-*tert*-butylnitrosyl radical (eq. 7.49).

$$\text{(7.49)}$$

Unlike the reaction of organomercuric halides with $NaBH_4$ and other metal hydrides, their reaction with Na/Hg appears to be stereospecific (7.42).

7.3.1.4 Transmetallation

Transmetallation represents yet another general mode of cleaving the Hg-C bond. The reaction can be induced by a variety of metal-containing compounds (Method XI) and metals (Method XII), as indicated by the following equations (eqs. 7.50 to 7.54).

$$R^1_2Hg \ + \ R^2Li \longrightarrow R^1HgR^2 \ + \ R^1Li \qquad (7.50)$$

$$(pK_a \text{ of } R^1H < pK_a \text{ of } R^2H)$$

$$ArHgCl \xrightarrow{BH_3} \left[ArB \Big\langle \right] \xrightarrow[NaOH]{H_2O_2} ArOH \qquad (7.51)$$

This reaction sequence provides an interesting route to phenols (7.43). The following transformations indicate that organomercuries can be converted to organocoppers (7.44) and organopalladiums (7.45).

$$(7.52)$$

$$(7.53)$$

$$n/2R_2Hg \ + \ M \longrightarrow n/2Hg \ + \ R_nM \qquad (7.54)$$

This oxidative-reductive transmetallation reaction (Sect. 2.12) takes place below 100°C with lithium and sodium. When less reactive metals, such as Mg, Zn, Cd, Al, and Sn, are used, higher temperatures (>100°C) are required.

All reactions discussed above except one are substitution reactions of the Hg-C bond. Organomercuries, however, can undergo other types of reactions as well.

7.3.1.5 Eliminations and Rearrangements

Although α- and β-heterosubstituted organomercuries are obtainable as isolable substances, they can undergo α- and β-elimination reactions, respectively. As discussed later in more detail, certain α-haloorganomercuries produce carbenes on thermolysis (7.11).

$$\underset{X}{\overset{|}{\underset{|}{C}}}\text{C-HgY} \xrightarrow{\Delta} \quad \overset{|}{\underset{|}{C}}: \quad + \text{ HgXY} \qquad (7.55)$$

β-Heterosubstituted organomercuries can undergo thermal or assisted β-elimination. Thus, for example, β-methoxy organomercuries give olefins on acid treatment, as shown in eq. 7.56 (7.46).

$$\text{(structure with HgI and OMe)} \xrightarrow{\text{H}^+} \text{(cyclohexene structure)} \qquad (7.56)$$

Although we concluded earlier that the intrinsic nucleophilicity of the Hg-C bond is low, α,β- and β,γ-unsaturated organomercuries should be reasonably reactive toward various electrophiles in light of their ability to participate well in the σ-π conjugation (Sect. 6.4.2.1). The carbocationic species produced by such reactions should readily undergo elimination (eqs. 7.57, 7.58).

$$\underset{|}{\overset{|}{C}}=\underset{|}{\overset{|}{C}}\text{-HgX} \xrightarrow{E^+} \overset{+}{\underset{|}{\overset{|}{C}}}-\underset{|}{\overset{\overset{HgX}{|}}{C}}-E \longrightarrow \underset{|}{\overset{|}{C}}=\underset{|}{\overset{|}{C}}-E \qquad (7.57)$$

$$\underset{|}{\overset{|}{C}}=\underset{|}{\overset{|}{C}}-\underset{|}{\overset{|}{C}}\text{-HgX} \xrightarrow{E^+} -\underset{|}{\overset{\overset{E}{|}}{C}}-\underset{|}{\overset{+}{\overset{|}{C}}}-\underset{|}{\overset{HgX}{|}}{C}- \longrightarrow \text{E-}\underset{|}{\overset{|}{C}}-\underset{|}{\overset{|}{C}}=\underset{|}{\overset{|}{C}} \qquad (7.58)$$

The overall transformation involves substitution of the mercury-containing group with an electrophilic group (E). Although such reaction schemes might prove general and useful, this possibility has not been fully tested. At present few rearrangement reactions involving cleavage of the Hg-C bond are known.

7.3.2   Synthetic Procedures Involving Organomercuries

7.3.2.1   The Markovnikov Hydration of Olefins and Related Reactions via Heteromercuration-Demercuration

A combination of oxymercuration (Sect. 7.2.2.1) and demercuration by reduction of β-oxyorganomercuries provides one of the most convenient procedures for the Markovnikov hydration of olefins, which nicely complements the anti-Markovnikov hydration via hydroboration-oxidation (eq. 7.59).

$$RCH=CH_2
\begin{cases}
\xrightarrow[\text{2. } NaBH_4]{\text{1. } HgX_2,\ H_2O} & RCHCH_3 \\
& \overset{|}{OH} \\
\\
\xrightarrow[\text{2. } H_2O_2\text{-}NaOH]{\text{1. } HB{<}} & RCH_2CH_2OH
\end{cases}$$

(7.59)

Although in some special cases the reduction of organomercuries with NaBH4 can be complicated by skelètal rearrangements and other side reactions, it is a highly dependable and general reaction. Thus the scope and limitations of the Markovnikov hydration via oxymercuration-demercuration are roughly comparable to those of the oxymercuration reaction discussed earlier.

Other oxymercuration and heteromercuration reactions can also be combined with the demercuration reaction via reduction with NaBH4 or other reducing agents to provide convenient procedures for the synthesis of ethers (7.25), amines (7.33), amides (7.34), and azides (7.35), as shown in eq. 7.60.

$$RCH=CH_2
\begin{cases}
\xrightarrow[\text{2. } NaBH_4]{\text{1. } Hg(O_2CCF_3)_2,\ HOR'} & RCHCH_3 \\
& \overset{|}{OH} \\
\\
\xrightarrow[\text{2. } NaBH_4]{\text{1. } Hg(OAc)_2,\ HNR_2^1} & RCHCH_3 \\
& \overset{|}{NR_2} \\
\\
\xrightarrow[\substack{\text{2. } H_2O \\ \text{3. } NaBH_4}]{\text{1. } Hg(NO_3)_2,\ CH_3CN} & RCHCH_3 \\
& \overset{|}{NHCOCH_3} \\
\\
\xrightarrow[\text{2. } NaBH_4]{\text{1. } Hg(OAc)_2,\ NaN_3} & RCHCH_3 \\
& \overset{|}{N_3}
\end{cases}$$

(7.60)

7.3.2.2   Carbene Transfer Reactions

α-Halomethylmercury compounds have been developed as carbene, or divalent carbon, transfer reagents mainly by Seyferth and his associates (7.11).

$$\text{PhHg}\overset{\text{X}}{\underset{\text{Z}}{\text{C}}}\text{-Y} \; + \; \text{\Large$>$C=C$\Large<$} \; \xrightarrow{\Delta} \; \text{\Large$>$C}\underset{\text{C\Large$<$}}{\overset{\overset{\text{Y}\diagdown\diagup\text{Z}}{\text{C}}}{\diagup\diagdown}} \; + \; \text{PhHgX} \quad (7.61)$$

In many cases the required reagents can be prepared according to the following equation (7.47).

$$\text{PhHgX} \; + \; \text{KOBu-}t \; + \; \text{HCXYZ} \xrightarrow{\text{THF}} \text{PhHgCXYZ} \; + \; \text{HBu-}t \; + \; \text{KX} \quad (7.62)$$

An extensive kinetic and product study by Seyferth (7.11) indicates that in most cases the α-halomethylmercury reagents appear to produce directly free carbenes via 7.6 without involving any carbanion intermediates (eq. 7.63).

$$\text{PhHgCXYZ} \rightleftharpoons \text{PhHg} \overset{:\text{X}}{\longrightarrow} \text{CYZ} \rightleftharpoons :\text{CYZ} \; + \; \text{PhHgX}$$

$$\underline{7.6}$$

$$:\text{CYZ} \; + \; \text{\Large$>$C=C$\Large<$} \longrightarrow \text{\Large$>$C}\underset{\text{C\Large$<$}}{\overset{\overset{\text{Y}\diagdown\diagup\text{Z}}{\text{C}}}{\diagup\diagdown}}$$

$$(7.63)$$

In some cases, however, other mechanisms appear to be operating. For example, the reaction of $Hg(CH_2Br)_2$ with olefins appears to involve a direct interaction between the two reactants in a manner analogous to the reaction of iodomethylzinc iodide (4.127).

Carbenes or divalent carbon species that have been successfully transferred include: CH_2, $CHCl$, $CHBr$, CF_2, CCl_2, CBr_2, $CFCl$, $CFBr$, $CClBr$, $CClCOOR$, $CBrCOOR$, $CClSO_2Ph$, $CBrSO_2Ph$, $CClSiMe_3$, and $CBrSiMe_3$. In accord with the mechanism shown in eq. 7.63, the most nucleophilic and least tightly bound halogen atom is transferred from carbon to mercury, the relative reactivity order of halogens being: $I > Br > Cl > F$. The relative reactivities of olefins toward phenyl(bromodichloromethyl) mercury are: $Me_2C=C(Me)Et > Et_2C=CHMe > Pr(Me)C=CH_2 >$ cyclohexene $>$ cis-$EtCH=CHPr >$ trans-$EtCH=CHPr > n$-$C_5H_{11}CH-CH_2$.

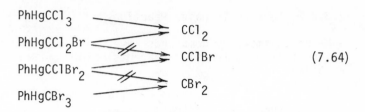

$$(7.64)$$

Both the order and the relative rate constants are essentially the same as those observed with CCl_3COONa, indicating that these two reactions involve the same reactive species. One significant difference between the α-halomethylmercury reagents and the more conventional reagents, such as CHX_3 + KOBu-t and CX_3COONa, is that the former reagents can produce carbenes thermally and directly, whereas the latter reagents first produce carbanion intermediates ($-CX_3$), which can induce undesirable side reactions. Although all of these reagents give satisfactory and comparable results in many cases, the α-halomethylmercury reagents are often superior to the more conventional reagents in terms of the product yield. For example, highly electrophilic olefins, such as $CH_2=CHCN$ and $CH_2=CHCOOR$, may undergo competitive conjugate addition reactions with carbanion intermediates ($-CX_3$). Such a complication does not appear to exist in the reaction of the α-halomethylmercury reagents. A detailed discussion of the synthetic utility of carbenes is beyond the scope of this monograph. Readers are referred to several extensive monographs on this subject (e.g., *3.11*).

7.3.2.3 Other Synthetic Procedures Involving Organomercuries

Transmetallation. As discussed earlier (Sect. 7.3.1), organomercuries can be converted via transmetallation into other organometallics, such as those containing Li, Na, Mg, Zn, Cd, B, Al, Sn, Cu, and Pd. These organometallics can, in turn, be utilized in organic synthesis, as discussed in their respective chapters.

Reactions with Carbon Electrophiles. As pointed out earlier (Sect. 7.3.1), unactivated Hg-C bonds are generally inert toward carbon electrophiles. The mercury atom, however, can activate the alkenyl, aryl, and alkynyl groups, which are either directly bonded to or separated by one carbon atom from the mercury atom, through σ-π conjugation. These α,β- and β,γ-unsaturated organomercuries are therefore considerably more reactive toward carbon

electrophiles. Although this property does not appear to have
been fully explored and exploited, the following examples point
to its potential synthetic utility. Arylmercuric chlorides
react with acyl halides to give aryl ketones (7.4).

$$ArHgCl \; + \; RCOCl \longrightarrow ArCOR \; + \; HgCl_2 \qquad (7.65)$$

Alkenylmercuric chlorides and acyl chlorides react at room temperature in the presence of $AlCl_3$ to give high yields of α,β-un-
saturated ketones with retention of configuration (7.4). As a
number of alkenylmercuric chlorides can readily be obtained as
stereodefined species via hydroboration-transmetallation of acetylenes, the reaction now provides a procedure for the conversion
of acetylenes into (E)-alkenyl ketones (eq. 7.66).

$$RC\equiv CH \quad \xrightarrow[\text{2. } HgX_2]{\text{1. } HB\diagdown} \quad \underset{H}{\overset{R}{>}}C=C\underset{HgX}{\overset{H}{<}} \quad \xrightarrow[AlCl_3]{R'COCl} \quad \underset{H}{\overset{R}{>}}C=C\underset{COR'}{\overset{H}{<}} \qquad (7.66)$$

7.4   REACTIONS OF ORGANOTHALLIUMS AND ORGANOLEADS

As mentioned in Sect. 7.1.2, organothallium(III) compounds of
the R_3Tl and R_2TlX types and organolead(IV) compounds of the
R_4Pb, R_3PbX, and R_2PbX_2 types have not found extensive uses in
organic synthesis despite their ready availability as isolable
species. Also very much limited is the application of organo-
thallium(I) and organolead(II) species in organic synthesis. On
the other hand, monoorganothalliums ($RTlX_2$), which are readily
obtainable as unstable intermediates either by thallation of aromatic compounds (Sect. 7.2.3) or by oxythallation of olefins and
acetylenes, have exhibited uniquely useful synthetic capabilities. Similarly, in some synthetically useful reactions of olefins and acetylenes with lead(IV) compounds, such as $Pb(OAc)_4$,
monoorganoleads of the $RPbX_3$ type appear to be formed as unstable
intermediates.

7.4.1   Organothalliums in Organic Synthesis

The reactions of organothalliums of current synthetic interest
may be classified as follows: (1) alkylation of cyclopenta-
dienylthallium(I), (2) reactions of arylthallium(III) derivatives, and (3) reactions of olefins and acetylenes with thallium-
(III) salts. There are several excellent reviews on these subjects (7.5, 7.6, 7.48).

7.4.1.1 Organic Synthesis Involving Organothallium(I) Compounds

Cyclopentadienylthallium(I) is readily obtainable in 90% yield as a stable organothallium(I) compound by adding freshly distilled cyclopentadiene to a suspension of Tl_2SO_4 in dilute KOH solution (7.49). The potential utility of cyclopentadienylthallium has been demonstrated by its conversion into 5-methoxymethylcyclopentadiene (7.7) which can be further converted to a useful intermediate in the synthesis of the primary prostaglandins (7.50).

$$\text{(structure)} \quad Tl^+ \quad \xrightarrow{ClCH_2OCH_3} \quad \text{(structure)}-CH_2OCH_3 \quad \xrightarrow[\text{2. KOH}]{\text{1. } CH_2=CCl(CN)} \quad \text{(structure)} $$

7.7

$$\text{(7.67)}$$

Isomerization of 7.7 to 1-methoxymethylcyclopentadiene, which is a serious side reaction in the corresponding reaction of the sodium or lithium salt, does not occur to any significant extent.

7.4.1.2 Organic Synthesis via Thallation of Aromatic Compounds

Reactions of Arylthalliums with Nucleophiles. As discussed in Sect. 7.2.3, the reaction of arenes with $Tl(O_2CCF_3)_3$ (TTFA) provides a convenient direct entry into arylthalliums of the $ArTlX_2$ type. Although arylthallium difluoroacetates thus obtained are relatively stable, exchange of the O_2CCF_3 group by other groups (X) tends to form unstable derivatives which may decompose to give substituted aromatic compounds. Most of the results can be accommodated by the following general scheme (eq. 7.68).

$$ArTl(O_2CCF_3)_2 \xrightarrow{MX} Ar-Tl\underset{X}{\overset{X}{\big<}} \longrightarrow ArX + TlX \qquad (7.68)$$

The reaction is regiospecific in that X Group ends up occupying the position originally held by the thallium-containing group. A variety of nucleophiles have been used as the MX reagent.

(a) Halogen nucleophiles. Aryl iodides can readily be prepared using KI (7.51).

$$\text{ArTl}(O_2CCF_3)_2 \xrightarrow{\text{2KI}} \text{ArTlI}_2 \longrightarrow \text{ArI} + \text{TlI} \quad (7.69)$$

ex.

$$(7.70)$$

79%

Aryl bromides, on the other hand, can be obtained more directly according to eq. 7.71 (7.52).

$$\text{ArH} + \text{Br}_2 + \text{Tl}(OAc)_3 \xrightarrow{\text{CCl}_4} \text{ArBr} \quad (7.71)$$

In contrast to the majority of electrophilic aromatic bromination techniques, this procedure usually gives a single, pure monobromo isomer. In the case of monosubstituted benzenes only para bromination takes place. The reaction, however, does not seem to involve the intermediacy of arylthallium species.

(b) Oxygen nucleophiles. Arylthallium derivatives can be converted to phenols either by treatment with Pb(OAc)$_4$ and PPh$_3$ followed by hydrolysis with aqueous NaOH (7.53) or by treatment with a borane reagent followed by oxidation (7.54).

$$(7.72)$$

ex.

$$(7.73)$$

39% (only isomer)

(c) Nitrogen nucleophiles. Photolysis of $ArTl(O_2CCF_3)_2$ in the presence of ammonia gives the corresponding anilines (7.5).

$$ArTl(O_2CCF_3)_2 \xrightarrow[NH_3]{h\nu} ArNH_2 \qquad (7.74)$$

(d) Carbon nucleophiles. Treatment of $ArTl(O_2CCF_3)_2$ with an excess of aqueous KCN gives $K[ArTl(CN)_3]$, which can be decomposed by irradiation to form aromatic nitriles (7.53).

$$ArTl(O_2CCF_3)_2 \xrightarrow{aq\ KCN} K[ArTl(CN)_3] \xrightarrow{h\nu} ArCN \qquad (7.75)$$

H_3C—⬡—CN (50%)   ⬡ with CH_2OMe and —CN (55%)

Photolysis of $ArTl(O_2CCF_3)_2$ in benzene produces the corresponding arylated benzenes in high yields (7.55).

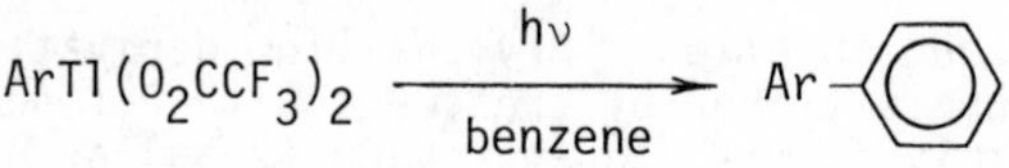

$$ArTl(O_2CCF_3)_2 \xrightarrow[benzene]{h\nu} Ar-⬡ \qquad (7.76)$$

CH_3—⬡—⬡ Cl—⬡—⬡

(91%, p/o = 95/5)   (87%, p/o = 92/8)

The intramolecular version of the biaryl synthesis provides a facile entry into certain alkaloids, such as ocoteine, as shown in eq. 7.77 (7.56).

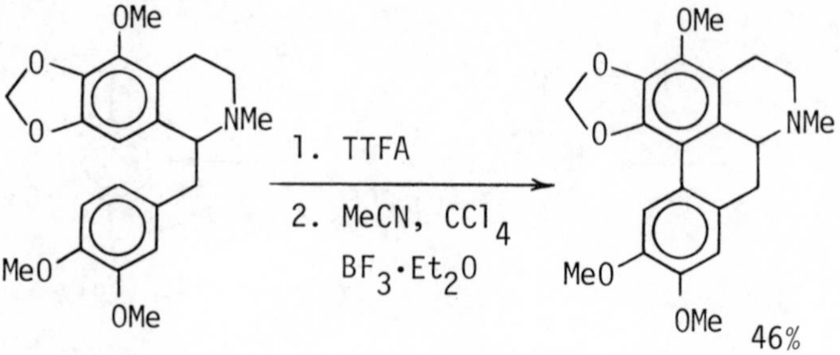

1. TTFA

2. MeCN, CCl_4
   $BF_3 \cdot Et_2O$

46%

(7.77)

(e) Hydrogen nucleophiles. The reaction of $ArTl(O_2CCF_3)_2$ with $LiAlH_4$ (or $LiAlD_4$) gives ArH (or ArD) (7.5).

$$ArTl(O_2CCF_3)_2 \xrightarrow{\text{LiAlH}_4(\text{or LiAlD}_4)} ArH(D) \qquad (7.78)$$

Reactions of Arylthalliums with Electrophiles. Arylthallium compounds are capable of reacting with electrophiles as indicated by their reactions with $Pb(OAc)_4$ and borane (eq. 7.72). The reaction of arylthalliums with Li_2PdCl_4 in the presence of olefins produces arylated olefins (7.57). The reaction presumably involves formation of arylpalladiums via transmetallation.

$$ArTlX_2 + CH_2=CHCOOMe \xrightarrow{\text{Li}_2\text{PdCl}_4} \begin{array}{c} Ar \\ H \end{array}C=C\begin{array}{c} H \\ COOMe \end{array} \qquad (7.79)$$

Arylthalliums also react with certain nonmetallic electrophiles. Their reaction with nitrosyl chloride gives nitrosoarenes (7.58).

$$ArTl(O_2CCF_3)_2 + NOCl \longrightarrow ArNO \qquad (7.80)$$

7.4.1.3 Organic Synthesis via Oxythallation

Various thallium(III) salts, such as $Tl(NO_3)_3$(TTN), $Tl(OAc)_3$, $Tl(O_2CCF_3)_3$(TTFA), $Tl_2(SO_4)_3$, and $Tl(ClO_4)_3$, react with olefins acetylenes to form the corresponding monoalkylthallium(III) derivatives (7.5). In marked contrast to the oxymercuration products, however, the monoalkylthallium(III) derivatives are generally quite unstable and decompose spontaneously under the oxythallation conditions in the great majority of cases, although some such derivatives have been isolated.

$$\begin{array}{c} C=C \end{array} + TlX_3 \quad (7.81)$$

The fate of the monoalkylthallium(III) derivatives depends on a number of factors. The results obtained to date can, however, be interpreted in terms of the following several mechanistic schemes suggested by Taylor and McKillop (7.5b).

Type 1 (s_N2)

$$Y-\overset{|}{\underset{|}{C}}-\overset{|}{\underset{|}{C}}-Tl\overset{X}{\underset{X}{\diagdown}} \longrightarrow Y-\overset{|}{\underset{|}{C}}-\overset{|}{\underset{|}{C}}-Y + TlX + X^- \qquad (7.82)$$

Type 2 ($s_N i$)

$$Y-\overset{|}{\underset{|}{C}}-\overset{|}{\underset{|}{C}}-Tl\overset{X}{\underset{X}{\diagdown}} \longrightarrow \;>\!C\!\!-\!\!-\!\!C\!<\; + \;^-Y \longrightarrow Y-\overset{|}{\underset{|}{C}}-\overset{|}{\underset{|}{C}}-Y \qquad (7.83)$$

Type 3 (s_N1)

$$Y-\overset{|}{\underset{|}{C}}-\overset{|}{\underset{|}{C}}-Tl\overset{X}{\underset{X}{\diagdown}} \longrightarrow Y-\overset{|}{\underset{|}{C}}-\overset{|}{\underset{|}{C}}{}^+- \overset{^-Y}{\longrightarrow} Y-\overset{|}{\underset{|}{C}}-\overset{|}{\underset{|}{C}}-Y \qquad (7.84)$$

Type 4 (s_N2 with rearrangement)

$$Y-\overset{R}{\underset{Y^-}{\overset{|}{\underset{|}{C}}}}-\overset{|}{\underset{|}{C}}-Tl\overset{X}{\underset{X}{\diagdown}} \longrightarrow Y-\overset{R}{\overset{|}{\underset{|}{C}}}-\overset{|}{\underset{Y}{\underset{|}{C}}}- + TlX + X^- \qquad (7.85)$$

Type 5 ($s_N i$ with rearrangement)

$$Y-\overset{R}{\overset{|}{\underset{|}{C}}}-\overset{|}{\underset{|}{C}}-Tl\overset{X}{\underset{X}{\diagdown}} \longrightarrow Y-\overset{R}{\overset{|}{\underset{|}{C}}}-\overset{|}{\underset{X}{\underset{|}{C}}}- + TlX \qquad (7.86)$$

Type 6 (s_N1 with rearrangement)

$$Y-\overset{R}{\overset{|}{\underset{|}{C}}}-\overset{|}{\underset{|}{C}}-Tl\overset{X}{\underset{X}{\diagdown}} \longrightarrow Y-\overset{R}{\underset{Y^-}{\overset{|}{\underset{|}{C}}}}-\overset{|}{\underset{|}{C}}{}^+- \longrightarrow Y-\overset{R}{\overset{|}{\underset{|}{C}}}-\overset{|}{\underset{Y}{\underset{|}{C}}}- \qquad (7.87)$$

It should be noted that the first three reactions bring about the same overall transformations and are different only in mechanistic details. Likewise, the Types 4 to 6 reactions involve basically the same overall transformations. Since we are primarily concerned with the synthetic aspects of these reactions, for the sake of convenience the Types 1 to 3 and Types 4 to 6 reactions will be termed the Type A and Type B reactions, respectively. Both Type A and Type B reactions involve reduction of thallium(III) to thallium(I), which proceeds via cleavage of the thermodynamically very weak (25 to 30 kcal/mole) Tl-C bond.

Unfortunately, the exact course of a given reaction is still quite difficult to predict, and is dependent on a number of reaction parameters. Mainly due to the systematic studies by Taylor and McKillop, however, we now can not only readily interpret the observed results but, in many cases, steer these reactions in the desired directions. For example, the reaction of cyclohexene with $Tl(OAc)_3$ in HOAc was reported to give a mixture of 7.8 and 7.9 along with other minor products (7.59).

The glycol and acetal derivatives must have arisen via Type A and Type B processes, respectively. If $Tl(NO_3)_3$ is used in place of $Tl(OAc)_3$, the Type B process predominates, producing cyclopentanecarboxaldehyde in 85% yield. Thus the Type B reaction can be favored over the Type A reaction by highly ionic thallium(III) compounds, such as $Tl(NO_3)_3$, $Tl(ClO_4)_3$, and $Tl_2(SO_4)_3$, which contain poor nucleophilic ligands.

The following examples indicate the synthetic scope of the reactions of olefins, acetylenes, and ketones with such thallium(III) compounds. A detailed discussion of each of these reactions is not presented here due to space limitations. Interested readers are referred to the original papers reporting these results. It should be clear, however, that all of these reactions involve the Type A and/or Type B processes as key steps. The Type B reaction is often accompanied by a skeletal rearrangement, which involves migration of a carbon group. On the other hand, if a hydrogen atom migrates, no skeletal rearrangement results.

1. Olefins (7.60):

$$\bigcirc\!\!-CH\!=\!CH_2 \xrightarrow{\ Tl_2(SO_4)_3\ } \bigcirc\!\!-CH_2CHO \qquad (7.89)$$
$$90\%$$

$$CH_3(CH_2)_3CH\!=\!CH_2 \xrightarrow{\ Tl_2(SO_4)_3\ } CH_3(CH_2)_3COCH_3 \qquad (7.90)$$
$$57\%$$

$$\begin{array}{c} Ar^1 \\ \\ Ar^2 \end{array}\!\!\!C\!=\!CH_2 \xrightarrow{\ TTN\ } Ar^1CH_2COAr^2 \qquad (7.91)$$
$$75 \text{ to } 95\%$$

$$\boxed{}\!\!-\!CH_3 \xrightarrow{\ Tl(ClO_4)_3\ } \triangleright\!\!-COCH_3 \qquad (7.92)$$
$$100\%$$

2. Acetylenes (7.61):

$$RC\!\equiv\!CH \xrightarrow[\ H_3O^+\ ]{\ 2 \text{ eq TTN}\ } RCOOH + CH_2O \qquad (7.93)$$

$$RC\!\equiv\!CR' \xrightarrow[\ H_3O^+\ ]{\ 1 \text{ eq TTN}\ } \overset{O}{\underset{}{R\overset{\|}{C}}}\!-\!\overset{OH}{\underset{}{\overset{|}{C}HR'}} \qquad (7.94)$$

$$ArC\!\equiv\!CAr' \nearrow \xrightarrow[\ H_3O^+\ ]{\ 1 \text{ eq TTN}\ } \overset{O}{\underset{}{Ar\overset{\|}{C}}}\!-\!\overset{OH}{\underset{}{\overset{|}{C}HAr'}}$$
$$\searrow \xrightarrow[\ H_3O^+\ ]{\ 3 \text{ eq TTN}\ } \overset{O}{\underset{}{Ar\overset{\|}{C}}}\!-\!\overset{O}{\underset{}{\overset{\|}{C}Ar'}} \qquad (7.95)$$

$$RC \equiv CAr \xrightarrow[\text{MeOH}]{1 \text{ eq TTN}} \begin{array}{c} Ar \\ R \end{array} CHCOOMe \qquad (7.96)$$

As indicated in the equations above, more than one equivalent of the thallium(III) reagent is required in some cases to observe a complete consumption of the acetylene used. The reaction of monoalkylacetylenes presumably involves a process consisting of oxythallation, tautomerization, Type A reaction, and TTN-promoted fragmentation (eq. 7.97).

$$RC \equiv CH \xrightarrow[\text{H}_2\text{O}]{\text{TTN}} \begin{array}{c} R \\ HO \end{array} C = C \begin{array}{c} H \\ Tl(NO_3)_2 \end{array} \longrightarrow \begin{array}{c} O \\ \parallel \\ RC \end{array} - CH_2 - Tl - NO_3$$

$$\qquad \qquad \qquad \qquad \qquad \qquad \qquad \qquad \qquad \qquad \qquad H_2O: \qquad NO_3$$

$$\qquad \qquad \qquad \qquad \qquad \qquad \qquad \qquad \qquad \qquad \qquad \qquad \qquad (7.97)$$

$$\longrightarrow \begin{array}{c} O \\ \parallel \\ RCCH_2OH \end{array} \xrightarrow{\text{TTN}} \begin{array}{c} O \\ \parallel \\ RC \end{array} - CH_2O - Tl - NO_3 \longrightarrow RCOOH + CH_2O$$

$$\qquad \qquad \qquad \qquad \qquad \qquad H_2O: \qquad NO_3$$

The fragmentation reaction is not competitive in the case of dialkylacetylenes (eq. 7.94). The reaction of monoarylacetylenes does not appear to have been investigated in detail. Benzoins, which can be obtained in varying yields from diarylacetylenes, readily enolize and react further with TTN to form benzils. A cyclic mechanism has been proposed, but the results can also be accommodated by a thallation-dethallation mechanism (eq. 7.98).

$$\begin{array}{c} O \quad OH \\ \parallel \quad \mid \\ ArC - CHAr' \end{array} \longrightarrow \begin{array}{c} HO \quad OH \\ \mid \quad \mid \\ ArC = CAr' \end{array} \begin{array}{c} \xrightarrow{\text{TTN}} \\ \\ \xrightarrow{\text{TTN}} \end{array} \begin{array}{c} TlNO_3 \\ O \quad O \\ \mid \quad \mid \\ ArC = CAr' \\ \\ H - O \quad O - H \\ \mid \quad \mid \\ ArC - CAr' \\ + Tl - NO_3 \\ NO_3 \end{array} \begin{array}{c} O \quad O \\ \parallel \quad \parallel \\ ArC - CAr' \quad (7.98) \end{array}$$

The results obtained with alkylarylacetylenes have been interpreted in terms of a process consisting of oxythallation, Type B reaction, and hydrolysis (7.61).

3. Chalcones:

$$\text{ArCH=CHCOAr'} \xrightarrow[\text{H}_3\text{O}^+]{\text{TTN}} \underset{\underset{\text{CHO}}{|}}{\text{ArCHCOAr'}} \xrightarrow{\text{H}_3\text{O}^+} \text{ArCH}_2\text{COAr'}$$

(7.99)

$$\xrightarrow[\text{H}_3\text{O}^+]{\text{TTN}} \underset{}{\overset{\text{HO   O}}{\text{ArCH-CAr'}}} \xrightarrow{\text{TTN}} \overset{\text{O   O}}{\text{ArC-CAr'}}$$

One equivalent of TTN converts chalcones to aryl benzyl ketones, whereas the use of three equivalents of TTN results in the formation of benzils via benzoins (7.62). When chalcones are appropriately substituted with a hydroxy group as shown in 7.10, they can readily be converted to isoflavones (7.63).

7.10

1. TTN, MeOH
2. H₃O⁺
3. H₂

sophorol   (7.100)

4. Ketones (pyrazolones):

1. TTN, HOAc
2. NaHCO₃

⟶ =O (95%)

(7.101)

1. TTN, HOAc
2. Δ (40°C)

⟶ —COOH (95%)

$$\text{ArCOCH}_3 \xrightarrow[\text{MeOH}]{\text{TTN}} \text{ArCH}_2\text{COOMe}$$

85 to 95%

(7.102)

$$\text{RCOCH}_2\text{COOR'} \xrightarrow{\text{N}_2\text{H}_4} \text{[pyrazolone]} \xrightarrow[\text{MeOH}]{\text{TTN}} \text{RC}\equiv\text{CCOOMe}$$

70 to 95%

(7.103)

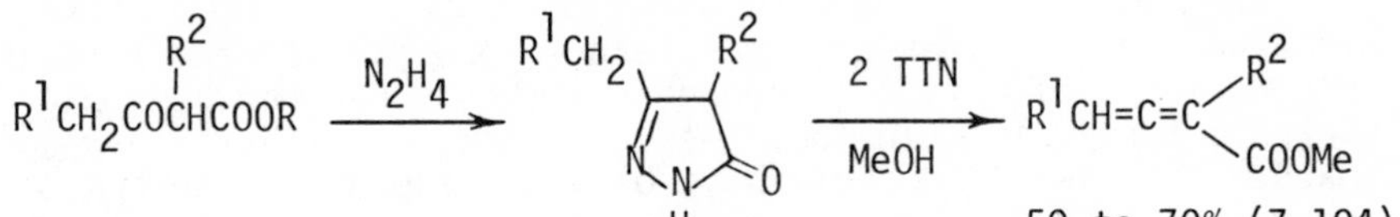

$$R^1CH_2COCHCOOR \xrightarrow{N_2H_4} \quad \text{(pyrazolone)} \xrightarrow[\text{MeOH}]{2\ TTN} R^1CH=C=C\begin{smallmatrix}R^2\\ \\COOMe\end{smallmatrix}$$

50 to 70% (7.104)

Enols and enamines most probably are the actual reactive species in these reactions. One can think of a number of different mechanistic paths for the reaction shown in eq. 7.101 (7.5). The Willgerodt-like reaction of acetophenones must involve a process consisting of enolization, oxythallation, and Type B reaction (7.64). The following schemes have been proposed for the reactions of pyrazolones (7.65).

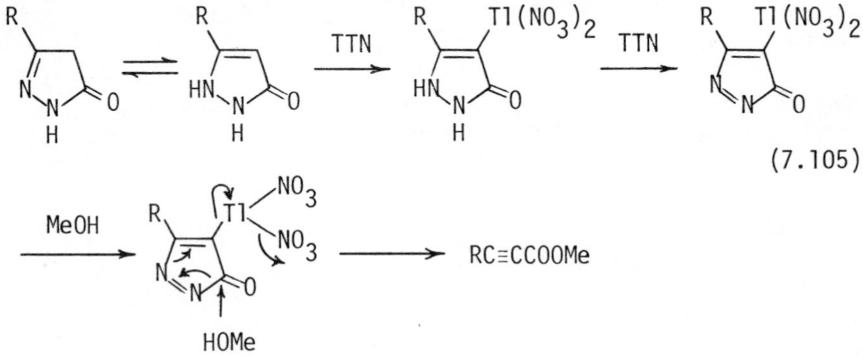

$$\text{(scheme)} \xrightleftharpoons{} \text{(scheme)} \xrightarrow{TTN} \text{(scheme)} \xrightarrow{TTN} \text{(scheme)}$$

(7.105)

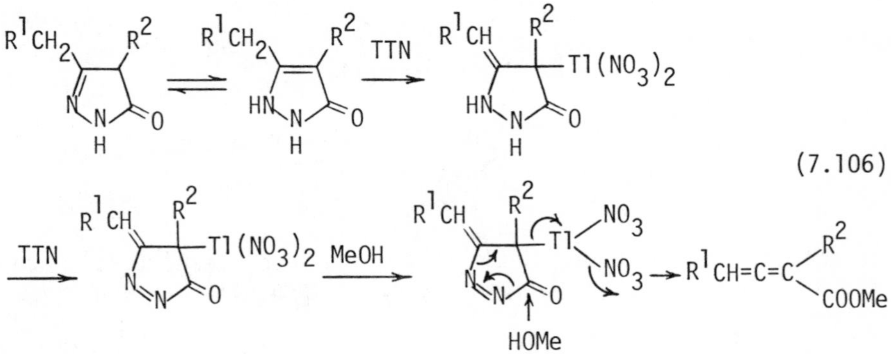

$$\xrightarrow{MeOH} \text{(scheme)} \longrightarrow RC{\equiv}CCOOMe$$

$$\begin{array}{c}R^1CH_2\\ \text{(scheme)}\end{array} \xrightleftharpoons{} \begin{array}{c}R^1CH_2\\ \text{(scheme)}\end{array} \xrightarrow{TTN} \begin{array}{c}R^1CH\\ \text{(scheme)}\end{array}$$

(7.106)

$$\xrightarrow{TTN} \text{(scheme)} \xrightarrow{MeOH} \text{(scheme)} \rightarrow R^1CH=C=C\begin{smallmatrix}R^2\\ \\COOMe\end{smallmatrix}$$

The corresponding reactions of aldehydes do not appear to have been well delineated.

7.4.2 Organoleads in Organic Synthesis

Although various types of organolead compounds are readily available, their application to organic synthesis has not been explored extensively. In fact, very few new reactions of organoleads of synthetic interest have been developed over the past several years. The Pb-C bond, however, is generally labile and reactive. Therefore, organoleads might turn out to be a useful class of reagents and intermediates in the future.

In this section, a few reactions of $Pb(OAc)_4$ with organic compounds which presumably proceed via unstable monoorganolead derivatives, will be briefly described.

The major driving forces of these reactions are: (a) the high polarizability and instability of the Pb(IV)-O bond, and (b) the ease with which the Pb-C bond is cleaved via one- and/or two-electron reduction (Sect. 7.1.1).

7.4.2.1 Olefins

The reaction of olefins with $Pb(OAc)_4$ can proceed in a manner analogous to the corresponding reaction with thallium(III) compounds as shown in eq. 7.107 (7.66).

$$(7.107)$$

In some favorable cases, the reaction proceeds cleanly, as shown in eq. 7.108 (7.67), although mixtures of products are usually obtained.

$$(7.108)$$

7.4.2.2 Carboxylic Acids

Carboxylic acids can be oxidized with $Pb(OAc)_4$ to produce decarboxylated carbocations presumably via monoorganolead species. The carbocations thus generated may undergo a variety of reactions characteristic of such species (7.68).

$$RCOOH \xrightarrow{Pb(OAc)_4} RCOOPb(OAc)_3 \longrightarrow RCOO\cdot \; + \; \cdot Pb(OAc)_3$$

$$\xrightarrow{-CO_2} R\cdot \; + \; \cdot Pb(OAc)_3 \longrightarrow R\!-\!\overset{\displaystyle \curvearrowright}{\underset{\overset{\displaystyle |}{OAc}}{Pb}}(OAc)_2 \longrightarrow R^+ \tag{7.109}$$

Typical examples follow (7.68).

$$\text{(7.110)}$$

$$30\%$$

$$HOOC(CH_2)_6COOEt \xrightarrow[\substack{Cu(OAc)_2 \\ Py,\ C_6H_6}]{\cdot Pb(OAc)_4} CH_2{=}CH(CH_2)_4COOEt \quad (7.111)$$

$$35\%$$

This procedure is applicable to the synthesis of α-methylenelactones (eq. 7.112) (7.69) and enones (eq. 7.113) (7.70).

$$\text{(7.112)}$$

$$50\%$$

The current scope of the use of organoleads in organic synthesis clearly is limited. Their ready availability and high reactivity, however, seem to warrant extensive future investigations in this area. For example, despite the fact that the Pd-C bond can greatly stabilize an adjacent carbocation center through σ-π conjugation (Sect. 6.4.1), little effort appears to have been made to exploit this attractive capability of lead. Explorations of the reactions of alkenyl-, aryl-, and alkynylleads might prove fruitful.

$$(7.113)$$

REFERENCES

7.1   Lee, A. G., *The Chemistry of Thallium*, Elsevier, Amsterdam, 1971, 336 pp.

7.2   Shapiro, H., and Frey, F. W., *The Organic Compounds of Lead*, John Wiley & Sons, New York, 1968, 468 pp.

7.3   Makarova, L. G., *Organometal. React.*, **1**, 119 (1970); **2**, 335 (1971).

7.4   Larock, R. C., *J. Organometal. Chem. Libr.*, **1**, 257 (1976).

7.5   (a) Taylor, E. C., and McKillop, A., *Acc. Chem. Res.*, **3**, 338 (1970); (b) McKillop, A., and Taylor, E. C., *Adv. Organometal. Chem.*, **11**, 147 (1973).

7.6   Lee, A. G., *Organometal. React.*, **5**, 1 (1975).

7.7   Pritchard, H. O., and Skinner, H. A., *Chem. Rev.*, **55**, 745 (1955).

7.8   Latimer, W., *Oxidation Potentials*, 2nd ed., Prentice-Hall, Englewood Cliffs, N. J., 1961.

7.9   Wittig, G., and Lehmann, G., *Chem. Ber.*, **90**, 875 (1957).

7.10 Makarova, L. G., and Nesmeyanov, A. N., *The Organic Chemistry of Mercury*, North-Holland Publishing Co., Amsterdam, 1967, 532 pp.

7.11 Seyferth, D., *Acc. Chem. Res.*, **5**, 65 (1972).

7.12 Larock, R. C., *Intra-Sci. Chem. Rep.*, **7**, 95 (1973).

7.13 Larock, R. C., and Brown, H. C., *J. Am. Chem. Soc.*, **92**, 2467 (1970).

7.14 Larock, R. C., *J. Organometal. Chem.*, <u>67</u>, 353 (1974); <u>72</u>, 35 (1974).

7.15 (a) Larock, R. C., and Brown, H. C., *J. Organometal. Chem.*, <u>36</u>, 1 (1972); (b) Larock, R. C., Gupta, S. K., and Brown, H. C., *J. Am. Chem. Soc.*, <u>94</u>, 4371 (1972); (c) Larock, R. C., *J. Org. Chem.*, <u>40</u>, 3237 (1975).

7.16 Brown, H. C., and Geoghegan, P. J., Jr., *J. Org. Chem.*, <u>35</u>, 1844 (1970); <u>37</u>, 1937 (1972).

7.17 Traylor, T. G., and Backer, A. W., *J. Am. Chem. Soc.*, <u>85</u>, 2746 (1963).

7.18 Pasto, D. J., and Gontarz, J. A., *J. Am. Chem. Soc.*, <u>92</u>, 7480 (1970); <u>93</u>, 6902 (1971).

7.19 (a) Traylow, T. G., and Baker, A. W., *Tetrahedron Lett.*, 14 (1959); (b) Tidwell, T. T., and Traylor, T. G., *J. Org. Chem.*, <u>33</u>, 2614 (1968).

7.20 (a) Brown, H. C., and Kawakami, J. H., *J. Am. Chem. Soc.*, <u>95</u>, 8665 (1973); (b) Brown, H. C., Kawakami, J. H., and Ikegami, S., *J. Am. Chem. Soc.*, <u>89</u>, 1525 (1967).

7.21 Brown, H. C., Geoghegan, P. J., Jr., Kurek, J. T., and Lynch, G. J., *Organometal. Chem. Synth.*, <u>1</u>, 7 (1970/1971).

7.22 Ford, D. N., Kitching, W., and Wells, P. R., *Aust. J. Chem.*, <u>22</u>, 1157 (1969).

7.23 Borisov, A. E., Vil'chevskaya, V. D., and Nesmeyanov, A. N., *Izv. Akad. Nauk SSSR, Otd. Khim.*, 1008 (1954).

7.24 Drefahl, G., Heublein, G., and Wintzer, A., *Angew. Chem.*, <u>70</u>, 166 (1958).

7.25 Brown, H. C., and Rei, M. H., *J. Am. Chem. Soc.*, <u>91</u>, 5464 (1969).

7.26 Sokolov, V. I., and Reutov, O. A., *J. Org. Chem. USSR*, <u>5</u>, 174 (1969).

7.27 Schmitz, E., Rieche, A., and Brede, O., *J. Prakt. Chem.*, <u>312</u>, 30 (1970).

7.28 Ballard, D. H., and Bloodworth, A. J., *J. Chem. Soc. C*, 945 (1971).

7.29 Bloodworth, A. J., and Griffin, I. M., *J. Chem. Soc.*, *Perkin I*, 195 (1975).

7.30 Olah, G. A., and Clifford, P. R., *J. Am. Chem. Soc.*, <u>93</u>, 1261 (1971); <u>95</u>, 6067 (1973).

7.31 (a) Brown, H. C., and Kawakami, J. H., *J. Am. Chem. Soc.*, 92, 201 (1970); (b) Brown, H. C., and Liu, K. T., *J. Am. Chem. Soc.*, 92, 3502 (1970); (c) Brown, H. C., Kawakami, J. H., and Liu, K. T., *J. Am. Chem. Soc.*, 95, 2209 (1973).

7.32 Brown, H. C., *Chem. Brit.*, 199 (1966).

7.33 (a) Hall, H. K., Jr., Schaefer, J. P., and Spanggord, R. J., *J. Org. Chem.*, 37, 3069 (1972); (b) DeBrule, R. F., and Hess, G. G., *Synthesis*, 197 (1974); (c) Aranda, V. G., Barluenga, J., Yus, M., and Asensio, G., *Synthesis*, 806 (1974); (d) Barluenga, J., Concellon, J. M., and Asensio, G., *Synthesis*, 467 (1975).

7.34 (a) Brown, H. C., and Kurek, J. T., *J. Am. Chem. Soc.*, 91, 5647 (1969); (b) Beger, J., and Vogel, D., *J. Prakt. Chem.*, 311, 737 (1969).

7.35 Heathcock, C. H., *Angew. Chem.*, 81, 148 (1969).

7.36 Bachman, G. B., and Witehouse, M. L., *J. Org. Chem.*, 32, 2303 (1967).

7.37 Taylor, E. C., Kienzle, F., Robey, R. L., McKillop, A., and Hunt, J. D., *J. Am. Chem. Soc.*, 93, 4845 (1971).

7.38 Jensen, F. R., and Rickborn, B., *Electrophilic Substitution of Organomercurials*, McGraw-Hill, New York, 1968, 203 pp.

7.39 Pike, P. E., Marsh, P. G., Erickson, R. E., Waters, W. L., *Tetrahedron Lett.*, 2679 (1970).

7.40 (a) Tobler, E., and Foster, D. J., *Z. Naturforsch.*, 17B, 135 (1962); (b) Foster, D. J., and Tobler, *J. Org. Chem.*, 27, 834 (1962).

7.41 (a) Pasto, D. J., and Gontarz, J. A., *J. Am. Chem. Soc.*, 91, 719 (1969); (b) Gray, G. A., and Jackson, W. R., *J. Am. Chem. Soc.*, 91, 6205 (1969); (c) Whitesides, G. M., and San Filippo, J., Jr., *J. Am. Chem. Soc.*, 92, 6611 (1970).

7.42 Jensen, F. R., Miller, J. J., Cristol, S. J., and Beckley, R. S., *J. Org. Chem.*, 37, 4341 (1972).

7.43 Breuer, S. W., Leatham, M. J., and Thorpe, F. G., *Chem. Commun.*, 1475 (1971).

7.44 Bergbreiter, D. E., and Whitesides, F. M., *J. Am. Chem. Soc.*, 96, 4937 (1974).

7.45 Larock, R. C., and Mitchell, M. A., *J. Am. Chem. Soc.*, 98, 6718 (1976).

7.46   (a) Kreevoy, M. M., and Kowitt, F. R., *J. Am. Chem. Soc.*,
       82, 739 (1960); (b) Kreevoy, M. M., and Ditsch, L. T., *J.
       Am. Chem. Soc.*, 82, 6127 (1960); (c) Schaleger, L. L.,
       Turner, M. A., Chamberlain, T. C., and Kreevoy, M. M., *J.
       Org. Chem.*, 27, 3421 (1962).

7.47   (a) Reutov, O. A., and Lovtsova, A. N., *Izv. Akad. Nauk.
       SSSR, Otd. Khim.*, 1716 (1960); (b) Seyferth, D., and Lam-
       bert, R. L., *J. Organometal. Chem.*, 16, 21 (1969).

7.48   (a) McKillop, A., *Proc. R. A. Welch Found. Conf. Chem.
       Res.*, 17, 153 (1974); (b) McKillop, A., *Pure Appl. Chem.*,
       43, 463 (1975).

7.49   Meister, H., *Angew. Chem.*, 69, 533 (1957).

7.50   Corey, E. J., Koelliker, U., and Neuffer, J., *J. Am. Chem.
       Soc.*, 91, 1489 (1971).

7.51   McKillop, A., Fowler, J. S., Zelesko, M. J., Hunt, J. D.,
       Taylor, E. C., and McGillivray, G., *Tetrahedron Lett.*,
       2427 (1969).

7.52   McKillop, A., Bromley, D., and Taylor, E. C., *Tetrahedron
       Lett.*, 1623 (1969).

7.53   Taylor, E. C., Altland, H. W., Danforth, R. H., McGilliv-
       ray, G., and McKillop, A., *J. Am. Chem. Soc.*, 92, 3520
       (1970).

7.54   Breuer, S. W., Pickles, G. M., Podesta, J. C., and Thorpe,
       F. G., *J. C. S. Chem. Comm.*, 36 (1975).

7.55   Taylor, E. C., Kienzle, F., and McKillop, A., *J. Am. Chem.
       Soc.*, 92, 6088 (1970).

7.56   Taylor, E. C., Andrade, J. G., and McKillop, A., *J. C. S.
       Chem. Comm.*, 538 (1977).

7.57   Spencer, T., and Thorpe, F. G., *J. Organometal. Chem.*, 99,
       C8 (1975).

7.58   Taylor, E. C., Danforth, R. H., and McKillop, A., *J. Org.
       Chem.*, 38, 2088 (1973).

7.59   Anderson, C. B., and Winstein, S., *J. Org. Chem.*, 28, 605
       (1963).

7.60   (a) McKillop, A., Hunt, J. D., Taylor, E. C., and Kienzle,
       F., *Tetrahedron Lett.*, 5275 (1970); (b) Byrd, J. E.,
       Cassar, L., Eaton, P. E., and Halpern, J., *J. C. S. Chem.
       Comm.*, 40 (1971).

7.61 McKillop, A., Oldenziel, O. H., Swann, B. O., Taylor, E. C., and Robey, R. L., *J. Am. Chem. Soc.*, <u>93</u>, 7331 (1971); <u>95</u>, 1296 (1973).

7.62 McKillop, A., Swann, B. P., and Taylor, E. C., *Tetrahedron Lett.*, 5281 (1970).

7.63 Farkas, L., Gottsegen, A., Nógrádi, M., and Antus, S., *J. C. S. Chem. Comm.*, 825 (1972).

7.64 McKillop, A., Swann, B. P., and Taylor, E. C., *J. Am. Chem. Soc.*, <u>93</u>, 4919 (1971).

7.65 Taylor, E. C., Robey, R. L., and McKillop, A., *Angew. Chem. Int. Ed. Engl.*, <u>11</u>, 48 (1972); *J. Org. Chem.*, <u>37</u>, 2797 (1972).

7.66 Wolfe, S., Campbell, P. G. C., and Palmer, G. E., *Tetrahedron Lett.*, 4203 (1966).

7.67 Moriarty, R. M., Walsh, H. G., and Gopal, H., *Tetrahedron Lett.*, 4363 (1966).

7.68 For a review, see Sheldon, R. A., and Kochi, J. K., *Org. React.*, <u>19</u>, 279 (1972).

7.69 (a) Divakar, K. J., Sane, P. P., and Rao, A. S., *Tetrahedron Lett.*, 399 (1974); (b) Ho, T.-L., and Wong, C. M., *Synthetic Comm.*, <u>4</u>, 133 (1974).

7.70 McMurry, J. E., and Blaszczak, L. C., *J. Org. Chem.*, <u>39</u>, 2217 (1974).

Appendix I

GENERAL REFERENCES

I GENERAL TEXTBOOKS AND MONOGRAPHS

0.1 Rochow, E. G., Hurd, D. J., and Lewis, R. N., *The Chemistry of Organometallic Compounds*, Wiley, New York, 1957, 344 pp.

0.2 Zeiss, H., Ed., *Organometallic Chemistry*, Reinhold, New York, 1960, 549 pp.

0.3 Pauson, P. L., *Organometallic Chemistry*, St. Martin's Press, New York, 1967, 202 pp.

0.4 Coates, G. E., Green, M. L. H., Powell, P., and Wade, K., *Principles of Organometallic Chemistry*, Methuen, London, 1968, 259 pp.

0.5 Swan, J. M., and Black, D. St. C., *Organometallics in Organic Synthesis*, Chapman and Hall, London, 1974, 158 pp.

0.6 Coates, G. E., and Wade, K., *Organometallic Compounds*, Vol. 1, *The Main Group Elements*, 3rd ed., Methuen, London, 1967, 573 pp.

0.7 Eisch, J. J., *The Chemistry of Organometallic Compounds*, *The Main Group Elements*, Macmillan, New York, 1967, 178 pp.

0.8 Nesmeyanov, A. N., and Kocheshkov, K. A., Eds., *Methods of Elemento-Organic Chemistry*, North-Holland, Amsterdam, Vol. 1, Nesmeyanov, A. N., and Sokolik, R. A., *The Organic Compounds of Boron, Aluminum, Gallium, Indium, and Thallium*, 1967, 635 pp; Vol. 2, Ioffe, S. T., and Nesmeyanov, A. N., *The Organic Compounds of Magnesium, Beryllium, Calcium, Strontium, and Barium*, 1967, 735; Vol. 3, Sheverdina, N. I., and Kocheshkov, K. A., *The Organic Compounds of Zinc and Cadmium*, 1967, 252 pp; Vol. 4, Makarova, L. G., and Nesmeyanov, A. N., *The Organic Compounds of Mercury*, 1967,

532 pp.

0.9 Bayer, O., Müller, E., and Ziegler, K., Eds., *Methoden der Organischen Chemie* (Houben-Weyl), Vol. 13: *Metallorganische Verbindungen.* Thieme, Stuttgart. Part 1 (1970), Li, Na, K, Rb, Cs, Cu, Ag, Au, 940 pp; Part 2a (1974), Be, Mg, Ca, Sr, Ba, Zn, Cd, 949 pp; Part 2b (1974), Hg, 483 pp; Part 4 (1970), Al, Ga, In, Tl, 430 pp; Part 7 (1975), Pb, Ti, Zr, Hf, V, Nb, Ta, Cr, Mo, W, 570 pp.

0.10 Matteson, D. S., *Organometallic Reaction Mechanisms*, Academic, New York, 1974, 353 pp.

0.11 Green, M. L. H., *Organometallic Compounds*, Vol. 2, *The Transition Elements*, Methuen, London, 1968, 376 pp.

0.12 King, R. B., *Transition-Metal Organometallic Chemistry*, Academic, New York, 1969, 204 pp.

0.13 Heck, R. F., *Organotransition Metal Chemistry*, Academic, New York, 1974, 338 pp.

II MONOGRAPHS AND COLLECTIONS OF REVIEWS ON SPECIFIC TOPICS

0.14 Fischer, E. O., and Werner, H., *Metal π-Complexes*, Vol. 1, Elsevier, Amsterdam, 1966, 246 pp; Herberhold, M., *Metal* π-Complexes, Vol. 2, Elsevier, Amsterdam, 1972, 643 pp.

0.15 Bird, C. W., *Transition Metal Intermediates in Organic Synthesis*, Academic, New York, 1967, 280 pp.

0.16 Ramsey, B. G., *Electronic Transitions in Organometalloids*, Academic, New York, 1969, 297 pp.

0.17 Denny, D. B., Ed., *Techniques and Methods of Organic and Organometallic Chemistry*, Dekker, New York, 1969, 232 pp.

0.18 Brilkina, T. G., and Shuchurov, V. A., *Reactions of Organometallic Compounds with Oxygen and Peroxides*, Chemical Rubber Co. Press, Cleveland, Ohio, 1969, 225 pp.

0.19 Cox, J. D., and Pilcher, G., *Thermochemistry of Organic and Organometallic Compounds*, Academic, New York, 1970, 643 pp.

0.20 Tsutsui, M., Levy, M. N., Nakamura, A., Ichikawa, M., and Mori, K., *Introduction to Metal π-Complex Chemistry*, Plenum, New York, 1970, 210 pp.

0.21 Tsutsui, M., Eds., *Characterization of Organometallic Compounds*, Wiley-Interscience, New York, Part 1, 1969, 371 pp; Part 2, 1971, 877 pp.

0.22 George, W. O., Ed., *Spectroscopic Methods in Organometallic Chemistry*, Butterworths, London, 1970, 224 pp.

0.23 Muetterties, E. L., Ed., *Transition Metal Hydrides*, Dekker, New York, 1971, 342 pp.

0.24 Shrauzer, G. N., Ed., *Transition Metals in Homogeneous Catalysis*, Dekker, New York, 1971.

0.25 Kepert, D. L., *The Early Transition Metals*, Academic, New York, 1972, 499 pp.

0.26 Crompton, T. R., *Chemical Analysis of Organometallic Compounds*, 3 vols., Academic, New York, Vol. 1, 1973, 258 pp; Vol. 2, 1974, 163 pp; Vol. 3, 1974, 211 pp.

0.27 Bernel, I., Ed., *Horizons in Organometallic Chemistry*, New York Academy of Science, New York, 1974, 321 pp.

0.28 Litzow, M. L., and Spalding, T. R., *Mass Spectrometry of Inorganic and Organometallic Compounds*, Elsevier, Amsterdam, 1974, 622 pp.

0.29 Ishii, Y., and Tsutsui, M., Eds., *Organotransition-Metal Chemistry*, Plenum, New York, 1975, 398 pp.

0.30 Tsuji, J., *Organic Synthesis by Means of Transition Metal Complexes*, Springer-Verlag, New York, 1975, 199 pp.

0.31 Alper, H., Ed., *Transition Metal Organometallics in Organic Synthesis*, Vol. 1, Academic, New York, 1976, 258 pp.

0.32 Maslowsky, E., Jr., *Vibrational Spectra of Organometallic Compounds*, John Wiley & Sons, New York, 1977, 528 pp.

Appendix II

PERIODIC TABLE AND ELECTRONEGATIVITIES OF THE ELEMENTS[a]

IA	IIA	IIIB	IVB	VB	VIB	VIIB	VIII	VIII	VIII	IB	IIB	IIIA	IVA	VA	VIA	VIIA	VIIA
H 2.2 / 2.20																	He – / –
Li 1.0 / 0.97	Be 1.5 / 1.47											B 2.0 / 2.01	C 2.5 / 2.50	N 3.0 / 3.07	O 3.5 / 3.50	F 4.0 / 4.10	Ne – / –
Na 0.9 / 1.01	Mg 1.2 / 1.23											Al 1.5 / 1.47	Si 1.8 / 1.74	P 2.1 / 2.06	S 2.5 / 2.44	Cl 3.0 / 2.83	Ar – / –
K 0.8 / 0.91	Ca 1.0 / 1.04	Sc 1.3 / 1.20	Ti 1.5 / 1.32	V 1.6 / 1.45	Cr 1.6 / 1.56	Mn 1.5 / 1.60	Fe 1.8 / 1.64	Co 1.8 / 1.70†	Ni 1.8 / 1.75†	Cu 1.9 / 1.75	Zn 1.6 / 1.66	Ga 1.6 / 1.82	Ge 1.8 / 2.02	As 2.0 / 2.20	Se 2.4 / 2.48	Br 2.8 / 2.74	Kr – / –
Rb 0.8 / 0.89	Sr 1.0 / 0.99	Y 1.2 / 1.11	Zr 1.4 / 1.22†	Nb 1.6 / 1.23†	Mo 1.8 / 1.30†	Tc 1.9 / 1.36†	Ru 2.2 / 1.42†	Rh 2.2 / 1.45†	Pd 2.2 / 1.35†	Ag 1.9 / 1.42	Cd 1.7 / 1.46	In 1.7 / 1.49	Sn 1.8 / 1.72	Sb 1.9 / 1.82	Te 2.1 / 2.01	I 2.5 / 2.21	Xe – / –
Cs 0.7 / 0.86	Ba 0.9 / 0.97	La* 1.1 / 1.08	Hf 1.3 / 1.23†	Ta 1.5 / 1.33†	W 1.7 / 1.40†	Re 1.9 / 1.46†	Os 2.2 / 1.52†	Ir 2.2 / 1.55†	Pt 2.2 / 1.44	Au 2.4 / 1.42†	Hg 1.9 / 1.44†	Tl 1.8 / 1.44†	Pb 1.8 / 1.55†	Bi 1.9 / 1.67†	Po 2.0 / 1.7†	At 2.2 / 1.90†	Rn – / –
Fr 0.7 / 0.86	Ra 0.9 / 0.97	Ac** 1.1 / 1.00															

* Lanthanides

Ce	Pr	Nd	Pm	Sm	Eu	Gd	Tb	Dy	Ho	Er	Tm	Yb	Lu
– / 1.08†	– / 1.07†	– / 1.07†	– / 1.07†	– / 1.07†	– / 1.01†	– / 1.11†	– / 1.10†	– / 1.10†	– / 1.10†	– / 1.11†	– / 1.11†	– / 1.06†	– / 1.14†

** Actinides

Th	Pa	U	Np	Pu	Am	Cm	Bk	Cf	Es	Fm	Md	No	Lw
1.3† / 1.11†	1.5† / 1.14†	1.7† / 1.22†	1.3† / 1.22†	1.3† / 1.22†	1.3† / 1.2†	1.3† / 1.2†	1.3† / 1.2†	1.3† / 1.2†	1.3† / 1.2†	1.3† / 1.2†	1.3† / 1.2†	1.3† / 1.2†	– / –

Key

H
2.2 ← Pauling electronegativity value
2.20 ← Allred-Rochow electronegativity value

a — The Pauling electronegativity values are taken from Pauling, L., *The Nature of the Chemical Bond*, 3rd ed., Cornell University Press, Ithaca, N. Y., 1960, p. 93. The Allred-Rochow electronegativity values are taken from Allred, A. L., and Rochow, E. G., *J. Inorg. Nucl. Chem.*, 5, 264 (1958) except for the numbers with daggers which are taken from Little, E. J., and Jones, M. M., *J. Chem. Educ.*, 37, 231 (1960) and Sanderson, R. T., *Inorganic Chemistry*, Van Nostrand-Reinhold, New York, 1967, p. 72.

Appendix III

ELECTRONIC CONFIGURATIONS OF THE ELEMENTS

Element		$1s$	$2s$	$2p$	$3s$	$3p$	$3d$	$4s$	$4p$	$4d$
1.	H	1								
2.	He	2								
3.	Li	2	1							
4.	Be	2	2							
5.	B	2	2	1						
6.	C	2	2	2						
7.	N	2	2	3						
8.	O	2	2	4						
9.	F	2	2	5						
10.	Ne	2	2	6						
11.	Na	2	2	6	1					
12.	Mg	2	2	6	2					
13.	Al	2	2	6	2	1				
14.	Si	2	2	6	2	2				
15.	P	2	2	6	2	3				
16.	S	2	2	6	2	4				
17.	Cl	2	2	6	2	5				
18.	Ar	2	2	6	2	6				
19.	K	2	2	6	2	6		1		
20.	Ca	2	2	6	2	6		2		
21.	Sc	2	2	6	2	6	1	2		
22.	Ti	2	2	6	2	6	2	2		
23.	V	2	2	6	2	6	3	2		
24.	Cr	2	2	6	2	6	5	1		
25.	Mn	2	2	6	2	6	5	2		
26.	Fe	2	2	6	2	6	6	2		
27.	Co	2	2	6	2	6	7	2		
28.	Ni	2	2	6	2	6	8	2		
29.	Cu	2	2	6	2	6	10	1		
30.	Zn	2	2	6	2	6	10	2		
31.	Ga	2	2	6	2	6	10	2	1	
32.	Ge	2	2	6	2	6	10	2	2	
33.	As	2	2	6	2	6	10	2	3	
34.	Se	2	2	6	2	6	10	2	4	
35.	Br	2	2	6	2	6	10	2	5	
36.	Kr	2	2	6	2	6	10	2	6	

Element	K	L	M	4s	4p	4d	4f	5s	5p	5d	5f	5g	6s	6p	6d
37. Rb	2	8	18	2	6			1							
38. Sr	2	8	18	2	6			2							
39. Y	2	8	18	2	6	1		2							
40. Zr	2	8	18	2	6	2		2							
41. Nb	2	8	18	2	6	4		1							
42. Mo	2	8	18	2	6	5		1							
43. Tc	2	8	18	2	6	6		1							
44. Ru	2	8	18	2	6	7		1							
45. Rh	2	8	18	2	6	8		1							
46. Pd	2	8	18	2	6	10									
47. Ag	2	8	18	2	6	10		1							
48. Cd	2	8	18	2	6	10		2							
49. In	2	8	18	2	6	10		2	1						
50. Sn	2	8	18	2	6	10		2	2						
51. Sb	2	8	18	2	6	10		2	3						
52. Te	2	8	18	2	6	10		2	4						
53. I	2	8	18	2	6	10		2	5						
54. Xe	2	8	18	2	6	10		2	6						
55. Cs	2	8	18	2	6	10		2	6				1		
56. Ba	2	8	18	2	6	10		2	6				2		
57. La	2	8	18	2	6	10		2	6	1			2		
58. Ce	2	8	18	2	6	10	2	2	6				2		
59. Pr	2	8	18	2	6	10	3	2	6				2		
60. Nd	2	8	18	2	6	10	4	2	6				2		
61. Pm	2	8	18	2	6	10	5	2	6				2		
62. Sm	2	8	18	2	6	10	6	2	6				2		
63. Eu	2	8	18	2	6	10	7	2	6				2		
64. Gd	2	8	18	2	6	10	7	2	6	1			2		
65. Tb	2	8	18	2	6	10	9	2	6				2		
66. Dy	2	8	18	2	6	10	10	2	6				2		
67. Ho	2	8	18	2	6	10	11	2	6				2		
68. Er	2	8	18	2	6	10	12	2	6				2		
69. Tm	2	8	18	2	6	10	13	2	6				2		
70. Yb	2	8	18	2	6	10	14	2	6				2		
71. Lu	2	8	18	2	6	10	14	2	6	1			2		
72. Hf	2	8	18	2	6	10	14	2	6	2			2		
73. Ta	2	8	18	2	6	10	14	2	6	3			2		
74. W	2	8	18	2	6	10	14	2	6	4			2		
75. Re	2	8	18	2	6	10	14	2	6	5			2		
76. Os	2	8	18	2	6	10	14	2	6	6			2		
77. Ir	2	8	18	2	6	10	14	2	6	7			2		
78. Pt	2	8	18	2	6	10	14	2	6	9			1		

Element		K	L	M	4s	4p	4d	4f	5s	5p	5d	5g	6s	6p	6d	7
79.	Au	2	8	18	2	6	10	14	2	6	10		1			
80.	Hg	2	8	18	2	6	10	14	2	6	10		2			
81.	Tl	2	8	18	2	6	10	14	2	6	10		2	1		
82.	Pb	2	8	18	2	6	10	14	2	6	10		2	2		
83.	Bi	2	8	18	2	6	10	14	2	6	10		2	3		
84.	Po	2	8	18	2	6	10	14	2	6	10		2	4		
85.	At	2	8	18	2	6	10	14	2	6	10		2	5		
86.	Rn	2	8	18	2	6	10	14	2	6	10		2	6		
87.	Fr	2	8	18	2	6	10	14	2	6	10		2	6		1
88.	Ra	2	8	18	2	6	10	14	2	6	10		2	6		2
89.	Ac	2	8	18	2	6	10	14	2	6	10		2	6	1	2
90.	Th	2	8	18	2	6	10	14	2	6	10		2	6	2	2
91.	Pa	2	8	18	2	6	10	14	2	6	10	2	2	6	1	2
92.	U	2	8	18	2	6	10	14	2	6	10	3	2	6	1	2
93.	Np	2	8	18	2	6	10	14	2	6	10	5	2	6		2
94.	Pu	2	8	18	2	6	10	14	2	6	10	6	2	6		2
95.	Am	2	8	18	2	6	10	14	2	6	10	7	2	6		2
96.	Cm	2	8	18	2	6	10	14	2	6	10	7	2	6	1	2
97.	Bk	2	8	18	2	6	10	14	2	6	10	8	2	6	1	2
98.	Cf	2	8	18	2	6	10	14	2	6	10	10	2	6		2
99.	Es	2	8	18	2	6	10	14	2	6	10	11	2	6		2
100.	Fm	2	8	18	2	6	10	14	2	6	10	12	2	6		2
101.	Md	2	8	18	2	6	10	14	2	6	10	13	2	6		2
102.	-	2	8	18	2	6	10	14	2	6	10	14	2	6		2
103.	Lw	2	8	18	2	6	10	14	2	6	10	14	2	6	1	2

Appendix IV

IONIZATION ENERGIES[a]

Element	Ionization Energies (eV)			Element	Ionization Energies (eV)		
	I	II	III		I	II	III
H	13.60			Ru	7.37	16.76	28.47
He	24.59	54.42		Rh	7.46	18.08	31.06
Li	5.39	75.64	122.45	Pd	8.34	19.43	32.93
Be	9.32	18.21	153.89	Ag	7.58	21.49	34.83
B	8.30	25.15	37.93	Cd	8.99	16.91	37.48
C	11.26	24.38	47.89	In	5.79	18.87	28.03
N	14.53	29.60	47.45	Sn	7.34	14.63	30.50
O	13.62	35.12	54.93	Sb	8.64	16.53	25.3
F	17.42	34.97	62.71	Te	9.01	18.06	27.96
Ne	21.56	40.96	63.45	I	10.45	19.13	33.00
Na	5.14	47.29	71.64	Xe	12.13	21.21	32.1
Mg	7.65	15.04	80.14	Cs	3.89	23.1	
Al	5.99	18.83	28.45	Ba	5.21	10.00	
Si	8.15	16.34	34.49	La	5.58	11.06	19.18
P	10.49	19.72	30.18	Ce	5.47	10.85	20.20
S	10.36	23.33	34.83	Pr	5.42	10.55	21.62
Cl	12.97	23.81	39.61	Nd	5.49	10.72	
Ar	15.76	27.63	40.74	Pm	5.55	10.90	
K	4.34	31.62	45.72	Sm	5.63	11.07	
Ca	6.11	11.87	50.91	Eu	5.67	11.25	
Sc	6.54	12.80	24.76	Gd	6.13	12.1	
Ti	6.82	13.58	27.49	Tb	5.85	11.52	
V	6.74	14.65	29.31	Dy	5.93	11.67	
Cr	6.77	15.50	30.96	Ho	6.02	11.80	
Mn	7.44	15.64	33.67	Er	6.10	11.93	
Fe	7.87	16.18	30.65	Tm	6.18	12.05	23.71
Co	7.86	17.06	33.50	Yb	6.25	12.17	25.2
Ni	7.64	18.17	35.17	Lu	5.43	13.9	
Cu	7.73	20.29	36.83	Hf	7.0	14.9	23.3
Zn	9.39	17.96	39.72	Ta	7.89		
Ga	6.00	20.51	30.71	W	7.98		
Ge	7.90	15.93	34.22	Re	7.88		
As	9.81	18.63	28.35	Os	8.7		
Se	9.75	21.19	30.82	Ir	9.1		
Br	11.81	21.08	36.00	Pt	9.0	15.56	
Kr	14.00	24.36	36.95	Au	9.22	20.5	
Rb	4.18	27.28	40.00	Hg	10.44	18.76	34.2
Sr	5.70	11.03	43.06	Tl	6.11	20.43	29.83
Y	6.38	12.24	20.52	Pb	7.42	15.03	31.94
Zn	6.84	13.13	22.99	Di	7.29	16.69	25.56
Nb	6.88	14.32	25.04	Po	8.42		
Mo	7.10	16.15	27.16	At			
Tc	7.28	15.26	29.54	Rn	10.75		

[a] The ionization energy values are taken from Moore, C. E., *Ionization Potentials and Ionization Limits Derived from the Analyses of Optical Spectra*, NSRDS-NBS 34, National Bureau of Standards, Washington, D. C., 1970. Most of the values cited in this table are rounded to the nearest hundredth.

Appendix V

COVALENT AND IONIC RADII (Å)[a]

Element	V_{cov}	V_{ion}	Element	V_{cov}	V_{ion}
H	0.37	2.08 (−1)	Rb	2.16	1.48 (+1)
Li	1.34	0.64 (+1)	Sr	1.91	1.13 (+2)
Be	1.25	0.31 (+2)	Y	1.62	0.93 (+3)
B	0.90	0.20 (+3)	Zr	1.45	0.80 (+4)
C	0.77	2.60 (−4)	Nb	1.34	0.70 (+5)
N	0.75	1.71 (−3)	Mo	1.30	0.62 (+6)
O	0.73	1.40 (−2)	Tc	1.27	0.63 (+4)
F	0.71	1.36 (−1)	Ru	1.25	0.62 to 0.70
Na	1.54	0.95 (+1)	Rh	1.25	0.62 to 0.67
Mg	1.45	0.65 (+2)	Pd	1.28	0.64 to 0.86
Al	1.30	0.50 (+3)	Ag	1.34	1.26 (+1)
Si	1.18	0.41 (+4)	Cd	1.48	0.97 (+2)
P	1.10	2.12 (−2)	In	1.44	0.81 (+3)
S	1.02	1.82 (−2)	Sn	1.40	0.71 (+4)
Cl	0.99	1.81 (−1)	Sb	1.43	0.62 (+5)
K	1.96	1.33 (+1)	Te	1.35	2.21 (−2)
Ca	1.74	0.99 (+2)	I	1.33	2.16 (−1)
Sc	1.44	0.81 (+3)	Cs	2.35	1.69 (+1)
Ti	1.32	0.68 (+4)	Ba	1.98	1.35 (+2)
V	1.22	0.59 (+5)	La	1.69	1.18 (+3)
Cr	1.18	0.52 (+6)	Hf	1.44	0.71 to 0.83 (+4)
Mn	1.17	0.46 (+7)	Ta	1.34	0.64 to 0.67
Fe	1.17	0.49 to 0.78	W	1.30	0.41 to 0.65
Co	1.16	0.53 to o.74	Re	1.28	0.40 to 0.63
Ni	1.16	0.56 to 0.70	Os	1.26	0.63 (+4)
Cu	1.17	0.96 (+1)	Ir	1.27	0.63 to 0.73
Zn	1.25	0.74 (+2)	Pt	1.30	0.60 to 0.99
Ga	1.26	0.62 (+3)	Au	1.34	1.37 (+1)
Ge	1.22	0.53 (+4)	Hg	1.49	1.10 (+2)
As	1.22	0.47 (+5)	Tl	1.48	0.95 (+3)
Se	1.17	0.42 (+6)	Pb	1.47	0.78 to 0.94 (+4)
Br	1.14	0.39 (+7)			

[a] The covalent and ionic radii listed in this table are taken from Huheey, J. E., *Inorganic Chemistry*, Harper & Row, New York, 1972, pp. 73 to 75, 184 to 185, and from *Table of Periodic Properties of the Elements* (Catalog No. S-18806). Sargent-Welch Scientific Co., Skokie, Ill.

Appendix VI
ACIDITY OF BRONSTED ACIDS

Table 1. Hydrocarbon Acids [a]

Compound	pK_a	Compound	pK_a
t-Bu-H	> 52	(⟨⟩-)₂CH₂	34
⟨⟩-H	51 to 52	(⟨⟩-)₃CH	31.5
Et-H	49 to 50	HC≡CH	25
▷-H	46	fluorene	23
H₂C=CH₂	44		
⟨⟩-H	43	indene	19 to 20
⟨⟩-CH₃	41		
H₂C=CHCH₃	40	cyclopentadiene	15 to 16

[a] These pK_a values are taken from: Streitwieser, A., Jr., and Heathcock, C.H., *Introduction to Organic Chemistry*, Macmillan, New York, 1976, 1279 pp.

Table 2. Hetero-Substituted Carbon Acids [a]

Compound	pK_a	Compound	pK_a
1,3-dithiane, 2-CH$_3$, 2-H (cyclic)	38 [b]	$(CH_3SO_2)_2CH_2$	14
PhSOCH$_3$	35 [c]	$(NC)_2CH_2$	11
1,3-dithiane, 2-H, 2-H (cyclic)	31 [b]	CH_3OC\\EtO_2C $>CH_2$	11
PhS\\PhS $>C(H)(H)$	31 [d]	O_2NCH_3	10
1,3-dithiane, 2-Ph, 2-H (cyclic)	29 [b]	NCH	9
RSO_2CH_3	23 to 27 [c]	$(CH_3OC)_2CH_2$	9
$NCCH_3$	25	NC\\MeO_2C $>CH_2$	9
Cl_3CH	25	$(CH_3OC)_3CH$	5.9
EtO_2CCH_3	24 to 25	$(O_2N)_2CH_2$	3.6
CH_3COCH_3	20	$(O_2N)_3CH$	0
		$SNCH$	-1.9

[a] Unless otherwise noted, the pK_a values are taken from the reference cited in footnote a of Table 1.

[b] Streitwieser, A., Jr., and Ewig, S. P., *J. Am. Chem. Soc.*, 97, 190 (1975).

[c] House, H. O., *Modern Synthetic Reactions*, 2nd ed., W. A., Benjamin, Menlo Park, Calif., 1972, p. 494. Bordwell, F. G., Mathews, W. S., Vanier, N. R., *J. Am. Chem. Soc.*, 97, 442 (1975).

[d] Bordwell, F. G., and Matthews, W. S., *J. Am. Chem. Soc.*, 96, 1214 (1974).

Table 3. Amines and Other Nitrogen Acids [a]

Compound	pK_a	Compound	pK_a
Me$\underline{N}$H$_2$	35		5.3
$\underline{N}$H$_3$	35		
Ph$\underline{N}$H$_2$	27 [b]		4.6
Ph$_2$$\underline{N}$H	23 [b]		1.0
Me$_2$$\overset{+}{N}$$\underline{H}$$_2$	11		
MeN$\overset{+}{}$$\underline{H}$$_3$	10.6	Me$_2$C=N$\overset{+}{}$$\underline{H}$OH	-1.9
Me$_3$$\overset{+}{N}$$\underline{H}$	10	MeC≡N$\overset{+}{}$$\underline{H}$	-10
$\overset{+}{N}$$\underline{H}$$_4$	9.2		
	7.0		-12

[a] See footnote a of Table 2.
[b] Hendrickson, J. B., Cram, D. J., and Hammond, G. S., *Organic Chemistry*, 3rd ed., MaGraw-Hill, New York, 1970, p. 304.

Table 4. Oxygen Acids [a]

Compound	pK_a	Compound	pK_a
$t\text{-BuO}\underline{H}$	18	$O_3N\underline{H}$	1.3
$EtO\underline{H}$	16		
$O\underline{H}_2$	15.7	$O_2N{-}\underset{NO_2}{\overset{NO_2}{\bigcirc}}{-}O\underline{H}$	0.25
$O_2\underline{H}_2$	11.6		
$PhO\underline{H}$	10	$CF_3COO\underline{H}$	0.2
$O_3B\underline{H}_3$	9.2	$Me_2\overset{+}{S}O\underline{H}$	0
$BrO\underline{H}$	8.6	$CH_4SO_3\underline{H}$	-1.2
$ClO\underline{H}$	7.5	$Et\overset{+}{O}\underline{H}_2$	-2.4
$O_2N{-}\bigcirc{-}O\underline{H}$	7.2	$t\text{-Bu}\overset{+}{O}\underline{H}_2$	-3.8
		$O_4S\underline{H}_2$	-5.2
$O_3C\underline{H}_2$	6.4	$CH_3\underset{OEt}{\overset{\|}{C}}{=}\overset{+}{O}\underline{H}$	-6.5
$CH_3COO\underline{H}$	4.7		
$O_2N{-}\underset{NO_2}{\bigcirc}{-}O\underline{H}$	4.1	$Ph\underset{CH_3}{\overset{+}{\underset{\|}{O}}}\underline{H}$	-6.5
		$Ph\overset{+}{O}\underline{H}_2$	-6.7
$O_2N{-}\bigcirc{-}COO\underline{H}$	3.4	$Me_2C{=}\overset{+}{O}\underline{H}$	-7
$O_2N\underline{H}$	3.2	$MeCH{=}\overset{+}{O}\underline{H}$	-8
$O_3P\underline{H}_3$	2.2	$O_4Cl\underline{H}$	-10
$O_3S\underline{H}_2$	1.8	$Me\overset{+}{N}O(O\underline{H})$	-12

[a] See footnote a of Table 2.

Table 5. Other Acids [a]

Compound	pK_a	Compound	pK_a
FH	3.2	SH$_2$	7
ClH	-7	Me$_2$S$^+$H	-5.2[b]
BrH	-9		
IH	-9.5	MeS$^+$H$_2$	-7

[a] See footnote a of Table 2.
[b] See footnote b of Table 3.

INDEX

Acetates, synthesis via
  oxymercuration, 446
  oxyplumbation, 490
  oxythallation, 485
Acetoacetic ester, alkylation
  of, 163, 187
Acetylenes, see Alkynes
Acetylene "zipper" reaction,
  251-253
Acid-base reactions, 21
  in the preparation of organo-
    metallics, 36
  in metal-hydrogen exchange,
    41, 43
Acidity of carbon acids
  factors effecting, 11, 12
  in metal-halogen exchange, 38
  in metal-hydrogen exchange,
    42, 136
  in alkylation of dianions,
    163
  pK_a values of, 506
Acyl anion equivalents, 151-160,
  438
Acyl halides, reaction with
  metal enolates, 210
  Group IA and II metals, 119
  organoaluminums, 356, 368

organoborons, 321, 345
organosilicons, 415
organotins, 413
Acyloins, from 1,2-glycols, 446
Acyloin reaction, 230, 231
1,2-Addition reactions to
  α,β-unsaturated carbonyls,
  127, 212, 372
1,4-Addition, see Conjugate
  addition
Alcohols
  protection of, 397
  synthesis of
    allylic, 136, 155, 325, 361,
      369, 409, 422, 429
    β-amino-, 438
    cyclopropanols, 176
    diols, 437
    α-halo-, 125
    α-keto-, 486
    phenol, 481
    symmetric, 120
    tertiary, 118, 119, 331,
      333, 337, 338, 440
  synthesis via
    Group IA and II organo-
      metallics, 98, 99, 118-
      125, 134, 136, 154, 231,
      242

511

Alcohols (cont.)
  organoaluminums, 361, 364, 369, 375-376
  organoborons, 243, 294, 304, 309, 311, 312, 318, 323, 325, 331, 333, 335-339, 345-349
  organomercuries, 476
  organosilicon, 409, 422, 426, 427, 429, 437
  organothalliums, 481, 486
  organotins, 422, 440, 447
  volatilization of, 438
Aldehydes
  deoxygenation of, 345
  epoxidation of, 172
  reactions with
    dianion of β-ketosulfoxides, 203
    lithium enolates, 202
    organoaluminums, 369
    organoborons, 77, 325
    organosilicons, 426, 429, 438
  synthesis of
    α-halo-, 431
    hydroxy-, 345
    α-keto-, 209
    α-silylated, 426
  synthesis via
    Group IA and II organometallics, 159, 201, 209
    organoaluminums, 378
    organoborons, 159, 311, 318, 333-335, 343, 345
    organothalliums, 485, 486
    organotins, 445, 447
Aldol reaction, 198-209, 433
Alkanes
  synthesis via
    hydrometallation of alkenes, 345, 357, 406-408, 410, 446
    reduction of organomercuries, 472
Alkenes

  activation of, 134
  basicity of, 51
  carbometallation of, 134, 341, 354
  hydrometallation of, 47, 294-298, 357, 358, 406, 410, 446
  inversion of configuration of, 414, 425
  substitution of, via addition-elimination, 68
  synthesis of
    arylated, 483
    disubstituted, 319, 352, 354, 355, 356, 365-368
    heterosubstituted, 168
    trans-, 233
    trisubstituted, 319, 355, 356, 368
  synthesis via
    Group IA and II organometallics, 161-171
    organoaluminums, 352, 354-356, 365-368
    organoborons, 304, 318, 319, 322, 323, 332
    organoleads, 491
    organosilicons, 167-169, 414, 417, 420, 421, 425-427
    organotins, 441-443
  See also Alkenyl halides
Alkenyl halides
  conversion to alkenyllithium, 96
  cross coupling of, 116
  hydroboration of, 331
  reduction of, 347
  synthesis of, 169, 315, 420, 427
  See also Allenyl halides
Alkenylation of enolates, 193
Alkenylmetals,
  as acyl anion equivalents, 153
  preparation of
    stereospecific, trans-, 55

Alkenylmetals (cont.)
  See also Hydrometallation
Alkoxide analogs, 433
Alkyl halides
  abstraction of, by free
    radials, 402
  cross coupling of, 105-118
  one-carbon homologation of,
    178
  reactivity of, 32, 37, 39, 65
  reduction of, 230, 238, 345,
    346, 349
  substitution reaction of, 107
  synthesis via
    organoaluminums, 364
    organoborons, 304, 313
    organomercuries, 471
    organosilicons, 419
  See also Allylic halides,
    Benzylic halides, and Pro-
    pargylic halides
Alkylation of
  aldehydes, 193-198, 370
  allylic organometallics, 163-
    166, 474
  carbinols, 370
  carboxylic acids, 370
  cyclopentadienylthallium(I),
    479, 480
  epoxides, 376
  esters, 163, 193-198
  α-heterosubstituted organo-
    metallics, 138, 154-161
  ketones, 162-163, 191, 370
  metal enolates, 28, 105, 185-
    197, 445
  nitriles, 193-198
  α,β-unsaturated ketones,
    371-373, 441
Alkylidene transfer, 177
Alkynes
  basicity of, 51
  carbometallation of, 133,
    134, 354-356, 341
  hydrometallation of, 47, 55,
    294, 298, 357, 359, 360,

    362, 377, 406, 408, 410,
    411, 446
  isomerization of, internal
    to terminal, 251
  oxymetallation of, 465, 479,
    483, 486
  protection of, 254
  synthesis via
    organoaluminums, 366, 376
    organoborons, 319
    organotins, 441, 442
  terminal, alkynylmetals
    from 98, 116
Alkynyl halides, cross
  coupling of, 116
Alkynylmetals, acyl anion
  equivalents, 153
Allenyl halides, reaction with
  n-butyllithium, 98
Alloy method, 33, 460
Allyl-allyl coupling, 107, 115,
  367
Allylic halides
  cross coupling of, 117
  cyclopropenes from, 55
  rearrangements of, 113
Allylic organometallics,
  alkylation of, 163-166
  rearrangement of, 108
Aluminum (Al), organometallics
  of
  alkenylaluminates, 370
  carboalumination, 354-356, 359
  cross-coupling reactions, 366-
    368
  halogenolysis of, 364
  hydrocyanation, 373-375
  oxidation of, 364
  preparations of, 353-356,
    362-363
  properties of, 350-352
  protonolysis of, 364
  reactions with
    acetylenes, 354, 355, 357,
      359
    acyl halides, 368

Aluminum (Al) (cont.)
  alcohols, 367
  alkenes, 354, 355, 357
  carbonyl compounds, 366,
    368, 369
  epoxides, 126, 366, 375
  organic halides, 65, 350,
    366, 368, 370
  propargylic alcohols, 361
  sulfonates, 366
  α,β-unsaturated ketones,
    68, 130, 133, 366, 371,
    372, 373
  transmetallation involving
    55-57, 301, 362
Amides
  reduction to aldehydes, 345
  synthesis of, from organo-
    mercuries, 476
Amination of organoborons, 316-
  317
Amines
  complexes with organoborons,
    289
  protection of, 367, 438
  reduction of, 340
  synthesis via
    Group IA and II organo-
      metallics, 201, 248
    organoborons, 294, 309, 316,
      345
    organomercuries, 476
    organotins, 444
  volatilization of, 438
Aminometallation, 50, 316
Amphophile
  definition, 65
  organoaluminums as, 352, 365,
    369, 376
Anilines, preparation of, 482
Annulation
  using masked carbonyl deri-
    vatives, 189
  via organoborons, 335
  See also Robinson annulation
    reaction

Anti-Markovnikov addition, 297,
  358, 476
Arenes
  basicity of, 51
  cross coupling of, 116
  deuteration of, 483
  reduction of, 234-237
  synthesis via
    organosilicons, 424
    organomercuries, 479
    organothalliums, 480-483
  See also Arylation
Arsenic (As), organometallics
  of, 33
Aryl halides
  cross coupling of, 116
  synthesis via
    organosilicons, 424
    organothalliums, 480, 481
Arylation of
  benzene, 482
  enolate anions, 193
  olefins, 483
  α,β-unsaturated ketones, 117
Asymmetric synthesis via
  asymmetric hydroboration,
    299
  oxazolines, 131, 221
  B-α-pinanyl-9-BBN, 349
"Ate" complexes, 37, 99, 249,
  302, 351
Aufbau principle, 4
Azides, synthesis of,
  in organomercuries, 476

Barium (Ba), organometallics of
  preparation of, 32
  transmetallation involving,
    249
Basicity
  definition, 91, 245
  of enolate anions, 212
  of organometallics, 41, 254
  solvent effects on, 42
Benzils, synthesis of, 487
Benzoins, synthesis of, 487

Benzylic halides
  carbenes from 175
  cross coupling of, 115, 117
  reduction of, 349
Benzyne, preparation of, 38
Berryllium (Be), organo-
    metallics of
  cross coupling involving, 107
  preparation of, 32, 249, 364
Bicyclic compounds, synthesis
    of, 214
Biellmann coupling, 1,5-diene
    synthesis, 114, 240
Bimetallic complexes, 401
Birch reduction, 234-237
Bismuth (Bi), organometallics
    of, 57
Bond angles, factors in, 18-20
Bond dissociation energy (DE),
    10
Bond energy (BE)
  effects of multiplicity on,
    14
  values for, 10, 396
Bond lengths, 10, 13
Bond polarization, 6, 62
p_π-d_π Bonding, 399
p_π-p_π Bonding, 289, 307, 314,
    352
9-Borabicyclo[3.3.1]nonane
    (9-BBN), 296
  intramolecular coupling
    reaction using, 319
  lithium dialkyl derivatives
    of, 343, 349
  reactions of, 325, 326
  reduction of
    acyl halides, 345
    enones, 345
Boron (B), organometallics of
  alkenylboranes, 159, 318
  alkenylborates, halogenation
    of, 67, 319
  allylboranes, 134, 341
  amination of, 316
  bond angles in, 19

boron-stabalized carbanions,
    340
carboboration, 341
carbonylation of, 250, 333-
    337
concerted reactions of, 305-
    306
conjugate addition reactions
    of, 325-328
copper borates, 322
cyanoborates, 337
enol borates, 203, 219, 324,
    325
fragmentation reactions of
    δ-heterosubstituted, 340
free radical reactions of,
    307, 308
halogenolysis of, 72, 307,
    312-316, 330-333
α-heterosubstituted, 138
hydroboration, see Hydro-
    metallation
hydrogenolysis, 310
intermolecular transfer
    reactions of, 303, 322
intramolecular transfer
    reactions of, 68, 304, 305,
    328, 329
ligand exchange, 81
one-electron transfer
    reactions of, 72, 84, 341
oxidation of, 304, 310-312
preparation of, 290-302
properties of, 57, 71, 285-
    290, 343
protonolysis of, 309, 310
reactions with
  acetylenes, 77, 134, 341
  acyl halides, 321
  alcohols, 288
  alkenes, 134, 291, 293, 316,
    341
  alkynes, 77, 134, 291, 293,
    332, 341
  arenes, 291
  carbonyl compounds, 4, 77,

Boron (cont.)
294, 295, 309, 321, 324,
328-330, 368
cyclopropanes, 291
epoxides, 288, 294, 321,
322
α-haloenolate anions, 328,
329
ketones, 4, 77, 288, 294,
324
nitriles, 294
nitro groups, 294
polyhalocarbanions, 338
α,β-unsaturated ketones,
113, 306
reduction with, 343-350
subvalent complexes of, 51
Bouveault-Blanc reduction, 228

Cadmium (Cd), organometallics of
preparation of, 33, 35
reaction with acyl halides,
119
Reformatsky reaction in-
volving, 104
transmetallation involving,
104, 301
Calcium (Ca), organometallics of
cross coupling involving, 108
preparation of, 33, 35
reaction with perfluro
olefins, 36
Carbanions
decomposition to carbenes,
173
organometallics as sources
of, 62, 109
structure of, 94
Carbenes, 46, 52
addition reactions of, 46,
74, 175, 436
organometallic analogs of,
402, 457
organometallic sources of,
62, 73-76, 173-177, 477,
478

preparation of, 137, 138,
173-175
reactions of 175-177
Carbocations
addition reactions of, 71
organometallic sources of,
62, 70, 457, 491
Carbometallation, 23, 48, 79,
82, 251
of Group IA and II organometallics, 133-136
of organoaluminums, 354,
355, 359
of organoborons, 341
Carbonyl compounds, masked,
151-161
Carbonyl compounds, α,β-un-
saturated
1,2-addition to, 127-133, 156
1,4-addition to, 65, 127-133,
138, 152-155, 156, 306,
325-327, 366, 371-373, 403,
413, 429, 430, 443
catalytic hydrogenation of,
232
reaction with bimetallic
complexes, 415, 416
reduction of, 182, 232, 345,
346, 348, 349, 409, 430, 443
synthesis of, 119, 171, 199,
244, 436, 479, 491, 492
titanium-promoted aldol reaction, 433
See also Conjugate addition
Carbonylation
of organoboranes, 250, 333-
338
of lithium amides, 154, 555
Carboranes, 287
Carboxylic acids
addition reaction with Group
IA and II organometallics,
118-120
addition reaction of
dianions of, 212
alkylation of, 371

Carboxylic acids (cont.)
 decarboxylation of, 491
 preparation of cyclic ketones
  from, 230
 protection of, 345
 protonolysis of organo-
  metallics with, 309, 419
 reduction of, 294, 377, 343,
  378
 synthesis via
  dihydro-1,3-oxazines, 131,
   201, 220
  organoaluminates, 370
  organoborates, 327
  organomercuries, 472
  organosilicons, 428
β-*trans*-Carotene, synthesis of,
 171
Catecholborane, in the preparation of alkenylmercuries,
 462
Cesium (Cs), organometallics of,
 91
Chalcones, reaction with
 thallium(III) salts, 488
Charge affinity inversion operations, 137, 160-161, 151-
 159
Charge transfer complexes, 19
Chemically induced dynamic
 nuclear polarization
 (CIDNP), detection of
 organometallic inter-
 mediates, 64, 111
Chromium (Cr), 33, 35
Claisen condensation, 185, 198,
 209-211
Claisen-Schmidt reaction, 201
Clemmensen reduction, 229
Cobalt (Co), 35
 organometallics of, 37, 85
Complexation, in the preparation of organometallics,
 50, 63, 81, 362
π-Complexes, 51, 76, 80, 85
σ-Complexes, 51

Conjugate addition
 chiral synthesis involving,
  131-132
 factors affecting, 128
 of alkali metal cyanides, 152
 of enolate anions, 211
 of free radicals, 128
 of Group IA and II organometallics, 127-133
 of lithium-S,S-acetal
  derivatives, 156
 of organoaluminums, 366, 371-
  373
 of organoborons, 306, 325-327
 of organocuprates, 65, 130,
  132, 325
 of organosilanes, 413, 430
 of organotins, 403, 429, 443
σ-π Conjugative effects, 68
 of silicon and tin, 398, 417
  418, 423, 429, 446, 457,
  475, 478, 491
Coordinative saturation, 16, 51
Coordinative unsaturation, 16,
  20, 37, 44, 50, 51, 79, 82,
  92, 94
Copper (Cu), organometallics of
 conjugate addition of, 65,
  130, 132, 325
 cross coupling involving, 69,
  113, 117, 118
 preparation of, 37, 474
 reaction with
  acyl halides, 368
  ketones, 124
Covalent radii, 10, 11, 505
Cram's rule, 123, 204
Cross aldol reaction, 433
Cross-coupling reaction
 nickel- or palladium-
  catalyzed, 118, 365
 with Group IA and II organometallics, 105-118
 with hindered alkyl halides,
  65
 with organoaluminums, 351, 366

Cyanides, metal derivatives of,
    42
  1,2-addition of, 130
  1,4-addition of, 130, 152
  as acyl anion equivalents, 152
Cyanohydrin, synthesis of
  via organosilicons,152
Cycloaddition reactions of
  silyloxysubstituted olefins,
  433-436
Cyclobutanes, synthesis of, 340
Cyclopropanation
  of silyl enol ethers, 436
  via Group IA and II organo-
    metallics, 75, 138, 173
  via organomercuries, 477, 478
Cyclopropane
  addition reaction with organo-
    boranes, 291
  synthesis via
    organolithiums, 134, 136
  See also Cyclopropyl halides
    organoborons, 339
    organomercuries, 477
Cyclopropene
  addition of organolithiums to,
    133
  synthesis of, 175
Cyclopropyl halides, 96
Cycloreversion, 75

Darzens glycidic ester conden-
    sation, 172, 201
Decarboxylation
  oxidative, 247
  via organoleads, 491
Dehalogenation of β-chloro-
    enones, 230, 238
Dehydrometallation, 46
  in ligand exchange, 81
  mechanism of, 76
Dehydroprogesterone, synthesis
    of, 199
Deoxygenation of ketones, 229,
    230

Desulfurization of S,S-acetals,
    156
Deuterium labeling, 223, 473,
    483
Dialkylative enone transposition,
    441
Dianionic metal complexes,
    formation of, 53, 54
α-Diazo carbonyl compounds,
    alkylation of, 329
Diborane, preparation of, 294
Dicarbonyl compounds
  alkylation of, 187
  synthesis of, 152, 210, 217
Dieckmann reaction, 209, 231
Diels-Alder reaction
  alkenyl- and alkynylboranes
    as dienophiles in, 342
  of silyl enol ethers, 433-436
Dienes
  reduction of, 232
  synthesis of,
    conjugated, 319, 332
    1,4-, 235
    1,5-, 114, 115
Dihydro-1,3-oxazines
  carbon-carbon bond formation
    via, 219-222
Diisobutylaluminum hydride
    (DIBAH), hydroalumination
    with, 357-360
  reduction of organic compounds
    with, 377
Discoordinate metal complexes
    (MX_2), 402
Disiamylborane,
  hydroboration with, 295, 297
  protection of carboxylic
    acids with, 345
  reductions with, 343, 345
Displacement reactions, as a
    preparative method for
    organoalanes, 353
    organoboranes, 292
Disproportionation
  in ligand exchange, 79, 80

Disproportionation (cont.)
  in substitution reactions of
    alkyllithiums, 65
  in transmetallation, 249, 250
  of free radicals, 73
  of Grignard reagents, 94
  of organoalanes, 353
  of organoboranes, 201, 292
  of organothalliums, 457
Dissociation of organoalanes,
    353
Dissolving metal reductions,
    223-240
Dithiane, acyl anion
    equivalent, 155
Diynes, conjugated
  synthesis via organoborates,
    319

Effective nuclear charge, 7, 24
Electron affinity (EA)
  definition of, 7
  in the reduction of organic
    compounds, 227
Electron-election repulsions, 19
Electron-paired species, organo-
    metallic sources of, 62
Electron rule, sixteen or
    eighteen, 15, 81
Electron spin resonance in the
    detection of organometallic
    intermediates, 64
Electron spin state of metals
  in organometallics prepared
    from metal vapors, 36
Electronegativity (EN)
  Allred-Rochow values, 6, 500
  definition of, 6
  effects on mechanistic path-
    ways, 65
  effects on pK_a values, 11
  in addition reactions of
    carbonyl compounds, 128
  in determining covalent bond
    lengths, 10

  in metal-halogen exchange, 38
  in transmetallation, 54, 99
  of orbitals, 12
  of organometallics, 41
  Pauling values, 6, 510
Electronic configurations of
    the elements, 4, 501
π-Electrons, complexes of, 36
Electrophilic aromatic sub-
    stitution
  mercuration, 468
  of arylsilanes, 424
  of metal-containing electro-
    philes, 44
  thallation, 468
Electrostatic interactions,
    types of, 20
α-Elimination reactions of α-
    heterosubstituted organo-
    metallics, 108, 137, 175-
    177, 474, 477
β-Elimination reactions
  in cross coupling reaction,
    107, 108, 351
  metal hydrides via, 84
  of β-heterosubstituted organo-
    metallics, 65, 66, 315, 319,
    420, 421, 425-427, 474
Enamines,
  in Michael reaction, 213
  reduction of, 345
Ene reaction, 342
Enolboranes, 325, 348
Enol ethers,
  hydroboration of, 298
  silyl, 427, 430-438
  stannyl, 443, 444
Enolate anions,
  alkenylation of, 193
  alkylation of, 28, 105
  analogs of, 443
  arylation of, 193
  carbon-carbon bond formation
    via, 178-222
  dienolates, 207

Enolate anions (cont.)
  halogenation of, 242
  α-heterosubstituted, 214-216,
    328
  hydroxylation of, 243
  kinetic vs. thermodynamic,
    184, 189, 203
  masked, 218-222
  preparation of
    from aldehydes, esters and
      nitriles, 120, 193-198
    from ketones, 181-184
  reactions with
    carbonyl compounds,
      198-216
    epoxides, 216-218
    organic halides, 28, 105,
      185-198
  selenenylation of, 245
  structure of, 181
  sulfenylation of, 245
  trapping of, 213
  use of preformed metal
    enolates, 210
  See also Enol ethers
Enynes, synthesis of, 319, 323
Epoxidation,
  mechanism of, 201
  of aldehydes, 172
  of alkenyl silanes, 424
  of ketones, 172
  via α-heterosubstituted
    organometallics, 138, 172,
    173
Epoxides,
  deoxygenation of, 414
  reactions with
    borohydrides, 294
    metal enolates, 216-218
    organoalkali metal compounds,
      125-127
    organoalanes, 375
    organoboranes, 288
    organosilicons, 413
    organostannanes, 413

reductions of, 238, 346, 347,
  349
Esters,
  conversion to ketones, 121
  cyclopropyl esters, 215
  enolization of, 185
  reduction of, 119, 228, 230,
    231, 294, 344, 370, 378
  synthesis via
    Group IA and II organo-
      metallics, 134, 168, 170,
      200, 209, 220, 244
    organoaluminums, 370
    organoborons, 328, 329
    organoleads, 490
    organomercuries, 465
    organosilicons, 429
    organothalliums, 488, 489
Ethers
  carbenes from, 175
  cleavage of, 100
  complexes of, 289
  reduction of, 238
  synthesis via
    organomercuries, 464, 476
    organosilicons, 429
Ethynylation, 98, 116

Farnesol, synthesis of, 361
Favorskii rearrangement, 171,
  444
Francium (Fr), organometallics
  of, 91
Friedel-Crafts acylation, 369
Free radicals
  detection of, using NMR, 112
  in cross coupling, 64, 112
  in hydrometallation, 48
  organometallic sources of,
    62, 71, 341, 451
Free radical reactions, 73
  of organolithiums, 40, 41, 112
  of organoborons, 289, 307
  of organomercuries, 473
  of organosilicons, 48, 402,
    407

Free radical reactions (cont.)
  of organotins, 48, 440, 445,
    446
  study of, using electron spin
    resonance, 112
Frontier orbital theory, 23

Gallium (Ga), organometallics
    of, 286, 364
Gegenion
  effects on alkylation of
    allylic organometallics, 165
  effects on metal-hydrogen
    exchange, 41
  effects on the ratio of
    kinetic vs. thermodynamic
    enolates, 184
Germanium (Ge)
  enolates containing, 181
  general properties of, 394
  metal vapor, 35
Gold (Ag)
  metal vapor, 35
  organometallics of, 35
Grignard reagents, see
    Magnesium, organometallics
    of
Grob-type fragmentation, 340
Grovenstein-Zimmerman re-
    arrangement, 253

Haloboranes
  halometallation of
    arenes, 291
    olefins, 291
Halogenolysis
  of Group IA and II organo-
    metallics, 241, 242
  of organoaluminums, 364, 361
  of organoborons, 312-316
  of organomercuries, 471
  of organotins, 416, 419, 439
  of silyl enol ethers, 431, 432
Halohydrins, epoxides from, 125
Halometallation, 48-50, 291, 467
Hard acids, 23, 25-27

Hard bases, 23, 25-27
Hard and soft acid and base
    principle (HSAB), 23-28, 186
Heats of formation, values for
    methyl derivatives of main
    group metals, 56
Heterometallation, 44, 48-50,
    82, 470
  aminomercuration, 467
  haloboration, 291
  halomercuration, 467
  oxymercuration, 49, 51, 463-
    467
  oxythallation, 479, 485, 486
α-Heterosubstituted organo-
    metalics, see Individual
    metals
Highest occupied molecular
    orbital, (HOMO), 23
Hund's rule, 4
Hybridization
  effects on acidity, 12
  effects on bond lengths, 13
Hydroalumination, see Hydro-
    metallation
Hydroboration, see Hydro-
    metallation
Hydrocyanation of enones, 373-
    375
Hydrogen abstraction by free
    radicals, 71, 73
Hydrogenolysis
  of acetylenes, 233
  of Group IA and II organo-
    metallics, 233
  of organoborons, 310
  of organotins, 419
Hydrometallation, 23, 45-48, 76,
    81, 82, 83, 84
  hydroalumination, 353, 357-362
  hydroboration, 290, 293-300,
    358
  asymmetric, 299
  factors effecting rate of, 20
  oxidation reaction, 287, 426
  organomercuries via, 461

Hydrometallation (cont.)
  scope of, 294, 331
  with triorganoborohydrides,
    294, 302, 342-350
  hydrogermation, 406
  hydroplumbation, 406
  hydrosilation, 45, 406, 410
  hydrostannation, 406, 410-412,
    446-447
  hydrozirconation, 45
Hydrosilation, see Hydro-
    metallation
Hydrostannation, see Hydro-
    metallation
Hydrozirconation, see Hydro-
    metallation
Hyperconjugation, 13, 417

Imines, lithium salts of, in
    cross aldol reaction, 202,
    219
Inductive effects
  due to heteroatoms, 11
  in hydroboration, 297
  of silicon and tin, 398
Indium (In)
  metal powder of, 35
  metal vapor of, 35
  organometallics of, 286
Infrared (IR), in enolate
    studies, 181
Insertion reactions
  of carbenes, 74
  of discoordinate organo-
    metallics, 402
Intramolecular transfer reac-
    tions of organoborates, 322
p_π-p_π Interactions, 400
  See also p_π-p_π Bonding
Interconversion of organo-
    metallic reagents, 63, 68,
    77, 78
  See Transmetallation
Intermolecular forces, 19
Intermolecular transfer reac-
    tions

  of organoalanes, 351
  of organoborates, 321
Ion pairs, in cross-coupling
    reaction, 109
Ionic reactions of
  organoboranes, 303
  organoborates, 304
Ionicity of bonds, 6-9, 15, 19,
    54, 62, 65, 107, 128, 181,
    186
Ionization energy, 6, 7, 32, 53,
    224
  table of, 504
Iridium (Ir), metal vapor, 35
Iron (Fe)
  π-complex of, in Michael
    reaction, 215
  in dissolving metal reductions,
    224
  organometallics of, reaction
    with acyl halides, 368
Isoflavones, preparation of, 488
Isomerization of alkylboranes,
    292

cis-Jasmone, synthesis of 199
Juvabione, synthesis of, 335
Juvenile hormone, synthesis of,
    361

Ketones
  alkylation of, 162-163,
    185-193, 230, 345
  charge affinity patterns of,
    151, 160
  deoxygenation of, 230, 345
  olefination reactions of,
    167-171
  reductive amination of, 345
  synthesis of
    (E)-alkenyl-, 479
    α-alkoxy-, 444
    β-amido-, 435
    α-amino-, 444
    β-amino-, 201
    bicyclic, 210

Ketones (cont.)
  cyclopropyl, 486
  diketones, 217, 486, 487
  α-halo-, 431, 443
  α-hydroxy-, 153, 154, 243,
    486
  β-hydroxy-, 119, 159, 199,
    433, 437
  γ-hydroxy-, 216
  keto acids, 437
  α-keto aldehydes, 209
  α-methyl-, 436
  α-oxy-, 472
  α,β-unsaturated, 158, 171,
    322, 323, 325, 479, 491,
    492
  synthesis via
    Group IA and II organo-
      metallics, 118-121, 153-
      155, 158-160, 168, 171,
      188, 209, 210, 211, 220
    organoaluminums, 368, 369
    organoborons, 160, 311,
      321-325, 335-338
    organoleads, 491, 492
    organomercuries, 472, 479
    organosilicons, 160, 409,
      422, 424, 425, 429, 431,
      432, 435, 436, 437
    organothalliums, 486, 487
    organotins, 422, 440, 443-
      446
  See also reduction of
Knoevenagel reaction, 200

β-Lactams, synthesis of, 435
Lactones
  α-methylene, preparation of,
    491
  reduction of, 345, 378
  synthesis via
    Group IA and II organo-
      metallics, 134, 220
    organoleads, 490, 491
Lasalocid A, synthesis of, 206
Lead (Pb), organometallics of
  carbocations from, 70, 491
  in cross coupling reaction,
    107
  oxymetallation reactions, 49
  preparation of, 364, 462-463,
    468
  properties of, 68, 91, 394,
    455, 490
  reaction with
    carboxylic acids, 490, 491
    enones, 492
    olefins, 489
  structure of, 457-459
  transmetallation involving, 99
Lewis acids, 15, 41, 44, 51,
    125, 289
Lewis bases, 16, 125
Lewis octet rule, 15, 81
Ligand abstraction, 72
Ligand displacement, 79, 80,
    81, 251
Ligand exchange, See Dispro-
    portionation
Ligand migration, 79, 82, 251
Limonene, selective
    metallation of, 45
Lindlar's catalyst, 233
Lithiodithiane, 155, 157, 158,
    244
Lithium aluminum hydride, 360,
    361
Lithium cuprates, 426
Lithium dialkyl-9-borabicyclo-
    [3.3.1]nonanates (9 BBN),
    343, 349
Lithium hydride, 302
Lithium trialkylborohydrides,
    343, 346
Lithium trisiamylborohydrides,
    347
Lithium (Li), organometallics of
  activation of organic
    compound with, 254
  allylic, alkylation of, 163-
    166
  ate complexes of, 302

Lithium (Li) (cont.)
  carbometallation reactions
    of, 133-136
  cationic 1,2-shifts of, 253
  cross-coupling reactions
    of, 69, 105-118
  dissolving metal reductions
    with, 224-240
  enolates of, 178-185
  halogenolyses of, 241
  α-heterosubstituted, 136-139
    as acyl anion equivalents,
      151-161
    alkylation reactions of,
      162-167
    cyclopropanation reactions
      of, 173-178
    epoxides via, 172, 173
    α-halo-, 139, 172
    α-nitrogen, 148
    olefination reactions of,
      167-172
    α-oxy-, 141, 161
    α-phosphorus, 149, 168
    α-selenium, 147, 161, 170
    sigmatropic reactions of,
      161-162
    α-silicon, 150, 168, 169
    α-sulfur, 142, 161, 169,
      170, 173, 178
  metal vapor, 35
  preparation of, 32, 39, 95-
    100, 136-137, 441
  reactions of, with
    acetylenes, 133-136
    acyl halides, 119
    alkyl halides, 65, 105-118
    alkenyl halides, 96
    cadmium and zinc halides,
      104
    carbonyl compounds, 120-
      125, 127-133, 172
    epoxides, 125-127
    esters, 119
    olefins, 133
    sulfonates, 110

  solvent effects on, 39
  structure of, 92, 93, 95
  transmetallation reaction, 55,
    99, 100, 249, 301, 404
  See also Enolate anions of
London forces, 16, 19
Lowest unoccupied molecular
    orbital (LUMO), 23, 53, 64,
    71, 227

Magnesium (Mg)
  catalyst in cross-aldol
    reaction, 202
  dissolving metal reductions
    with, 224
  metal powder, 33
  metal vapor, 35
Magnesium (Mg), organometallics
    of
  allylic rearrangements of, 251
  ate complexes of, 302
  conjugate addition reactions
    of, 127-136
  cross-coupling reaction of,
    69, 105-118
  di-Grignard reagents, 101
  enolates of, 178-185
  oxidation of, to radical
    cations, 110, 111
  preparation of, 32, 34, 36,
    101-103, 249
  reactions with
    acetone, 36
    acetylenes, 102
    acyl halides, 119
    alkyl halides, 105-118
    cadmium and zinc halides,
      104
    carbonyl compounds, 120-
      125, 127-133
    enolates, 208
    epoxides, 125-127
    esters, 119
    ketones, 60, 65, 122-124
    trimethylsilylethylene
      oxide, 426

Magnesium (Mg) (cont.)
  structure of, 93-95
  transmetallation of, 55,
    249. 301, 404
Malonic esters, alkylation of,
    163, 187
Manganese (Mn)
  manganese hydride, 17
  metal vapor of, 35
Mannich bases, in Michael
    reaction, 213
Mannich reaction, 200
Markovnikov functionalization
    of olefins, 318, 463, 464
Mechanism of
  addition reactions of allyl-
    and enolboranes, 325
  amination of organoboranes,
    317
  anionotropic rearrangements
    of organosilicons, 80, 428
  bromination of organoborons,
    307, 313, 314
  bromination of organotins,
    440
  carbene formation, 76
  carbometallation, 76, 82, 342
  carbonylation of organo-
    boranes, 334
  cationic 1,2-shifts, 253
  conjugate addition reactions
    of Group IA and II organo-
      metallics, 128-130
    of organoalanes, 371
    of organoboranes, 306, 325
  cross-coupling reactions of
    Group IA and II organo-
      metallics, 63, 69-70,
      108-112
    organoaluminums, 367
  cyclopropanation by α-halo-
    organometallics, 173-174,
    477
  Grignard additions, 124-125,
    126-127
  hydroboration, 297, 347

  hydrosilation, 407
    transition metal catalyzed,
      410
  hydrostannation, 412, 446
  1,2-migration reactions, 82,
    303
  oxidation of
    organoborons, 311
    organotins, 440, 445
  oxymetallation with
    mercury, 466
    thallium, 483-489
  protonolysis of organoboranes,
    309
  scission of metal-carbon
    bonds, 75-77
  substitution reactions, 63-71
    addition elimination, 67,
      135
    one-electron transfer, 64
    oxidative addition-reduc-
      tive elimination, 69
    S_E2, 66
    S_H2, 71
    S_N1, 65, 351
    S_N2, 38, 63, 67, 109, 127
  transfer reaction of
    organoaluminums, 351
    organoborons, 303-305
Mercury (Hg)
  mercuric salts, reaction with
    acetylenes, 463-468
    alkenes, 463, 468
    arenes, 468-470
    Grignard reagents, 460
    organoboranes, 461, 462
    organolithiums, 460
Mercury (Hg), organometallics of
  carbenes from, 74, 76, 137,
    477, 478
  elimination reactions of,
    475
  enolates of, 181
  halogenolysis of, 471, 472
  heteromercuration, 49-51,
    463-467

Mercury (Hg) (cont.)
  α-heterosubstituted, 74, 76,
    474, 475
  β-heterosubstituted, 463,
    465, 474, 475
  ligand abstraction of, 72, 73
  oxidation of, 472-473
  preparation of, 44, 57, 91,
    395, 455-457
  protonolysis of, 223, 471
  reactions with
    acetylenes, 479
    acyl halides, 479
    olefins, 475-478
    trialkylhalosilanes, 430
  rearrangement of, 475
  reductive cleavage of, 72,
    472
  structure of, 457-460
  thermolysis of, 472
  transmetallation of, 54, 57,
    99, 100, 363, 473, 474, 478
  α,β-unsaturated, reaction
    with electrophiles, 470
Metal cations, 53
Metal-halogen exchange
  in cross-coupling reaction,
    107, 108, 112, 116
  in ligand displacement, 81
  in the preparation of organo-
    metallics, 38-40, 96-98
Metal-hydrogen exchange
  in ligand displacement, 81
  in the preparation of organo-
    metallics, 41-45, 57, 95,
    98, 99, 136, 291, 468-470
Metal powders, 33
  in the preparation of
    Grignard reagents, 102
  in the preparation of
    organozincs, 104
Metal vapors, 35-37
  in the preparation of
    Grignard reagents, 102
Metallate anions, 37, 38
  one-electron transfer

reaction of, 72
Metallocycles, 52, 75
Metallotropy, 22, 23, 79, 80, 82
Methyl methylthiomethyl
  sulfoxide, 157, 158
Michael reaction, 198, 211-216,
    433
1,2-Migration reactions, 82
  of organoaluminums, 352
  of organoboranes, 289, 291,
    303, 305, 317, 318, 324,
    332, 337, 339
  of silicon, 422
Molecular orbital theory, 235
Molybdenum (Mo)
  metal vapor, 35
  organometallics of, hydro-
    metallation, 47
Monolithioacetylene,
  preparation of, 98
Multicenter bonding, 15, 16,
    17, 94

Nagata hydrocyanation
  reaction, 373
Nef reaction, 152
Neutral free radicals, 64
Nickel (Ni)
  in cross coupling reaction,
    69, 106, 118
  in conjugate addition, 133
  metal powder, 33
  metal vapor, 35, 36, 37, 69
  organometallics of, 33, 37,
    364
Niobium (Nb), metal vapor of,
    35
Nitriles
  aromatic, preparation of, 482
  one carbon homologation of,
    341
  reduction of, 294, 370, 378
Nitro compounds,
  aromatic, preparation of, 483
  one carbon homologation of,
    341

Nitro compounds (cont.)
  metal derivatives of, 152
  reactivity toward boro-
    hydrides, 294
Non-bonding electrons, 36
Norbornene
  addition of organolithiums
    to, 133
  oxymetallation of, 49
19-Norsteriods, synthesis of,
  232
Nuclear magnetic resonance
  detection of metallotropic
    reactions by, 80
  detection of organometallic
    intermediates by, 64
  in enolate studies, 181

Ocoteine, preparation of, 482
Olefins, see Alkenes
Olefin metathesis, 75
Olefination reaction, 138, 167-
  173, 201, 245-246, 442
One electron processes, 40, 53,
  64, 72, 96, 109, 110, 125,
  129, 130, 223, 341
Oxazolines, carbon-carbon bond
  formation via, 219-222
Oxidation of
  acetylenes, 486
  alcohols, 445
  olefins, 486
  organic compounds with lead
    tetraacetate, 459
  organoalkali and alkaline
    earth metals, 242-243
  organoaluminums, 364
  organoboranes, 243, 310-312,
    318, 333, 476
  organomercuries, 472
  organosilicons, 422
  organotins, 440, 441, 445
  selenides, 245
  silyl enol ethers, 243, 436
Oxidative addition - reductive
  elimination, 69, 77

Oxidative displacement, in the
  preparation of organo-
  metallics, 30, 37-38
Oxidative coupling, in the
  preparation of organo-
  metallics, 51, 52
Oxidative metallation, in the
  preparation of organo-
  metallics, 40, 41, 95, 101,
  103, 356, 403
Oxidation potentials of alkali
  and alkaline earth metals,
  224
Oxidative-reductive transmetallation, 56
Oximes, reduction of, 345
Oxygenolysis, see Oxidation
Oxymetallation, 48, 49, 51,
  463-467, 479, 485, 486

Palladium (Pd)
  in cross-coupling reaction, 69
  metal powder of, 33
  organometallics of, 287, 474
Pauli exclusion principal, 4
Pericyclic reactions, 21, 75,
  77, 305, 325
Periodic table, 500
Perkin reaction, 201
Peterson reaction, 168, 169, 201
Phase transfer catalysts, 152
Phenols, preparation of, 481
Phenyllithium, 96
Phosphines
  reduction of, 240
  synthesis of, 248
Phosphorous (P), organic com-
  pounds of, 33
Photochemical reactions, 21, 72
Pinacols, preparation of, 231
Platinum (Pt)
  metal powder, 33
  metal vapor, 35
  organometallics of, 287
    carbocations from, 70
    catalyst in hydrosilation,
      45, 47

Polarizability
  in carbocation formation, 70
  in charge controlled
    reactions, 24
  in oxymetallation, 49
  in substitution reactions, 67
  of organolead and mercuries,
    456
Polarographic half-wave
  reduction potentials,
  226-227
Polymerization
  of conjugated dienes, 135
  of epoxides, 125
Potassium (K)
  metal powders of, 33
  organometallics of,
    acetylene "zipper" reac-
      tion involving, 251-253
    preparation of, 32, 39,
      100-101, 249
    stability of, 95
  See Lithium for reactions
    common to Group IA metals
Preparation of organometallics,
  30-59
Principle of conservation of
  orbital symmetry, 21, 23
Propargylic
  halides, 110, 115, 117
  organometallics,
    protection of, 113
    rearrangement of, 108
Propyne, dilithiated, 98
Prostaglandins, synthetic
  intermediates of, 480
Protection of
  active hydrogen containing
    functionalities, 113,
    438
  by organometallics, 84, 85
Protonolysis of
  organoaluminums, 363, 364
  organoborons, 310
  organomercuries, 471
  organothallium, 459

Pyrazolones, reaction with
  thallium(III) salts, 488,
  489

Radical coupling, 73, 110
Radical anions, 53
Radium (Ra), 91
Ramberg-Bachlund reaction, 171
Raney nickel, 230
Redox reactions, 56, 456
Reduction of
  alcohols, 238, 239, 240, 367
  aldehydes, 228, 229, 294, 346,
    349
  alkanes, 224
  alkenes, 224, 232, 233, 357,
    358, 377
  alkenyl halides, 347
  alkyl halides, 230, 238, 345,
    346, 349
  alkynes, 47, 55, 224, 233,
    234, 294, 298, 357, 359,
    360, 362, 377, 406, 410,
    411, 4-6
  amides, 345
  amines, 289
  arenes, 234-237
  aryl halides, 347
  benzylic halides, 349
  carboxylic acids, 294, 343,
    377, 378
  conjugated dienes, 232
  cyano group, 152
  esters, 228
  ethers, 238
  ketones, 60, 120, 182, 228-
    231, 346, 347, 349
  lactones, 345, 378
  metal enolates, 239
  organomercuries, 472, 473
  phenols, 239
  phosphonium compounds, 240
  quatenary amines, 240
  sulfides, 240
  sulfones, 115
  sulfoxides, 115

Reduction of (cont.)
  α,β-unsaturated ketones,
    182, 232, 345, 346, 348,
    349, 409, 430, 443
Reduction potentials for
  organic compounds, 226-227
Reductive cleavage, 72
Reductive elimination, 69, 70,
  77
Reformatsky reaction, 32, 104,
  200, 286
Resonance effects, 11, 12, 297,
  431
Rhenium (Re), hydrometallation
  using, 47
Rhodium (Rh), as a catalyst in
  hydrosilation, 409
Robinson annulation reaction,
  189, 211, 212, 215
Rubidium (Rb), 91
Ruthenium (Ru), 35, 364

α-Santalol, 361
Schlenk equilibrium, 94, 249
Selenium (Se), organic com-
    pounds of,
  activation of olefins, 134
  conversion to olefins, 243,
    244, 245
  synthesis of, 243
  transmetallation involving,
    100
Selenoxides, rearrangement of,
  170
Sesquihalides, 32, 356
Sigmatropic shifts of
  α-heterosubstituted organo-
    metallics, 161, 162
Silicon (Si), organometallics
    of
  alkenylsilanes, 424
    hydroboration-oxidation of,
    426, 427
  allylsilanes, addition
    reactions of, 429-430

bimetallic complexes of,
  412-416
cross coupling of, 107
disproportionation of, 81
electrophilic substitution
  involving,   418, 423
elimination reactions of,
  420, 425, 426
β-heterosubstituted,
  424-430
hydrosilation, 47, 406-410
preparation of, 33, 35, 37,
  364, 403-416, 425, 427
properties of, 19, 57, 395,
  416
protection of functional
  groups by, 85, 113, 396,
  397, 438
reactions with
  acetylenes, 410, 41
  aldehydes, 413, 414, 426
  alkyl halides, 413
  epoxides, 413, 414, 425
  halides, 413
  ketones, 413, 414, 426
  olefins, 410, 413
  α,β-unsaturated ketones,
    413, 415
rearrangements of, 422
stabilizing effects on
  anions, cations and free
    radicals 398, 426
structure of, 299-402
subvalent complexes of, 51,
  402, 409, 417
Silyl ethers, α-functional
  reaction with aldehydes
  and ketones, 438
  synthesis of β-amino
    alcohols, 438
Silyl enol ethers
  alkylation of, 191
  conversion to enolate anion,
    430, 431
  cycloadditions of, 433-436

Silyl enol ethers (cont.)
  cyclopropanation of, 436
  halogenation of, 431, 432
  oxidation of, 436, 437
  preparation of, 409, 428, 430
  reaction with
    aldehydes, 433
    azides, 434
    isocyanates, 435
    ketones, 433
    α,β-unsaturated carbonyl
      compounds, 433, 434
  structure of, 181, 205
  titanium-promoted reactions
    of, 203, 204, 432, 433
Silylenes, carbene analog, 402,
  406
Simmons-Smith reaction, 74, 175
  Furukawa modification of, 177
Slater's rules, 6
Sodium borohydrides, 220, 344
  insitu generation of tin-
    hydrides, 474
  reduction of organomercuries,
    473
Sodium cyanoborohydride
  reductions with, 345, 346
Sodium (Na), organometallics of
  ate complexes with, 302
  cross-coupling reaction of,
    105, 106, 107, 109, 110, 110
  cyclopropanation reactions
    involving, 174, 175
  dissolving metal reductions
    with, 224-240
  enolates of, 178-185
    reactions of, 188, 195, 203,
      208, 209, 210, 213, 216
  α-heterosubstituted, 136-157
  preparations of, 32, 39, 95,
    100, 101, 249
  properties of, 95
  protection of acetylenes, 255
  olefination reaction, 168
  reaction with
    epoxides, 126

  ketones, 120-125
  α,β-unsaturated ketones, 129
  transmetallation of, 244, 301
Soft acids, 23, 25-27
Soft bases, 23, 25-27
Softness parameter ($E^{\ddagger}$), 27
Solvation effects on organometallics, 20, 64, 94, 166,
  168
Spin relaxation, 111
Steroids, synthetic intermediate for, 335
Stevens rearrangement, 422
Stobbe reaction, 200
Strontium (Sr), organometallics
  of, 32, 249
Substitution reactions, 63-71
Sulfides
  alkylation of allylic
    sulfides, 164-166
  α-halocarbenes from, 175
  olefination reactions of,
    171, 245, 246
  oxidation of, 245
  reduction of, 247
  synthesis of, 243-247, 309
Sulfonates
  allylic, 2,3-shifts of, 161
  cross coupling of, 63, 65,
    110, 366
  reactivity of organoborons
    toward, 288, 160
  reduction of, 345, 346
  synthesis via
    organoaluminums, 364, 366
    organosilicons, 423
Sulfones
  alkylation of, 162, 164
  synthesis of, 247
Sulfoxides
  allylic, 2,3-shifts of, 161
  alkylation of, 162, 164, 165
  olefins from, 170, 244
  synthesis of, 245
Sulfur ylides, alkylation of,
  329

Tetramethylethylenediamine
  (TMEDA), 42
Tetramethylsilane, 68, 67
Thallium (Tl),
  arylthalliums reaction with
    electrophiles, 483
    nucleophiles, 480-483
  in enolate alkylation, 187
  metal-hydrogen exchange in-
    volving, 44
  oxythallation, 483-489
  preparation of, 33, 44, 55,
    462-463
  properties of, 395, 455-457
  reaction with
    acetylenes, 479, 483, 486
    aldehydes, 489
    arenes, 468, 469, 470, 480
    chalcones, 488
    ketones, 488
    olefins, 479, 483, 486
    pyrazolones, 489
    α,β-unsaturated ketones, 488
  structure of, 457-460
  rearrangements of mono-
    alkylthalliums, 484-485
  transmetallation to boron,
    155, 48, 481
Thermal decomposition of π-
  complexes, 76
Thexylborane, 296, 319, 335,
  343
α-Thiophenoxycyclopropylli-
  thium, 178, 179
Thope-Ziegler reaction, 209
Tiffeneau-Dimjanov reaction,
  438
Tin (Sn), metal vapor, 35
Tin (Sn), organometallics of
  alkenyltins, reactions with
    electrophiles, 441, 442
  allyltins, addition
    reaction of, 442
  bimetallic complexes of,
    412, 413, 441
  cross coupling involving,

  107,108
  dissolving metal reductions
    with, 224
  elimination of, 420
  enolates of, 181, 444, 445
  enone transposition, 441
  halogenolysis of, 419
  α-heterosubstituted, 414, 427
  β-heterosubstituted, 442
  hydrogenolysis of, 419
  hydrostannation, 47, 84, 443,
    446, 447
  oxides of, 443, 445
  oxidation of, 422, 440
  preparation of, 33, 37, 364,
    403-416
  properties of, 394, 416, 439
  protonolysis of, 419
  reactions with
    acetylenes, 442
    epoxides, 443
    1,4-glycols, 446
    halogens, 440
    α-halo ketones, 444
    ketones, 445
    α,β-unsaturated ketones,
      441, 442, 443
  structure of, 399-403
  transmetallation involving,
    46, 57, 99, 301, 441
Titanium (Ti)
  tetrachloride
    in the addition of
      allylsilanes to carbonyl
      compounds, 429
    in the alkylation of silyl
      enol ethers, 191
    in the cross aldol reaction
      of silyl enol ethers, 433
  trichloride
    in the conversion of nitro-
      alkanes to ketones, 152,
      153
Tosylates, reduction of, 347
Transition metals
  acyltransition metal com-

Transition metals (cont.)
    plexes, 155
    carbometallation by, 82
    σ-complexes of, 51
    π-complexes of, 51
    complexes with divalent
        metals, 402
    hydrides of, 46
    masked enolate anions of, 219
    metallate anions of, 37
    oxidative coupling involving,
        52
Transmetallation, in the prepa-
        ration of organometallics,
        33, 47, 54, 55, 63, 69, 99,
        100, 249, 250, 300, 362,
        364, 441, 470, 474, 478
Tricyclic compounds, synthesis
    of, 241
Trihaloacetic acids, carbenes
    from, 174
Triisobutylaluminum   337
Two electron rule, 15

Uranium (U), preparation of
    metal powder, 33
    metal vapor, 35

Valence orbitals, in ligand-
    ligand exchange, 80
Van der Waals forces, 19
Van der Waals radii, 19
Vanadium (V), preparation of
    metal vapor, 35
Vinylic halides, preparation of,
    472
Vinylic hydrogen, substitution
    reaction of, 54
Vitamin A, synthesis of, 171
Vitamin B_{12}, metal-carbon
    bonds in, 60

Wadsworth-Emmons reaction, 168,
    169, 201
Wagner-Meerwein rearrangement,
    305, 466

Wittig reaction, 167, 168, 426
Wittig rearrangement, 253, 422
Wolf-Kishner reduction, 205, 345
Woodward-Hoffman rules, 21, 23,
    45, 76, 80
Wurtz coupling reaction, 32, 39,
    96, 102, 104, 107

Ylides
    conjugate addition to, 131
    epoxidation via sulfur
        ylides, 172, 173

Ziegler-Natta catalysts, 345
Ziegler-Natta polymerization,
    364
Zinc (Zn)
    addition to acyl halides, 119
    amalgam, 229
    as catalyst in cross aldol
        reaction, 202
    dissolving metal reductions
        with, 224, 233
    powder, 32, 33, 34, 104
    vapor, 35
Zinc (Zn), organometallics of
    carbenes from α-halo-
        derivatives of, 74
    cross coupling with, 69, 107,
        108
    cyclopropoanation reaction,
        104, 177, 175-177
    enolates of, in aldol
        reaction, 204
    preparation of, 103-104, 364
    transmetallation involving,
        55, 104, 301
Zirconium (Zr)
    metal vapor, 35
    organometallics of, 356, 364